AF732194

Solar Cells: Materials, Manufacture and Operation

Solar Cells: Materials, Manufacture and Operation

Edited by:
Tom Markvart
University of Southampton, UK and

Luis Castañer
Universidad Politecnica de Catalunya, Barcelona, Spain

ELSEVIER

Elsevier
The Boulevard, Langford Lane, Kidlington, Oxford OX5 1GB, UK
Radarweg 29, PO Box 211, 1000 AE Amsterdam, The Netherlands

First edition 2005
Reprinted 2006

Copyright © 2005 Elsevier Limited. All rights reserved

No part of this publication may be reproduced, stored in a retrieval system or transmitted in any form or by any means electronic, mechanical, photocopying, recording or otherwise without the prior written permission of the publisher

Permissions may be sought directly from Elsevier's Science & Technology Rights Department in Oxford, UK: phone: (+44) (0) 1865 843830, fax: (+44) (0) 1865 853333, e-mail: permissions@elsevier.com. Alternatively you can submit your request online by visiting the Elsevier web site at http://elsevier.com/locate/permissions, and selecting *Obtaining permission to use Elsevier material*

Notice
No responsibility is assumed by the publisher for any injury and/or damage to persons or property as a matter of products liability, negligence or otherwise, or from any use or operation of any methods, products, instructions or ideas contained in the material herein. Because of rapid advances in the medical sciences, in particular, independent verification of diagnosis and drug dosages should be made

British Library Cataloguing in Publication Data
A catalogue record for this book is available from the British Library

Library of Congress Cataloguing in Publication Data
A catalog record for this book is available from the Library of Congress

ISBN 1856174573

For information on all Elsevier publications
visit our website at books.elsevier.com

Transferred to digital printing 2007

Working together to grow
libraries in developing countries

www.elsevier.com | www.bookaid.org | www.sabre.org

ELSEVIER BOOK AID International Sabre Foundation

Publisher's Note

In response to a perceived demand for a concise account of solar cell technology, this volume 'Solar Cells: Materials, Manufacture and Operation' has been abstracted virtually without change from the 'Practical Handbook of Photovoltaics: Fundamentals and Applications', also edited by Dr Markvart and Professor Castañer and published in 2003. (ISBN 185617 3909)

A new chapter on 'Low Cost Industrial Manufacture of Crystalline Silicon Solar Cells' by Szlufcik et al has been provided as the original chapter was taken from a proceedings volume and was in need of updating.

To save time and costs, the original Section and Chapter numbering has been retained, although a new index has been provided.

Contents

APPENDICES

EDITORIAL INDEX

List of Contributors

Vyacheslav M. Andreev, Ioffe Physico-Technical Institute, 26 Polytekhnicheskaya str., St. Petersburg 194021, Russia
email: vmandreev@mail.ioffe.ru

Allen Barnett, 19 Nivin Lane, Landenberg, PA 19350, USA
email: ambarnett@aol.com

Dieter Bonnet, Breslauer Ring 9a, D-61381 Friedrichsdorf, Germany
email: DieterBonnet@aol.com

David E.Carlson,BP Solar, 989 Corporate Boulevard, Linthicum, MD 21090, USA
email: carlsde@bp.com

Luis Castañer, GDS, Modulo C4 Campus Nord, Universidad Politecnica de Catalunya, Calle Jordi Girona 1, 08034 Barcelona, Spain
email: castaner@eel.upc.es

Andres Cuevas, Department of Engineering, Australian National University, Canberra, Australia 0200
email: Andres.Cuevas@faceng.anu.edu.au

Emilio Fernandez Lisbona, Solar Array Section, ESA-Estec, Keplerlaan 1, 2200 AG Noordwijk, The Netherlands
email: Emilio.Fernandez.Lisbona@esa.int

Francesca Ferrazza, Eurosolare S.p.A. Via Augusto D'Andrea 6, 00048 Nettuno, Italy
email: francesca.ferrazza@eurosolare.agip.it

Martin A. Green, Centre of Excellence for Advanced Silicon Photovoltaics and Photonics, University of New South Wales, Sydney NSW 2052, Australia
email: m.green@unsw.edu.au

Peter T. Landsberg, Faculty of Mathematical Studies, University of Southampton, Southampton SO17 1BJ, UK
email: ptl@maths.soton.ac.uk

Tom Markvart, School of Engineering Sciences, University of Southampton, Southampton SO17 1BJ, UK
email: t.markvart@soton.ac.uk

Michael Mauk, AstroPower, Inc., 300 Executive Drive, Newark, DE 19702-3316, USA
email: mauk@AstroPower.com

Augustin McEvoy, Institute for Molecular and Biological Chemistry, Faculty of Basic Sciences, Ecole Polytechnique Fédérale de Lausanne, CH-1015 Lausanne, Switzerland.
email: augustin.mcevoy@epfl.ch

Robert P. Mertens, Interuniversity Microelectronic Center (IMEC), Kapeldreef 75, 3001 Leuven, Belgium
email: robert.mertens@imec.be

Jenny Nelson, Centre for Electronic Materials and Devices, Department of Physics, Imperial College, London SW72BW, UK
email: jenny.nelson@ic.ac.uk

Johan F. Nijs, Photovoltech, rue de l'Industrie 52 Nijverheidsstraat, 1040 Brussels, Belgium
email: johan.nijs@photovoltech.be

Carl R. Osterwald, NREL, 1617 Cole Boulevard, Golden, Colorado 80401-3393, USA
email: carl_osterwald@nrel.gov

Roger van Overstraeten (deceased), formerly of Interuniversity Microelectronic Center (IMEC), Kapeldreef 75, Leuven B-3001, Belgium

James Rand, AstroPower, Inc., 300 Executive Drive, Newark, DE 19702-3316, USA
email: jimrand@AstroPower.com

Uwe Rau, Institute of Physical Electronics, University of Stuttgart, Pfaffenwaldring 47, D-70569 Stuttgart, Germany
email: uwe.rau@ipe.uni-stuttgart.de

Hans Werner Schock, Institute of Physical Electronics, University of Stuttgart, Pfaffenwaldring 47, D-70569 Stuttgart, Germany
email: schock@ipe.uni-stuttgart.de

Paul Sims, AstroPower, Inc., 300 Executive Drive, Newark, DE 19702-3316, USA
email: pesims@AstroPower.com

Ronald A. Sinton, Sinton Consulting, Inc., 1132 Green Circle, Boulder, CO 80305, USA
email: Ron@Sintonconsulting.com

Siva Sivoththaman, University of Waterloo, Faculty of Electrical and Computer Engineering, 200 University Ave, Waterloo, Ontario, Canada N2L 3G1
email: sivoththman@uwaterloo.ca

Jozef Szlufcik, Photovoltech, rue de l'Industrie 52 Nijverrheidsstraat, 1040 Brussels, Belgium
email: Jozef.Szlufcik@imec.be

Pierre J. Verlinden, Origin Energy, GPO Box 1097, Adelaide, SA 5001, Australia
email: Pjverlinden@aol.com

Christopher R. Wronski, 215 Electrical Engineering West, The Pennsylvania State University, University Park, PA 16802, USA
email: crwece@engr.psu.edu

Kenji Yamamoto, Kaneka Corporation, 2-1-1, Hieitsuji, Otsu, Shiga, 520-0104, Japan
email: yamamoto@pv.kaneka.co.jp

Part II

Solar Cells

Part IIa

Introduction

IIa-1

Principles of Solar Cell Operation

Tom Markvart, School of Engineering Sciences,
University of Southampton, UK
Luis Castañer, Universidad Politecnica de Catalunya,
Barcelona, Spain

1 Introduction

Photovoltaic energy conversion in solar cells consists of two essential steps. First, absorption of light generates an electron–hole pair. The electron and hole are then separated by the structure of the device – electrons to the negative terminal and holes to the positive terminal – thus generating electrical power.

This process is illustrated in Figure 1 which shows, at a glance, the principal features of the typical solar cells in use today. Each cell is depicted in two ways. One diagram shows the physical structure of the device and the dominant electron transport processes that contribute to the energy conversion process. The same processes are shown on the band diagram of the semiconductor, or energy levels in the molecular devices.

The diagrams in Figure 1 are schematic in nature, and a word of warning is in place regarding the differences in scale: whilst the thickness of crystalline silicon cells (shown in Figures (a) and (f)) is of the order of a hundred micrometres or more, the thickness of the various devices in Figures (b)–(e) (thin film and GaAs based cells) might be several micrometres or less. The top surface of the semiconductor structures shown in Figure 1 would normally be covered with antireflection coating. The figure caption can also be used to locate the specific chapter in this book where full details for each type of device can be found.

2 Electrical Characteristics

2.1 The Ideal Solar Cell

An ideal solar cell can be represented by a current source connected in parallel with a rectifying diode, as shown in the equivalent circuit of Figure 2. The corresponding I–V characteristic is described by the Shockley solar cell equation

$$I = I_{ph} - Io\left(e^{\frac{qV}{k_B T}} - 1\right) \tag{1}$$

where k_B is the Boltzmann constant, T is the absolute temperature, q (>0) is the electron charge, and V is the voltage at the terminals of the cell. I_0 is well known to electronic device engineers as the diode saturation current (see, for example, [1]) serving as a reminder that a solar cell in the dark is simply a semiconductor current rectifier, or diode. The photogenerated current I_{ph} is closely related to the photon flux incident on the cell and its dependence on the wavelength of light is frequently discussed in terms of the quantum efficiency or spectral response (see Section 2.3). The photogenerated current is usually independent of the applied voltage with possible exceptions in the case of a-Si and some other thin film materials [2–4].

Figure 3(a) shows the I–V characteristic (Equation (1)). In the ideal case, the short circuit current I_{sc} is equal to the photogenerated current I_{ph}, and the open circuit voltage V_{oc} is given by

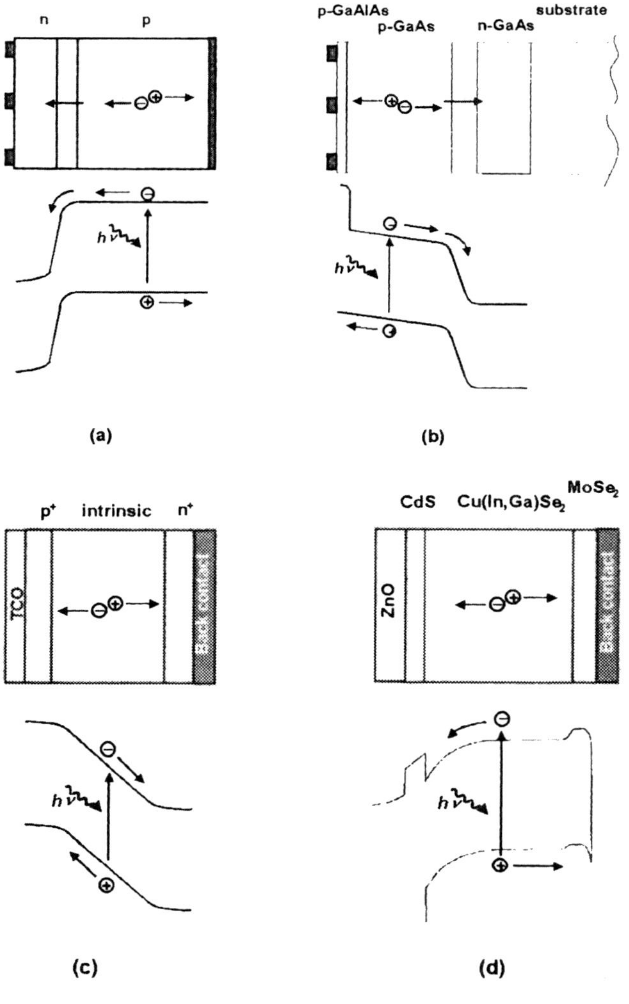

Figure 1 (a) The structure of crystalline silicon solar cell – the typical solar cell in use today. The bulk of the cell is formed by a thick p-type base where most of the incident light is absorbed and most power is generated. After light absorption, the minority carriers (electrons) diffuse to the junction where they are swept across by the strong built-in electric field. The electrical power is collected by metal contacts to the front and back of the cell (Chapters IIb-2 and -5). (b) The typical gallium arsenide solar cell has what is sometimes called a heteroface structure, by virtue of the thin passivating GaAlAs layer which covers the top surface. The GaAlAs 'window' layer prevents minority carriers from the emitter (electrons) to reach the surface and recombine but transmits most of the incident light into the emitter layer where most of the power is generated. The operation of this p–n junction solar cell is similar in many respects to the operation of the crystalline silicon solar cell in (a) but the substantial difference in thickness should be noted. (Chapter IId-1). (c) The structure of a typical single-junction amorphous silicon solar cells. Based on p–i–n junction, this cell contains a layer of intrinsic semiconductor which separates two heavily doped p and n regions near the contacts. Generation of electrons and holes occurs principally within the space-charge region, with the advantage that charge separation can be assisted by the built-in electric field, thus enhancing the collection efficiency. The contacts are usually formed by a transparent conducting oxide (TCO), at the top of the cell, and a metal contact at the back. Light trapping features in TCO can help reduce the thickness and reduce degradation. The thickness of a-Si solar cells ranges typically from a fraction

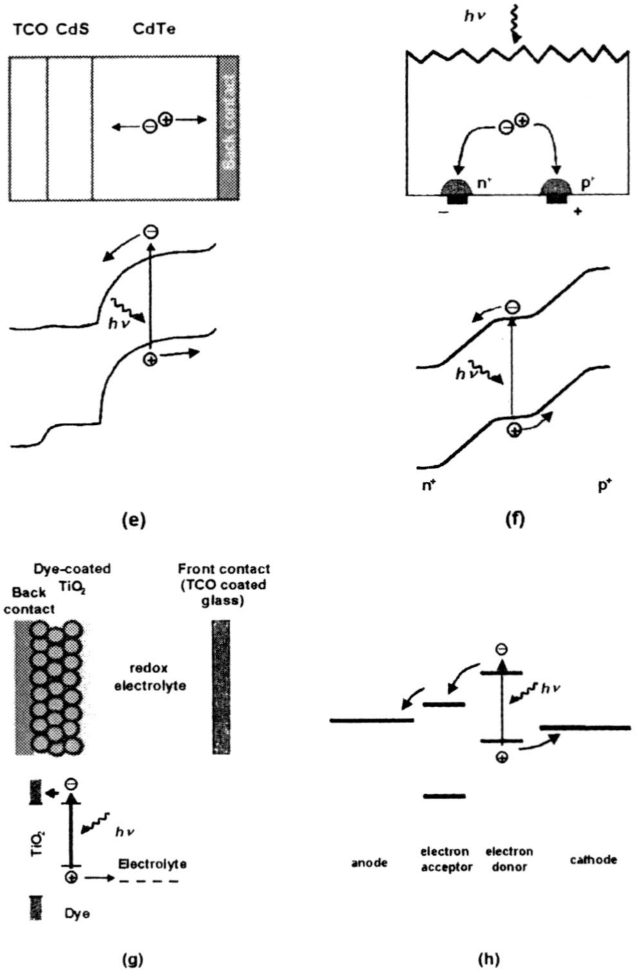

of a micrometer to several micrometers. (Chapter IIc-1). (d), (e) The typical structures of solar cells based on compound semiconductors copper indium–gallium diselenide (d) and cadmium telluride (e). The front part of the junction is formed by a wide band gap material (CdS 'window') which transmits most of the incident light to the absorber layer ($Cu(In,Ga)Se_2$ or CdTe) where virtually all electron–hole pairs are produced. The top contact is formed by a transparent conducting oxide. These solar cells are typically a few micrometers thick. (Chapters IIc-3 and -4). (f) Contacts can be arranged on the same side of the solar cell, as in this point contact solar cell. The electron–hole pairs are generated in the bulk of this crystalline silicon cell which is near intrinsic, usually slightly n-type. Slightly thinner than the usual crystalline silicon solar cell, efficient light absorption is aided here by light trapping: a textured top surface and a reflecting back surface (Chapter IId-2). (g), (h) The most recent types of solar cell are based on molecular materials. In these cells, light is absorbed by a dye molecule, transferring an electron from the ground state to an excited state, rather than from the valence band to the conduction band, as in the semiconductor cells. The electron is subsequently removed to an electron acceptor and the electron deficiency (hole) in the ground state is replenished from an electron donor. A number of choices exist for the electron acceptor and donor. In the dye sensitised cell (g, Chapter IIe-1), the electron donor is a redox electrolyte and the role of electron acceptor is the conduction band of titanium dioxide. In plastic solar cells (h, Chapter IIe-2), both electron donor and electron acceptor are molecular materials.

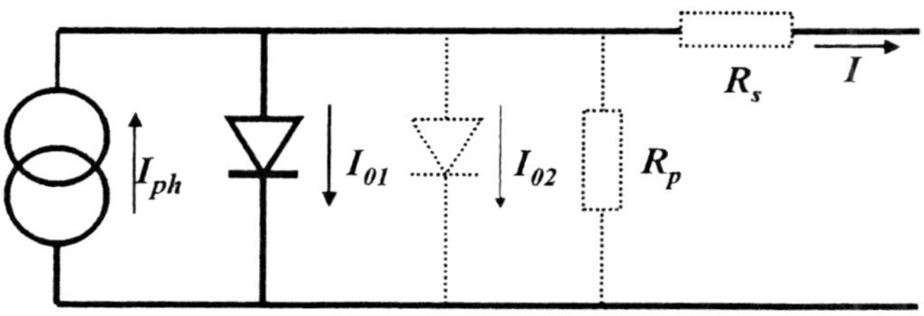

Figure 2 The equivalent circuit of an ideal solar cell (full lines). Non-ideal components are shown by the dotted line.

$$V_{oc} = \frac{k_B T}{q} \ln\left(1 + \frac{I_{ph}}{I_0}\right) \tag{2}$$

The maximum theoretically achievable values of the short circuit current density J_{ph} and of the open circuit voltage for different materials are discussed and compared with the best measured values in Chapter IIa-3.

The power $P = IV$ produced by the cell is shown in Figure 3(b). The cell generates the maximum power P_{max} at a voltage V_m and current I_m, and it is convenient to define the fill factor FF by

$$FF = \frac{I_m V_m}{I_{sc} V_{oc}} = \frac{P_{max}}{I_{sc} V_{oc}} \tag{3}$$

The fill factor FF of a solar cell with the ideal characteristic (1) will be furnished by the subscript 0. It cannot be determined analytically but it can be shown that FF_0 depends only on the ratio $v_{oc} = V_{oc}/k_B T$. FF_0 is determined, to an excellent accuracy, by the approximate expression [5]

$$FF_0 = \frac{v_{oc} - \ln(v_{oc} + 0.72)}{v_{oc} + 1}$$

The I–V characteristics of an ideal solar cell complies with the *superposition principle*: the functional dependence (1) can be obtained from the corresponding characteristic of a diode in the dark by shifting the diode characteristic along the current axis by I_{ph} (Figure 4).

2.2 Solar Cell Characteristics in Practice

The I–V characteristic of a solar cell in practice usually differs to some extent from the ideal characteristic (1). A two-diode model is often used to fit an observed curve, with the second diode containing an 'ideality factor' of 2 in the denominator of the argument of the exponential term. The solar cell (or circuit) may also contain series (R_s) and parallel (or shunt, R_p) resistances, leading to a characteristic of the form

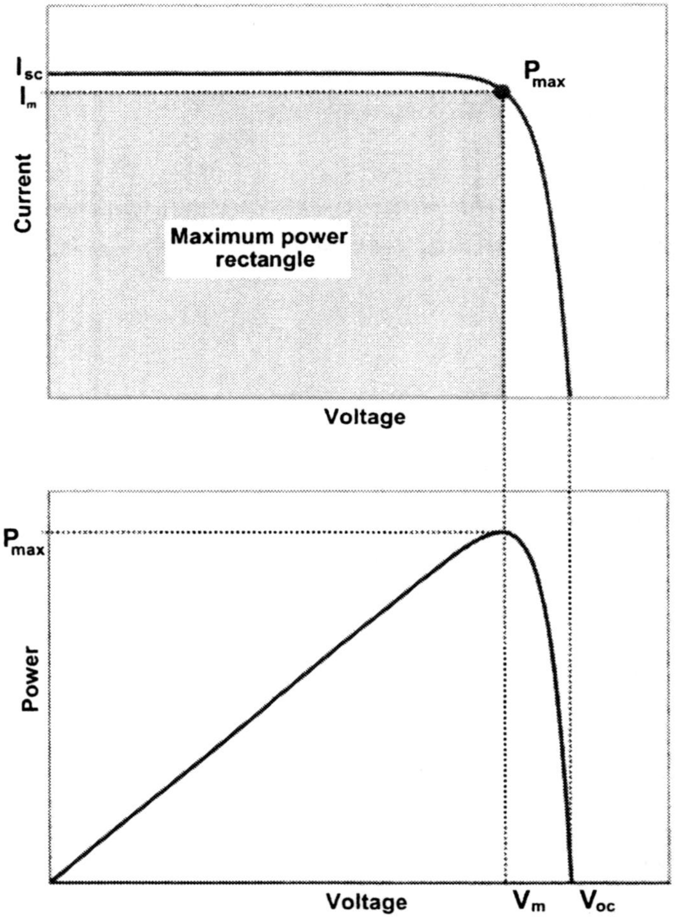

Figure 3 The I–V characteristic of an ideal solar cell (a) and the power produced by the cell (b). The power generated at the maximum power point is equal to the shaded rectangle in (a).

$$I = I_{ph} - I_{o1}\left\{exp\left(\frac{V + IR_s}{k_B T}\right) - 1\right\} - I_{o2}\left\{exp\left(\frac{V + IR_s}{2k_B T}\right) - 1\right\} - \frac{V + IR_s}{R_p} \quad (4)$$

where the light-generated current I_{ph} may, in some instances, depend on the voltage, as we have already noted. These features are shown in the equivalent circuit of Figure 2 by the dotted lines. The effect of the second diode, and of the series and parallel resistances, on the I–V characteristic of the solar cell is shown in Figures 5 and 6, respectively; further information about these parameters can be obtained from the dark characteristic (Figure 7). The effect of the series resistance on the fill factor can be allowed for by writing

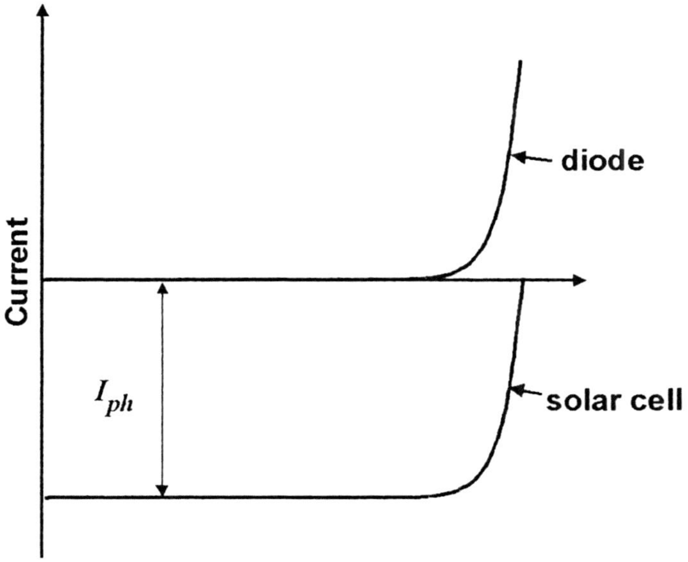

Figure 4 The superposition principle for solar cells.

$$FF = FF_0(1 - r_S) \tag{5}$$

where $r_s = R_s\, I_{sc}/V_{oc}$. An analogous expression exists also for the parallel resistance (see Chapter IIc-4). Instead of the two-diode equation (4), an empirical non-ideality factor n_{id} can be introduced in the single-diode equation (1) which usually lies between 1 and 2. Two among a number of possible sources of non-ideal behaviour – recombination in the depletion region and series resistance – are discussed in Sections 4.1.1 and 4.4.

2.3 The Quantum Efficiency and Spectral Response

The quantum efficiency of a solar cell is defined as the ratio of the number of electrons in the external circuit produced by an incident photon of a given wavelength. Thus, one can define external and internal quantum efficiencie (denoted by $EQE(\lambda)$ and $IQE(\lambda)$, respectively). They differ in the treatment of photons reflected from the cell: all photons impinging on the cell surface are taken into account in the value of the *EQE* but only photons that are not reflected are considered in the value of *IQE*.

If the internal quantum efficiency is known, the total photogenerated current is given by

$$I_{ph} = q\int_{(\lambda)} \Phi(\lambda)\{1 - \mathcal{R}(\lambda)\}IQE(\lambda)d\lambda \tag{6}$$

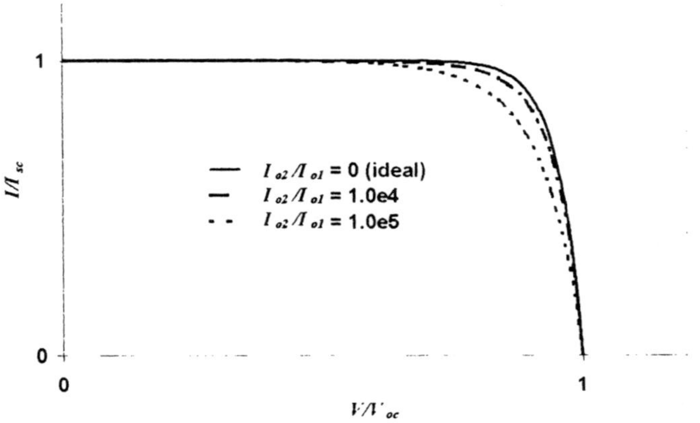

Figure 5 *The I–V characteristic of the solar cell in the two diode model for three values of the ratio* I_{02}/I_{01}.

where $\Phi(\lambda)$ is the photon flux incident on the cell at wavelength λ, $\mathcal{R}(\lambda)$ is the reflection coefficient from the top surface (see Section 3.1), and the integration is carried out over all wavelength λ of light absorbed by the solar cell. The values of the internal and external quantum efficiency are routinely measured to assess the performance of a solar cell by using interference filters or monochromators.

The *spectral response* (denoted by $SR(\lambda)$, with the units A/W) is defined as the ratio of the photocurrent generated by a solar cell under monochromatic illumination of a given wavelength, to the value of the spectral irradiance at the same wavelength. Since the number of photons and irradiance are related, the spectral response can be written in terms of the quantum efficiency as (see, for instance, [6])

$$SR(\lambda) = \frac{q\lambda}{hc} QE(\lambda) = 0.808 \cdot \lambda \cdot QE(\lambda) \tag{7}$$

where λ is in micrometres. Spectral response in (7) can be either internal or external, depending on which value is used for the quantum efficiency.

3 Optical Properties

3.1 The Antireflection Coating

Most solar cells rely on a thin layer of a dielectric (an antireflection coating) to reduce the reflection of light from the front surface of the cell. This section gives a brief description of the reflection of light from a bare semiconductor, and from a semiconductor with a single-layer antireflection coating. The discussion is confined to the case of normal incidence of light onto a smooth planar surface.

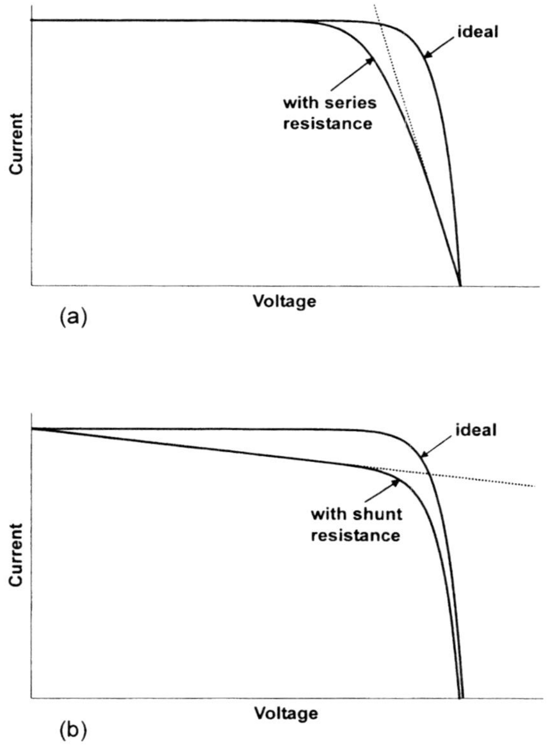

Figure 6 The effect of series (a) and parallel (b) resistance on the I–V characteristic of the solar cell.

The reflection coefficient from bare silicon for light incident from air is given by

$$\mathcal{R} = \frac{(\mathbf{n}-1)^2+\kappa^2}{(\mathbf{n}+1)^2+\kappa^2} \tag{8}$$

where $\mathbf{n}$ and κ are the refractive index and extinction coefficient of the semiconductor, both in general functions of the wavelength λ of light in vacuum. The extinction coefficient is related to the absorption coefficient α by

$$\kappa = \frac{\alpha\lambda}{4\pi\mathbf{n}} \tag{9}$$

For single-layer antireflection coating of refractive index $\mathbf{n}_{ar}$ between a top medium of refractive index $\mathbf{n}_0$ (for example, glass or air) and semiconductor, the reflection coefficient becomes, neglecting light absorption in the semiconductor

$$\mathcal{R} = \frac{r_0^2 + r_{sc}^2 + 2r_0 r_{sc}\cos 2\beta}{1 + r_0^2 r_{sc}^2 + 2r_0 r_{sc}\cos 2\beta} \tag{10}$$

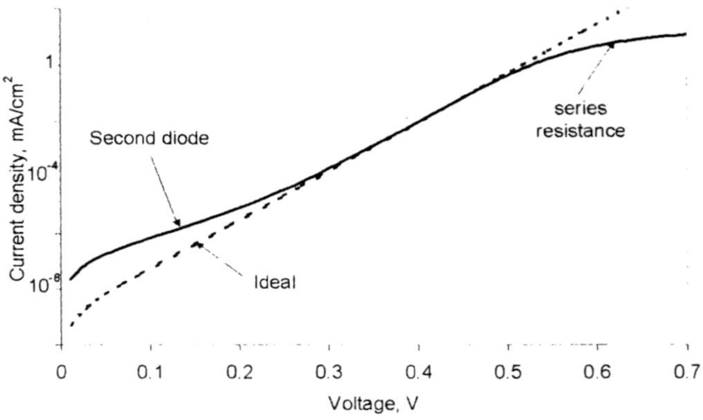

Figure 7 The dark I–V characteristic of a solar cell for the two-diode model including the series resistance. The shunt resistance has a similar effect to the second diode.

where

$$r_0 = \frac{\mathbf{n}_{ar} - \mathbf{n}_0}{\mathbf{n}_{ar} + \mathbf{n}_0}; \qquad r_{sc} = \frac{\mathbf{n}_{sc} - \mathbf{n}_{ar}}{\mathbf{n}_{sc} + \mathbf{n}_{ar}} \qquad \beta = \frac{2\pi}{\lambda}\mathbf{n}_{ar}d$$

and d denotes the thickness of the coating. The transmission coefficient is, in both cases, simply

$$\mathcal{T} = 1 - \mathcal{R} \tag{11}$$

In most cases of interest, both r_{sc} and r_0 are positive and $\mathcal{R}$ vanishes when

$$d = \frac{\lambda}{4\mathbf{n}_{ar}}; \qquad \frac{3\lambda}{4\mathbf{n}_{ar}}; \qquad \frac{5\lambda}{4\mathbf{n}_{ar}}; \ldots \tag{12}$$

and

$$\mathbf{n}_{ar} = \sqrt{\mathbf{n}_0\mathbf{n}_{sc}} \tag{13}$$

The first value of d in (12) is often used in practice under the name of *quarter-wavelength rule* since $\lambda/\mathbf{n}_{ar}$ is the wavelength of light in the antireflection coating.

Reflection from the top surface can be reduced further by the use of a multilayer coating. The details of such coatings as well as a general theory for an oblique incidence of light can be found, for example, in [7]. Figure 8 compares the reflection coefficients for a smooth bare silicon surface, a smooth surface covered with antireflection coating, and a textured surface with antireflection coating.

3.2 Light Trapping

In solar cells with a simple geometry, light rays enter the cell through the front surface and, if not absorbed, leave through the rear surface of the cell. More sophisticated arrangements exist which extend the path of light inside the

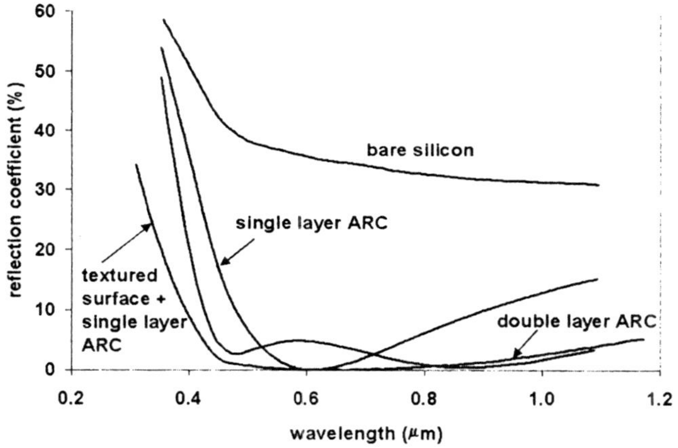

Figure 8 The reflection coefficient from polished bare silicon, and from a polished silicon surface covered with a single and double layer antireflection coating (after [31, 33]). The reflection coefficient for a textured surface is also shown.

cell, and are usually referred to as optical confinement or light trapping. In crystalline or amorphous silicon solar cells, light trapping is used to reduce the thickness of the cell without lowering the light absorption within the cell. Light trapping can also be used to enhance the open circuit voltage [8, 9].

The most common light trapping features include a textured top surface combined with an optically reflecting back surface (Figure 9). In the ideal case, Yablonovich [10, 11] (see also [12]) has shown that a randomly textured (so called Lambertian) top surface in combination with a perfect back-surface reflector produces a light trapping scheme which enhances the light intensity inside the cell by a factor of $\mathbf{n}_{sc}^2$ where, as in Section 3.1, $\mathbf{n}_{sc}$ is the refractive index of the solar cell material. This arrangement also increases the average path length of light rays inside the cell from $2W$, in the case of single pass through the cell, to $4\mathbf{n}_{sc}^2W$ in the case of complete light trapping, where W the cell thickness. Schemes have been developed to enhance the operation of practical devices including crystalline, polycrystalline and amorphous silicon cells (discussed in Chapters IIb-2, -3 and -5, and Chapters IIc-1 and -2). With application to the latter cells, Schropp and Zeman [13] consider the trapping and scattering of light at rough interfaces in some detail. In gallium arsenide cells, multilayer Bragg reflectors (in place of the back-surface reflector) have been used with success (see Chapter IId-1).

4 Typical Solar Cell Structures

4.1 The p–n Junction Solar Cell

The planar p–n junction solar cell under low injection is usually singled out for special analysis since realistic approximations exist that allow analytic solutions

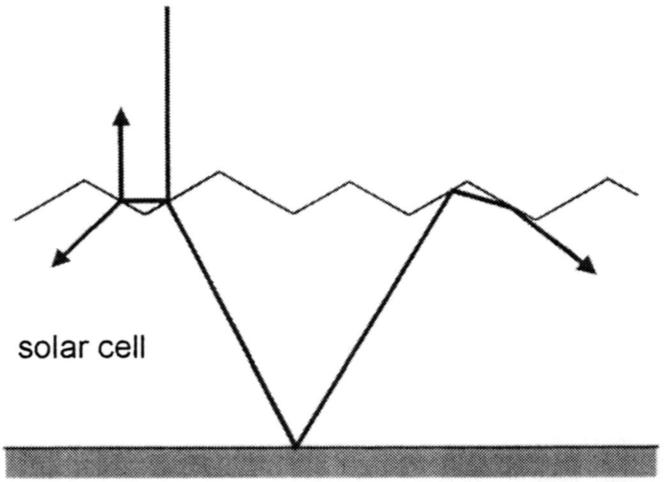

Figure 9 The textured top surface reduces reflection from the solar cell and, when combined with a reflecting back surface, helps to confine or 'trap' light within the cell.

to be developed and used successfully for the description of practical devices. The success of this model is due, to a large extent, on the clear way the cell can be divided into three regions – emitter, junction region and base – which serve a different purpose in solar cell operation.

The emitter and base – which remain largely neutral during the cell operation – absorb the main part of the incident light and transport the photogenerated minority carriers to the junction. The p–n junction – which contains a strong electric field and a fixed space charge – separates the minority carriers that are collected from the emitter and base. The junction is effectively devoid of mobile charge carriers and is sometimes called the depletion region.

4.1.1 The p–n Junction

Figure 10 shows the principal parameters of a p–n junction in equilibrium along the spatial coordinate perpendicular to the junction. In operation, the Fermi level E_F splits into two quasi-Fermi levels E_{Fn} and E_{Fp}, one each for the electrons and holes, with the corresponding potentials $\phi_n = -q/E_{Fn}$ and $\phi_p = -q/E_{Fp}$. Near the open circuit, the quasi-Fermi levels are parallel in the junction, their gradients are small, and their splitting is equal to the observed voltage at the junction (Figure 11). The charge carrier statistics in terms of the quasi-Fermi levels is discussed in Section 3 of Chapter IIa-2.

Under illumination or under applied bias in the dark, the electrostatic potential difference $\Delta\psi$ between the two sides of the junction is a difference of two terms: the equilibrium built-in voltage V_{bi} and the voltage V at the junction edges:

$$\Delta\psi = V_{bi} - V \qquad (14)$$

$$qV_{bi} = k_B T \ln\left(\frac{N_D N_A}{n_i^2}\right) \tag{15}$$

where N_A and N_D are the acceptor and donor concentrations on the p- and n-sides of the junction, respectively. In the absence of resistive losses, V is equal to the voltage measured at the terminals of the cell. The junction width W_j is given by

$$W_j = L_D \sqrt{\frac{2q\Delta\psi}{k_B T}} \tag{16}$$

Here, L_D is the Debye length,

$$L_D = \frac{\sqrt{\varepsilon k_B T}}{q^2 N_B} \tag{17}$$

where ε is the static dielectric constant and

$$N_B = \frac{N_A N_D}{N_A + N_D}$$

In an ideal p-n junction solar cell, the junction (or depletion) region serves as a lossless mechanism for extracting and separating the minority carriers from the quasi-neutral regions – the base and the emitter. The function of the junction can then be summarised in the form of boundary conditions which link the majority carrier concentration on one side of the junction with the minority carrier concentration on the other. For an n-type emitter and p-type base, for example, the following relations hold:

$$n(base) = n_0(base)e^{qV/k_B T} = n_0(emitter)e^{q(V - V_{bi})/k_B T} \tag{18}$$

Equation (18) relates the electron concentration n(base) at the edge of the depletion region of the base to its equilibrium value n_0(base), and to the equilibrium electron concentration n_0(emitter) at junction edge of the emitter. A similar relationship exists for the hole concentration at the junction edge of the emitter and base:

$$p(emitter) = p_0(emitter)e^{qV/k_B T} = p_0(base)e^{q(V - V_{bi})/k_B T} \tag{19}$$

Equations (18) and (19) are the boundary conditions for an analytical solution of the transport equations (discussed in Chapter IIa-2, Section 4) in the quasi-neutral regions. A rigorous discussion of this *depletion approximation* which forms the basis for the analytical treatment can be found in reference [14].

The photogenerated and dark saturation currents for the cell are obtained by adding the relevant quantities for the base and the emitter:

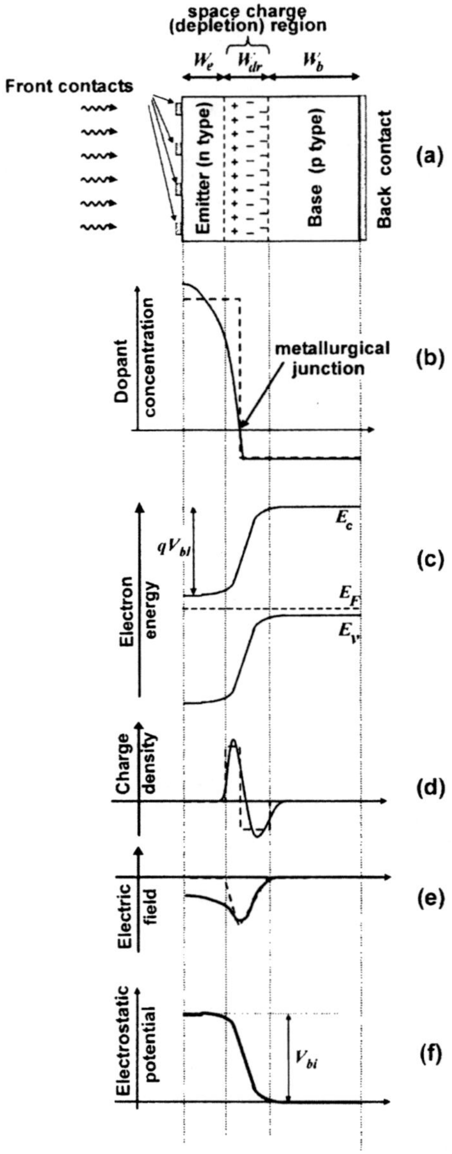

Figure 10 The p–n junction solar cell in equilibrium. (a) The physical layout (not to scale); (b) the difference of dopant concentrations $N_D - N_A$*; (c) the band diagram; (d) charge density; (e) electric field; (f) electrostatic potential. The quantities shown by the dashed line correspond to an idealised abrupt junction with constant dopant concentrations in the base and in the emitter; the full line corresponds to a typical industrial solar cell with a diffused emitter.*

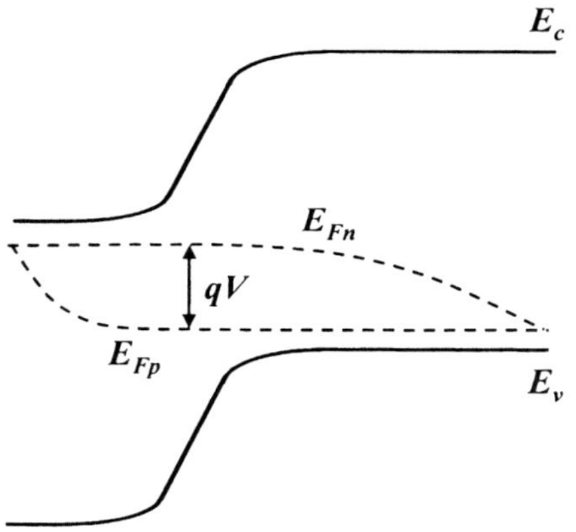

Figure 11 The p–n junction at open circuit.

$$\begin{aligned} I_{ph} &= I_{phb} + I_{phe} \\ I_0 &= I_{0b} + I_{0e} \end{aligned} \tag{20}$$

A similar results holds also for the quantum efficiency. It is sometimes convenient to define the collection efficiency ϑ_i for a region i (where i stands for the base, emitter or the depletion region) as the probability that an electron–hole pair generated in this region reaches the junction:

$$EQE_i(\lambda) = a_i(\lambda)\vartheta_i(\lambda) \tag{21}$$

where $a_i(\lambda)$ is the (fractional) number of electron–hole pairs generated by each photon of incident light in region i.

No recombination occurs in an ideal p–n junction but the (small) light generated current produced here can be added to the first Equation (20). Recombination is included in more realistic analytical theories: the original treatment by Sah et al. [15] uses the Shockley–Read–Hall model of recombination via defects (see Chapter IIa-2, Section 7) with the principal result that the current in (1) is reduced by a term of the form

$$I_{02}\left(e^{\frac{qV}{2k_BT}} - 1\right) \tag{22}$$

In other words, recombination in the depletion region gives rise to an additional dark current corresponding to the second diode in the I–V characteristic (4), as already discussed in Section 2.2.

4.1.2 Uniform Emitter and Base

Analytical expressions for the photogenerated and dark saturation current densities for the emitter or base can be obtained if the dopant concentration and all other parameters are assumed constant. To this end we define

$$\zeta = \frac{SL}{D} \tag{23}$$

$$\gamma_+ = (\zeta + 1)e^{W/L} + (\zeta - 1)e^{-W/L} \tag{24}$$

$$\gamma_- = (\zeta + 1)e^{W/L} - (\zeta - 1)e^{-W/L} \tag{25}$$

where S is the surface recombination velocity at external surface (front surface in the case of emitter and rear surface in the case of base), W is the width of the relevant region (W_e for the emitter and W_b for the base), $L = \sqrt{D\tau}$ is the minority-carrier diffusion length, τ is the minority carrier lifetime, and D is the minority carrier diffusion constant. The photogenerated and dark saturation currents for each region are then given by

$$J_0 = \frac{qD}{L}\frac{n_i^2}{N_{dop}}\frac{\gamma_+}{\gamma_-} \tag{26}$$

where N_{dop} is the dopant concentration N_A or N_D appropriate for the relevant region. The internal quantum efficiency for each region is given in Table 1.

4.1.3 Diffused Emitter

In practical silicon solar cells the emitter is generally fabricated by diffusion of impurities into the semiconductor wafer. This creates a thin layer where the impurity gradient is very high and the approximation of constant doping concentration does not hold. Simultaneously, the continuity and current equations do not combine into a second-order differential equation with constant coefficients, and a simple analytical solution cannot be found. Several approaches have been followed besides the numerical integration of the equations [16] to reach a reasonably simple analytical or truncated series solutions. Analytical solutions were reviewed in reference [17] where the errors have been estimated for the transparent emitter [18] and quasi-transparent emitter [19] solutions. An emitter is considered transparent when the recombination inside the emitter bulk is negligible and quasi-transparent when this recombination can be considered as a perturbation to the transparent solution. Solutions based on an infinite series which can be truncated to provide different order approximations were proposed in reference [20] and extended as a succession of asymptotic expansions in [21]. One of the simplest yet accurate solutions is given in [22] based on the superposition of a zero-input and a zero-state solutions of the continuity equation with a boundary condition at the surface given by a surface recombination velocity S as follows:

$$J_0 = \frac{qn_i^2}{\int_0^{W_e} \frac{N_{eff}}{D}dx + \frac{N_{eff}(W_e)}{S}} + qn_i^2 \int_0^{W_e} \frac{dx}{N_{eff}\tau} \tag{27}$$

Table 1 The internal quantum efficiency for the emitter and base in the uniform doping model. The subscripts *e* or *b* of γ, ζ, *L* and *W* refers to the emitter or base, respectively. In the case of base, IQE is understood per unit photon entering from the junction

	$IQE(\lambda)$	
Base	$\frac{\alpha L_b}{\gamma_{b-}}\left\{\frac{\zeta_b+1}{1+\alpha L_b}\left(e^{W_b/L_b}-e^{-\alpha W_b}\right)+\frac{\zeta_b-1}{1-\alpha L_b}\left(e^{-W_b/L_b}-e^{-\alpha W_b}\right)\right\} \rightarrow \frac{\alpha L_b}{1+\alpha L_b}$	*for an infinite base* ($W_b \rightarrow \infty$)
Emitter	$\frac{\alpha L_b e^{-\alpha W_e}}{\gamma_{e-}}\left\{\frac{\zeta_e+1}{1+\alpha L_e}\left(e^{-W_e/L_e}-e^{+\alpha W_e}\right)+\frac{\zeta_e-1}{1-\alpha L_e}\left(e^{+W_e/L_e}-e^{+\alpha W_e}\right)\right\}$	

where $N_{eff}(x)$ is the effective doping concentration at depth x taking into account the effect of band gap narrowing. A systematic and general formulation of the several approximations is given in [23]. An elegant formalism to deal with inhomogeneously doped emitters can be found in [24].

When the emitter in illuminated, the problem can be solved using the same approaches as used in the dark, computing the emitter collection efficiency ϑ_{em} equal, as in (21), to the ratio of the photogenerated current at the emitter boundary of the space charge region divided by the integrated carrier generation in the emitter. Bisschop et al [25] extended Park's solution in the dark [20] to illuminated emitters. Cuevas et al. [23] provided a formulation in terms of a series expansion and Pons et al. extended the dark superposition model [22]. The first-order result for the photocurrent is given by (see, for instance, [23])

$$J_{ph} = \frac{q\int_0^{W_e} g(x)dx}{1 + \frac{S}{N_{eff}(W_e)}\int_0^{W_e}\frac{N_{eff}}{D}dx + \frac{N_{eff}(W_e)}{S}} \tag{28}$$

where $g(x)$ is the generation rate (see Section 6.1, Chapter IIa-2).

4.2 Heterojunction Cells

Heterostructures represent an opportunity to manufacture efficient solar cells from highly absorbing thin-film materials without substantial losses through electron–hole recombination at the front surface. This is illustrated by the structures of the CdS/CdTe and CdS/CIGS solar cells where a wide band gap semiconductor (here, CdS) serves as a 'window' partner to a lower band gap 'absorber' where most of the power is generated.

An important consideration in the heterojunction design includes the band gap line-ups at the interface between the two semiconductors. Figure 12 shows the equilibrium band diagrams of typical heterojunctions between a wide-gap window A and an absorber B. The band diagram corresponds the usual situation encountered in CdTe and CIGS solar cells where an n-type wide-gap window and a p-type emitter are the most common arrangements. Similarly to the p–n junction, the built-in potentials $V_{bi}(A)$ and $V_{bi}(B)$ on the two sides of the junction

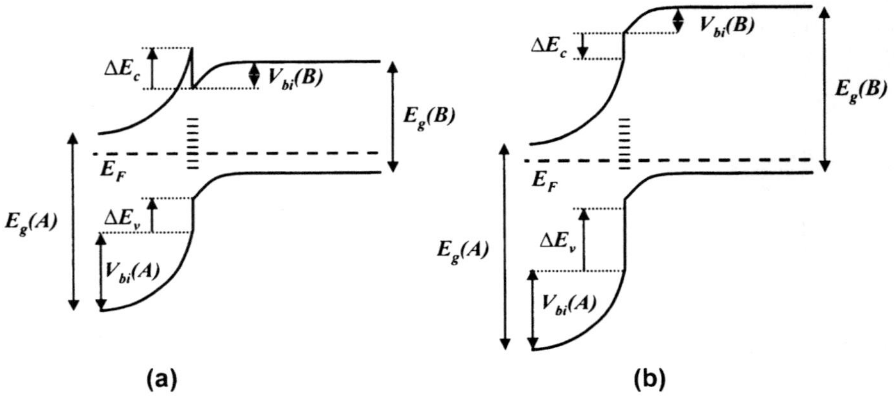

Figure 12 The band diagrams of typical heterojunction solar cells consisting of a window layer of wide-gap n-type semiconductor A, and an absorber of p-type semiconductor B. The energy levels of an interface defect layer, and the band-bending potential $V_{bi}(A)$ and $V_{bi}(B)$ are also shown. A spike in the conduction band occurs for positive ΔE_c, as shown in the structure (a).

can be determined by solution of the Poisson equation (see Equation (7) in Chapter IIa-2). The band gap discontinuities ΔE_c and ΔE_v have been subject to much discussion over the years, and a number of theories have evolved that provide an understanding in terms of electron affinities and the electron dipole moments at the interface. The discontinuity in the conduction band edge, for example, can be written in the form

$$\Delta E_c = \chi_B - \chi_A + \text{interface dipole terms}$$

where χ_A and χ_B are the electron affinities of semiconductors A and B [26].The classical Shockley–Anderson model [27] neglects the interface dipole terms. Its limited validity has been discussed extensively (see, for example, [28]) although it does seem to provide a reasonable description for some heterojunctions (Figure 13). In the application to solar cells, a full understanding of the problem is hindered further by the polycrystalline nature of the materials, and frequently the presence of more than two layers that need to be considered in the analysis.

On account of the wide band gap and weak generation in the window material, both the dark and photogenerated currents from the emitter are significantly smaller that the corresponding quantities from the base. In addition, dark current may contain a component due to recombination at the interface defect states; less frequently, these states also reduce the collection probability and the photogenerated current.

Because of the short-minority carrier diffusion lengths, it is desirable to ensure that minority carriers are generated predominantly in a region where electric field assists collection through drift rather than diffusion. As in the homojunction cell, this can be achieved by employing sufficiently low doping concentrations in the absorber to obtain a wide depletion region; a similar philosophy is also employed in amorphous silicon solar cells, as discussed in

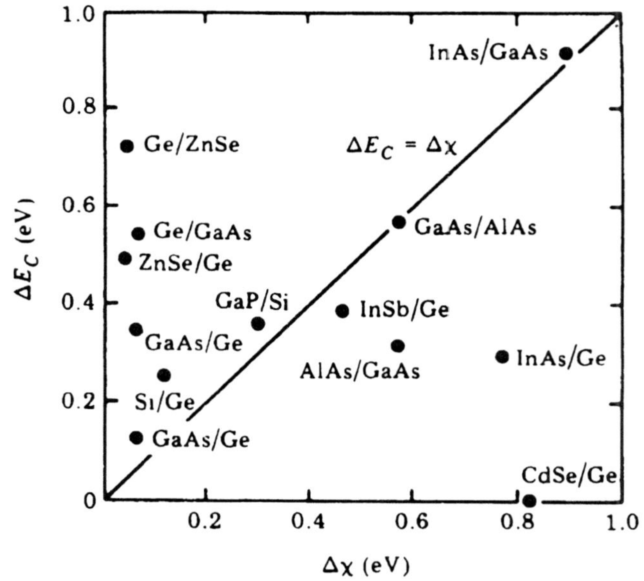

Figure 13 The conduction band discontinuity ΔE_c and the difference of electron affinities $\Delta\chi$ for a number of heterojunctions.(reference [26], p. 391). © Elsevier. Reprinted with permission.

Section 4.3 and Chapter IIc-1. More detailed description of heterojunctions in application to practical solar cells will be found in Chapters IIc-3 and -4.

Somewhat similar to heterojunction is the *heteroface* solar cell; a common structure in GaAs solar cells where a thin layer of wide-gap GaAlAs is deposited to reduce recombination at the top surface. It is more convenient, however, to describe this structure as a homojunction cell with surface passivation which can be treated by the methods described in Section 4.1.

4.3 The p–i–n Structure

The analysis of p–i–n junction solar cells is of considerable importance for the understanding of operation of amorphous silicon solar cells. Furthermore, similar principles have been invoked in the description of other thin film solar cells where carrier diffusion is ineffective and electric field is used to enhance carrier transport and collection. Despite this importance, however, the theoretical understanding of these structures is limited, hampered by the fundamental complexity of the problem. Indeed, the less-than-complete knowledge of the parameters of amorphous or polycrystalline material is compounded by mathematical difficulties arising principally from the need to solve the non-linear transport equations. Although a detailed description is possible only with the use of numerical computational techniques, a broad understanding can be gained through judicious approximations based on a physical insight [29].

A schematic band diagram of a p–i–n structure is shown in Figure 14. Noting that the carrier transport dominant chiefly by drift in the electric field of the junction rather than by diffusion, carrier collection will be described by the drift lengths ℓ_n and ℓ_p rather than by the diffusion lengths L_n and L_p (see Chapter IIa-2, Section 7.3). The recombination can be conveniently approximated with the use of minority carrier lifetimes τ_n and τ_p on the two sides of the junction where electrons and holes are minority carriers, respectively. The use of constant electric field $\mathcal{E}$ is obviously an approximation, but it is usually a good one if carrier injection is not too high. A reasonably simple analysis is then possible which, in the limit of weak absorption, results in an analytical expression in the form

$$J = qg\ell_c(1 - e^{-W_i/\ell_c}) \tag{29}$$

for the current density produced by illumination. In Equation (29), W_i is the width of the intrinsic region, and

$$\ell_c = \ell_n + \ell_p \tag{30}$$

is the collection length, d is the width of the i layer and g is the generation function which is assumed here to be constant. Equations similar to (29) have been used with success to interpret various characteristics of p–i–n solar cells (see, for example, [4]).

An extension of this theory was later proposed which allows for the three charge states of the dangling bonds in amorphous silicon rather than the two charge states usually considered in the Shockley–Read–Hall theory [30].

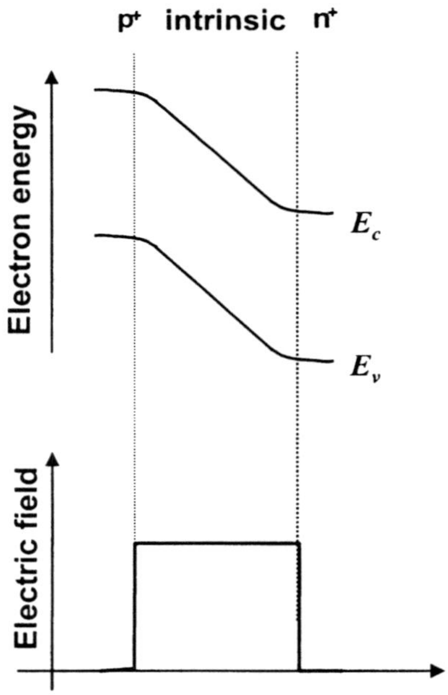

Figure 14 An idealised model of a p–i–n junction amorphous silicon solar cell with a constant electric field in the intrinsic region.

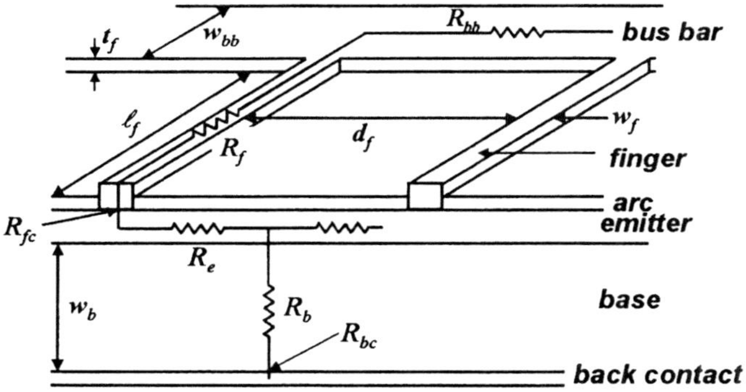

Figure 15 Components of the series resistance in a p–n junction solar cell (after [32] and [33]).

Table 2 Expressions for the various components of the series resistance; the bus bar resistance assumes that connection is made at one end. Here, R_{sp} is the sheet resistance of the emitter layer (in ohm/square), ρ_{cf} and ρ_{cr} are the contact resistances (in ohm/cm^2) of the front and rear contact, respectively, ρ_b is the base resistivity, and ρ_m is the resistivity of the front metallisation. The geometrical dimensions are defined in Figure 15

Component of resistance	Notation	Expression
Emitter resistance	R_e	$R_e = \dfrac{R_{sp}d_f}{7\ell_f}$
Resistance of the base	R_b	$R_b = AW_b\rho_b$
Contact resistance: front contact	R_{fc}	$R_{fc} = \dfrac{\sqrt{R_{sp}\rho_{cf}}}{\ell_f} coth\left(W_f\sqrt{\dfrac{R_{sp}}{\rho_{cf}}}\right)$
Contact resistance: rear contact	R_{bc}	$R_{bc} = A\rho_{cr}$
Resistance of the finger contact	R_f	$R_f = \dfrac{\ell_f\rho_m}{3t_fW_f}$
Resistance of the collecting busbar (per unit length)	R_{bb}	$R_{bb} = \dfrac{\rho_m}{3t_fW_{bb}}$

4.4 Series Resistance

Considerations regarding series resistance form an important part of the solar cell design. The main components of the series resistance of a typical crystalline silicon solar cell are shown in Figure 15 and expressions given in Table 2.

References

[1] Sze, S.M., 1981. *Physics of Semiconductor Devices* (2nd edition), John Wiley & Sons, New York.

[2] Hishikawa, Y., Imura, Y. and Oshiro, T., 2000. Irradiance dependence and translation of the I-V characteristics of crystalline silicon solar

cells, *Proc. 28th IEEE Photovoltaic Specialists Conf.*, Anchorage, pp. 1464–1467.

[3] Philips, J.E., Titus J. and Hofmann, D., 1997. Determining the voltage dependence of the light generated current in $CuInSe_2$-based solar cells using I-V measurements made at different light intensities, *Proc. 26th IEEE Photovoltaic Specialist Conf.*, Anaheim, pp. 463–466.

[4] Hegedus, S.S., 1997. Current-voltage analysis of a-Si and a-SiGe solar cells including voltage-dependent photocurrent collection, *Prog. Photovolt: Res. Appl.*, Vol. 5, pp. 151–168.

[5] Green, M.A., 1995. *Silicon Solar Cells: Advanced Principles and Practice.* Centre for Photovoltaic Devices and Systems, University of New South Wales.

[6] Castañer L. and Silvestre, S., 2002. *Modelling Photovoltaic Systems Using Pspice*, John Wiley & Sons, Chichester, 2002.

[7] Born, M. and Wolf, E., 1999. Principles of Optics (7th edition), Cambridge University Press, Cambridge, Section 1.6.

[8] Brendel, R. and Queisser, H.J., 1993. On the thickness dependence of open circuit voltages of p-n junction solar cells, *Sol. Energy Mater. Sol. Cells*, Vol. 29, p. 397.

[9] Markvart, T., 2000. Light harvesting for quantum solar energy conversion, *Prog. Quantum Electronics*, Vol. 24, p. 107.

[10] Yablonovich, E., 1982. Statistical ray optics. *J. Opt. Soc. Am.*, Vol. 72, p. 899.

[11] Yablonovich, E. and Cody, G.C., 1982. Intensity enhancement in textured optical sheets for solar cells, *IEEE Trans. Electron Devices*, Vol. ED-29, p. 300.

[12] Miñano, J.C., 1990. Optical confinement in photovoltaics, in A. Luque and G.L. Araujo Eds., *Physical Limitations to Photovoltaic Energy Conversion*, Adam Hilger, Bristol, p. 50.

[13] Schropp, R. and Zeman, M., 1998. *Amorphous and Microcystalline Silicon Solar Cells: Modelling, Materials and Device Technology*, Kluwer, Boston.

[14] Selberherr, S., 1984. *Analysis and Simulation of Semiconductors Devices*, Springer, Vienna, New York, 1984, pp. 141-146.

[15] Sah, C.T., Noyce, R.N. and Shockley, W., 1957. Carrier generation and recombination in p-n junctions and p-n junction characteristics, *Proc. IRE*, Vol. 45, p. 1228.

[16] Rover, D.T., Basore, P.A. and Thorson, G.M., 1985. *Proc. 18th IEEE Photovoltaic Specialist Conf.*, Las Vegas, pp. 703–709.

[17] Cuevas, A. and Balbuena, M., 1989. Review of analytical models for the study of highly doped regions of silicon devices. *IEEE Trans. Electron Devices*, Vol, ED-31, pp. 553–560.

[18] Shibib, M.A., Lindholm, F.A. and Therez, F., 1978. *IEEE Trans. Electron Devices*, Vol. ED-26, p.958.

[19] del Alamo, J.A. and Swanson, R.M., 1984. *Proc. 17th IEEE Photovoltaic Specialist Conf.*, Orlando, pp. 1303–1308.

[20] Park, J.S., Neugroschel, A. and Lindholm, F.A., 1986. *IEEE Trans. Electron Devices*, Vol. ED-33, p. 240.
[21] Rinaldi, N., 1993. Modelling of minority carrier transport in non-uniformly doped silicon regions with asymptotic expansions, *IEEE Trans. Electron Devices*, Vol. ED-40, pp. 2307–2317.
[22] Alcubilla, R., Pons J. and Castañer, L., 1992. Superposition solution for minority carrier current in the emitter of bipolar devices, *Solid State Electronics*, Vol. 35, pp. 529–533.
[23] Cuevas, A., Merchan, R. and Ramos, J.C., 1993. On the systematic analytical solutions for minority carrier transport in non-uniform doped semiconductors: application to solar cells, *IEEE Trans. Electron Devices*, Vol. ED-40, pp. 1181–1183.
[24] del Alamo, J.A. and Swanson, R.M., 1984. The physics and modelling of heavily doped emitters, *IEEE Trans. Electron Devices*, Vol. ED-31, p. 1878.
[25] Bisschop, F.J., Verhoef, L.A. and Sinke, W.C., 1990. An analytical solution for the collection efficiency of solar cell emitters with arbitrary doping profile. *IEEE Trans. Electron Devices*, Vol. ED-37, pp. 358–364.
[26] Brillson, L.J., 1992. Surfaces and interfaces: atomic-scale structure, band bending and band offsets. In: P.T. Landsberg Ed., *Handbook of Semiconductors*, Vol. 1, Elsevier, pp. 281–417.
[27] Anderson, R.L., 1962. Experiments on Ge–As heterojunctions, *Solid State Electronics*, Vol. 5, pp. 341–351.
[28] Kroemer, H., 1983. Heterostructure devices: a device physicist looks at interfaces, *Surface Science*, Vol. 132, pp. 543–576.
[29] Crandall, R.S., 1983. Modelling of thin film solar cells: uniform field approximation, *J. Appl. Phys.* Vol. 54, p. 7176.
[30] Hubin, J. and Shah, A.V., 1995. Effect of the recombination function on the collection in a p-i-n solar cell, *Phil. Mag.*, Vol. B72, p. 589.
[31] Zweibel, K. and Hersch, P. 1984. *Basic Photovoltaic Principles and Methods*, Van Nostrand Reinhold, New York.
[32] Overstraeten, R. van and Mertens, R.P., 1986. *Physics, Technology and Use of Photovoltaics*, Adam Hilger, Bristol.
[33] Goetzberger, A., Knobloch, J. and Voss, B. 1998. *Crystalline Silicon Solar Cells*, John Wiley & Sons, Chichester.

IIa-2

Semiconductor Materials and Modelling

Tom Markvart, School of Engineering Sciences,
University of Southampton, UK
Luis Castañer, Universidad Politecnica de Catalunya,
Barcelona, Spain

1 Introduction

Solar cell modelling is fundamental to a detailed understanding of the device operation, and a comprehensive model require a detailed knowledge of the material parameters. A brief overview of the semiconductor properties relevant to solar cell operation is given in this chapter, including the semiconductor band structure and carrier statistics, transport and optical properties, recombination processes, material aspects of radiation damage to solar cells in space and effects observed under heavy doping, with special attention given to the properties of hydrogenated amorphous silicon. The principal semiconductor parameters encountered in photovoltaic applications are summarised in Tables 1 and 2. The refractive indices of materials used for antireflection coating can be found in Table 3.

Numerous computer programs that use the material parameters to model solar cell operation have been developed over the years, and several are now available commercially:

- PC1D developed by P.A. Basore and colleagues at the University of New South Wales, Australia, is the standard one-dimensional simulator used by the PV community.
- ATLAS, a Device Simulation Software by SILVACO International, uses physical models in two and three dimensions. It includes the Luminous tool which computes ray tracing and response of solar cells. It allows the use of monochromatic or multi-spectral sources of light.
- MEDICI by Technology Modelling Associates models the two-dimensional distribution of potential and carrier concentration in a semiconductor device. It also includes an Optical device advanced application module where photogeneration can be computed for multi-spectral sources.

Some programs are available free of charge. These include, for example, SimWindows which can be downloaded from http://www-ocs.colorado.edu/SimWindows/simwin.html.

A discussion of the main principles of the numerical techniques can be found in specialised texts which deal with the modelling of solar cells (see, for example, the review [1]) or with the modelling of semiconductor devices in general [2].

2 Semiconductor Band Structure

The energy gap (or band gap) E_g and its structure as a function of the wave vector are key characteristics of the semiconductor material and of fundamental importance to the operation of the solar cell (see Figure 1). The principal features of interest are the temperature variation of the band gap energy E_g and the magnitude of wave vector associated with low-energy transitions.

The variation of the band gap with temperature can be described by an expression originally suggested by Varshni [10]

Table 1 Properties of the principal semiconductors with photovoltaic applications (all at 300 K). Bandgap: d = direct; i = indirect. Crystal structure: dia = diamond, zb = zinc blende, ch = chalcopyrite. The refractive index is given at the wavelength 590 nm (2.1 eV) unless otherwise stated. Principal sources of data: c-Si [20]; GaAs [3]; InP [4, 41]; a-Si [5] and Section 10; CdTe [6]; CIS [7]; $Al_xGa_{1-x}As$ [9]. Details of the absorption coefficient and refractive index, including the wavelength dependence, can be found in references [27] and [28]. χ is the electron affinity and TEC stands for thermal expansion coefficient

	E_g (eV)	Crystal strucure	ε	**n**	χ (eV)	Lattice const. (Å)	Density (g/cm^{-3})	TEC ($10^{-6}\,K^{-1}$)	Melting point (K)	Comments
c-Si	1.12 (i)	dia	11.9	3.97	4.05	5.431	2.328	2.6	1687	Cell material (Part IIb)
GaAs	1.424 (d)	zb	13.18	3.90	4.07	5.653	5.32	6.03	1510	Cell material (Ch. IId-1)
InP	1.35 (d)	zb	12.56	3.60	4.38	5.869	4.787	4.55	1340	Cell material (Ch. IId-1)
a-Si	~1.8 (d)	–	~11	3.32						Cell material, invariably Si:H alloy; sometimes also alloyed with germanium or carbon (Chs. IIc-1, -2)
CdTe	1.45–1.5 (d)	zb	10.2	2.89*	4.28	6.477	6.2	4.9	1365	Cell material (Ch. IIc-3)
$CuInSe_2$ (CIS)	0.96–1.04 (d)	ch			4.58			6.6	~1600	Cell material, often alloyed with gallium (Ch. IIc-4)
$Al_xGa_{1-x}As$ ($0 \leqslant x \leqslant 0.45$)	$1.424 + 1.247x$ (d)	zb	$13.18 - 3.12x$		$4.07 - 1.1x$	$5.653 + 0.0078x$	$5.36 - 1.6x$	$6.4 - 1.2x$		Window layer for GaAs solar cells
($0.45 < x \leqslant 1$)	$1.9 + 0.125x + 0.143x^2$ (i)				$3.64 - 0.14x$					

* At 600 nm.

Table 2 The energy gap E_g, refractive index n and the electron affinity χ of transparent conducting semiconductors used as window layers in thin-film solar cells

Material	E_g (eV)	n	χ
CdS	2.42	2.5	4.5
ZnS	3.58	2.4	3.9
$Zn_{0.3}Cd_{0.7}S$	2.8		4.3
ZnO	3.3	2.02	4.35
In_2O_3:Sn	3.7–4.4		4.5
SnO_2:F	3.9–4.6		4.8

Table 3 The refractive index at 590 nm (2.1 eV) of the common materials used for antireflection coating. The full wavelength dependence of most of these substances can be found in references [27] and [28]

Material	n
MgF_2	1.38
SiO_2	1.46
Al_2O_3	1.76
Si_3N_4	2.05
Ta_2O_5	2.2
ZnS	2.36
SiO_x	1.8–1.9
TiO_2	2.62

$$E_g(T) = E_{g0} - \frac{\alpha T^2}{T + \beta} \tag{1}$$

where T is the absolute temperature and the parameters α and β are given in Table 4.

The current produced by solar cells is generated by optical transitions across the band gap. Two types of such transitions can be distinguished: direct transitions where the momentum of the resulting electron–hole pair is very close to zero, and indirect transitions where the resulting electron hole pair has a finite momentum. The latter transitions require the assistance of a phonon (quantum of lattice vibration). Thus, there are two types of semiconductors:

- ***Direct gap semiconductors*** where the top of the valence band and the bottom of the conduction band lie at the Γ point of the first Brillouin zone (i.e., at zero wave vector $\mathbf{k} = 0$).
- ***Indirect gap semiconductors*** where the minima of the conduction band (in general, more than one) lie at a another point of the first Brillouin zone, with a different value of the wave vector $\mathbf{k}$.

	$E_{\Gamma 1}$	$E_{\Gamma 2}$	E_X	E_L	E_{SO}	n_c
Si	3.4	4.2	1.12	1.9	0.035	6
GaAs	1.42		1.90	1.71	0.34	1
InP	1.34		2.19	1.93	0.11	1

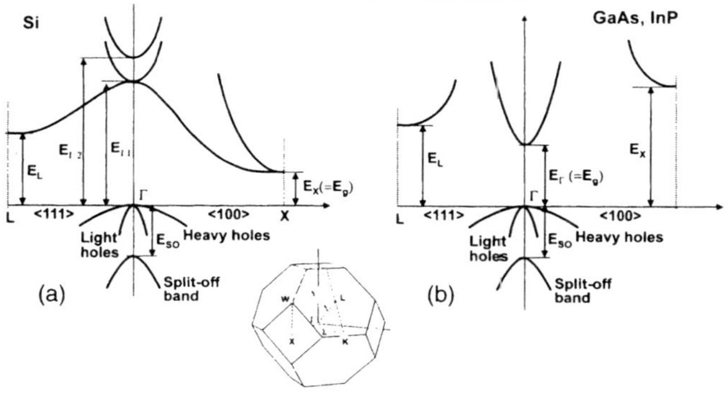

*Figure 1 The energy gaps in Si (a), and GaAs and InP (b) as functions of the wave vector **k**. The inset shows the Brillouin zone of the corresponding face-centred cubic crystal lattice.*

Table 4 The parameters E_{g0}, α and β in Equation (1). Sources of data: (a) Thurmond [11]; (b) Varshni [10]

	E_g ($T = 0$ K), eV	$\alpha \times 10^{-4}$, eV/K^2	β, K
Si (a)	1.17	4.730	636
GaAs (a)	1.52	5.405	204
InP (b)	1.42	4.906	327

The optical absorption in indirect-gap semiconductors is considerably weaker than in direct gap semiconductors, as shown in Figure 2 on the example of silicon and gallium arsenide.

3 Carrier Statistics in Semiconductors

In thermal equilibrium the temperature and electrochemical potential (the Fermi level, denoted by E_F) are constant throughout the device. The product of the electron concentration n and the hole concentration p is then independent of doping and obeys the mass action law

$$np = n_i^2 = \mathcal{N}_c \mathcal{N}_v \, exp\left(-\frac{E_g}{k_B T}\right) \tag{2}$$

where k_B is the Boltzmann constant, n_i is the electron (or hole) concentration in the intrinsic semiconductor, and the effective densities of states $\mathcal{N}_c$ and $\mathcal{N}_v$ are given by

$$\mathcal{N}_c = 2\left(\frac{2\pi m_e k_B T}{h^2}\right)^{3/2} \qquad \mathcal{N}_v = 2\left(\frac{2\pi m_h k_B T}{h^2}\right)^{3/2} \tag{3}$$

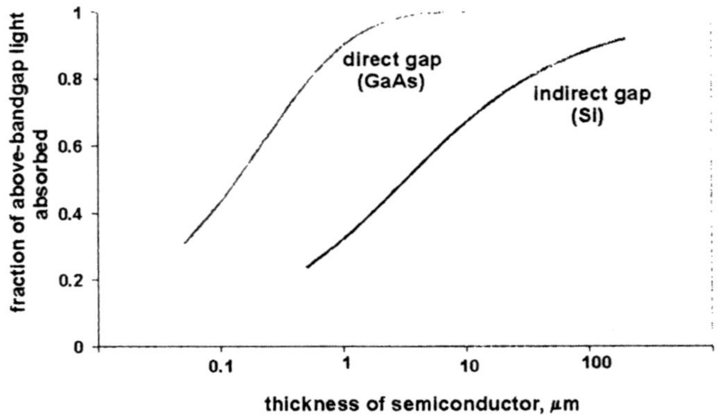

Figure 2 A comparison of the difference in strength of the optical absorption in direct and indirect semiconductors, illustrated on the examples of silicon and gallium arsenide.

Here, h is the Planck constant and m_e and m_h are the electron and hole density-of-states effective masses (see Table 6). In Equation (3), the effective mass m_e includes a factor which allows for several equivalent minima of the conduction band in indirect-gap semiconductors (see Figure 1).

Equation (2) does not hold for a solar cell in operation. Current flow as well as electron transitions between different bands and/or other quantum states are induced by differences and gradients of the electrochemical potentials, and temperature gradients may also exist. It is then usual to assign a quasi-Fermi level to each band which describes the appropriate type of carriers. Thus, electrons in the conduction band will be described by the quasi-Fermi level E_{Fn}, and holes by E_{Fp}. It is convenient to define also the appropriate potentials ϕ_n and ϕ_p by

$$\begin{aligned} E_{Fn} &= -q\phi_n \\ E_{Fp} &= -q\phi_p \end{aligned} \tag{4}$$

The use of quasi-Fermi levels to describe solar cell operation was already mentioned in Section 4.1 of Chapter IIa-1. The formalism leads to expressions for electron and hole concentrations which are summarised in Table 5.

4 The Transport Equations

The electron and hole current densities $\boldsymbol{J}_n$ and $\boldsymbol{J}_p$ are governed transport by the transport equations

$$\begin{aligned} \boldsymbol{J}_n &= q\mu_n n\mathcal{E} + qD_n\nabla n \\ \boldsymbol{J}_p &= q\mu_p p\mathcal{E} - qD_p\nabla p \end{aligned} \tag{5}$$

Table 5 Carrier concentration in degenerate and non-degenerate semiconductors. $F_{1/2}$ denotes the integral $F_{1/2}(z) = \frac{2}{\sqrt{\pi}}\int_0^{\infty} \frac{x^{1/2}}{1+exp(x-z)}dx$

	Non-degenerate semiconductors		General expressions
	In terms of the band parameters	In terms of the parameters of intrinsic semiconductor	
n	$\mathcal{N}_c exp\left(\frac{E_{Fn}-E_c}{k_B T}\right)$	$n_i exp\left\{\frac{q(\psi-\varphi_n)}{k_B T}\right\}$	$\mathcal{N}_c F_{1/2}\left(\frac{E_{Fn}-E_c}{k_B T}\right)$
p	$\mathcal{N}_v exp\left(\frac{E_v-E_{Fp}}{k_B T}\right)$	$n_i exp\left\{\frac{q(\varphi_p-\psi)}{k_B T}\right\}$	$\mathcal{N}_v F_{1/2}\left(\frac{E_v-E_{Fp}}{k_B T}\right)$

Table 6 The densities of states in the conduction and valence band ($\mathcal{N}_c$ and $\mathcal{N}_v$, respectively), the intrinsic carrier concentration n_i and the density-of-states effective masses m_e and m_h (all at 300 K) in Si, GaAs and InP. m_o denotes the free-electron mass. Data from references [20, 35]; intrinsic carrier concentration in Si from reference [12]

	$\mathcal{N}_c$ (cm^{-3})	$\mathcal{N}_v$ (cm^{-3})	n_i (cm^{-3})	m_e/m_o	m_h/m_o
Si	2.8×10^{19}	1.04×10^{19}	1.00×10^{10}	1.08	0.55
GaAs	4.7×10^{17}	7.0×10^{18}	1.79×10^{6}	0.063	0.53
InP	5.7×10^{17}	1.1×10^{19}	1.3×10^{7}	0.08	0.6

where n and p are the electron and hole the concentrations, μ_n and μ_p are the electron and hole mobilities, D_n and D_p are the electron and hole diffusion constants, and $\boldsymbol{\mathcal{E}}$ is the electric field. The first term in each equation is due to drift in the electric field $\boldsymbol{\mathcal{E}}$, and the second term corresponds to carrier diffusion. With the help of the quasi-Fermi levels, Equations (5) can be written as

$$\begin{aligned} \boldsymbol{J}_n &= -q\mu_n n \nabla\phi_n \\ \boldsymbol{J}_p &= -q\mu_p n \nabla\phi_p \end{aligned} \tag{6}$$

Equations (6) are valid for a semiconductor with a homogeneous composition but position-dependent dopant concentration is included. A generalisation to semiconductors with position-dependent band gap can be found, for example, in [13].

In a region where space charge exists (for example, in the junction), the *Poisson equation* is needed to link the electrostatic potential ψ with the charge density ρ:

$$\nabla \cdot \boldsymbol{\mathcal{E}} = \frac{\rho}{\varepsilon} \tag{7}$$

with

$$\rho = q(p - n + N_D - N_A) \tag{8}$$

$$\mathcal{E} = -\nabla \cdot \psi \tag{9}$$

where ε is the static dielectric constant of the semiconductor (see Table 1).

In non-degenerate semiconductors, the diffusion constants are related to mobilities by the *Einstein relations*

$$D_n = \frac{kT}{q}\mu_n; \qquad D_p = \frac{kT}{q}\mu_p \tag{10}$$

The generalisation of the Einstein relations to degenerate semiconductors is discussed, for example, in [14].

The conservation of electrons and holes is expressed by the *continuity equations*

$$\begin{aligned} \frac{\partial n}{\partial t} &= G - U + \frac{1}{q}\nabla \cdot \boldsymbol{J}_n \\ \frac{\partial p}{\partial t} &= G - U + \frac{1}{q}\nabla \cdot \boldsymbol{J}_p \end{aligned} \tag{11}$$

where G and U are the generation and recombination rates which may be different for electrons and holes if there are transitions into or from localised states.

5 Carrier Mobility

In weak fields, the drift mobilities in Equations (5) represent the ratio between the mean carrier velocity and the electric field. The mobilities – which are generally different for majority and minority carriers – depend on the concentration of charged impurities and on the temperature. For silicon, these empirical dependencies are generally expressed in the Caughey–Thomas form [15]:

$$\mu = \mu_{min} + \frac{\mu_0}{1 + \left(\dfrac{N}{N_{ref}}\right)^{\alpha}} \tag{12}$$

The values of the various constants for majority and minority carriers are given in Tables 7 and 8. A full model which includes the effects of lattice scattering, impurity scattering, carrier-carrier scattering and impurity clustering effects at high concentration is described in [16] where the reader can find further details.

In strong fields, the mobility of carriers accelerated in an electric field parallel to the current flow is reduced since the carrier velocity saturates. The field dependence of the mean velocity v (only quoted reliably for majority carriers) is given by

Table 7 The values of parameters in Equation (12) for majority-carrier mobility in silicon (from reference [17]; $T_n = T/300$)

	$\mu_{min} = AT_n^{-\beta}$		$\mu_0 = BT^{-\beta_2}$		$N_{ref} = CT_n^{\beta_3}$		$\alpha = DT_n^{-\beta_4}$	
	A	β_1	B	β_2	C	β_3	D	β_4
Electrons	88	0.57	7.4×10^8	2.33	1.26×10^{17}	2.4	0.88	0.146
Holes	54.3		1.36×10^8		2.35×10^{17}			

Table 8 The values of parameters in Equation (12) for minority-carrier mobility in silicon (from references [18, 19])

	μ_{min}	μ_0	N_{ref}	α
Electrons	232	1180	8×10^{16}	0.9
Holes	130	370	8×10^{17}	1.25

$$v = \frac{\mu_{lf}}{\left(1 + \left(\frac{\mu_{lf}\mathcal{E}}{v_{sat}}\right)^{\beta}\right)^{1/\beta}} \tag{13}$$

where μ_{lf} is the appropriate low field value of the mobility (Equation (12)), the parameter $\beta = 2$ for electrons and $\beta = 1$ for holes, and v_{sat} is the saturation velocity, identical for electrons and holes [20]:

$$v_{sat} = \frac{2.4 \times 10^7}{1 + 0.8e^{T/600}} \tag{14}$$

In gallium arsenide, the empirical fitting of the mobility data is of a more complex nature, and the reader is referred to the reference [21] for more information regarding the majority-carrier mobility. The temperature and dopant concentration dependences of the minority-carrier mobility are shown in Figures 3 and 4.

The electric field dependence of carrier mobility in GaAs is different from silicon as the velocity has an 'overshoot' which is normally modelled by the following equation [24]:

$$\mu_n = \frac{\mu_{lf}\mathcal{E} + v_{sat}(\mathcal{E}/\mathcal{E}_0)^{\beta}}{1 + (\mathcal{E}/\mathcal{E}_0)^{\beta}} \tag{15}$$

where μ_{lf} is the appropriate low-field mobility, $\mathcal{E}_0 = 4 \times 10^3$ V/cm, β equals 4 for electrons and 1 for holes, and v_{sat} is given by [25]

$$v_{sat} = 11.3 \times 10^6 - 1.2 \times 10^4 T \tag{16}$$

where T is the temperature in K.

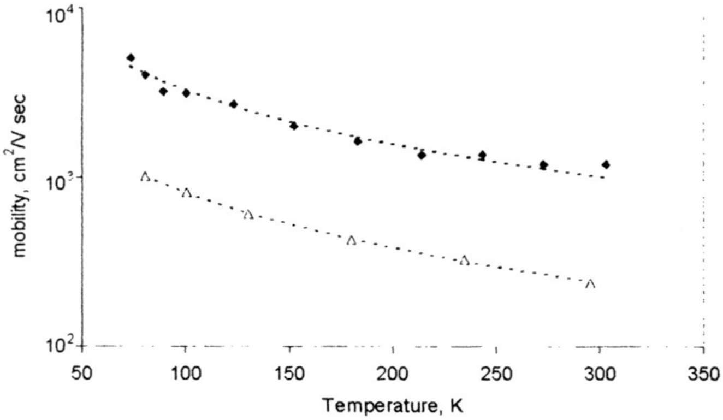

Figure 3 The temperature dependence of the minority-carrier mobility in GaAs. Full symbols: electron mobility in p-type GaAs ($N_A = 4 \times 10^{18}$ cm^{-3}); data from [22]. Empty symbols: hole mobility in n-type GaAs ($N_D = 1.8 \times 10^{18}$ cm^{-3}) [23]. The dashed lines show fits to the data with expressions

$$\mu_n = \frac{337\,100}{T} - 116; \qquad \mu_p = \frac{85\,980}{T} - 49.84;$$

where T is the temperature in K.

The majority carrier mobility in indium phosphide reported in reference [26] is shown in Figure 5.

6 Carrier Generation by Optical Absorption

6.1 Band-to-Band Transitions

The principal means of carrier generation in solar cells is the absorption of light. For a planar slab (Figure 6), a photon which enters the semiconductor generates $g(x)\,\delta x$ electron hole pairs in a thin layer at depth $x \rightarrow x + \delta x$. The generation function $g(x)$ is given by

$$g(x) = \alpha(\lambda) exp\{-\alpha(\lambda)x\} \qquad (17)$$

where $\alpha(\lambda)$ is the absorption coefficient, shown in Figure 7 for a number of semiconductors with photovoltaic applications. The generation rate G per unit volume which appears in Equation (11) is related to the generation function g in Equation (17) by $G = g/A$, where A is the illuminated area of the sample.

A useful formula for the absorption coefficient of silicon is provided by the expression of Rajkanan et al. [29]:

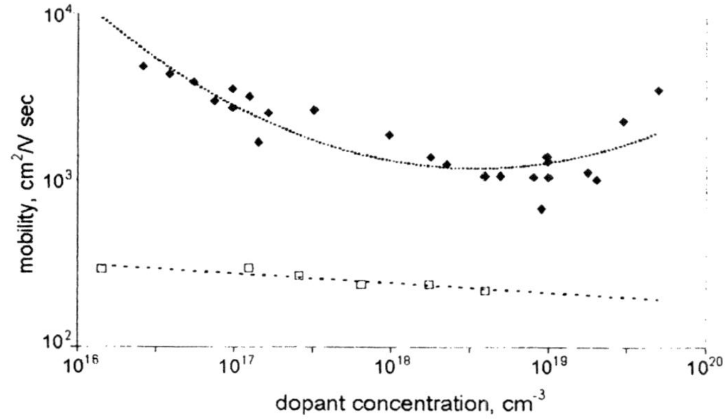

Figure 4 The dopant concentration dependence of the room-temperature minority-carrier mobility in GaAs. Full symbols: electron mobility in p-type GaAs; data from [21]. Empty symbols: hole mobility in n-type GaAs [23]. The dashed lines show fits to the data with the expressions

$$log\ \mu_n = 0.16(log\ N_A)^2 - 5.93 log\ N_A + 58$$
$$log\ \mu_p = -0.0575 log\ N_D + 3.416$$

where N_D *and* N_A *are the donor and acceptor concentrations in* cm^{-3}.

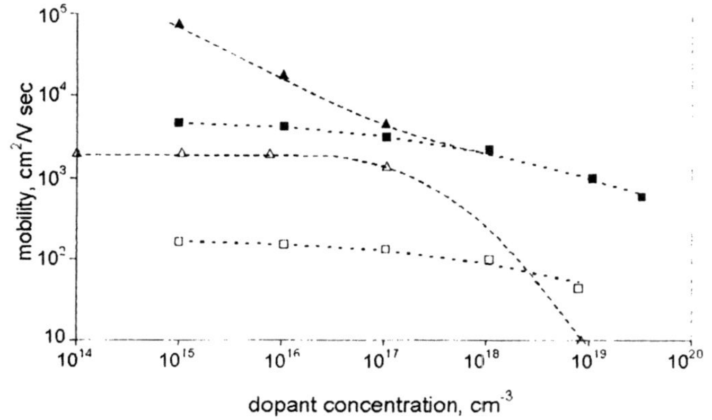

Figure 5 The majority-carrier mobilities in InP [26]. Full symbols correspond to electron mobility, empty symbols to hole mobility. Room temperature (300 K) data are shown by squares; data at 77 K by triangles. The room-temperature data were fitted with a Caughey–Thomas-type expression with parameters shown in Table 9.

$$\alpha(T) = \sum_{\substack{i=1,2\\j=1,2}} C_i A_j \left[\frac{\{h\nu - E_{gj}(T) + E_{pi}\}^2}{\{exp(E_{pi}kT) - 1\}} + \frac{\{h\nu - E_{gj}(T) - E_{pi}\}^2}{\{1 - exp(-E_{pi}kT)\}} \right] + A_d\{h\nu - E_{gd}(t)\}^{1/2} \tag{18}$$

Table 9 Parameters in Equation (12) which were used to fit the room-temperature electron and hole mobilities in InP

	μ_{min}	μ_O	N_{ref}	α
Electrons	0	4990	4.02×10^{17}	0.433
Holes		168	1.24×10^{18}	

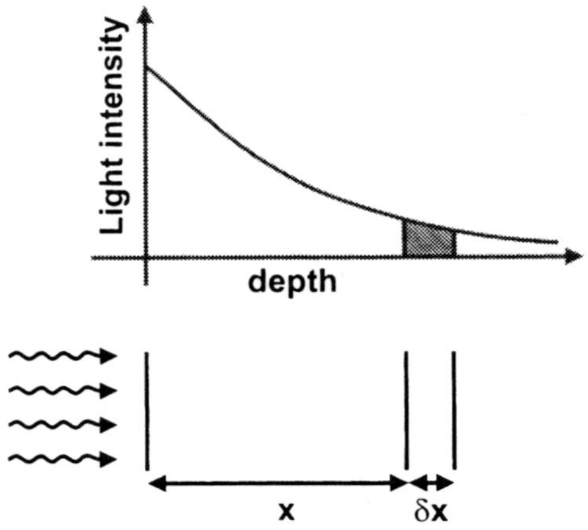

Figure 6 The geometry used to discuss the light absorption in semiconductors.

where $h\nu$ is the photon energy, $E_{g1}(0) = 1.1557$ eV, $E_{g2}(0) = 2.5$ eV and $E_{gd}(0) = 3.2$ eV are the two lowest indirect and the lowest direct band gap, respectively (used here as parameters to obtain a fit to the spectrum), $E_{p1} = 1.827 \times 10^{-2}$ eV and $E_{p2} = 5.773 \times 10^{-2}$ eV are the Debye frequencies of the transverse optical and transverse acoustic phonons, respectively, $C_1 = 5.5$, $C_2 = 4.0$, $A_1 = 3.231 \times 10^2$ cm^{-1} eV^{-2}, $A_2 = 7.237 \times 10^3$ cm^{-1} eV^{-2} and $A_1 = 1.052 \times 10^6$ cm^{-1} eV^{-2}. The temperature variation of the band gaps is described by Equation (1) where the original Varshni coefficients $\alpha = 7.021 \times 10^{-4}$, eV/K^2 and $\beta = 1108$ K are used for all three band gaps E_{g1}, E_{g2} and E_{gd}.

6.2 Free-Carrier Absorption

In regions with high-carrier concentration (due to doping or strong illumination, for example) photon absorption can also occur by electron transitions with initial and final states inside the same band. This free-carrier absorption does not generate electron–hole pairs and competes with the band-to-band transitions that produce the photogenerated current which were discussed above. Free carrier absorption might be significant in the case of photon energies near the band gap, and should not be included in the absorption coefficient α in front of the exponential in Equation (17). Figure 8 illustrates the different phenomena that occur near the band edge for high doping concentrations.

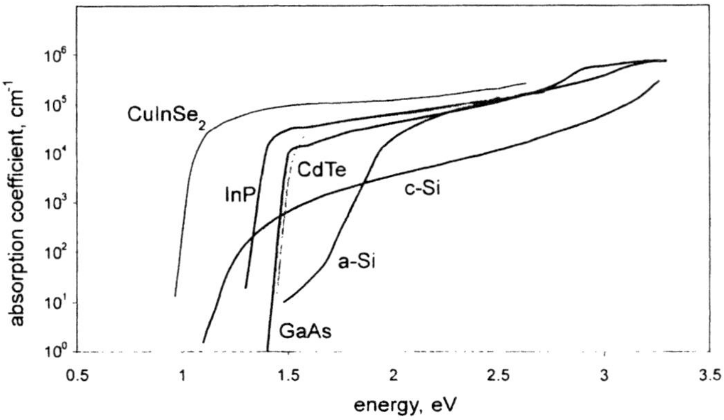

Figure 7 The absorption coefficients of the principal semiconductors used in the manufacture of solar cells. Full details of these and other semiconductors can be found in references [27] and [28]. Further details of silicon absorption are given in the text.

Table 10. The constants for the free carrier absorption coefficient in Equation (19)

	K_1	a	K_2	b
Si	2.6×10^{-27}	3	2.7×10^{-24}	2
GaAs	4×10^{-29}	3	–	–
InP	5×10^{-27}	2.5	–	–

Based on experimental data of [30] and [31], the PC1D model, for example, uses the following expression for the absorption coefficient due to free carrier absorption:

$$\alpha_{FC} = K_1 n \lambda^a + K_2 p \lambda^b \tag{19}$$

where λ is in nanometres and the empirically determined constants K_1, K_2, a, and b are given in Table 10.

7 Recombination

Recombination processes can be classified in a number of ways. Most texts distinguish between bulk and surface recombination, and between band-to-band recombination as opposed to transitions with the participation of defect levels within the band gap. Recombination processes can also be classified according to the medium which absorbs the energy of the recombining electron–hole pair: radiative recombination (associated with photon emission), or the two principal non-radiative mechanisms by Auger and multi-phonon transitions, where the recombination energy is absorbed by a free charge carrier or by lattice vibrations, respectively. An opposite process to Auger recombination (where an electron hole pair is generated rather than consumed) is called impact ionisation.

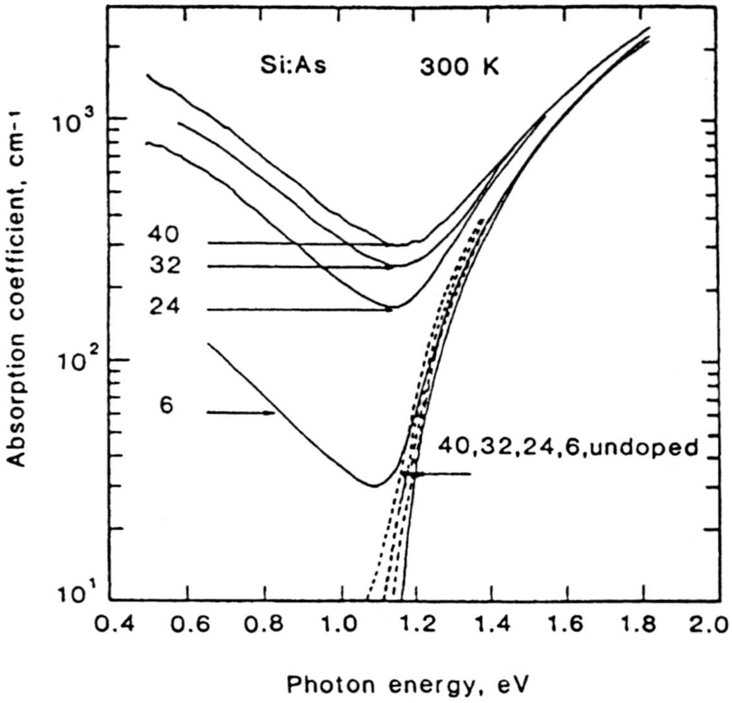

Figure 8 The effects observed near the absorption edge at high doping concentrations, illustrated here on the example of n-type silicon (from reference [30], as adapted by M.A. Green, Silicon Solar Cells – See Bibliography).

The following sections give a brief summary of the main recombination mechanisms with relevance to solar cell operation which are summarised schematically in Figure 9.

7.1 Bulk Recombination Processes

A detailed discussion of the variety of the bulk recombination mechanisms can be found in [32]. Here, we confine ourselves to a brief overview of the radiative, Auger and defect-assisted recombination processes which are most frequently encountered in practical operation of solar cells. These processes are depicted schematically in Figure 9 which also shows the notation used to describe the relevant parameters.

The rate of *band-to-band radiative recombination* can be written in the form

$$U_{rad} = B(np - n_i^2) \tag{20}$$

where the coefficient B is sometimes written as $R/n_i{}^2$. Radiative transitions between a free electron and a localised state within the band gap may also be important in certain situations, for example in novel concepts such as the

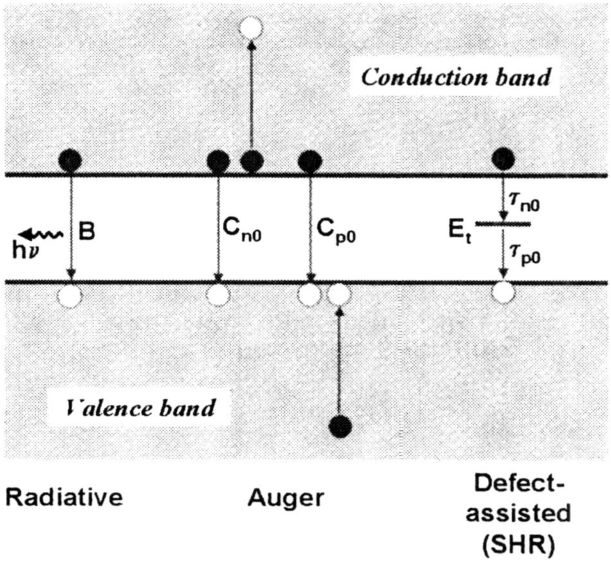

Figure 9 A schematic diagram of the principal recombination processes in semiconductors, and the notation for the rate constant adopted in this book. The direction of arrows indicates electron transitions.

impurity photovoltaic effect. The values of B for silicon, gallium arsenide and indium phosphide can be found in Table 11. In other situations, the radiative recombination coefficient B can be determined from optical absorption using the detailed balance argument due to van Roosbroeck and Shockley [33].

The rate of *band-to-band Auger recombination* can be written as

$$U_{Auger} = (C_{p0}p + C_{n0}n)(np - n_i^2) \tag{21}$$

where the first bracket gives the two, usually most important, Auger terms. Table 11 gives values of the coefficients C_{po} and C_{no} for Si, GaAs and InP.

The *recombination rate via defects* of concentration N_t with a level at energy E_t within the band gap is described by the Shockley–Read–Hall formula [36, 37]:

$$U_{SHR} = \frac{np - n_i^2}{\tau_p(n + n_1) + \tau_n(p + p_1)} \tag{22}$$

where

$$n_1 = n_i\, exp\left(\frac{E_t - E_i}{k_B T}\right) \qquad p_1 = p_i\, exp\left(\frac{E_t - E_i}{k_B T}\right) \tag{23}$$

and τ_n, τ_p are parameters, proportional to the defect concentration N_t, which are characteristic for the particular defect and energy level. At low injection, τ_n and τ_p assume the meaning of minority carrier lifetimes. With an appropriate

Table 11 The coefficients B of the radiative recombination rate (Equation (20)) and C_{no} and C_{po} of the Auger recombination rate (Equation (21)). Sources of data: (a) [34]; (b) [32]; (c) [35]; (d) [26]

	B, cm^3 s^{-1}	C_{no}, cm^6 s^{-1}	C_{po}, cm^6 s^{-1}
Si	1.8×10^{-15} (a)	2.8×10^{-31} (b)	0.99×10^{-31}(b)
GaAs	7.2×10^{-10} (a)	$\sim 10^{-30}$ (c)	
InP	6.25×10^{-10} (d)	$\sim 9 \times 10^{-31}$ (c)	

dependence on the doping concentration and temperature, τ_n and τ_p are also used extensively in material and device modelling (see reference [20], Section 1.5.3 and Section 7.3 below for a fuller discussion).

7.2 Surface Recombination

Surface recombination velocity is an important parameter which affects the dark saturation current and the quantum efficiency of solar cells. Similarly to dislocations and planar defects such as grain boundaries, surfaces (and interfaces in general) introduce band of electronic states in the band gap which can be ascribed to broken (or strained) bonds and impurities. A complete characterisation of surface recombination must also take into account the surface charge which may give rise to band bending. To achieve optimal operation, surface recombination is reduced by a passivating or window layer which prevents minority carriers from reaching the surface. Passivation of silicon surface by an oxide layer, or the deposition of a thin 'window' layer of GaAlAs on GaAs solar cells are just two examples of such practice.

For an oxidised silicon surface, surface recombination velocity is strongly dependent on the surface roughness, contamination, ambient gases used during oxidation and the annealing conditions. Under identical process parameters, however, one can identify trends in the dependence of the surface recombination velocity on the surface doping concentration. Cuevas et al. [38] proposed the following analytical relationship between surface recombination velocity and doping concentration:

$$S = 70 \text{ cm/s} \quad \text{for } N < 7 \times 10^{17} \text{cm}^{-3}$$

$$S = 70\left(\frac{N}{7 \times 10^{17}}\right) \quad \text{for } N > 7 \times 10^{17} \text{ cm}^{-3} \tag{24}$$

Equation (24) models several experimental results such as those reported in [39].

In gallium arsenide, the surface recombination velocity is very high (of the order of 10^6 cm/s). The deposition of a thin layer of GaAlAs, however, reduces the recombination velocity at the interface to 10–10^3 cm/s (see reference [40], p. 41). Coutts and Yamaguchi [41] quote the values of 10^3–2×10^4 cm/s and 1.5×10^5 cm/s for the surface recombination velocity in n- and p-type InP, respectively.

7.3 Minority-Carrier Lifetime

Under low injection – a regime of particular importance for solar cell operation – the majority-carrier concentration can be assumed excitation independent, and the effect of recombination can be discussed in terms of minority-carrier lifetime. In p-type material, for example, and the recombination rate can be written as

$$U = \frac{1}{\tau_n}(n - n_0) \tag{25}$$

where τ_n is the minority-carrier (electron) lifetime. An analogous equation can be written for the hole lifetime τ_p in n-type material. The inverse of the lifetime – the rate constant – is a sum of the different contributions to the lifetime:

$$\frac{1}{\tau} = \frac{1}{\tau_{rad}} + \frac{1}{\tau_{Auger}} + \frac{1}{\tau_{SRH}} \tag{26}$$

where τ stands for τ_n or τ_p, as appropriate. This additive nature of the recombination rate constant is also useful when discussing the radiation damage (see Section 8).

The effect of lifetime on transport properties by carrier diffusion can be discussed in terms of the diffusion length defined by

$$L = \sqrt{D\tau} \tag{27}$$

where D is the diffusion constant for the minority carriers in question. If, however, drift in electric field $\mathcal{E}$ is the dominant transport mechanism, it is appropriate to define the drift length as

$$\ell_n = \mathcal{E}\tau_n\mu_n; \qquad \ell_p = \mathcal{E}\tau_p\mu_p \tag{28}$$

for electrons and holes as minority carriers. This parameter plays an important role in the analysis of p–i–n junction solar cells (see Section 4.3 in Chapter IIa-1).

The wealth of available data for crystalline silicon has made it possible to arrive at a consensus as to the magnitude as well as the temperature and doping concentration dependence of the contributions (26) to minority carrier lifetime [42] The contribution to lifetime due to defects, when combined with recombination in intrinsic material, has been empirically observed to follow the equations

$$\begin{aligned}
\frac{1}{\tau_{n,SRH}} &= \left(\frac{1}{2.5 \times 10^{-3}} + 3 \times 10^{-13} N_D\right)\left(\frac{300}{T}\right)^{1.77} \\
\frac{1}{\tau_{p,SRH}} &= \left(\frac{1}{2.5 \times 10^{-3}} + 11.76 \times 10^{-13} N_A\right)\left(\frac{300}{T}\right)^{0.57}
\end{aligned} \tag{29}$$

where the first term in the brackets applies for recombination in intrinsic semiconductor. Similarly, the contribution to Equation (26) by Auger recombination can be described by the expressions

$$\frac{1}{\tau_{n,Auger}} = 1.83 \times 10^{-31} p^2 \left(\frac{T}{300}\right)^{1.18}$$
$$\frac{1}{\tau_{p,Auger}} = 2.78 \times 10^{-31} n^2 \left(\frac{T}{300}\right)^{0.72} \qquad (30)$$

Although the concept of minority carrier lifetime is most commonly applied to bulk recombination, a similar notion can be relevant for surface processes. For example, the effective lifetime observed of minority carriers with uniform concentration in a wafer can be written as

$$\frac{1}{\tau_{eff}} = \frac{1}{\tau_{SRH}} + \frac{2W}{A} S \qquad (31)$$

where S is the value of the recombination velocity, W is the wafer thickness and A is the area of the sample.

8 Radiation Damage

Solar cells which operate on board of a satellite in an orbit which passes through the van Allen belts are subjected to fluxes of energetic electrons and protons trapped in the magnetic field of the earth and by fluxes of particles associated with high solar activity [43, 44] (see Chapter IIId-2). When slowed down in matter, most of the energy of the incident proton or electron is dissipated by interaction with the electron cloud. A relatively small fraction of this energy is dissipated in collisions with the nuclei, resulting in the formation of a lattice defect when energy in excess of a minimum threshold value is transferred to the nucleus [45, 46]. A proton with energy in excess of this threshold causes the most damage (typically, of the order of 10–100 atomic displacements) near the end of its range in the crystal. At high energy, on the other hand, a proton creates simple point defects, and the displacement rate decreases with increasing proton energy. In contrast, the displacement rate by electrons increases rapidly with energy and approaches a constant value at higher energies. Electron damage can usually be assumed to be uniform throughout the crystal.

Some defects which are thus created act as recombination centres and reduce the minority-carrier lifetime τ. Thus, it is the lifetime τ (or, equivalently, the diffusion length L) which is the principal quantity of concern when the cell is subjected to the particle radiation in space. Under low injection, the reduction of the diffusion length L is described by the Messenger–Spratt equation (see, for example, [47], p. 151):

$$\frac{1}{L^2} = \frac{1}{L_0^2} + K_L\phi \tag{32}$$

where L_o is the diffusion length in the unirradiated cell, ϕ is the particle fluence (integrated flux), and K_L is a (dimensionless) diffusion-length damage constant characteristic for the material and the type of irradiation.

The damage constant K_L generally depends on the type of dopant and, even in single-junction cells, a different damage constant should therefore be introduced for each region of the cell. Moreover, care should be exercised when dealing with low-energy protons (with energy of the order of 0.1–1 MeV) which are stopped near the surface and may create non-uniform damage near the junction. Figures 10 and 11 show the doping concentration and energy dependence of the damage constant K_L for electrons and protons in p-type silicon. Details of the damage constants for other materials can be found in references [43], [44], [48] and [49].

Damage constants are usually quoted for 1 MeV electrons, and converted to other energies and particles (such as protons) by using the concept of damage equivalent. Conversion tables are available in the Solar Radiation Handbooks which will suit most circumstances. Radiation damage equivalence works well when applied to uniform damage throughout the cell but care should be exercised in the case of non-uniform damage, for example, for low-frequency protons.

The reduction of the diffusion length describes usually the most significant part of the damage but changes in other parameters on irradiation also sometimes need to be considered. The dark diode current I_o may increase as a result of compensation by the radiation-induced defects. This, however, occurs

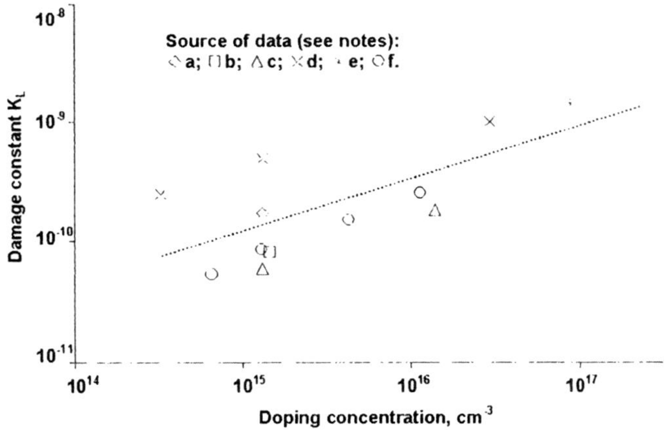

Figure 10 The radiation damage constant K_L for 1 MeV electrons in p-type silicon as a function of the dopant concentration. Source of data: (a) [50]; (b) [51]; (c) [52]; (d) [53]; (e) [54]; (f) [55]. The dashed line shows a fit to the data with the formula $K_L = 3.43 \times 10^{-17} N_A^{0.436}$, where N_A is the acceptor concentration in cm^{-3}.

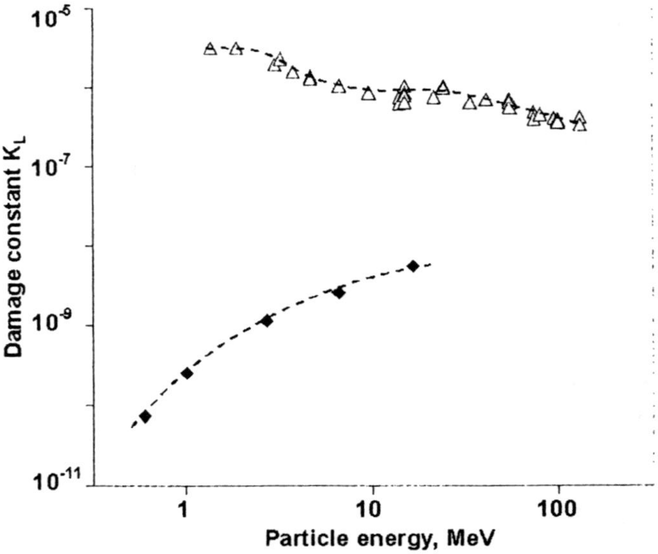

Figure 11 The energy dependence of the damage constant K_L *in 1 Ω cm p-type silicon by electron (full symbols, [55]) and protons (empty symbols, [56]).*

usually at a fluence orders of magnitude higher than that which degrades τ or L. Other source of degradation may be defects created in the depletion region, at the interfaces between different regions of the cell, or at the external surfaces.

9 Heavy Doping Effects

At high doping densities the picture of independent electrons interacting with isolated impurities becomes insufficient to describe the electron properties of semiconductors. As doping concentration increases, an impurity band is formed from the separate Coulomb wave functions located at individual doping impurities. This band gradually merges with the parent band and eventually gives rise to a tail of localised states. Electron–electron interaction becomes important, and the exchange and correlation energy terms have to be taken into account for a satisfactory description of the optical parameters and semiconductor device operation. An early overview of this multi-faceted physical problem can be found in [57].

In practical situations, this complex phenomenon is usually described in terms of band-gap narrowing, with a possible correction to effective masses and band anisotropy. Different manifestations of this effect are normally found by optical measurements (including absorption and low-temperature luminescence) and from data which pertain to device operation. Parameters which have been inferred from the latter are usually referred to as apparent band gap narrowing denoted by ΔE_g. For n-type material, for example,

$$\Delta E_g = kT \ln\left(\frac{p_O N_D}{n_i^2}\right) \tag{33}$$

where p_O is the minority-carrier (hole) concentration and N_D is the dopant (donor) concentration.

The experimental data obtained by various methods have been reviewed in [58] where the reader can find references to much of the earlier work. Jain et al. [59, 60] obtain a fit for band-gap narrowing in various materials using a relatively simple and physically transparent expression which, however, is less easy to apply to device modelling. Values for silicon which are now frequently used in semiconductor devices modelling are based on the work of Klaasen et al. [61] After reviewing the existing experimental data and correcting for the new mobility models and a new value for the intrinsic concentration, Klaasen et al. show that the apparent band gap narrowing of n- and p-type silicon can be accurately modelled by a single expression:

$$\Delta E_g(meV) = 6.92\left[\ln\left(\frac{N_{dop}}{1.3\times10^{17}}\right) + \sqrt{\left(\ln\frac{N_{dop}}{1.3\times10^{17}}\right)^2 + 0.5}\right] \tag{34}$$

where N_{dop} is the dopant concentration

For gallium arsenide, Lundstrom et al. [62] recommend the following formula based on fitting empirical data:

$$\Delta E_g = AN_{dop}^{1/3} + k_B T \ln\{F_{1/2}(E_F)\} - E_F$$

where E_F is the Fermi energy, the function $F_{1/2}$ is defined in the caption to Table 5, and

$$A = \left\langle \begin{matrix} 2.55\times10^{-8}eV & (p - GaAs) \\ 3.23\times10^{-8}eV & (n - GaAs) \end{matrix} \right.$$

10 Properties of Hydrogenated Amorphous Silicon[1]

In hydrogenated amorphous silicon (a-Si:H), the effective band gap between the conduction and valence band edges is around 1.8 eV but a thermal shrinking of the band gap with temperature has been reported [63]:

$$E_g = E_{go} - \gamma(T - T_o) \qquad \gamma \approx 5k_B \tag{35}$$

[1] A rigorous description of the band structure and charge carrier transport in amorphous silicon is a complex matter well beyond the scope of this Handbook. This section gives a simplified picture in terms of effective parameters which has been used in success in semiconductor device modelling.

For statistical calculations, the corresponding effective densities of states can be approximated by $\mathcal{N}_c \approx \mathcal{N}_V \approx 4 \times 10^{19}$ cm^{-3}. In contrast to crystalline silicon, the conduction and valence bands show evidence of exponential tails of localised states within the band gap [64]:

$$\begin{aligned} D_{Ct} &= D_{CO}\, exp\left(\frac{E - E_C}{kT_C}\right) \quad D_{CO} \approx 0.8 \times 10^{21} cm^{-3} \quad kT_C \approx 30\ meV \\ D_{Vt} &= D_{VO}\, exp\left(\frac{E_V - E}{kT_V}\right) \quad D_{VO} \approx 1.1 \times 10^{21} cm^{-3} \quad kT_V \approx 50\ meV \end{aligned} \tag{36}$$

In addition, there is a Gaussian distribution of dangling bond states (states corresponding to non-saturated silicon bonds) around mid gap (Figure 12):

$$D_{DB} = \frac{N_{DB}}{\sqrt{2\pi w^2}}\, exp\left(-\frac{(E - E_{DB})^2}{2w^2}\right) \quad w \approx 100\ meV \quad E_{DB} = \frac{E_C + E_V}{2} \tag{37}$$

In device-quality intrinsic a-Si:H films, the density of dangling bonds N_{DB} ranges from 10^{15} to 10^{16} cm^{-3}.

The carrier mobility in the localised states (band tails and dangling bonds) is negligible. The accepted values for extended states in the valence and conduction band are μ_p = 0.5 cm^2/V-s and μ_n = 10 cm^2/V-s, respectively. As in other

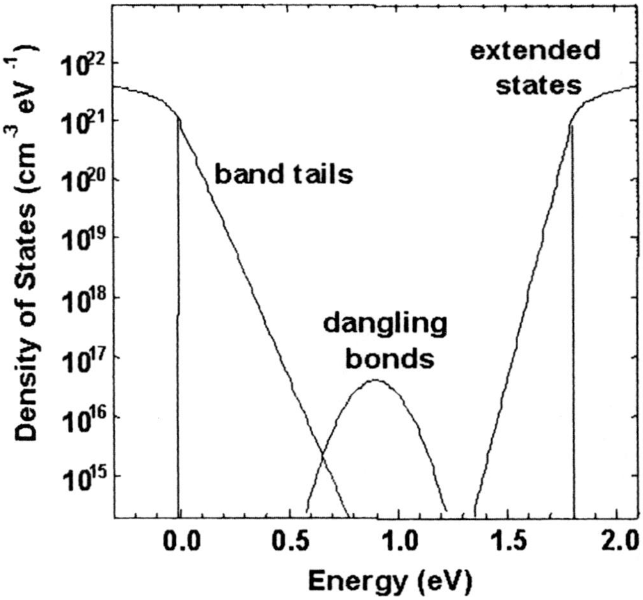

Figure 12 Typical density of states in a-Si:H. The states in the band tails behave as carrier traps whereas dangling bonds are recombination centres. Only those carriers in the extended states contribute to the electron transport [65].

semiconductors, the electron transport is given by the drift-diffusion equations (5) where the Einstein relations (10) apply.

The optical absorption coefficient shows three different zones (Figure 13):

- Band-to-band transitions (Tauc plot) for photon energies in excess of the band gap (> 1.8 eV).
- Transitions involving tail states (Urbach's front) for photon energies in the range 1.5–1.8 eV.
- Transitions involving dangling bonds for photon energies below 1.5 eV.

Once the optical absorption coefficient is known, the carrier generation can be easily calculated according to Equation (17) but it is important to note that only photons with energies in excess of the band gap yield useful electron-hole pairs for photovoltaic conversion.

Finally, the dominant recombination mechanism in intrinsic a-Si:H is given by the modified Shockley–Read–Hall equation, as applied to the amphoteric distribution of dangling bonds:

$$U = v_{th}\left(n\sigma_n^o + p\sigma_p^o\right)\frac{N_{DB}}{1 + \frac{p}{n}\frac{\sigma_p^o}{\sigma_n^+} + \frac{n}{p}\frac{\sigma_n^o}{\sigma_p^-}} \tag{38}$$

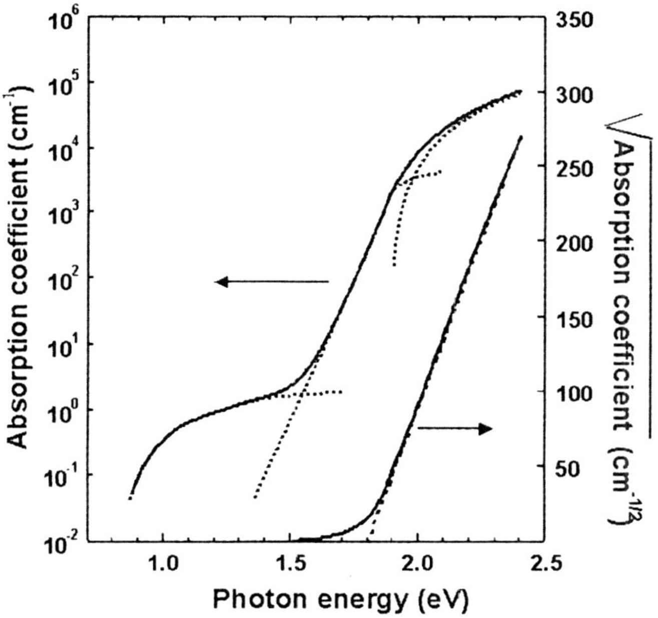

Figure 13 Typical optical absorption coefficient in a-Si:H. Three different regions can be observed corresponding to band-to-band absorption, transitions involving tail states and defect-associated absorption [65].

Table 12 Recombination parameters in amorphous silicon

$v_{th}\,(\mathrm{cm\cdot s^{-1}})$	$\sigma_n^o\,(\mathrm{cm^2})$	$\sigma_p^o\,(\mathrm{cm^2})$	$\sigma_n^+\,(\mathrm{cm^2})$	$\sigma_p^-\,(\mathrm{cm^2})$
10^7	10^{-15}	$\sigma_n^o/3$	$50\sigma_n^o$	$50\sigma_p^o$

where the parameters are given in Table 12 [66].

Acknowledgements

We are grateful to J. Puigdollers and C. Voz for providing graphs of the absorption coefficient and density of states in amorphous silicon.

References

[1] Schwartz, R.J. 1990. The application of numerical techniques to the analysis and design of solar cells, in T.J. Coutts and J.D. Meakin, Eds., *Current Topics in Photovoltaics*, Vol. 4, p. 25.

[2] Selberherr, S. 1984. *Analysis and Simulation of Semiconductor Devices*, Springer, Vienna, New York.

[3] Brozel, M.R. and Stillman, G.E., Eds. 1996. *Properties of Gallium Arsenide* (3rd edition), IEE/INSPEC, Institution of Electrical Engineers, London.

[4] Pearsall, T.P., Ed. 2000. *Properties, Processing and Applications of Indium Phosphide*, INSPEC/IEE, London.

[5] Shur, M. 1990. *Physics of Semiconductor Devices*, Prentice Hall, Englewood Cliffs, NJ.

[6] Chu, T.L. 1988. Cadmium telluride solar cells, in T.J. Coutts and J.D. Meakin, Eds., *Current Topics in Photovoltaics*, Vol. 3, p. 235.

[7] Kazmerski, L.L. and Wagner, S. 1985. Cu-ternary chalcopyrite solar cells, in Coutts, T.J. and Meakin, J.D., Eds., *Current Topics in Photovoltaics*, Vol. 1, p. 41.

[8] *Handbook of Physics and Chemistry of Solids* (82nd edition), CRC Press, Boca Raton, 2001.

[9] Adachi, S. 1985. GaAs, AlAs and $Al_xGa_{1-x}As$: material parameters for use in research and device applications, *J. Appl. Phys.*, Vol. 58, p. R1.

[10] Varshni, Y.P. 1967. Temperature dependence of the energy gap in semiconductors, *Physica*, Vol. 34, p. 149.

[11] Thurmond, C.D. 1975. The standard thermodynamic functions for the formation of electrons and holes in Ge, Si, GaAs and GaP, *J. Electrochem. Soc.*, Vol. 122, p. 1133.

[12] Sproul A.B. and Green, M.A. 1993. Intrinsic carrier concentration and minority-carrier mobility of silicon from 77-K to 300-K, *J. Appl. Phys.*, Vol. 73, pp. 1214–1225.

[13] Marshak A.H. and van Vliet, K.M. 1978. Electrical currents in solids with position dependent band structure, *Solid-State Electron.*, Vol. 21, p. 417.

[14] Smith, R.R. 1978. *Semiconductors* (2nd edition), Cambridge University Press, Cambridge.

[15] Caughey D.M. and Thomas, R.E. 1967. Carrier mobilities in silicon empirically related to doping and field, *Proc. IEEE*, Vol. 55, p. 2192.

[16] Klaassen, D.B.M. 1992. A unified mobility model for device simulation – I. Model equations and concentration dependence, *Solid-State Electron.*, Vol. 35, pp. 953–959.

[17] Arora, N.D., Hauser, T.R. and Roulston, D.J. 1982. Electron and hole mobilities in silicon as a function of concentration and temperature, *IEEE Trans. Electron Devices*, Vol. ED-29, p. 292.

[18] Swirhun, S.E., Kwark Y.-H. and Swanson, R.M. 1986. Measurement of electron lifetime, electron mobility and band-dap narrowing in heavily doped p-type silicon, *IEDM'86*, pp. 24–27.

[19] del Alamo, J., Swirhun, S. and Swanson, R.M. 1985. Simultaneous measurement of hole lifetime, hole mobility and band-gap narrowing in heavily doped n-type silicon, *IEDM'85*, pp. 290–293.

[20] Sze, S.M. 1981. *Physics of Semiconductor Devices* (2nd edition), John Wiley & Sons, New York.

[21] Lancefield, D. 1996. Electron mobility in GaAs: overview, in Brozel, M.R. and Stillman, G.E., Eds., *Properties of Gallium Arsenide* (3rd edition), IEE/INSPEC, Institution of Electrical Engineers, London, p. 41.

[22] Harmon, E.S., Lovejoy, M.L., Lundstrom M.S. and Melloch, M.R. 1996. Minority electron mobility in doped GaAs, in Brozel, M.R. and Stillman, G.E., Eds., *Properties of Gallium Arsenide* (3rd edition), IEE/INSPEC, Institution of Electrical Engineers, London, p. 81.

[23] Lovejoy, M.L., Melloch, M.R. and Lundstrom, M.S. 1996. Minority hole mobility in GaAs, in Brozel, M.R. and Stillman, G.E., Eds., *Properties of Gallium Arsenide* (3rd edition), IEE/INSPEC, Institution of Electrical Engineers, London, p. 123.

[24] Barnes, J.J., Lomax R.J. and Haddad, G.I. 1976. Finite element simulation of GaAs MESFET's with lateral doping profiles and submicron gates, *IEEE Trans. Electron Devices*, Vol. ED-23, p. 1042.

[25] Littlejohn, M.A., Hauser, J.R. and Glisson, T.H. 1977. Velocity-field characteristics of GaAs with $\Gamma_6^c - L_6^c - X_6^c$ ordering, *J. Appl. Phys.*, Vol. 48, p. 4587.

[26] Ahrenkiel, R.K. 2000. Minority carrier lifetime in InP, in T.P. Pearsall, Ed., *Properties, Processing and Applications of Indium Phosphide*, INSPEC/IEE, London.

[27] Palik, E.D., Ed., 1985. *Handbook of Optical Constants of Solids*, Academic Press Handbook Series, Orlando.

[28] Palik, E.D., Ed., 1991. *Handbook of Optical Constants of Solids II*, Academic Press, San Diego.

[29] Rajkanan, K., Singh, R., and Shewchun, J. 1979. Absorption coefficient of silicon for solar cell calculations, *Solid-State Electron.*, Vol. 22, p. 793.

[30] Schmid, P.E. 1981. Optical absorption in heavily doped silicon, *Phys. Rev.*, Vol. B23, p. 5531.

[31] Fan, H.Y. 1967. In R.K. Willardson and A.C. Beer, Eds., *Semiconductors and Semimetals*, Vol. 3, Academic Press, p. 409.
[32] Landsberg, P.T. 1991. *Recombination in Semiconductors*, Cambridge University Press.
[33] van Roosbroeck, W. and Shockley, W. 1954. Photon-radiative recombination of electrons and holes in germanium, *Phys. Rev.*, Vol. 94, p. 1558.
[34] Pilkuhn, M.H., Light emitting diodes, in Moss, T.S., Ed., Vol. 4, *Handbook of Semiconductors*, North Holland, p. 539.
[35] Levinstein, M., Rumyantsev, S. and Shur, M., Eds., 1996, 1999. *Handbook Series on Semiconductor Parameters*, Vols. 1 and 2, World Scientific, London. See also http://www.ioffe.rssi.ru/SVA/NSM/Semicond/.
[36] Shockley, W. and Read, W.T. 1952. Statistics of the recombination of holes and electrons, *Phys. Rev.*, Vol. 87, p. 835.
[37] Hall, R.N. 1952. Electron hole recombination in germanium, *Phys. Rev.*, Vol. 87, p. 387.
[38] Cuevas, A., Giroult-Matlakowski, G., Basore, P.A., du Bois C., and King, R. 1994. Extraction of the surface recombination velocity of passivated phosphorus doped emitters, *Proc. 1st World Conference on Photovoltaic Energy Conversion*, Hawaii, pp. 1446–1449.
[39] King, R.R., Sinton, R.A. and Swanson, R.M. 1990. Studies of diffused emitters: saturation current, surface recombination velocity and quantum efficiency, *IEEE Trans. Electron Devices*, Vol. ED-37, p. 365.
[40] Andreev, V.M., Grilikhes, V.A. and Rumyantsev, V.D. 1997. *Photovoltaic Conversion of Concentrated Sunlight*, Wiley, Chichester.
[41] Coutts, T.J. and Yamaguchi, M. 1988. Indium phosphide based solar cells: a critical review of their fabrication, performance and operation, in T.J. Coutts and J.D. Meakin, Eds., *Current Topics in Photovoltaics*, Vol. 3, p. 79.
[42] Klaassen, D.B.M. 1992. A unified mobility model for device simulation – II. Temperature dependence of carrier mobility and lifetime, *Solid-State Electron.*, Vol. 35, p. 961.
[43] Tada, H.Y., Carter, J.R., Anspaugh B.E. and Downing, R.G. 1982. *Solar-Cell Radiation Handbook*, JPL Publication 82-69, Jet Propulsion Laboratory, California Institute of Technology, Pasadena, CA.
[44] Anspaugh, B.E. 1996. *GaAs Solar Cell Radiation Handbook*, JPL Publication 96-9, Jet Propulsion Laboratory, California Institute of Technology, Pasadena, CA.
[45] Kinchin, G.H. and Pease, R.S. 1955. The displacement of atoms in solids by radiation, *Rep. Prog. Phys.*, Vol. 18, p. 1.
[46] Seitz, F. and Koehler, J.S. 1956. Displacement of atoms during irradiation, *Solid State Physics*, Vol. 2, p. 307.
[47] Hovel, H.J. 1975. Semiconductor solar cells, in R.K. Willardson and A.C. Beer, Eds., *Semiconductors and Semimetals*, Vol. 11, Academic Press, New York.

[48] Markvart, T. 1990. Radiation damage in solar cells, *J. Materials Science: Materials in Electronics*, Vol. 1, p. 1.
[49] Yamaguchi, M. and Ando, K. 1988. Mechanism for radiation resistance of InP solar cells, *J. Appl. Phys.*, Vol. 63, p. 5555.
[50] Smits, F.M. 1963. *IEEE Trans. Nucl. Sci.*, Vol. NS-10, p. 88.
[51] Meulenberg, A. and Treble, F.C. 1973. Damage in silicon solar cells from 2 to 155 MeV protons, *Proc. 10th IEEE Photovoltaic Specialists Conf.*, Palo Alto, p. 359.
[52] Rosenzweig, W. 1962. Diffusion length measurement by means of ionizing radiation, *Bell. Syst. Tech. J.*, Vol. 41, p. 1573–1588.
[53] Wilsey, N.D. 1972. *Proc. 9th IEEE Photovoltaic Specialists Conf.*, Silver Springs, p. 338.
[54] Minahan, J.A. and Green, M.J. 1985. *Proc. 18th IEEE Photovoltaic Specialists Conf.*, p. 350.
[55] Downing, R.G, Carter, J.R. Jr. and Denney, J.M. 1964. The energy dependence of electron damage in silicon, *Proc. 4th IEEE Photovoltaic Specialists Conf.*, Vol. 1, p. A-5-1.
[56] Rosenzweig, W., Smits, F.M., and Brown, W.L. 1964. Energy dependence of proton irradiation damage in silicon, *J. Appl. Phys.*, Vol. 35, p. 2707.
[57] Keyes, R.W. 1977. The energy gap of impure silicon, *Comm. Solid State Phys.*, Vol. 7(6), p. 149.
[58] Wagner, J. and delAlamo, J.A. 1988. Band-gap narrowing in heavily doped silicon: a comparison of optical and electrical data, *J. Appl. Phys.*, Vol. 63, p. 425.
[59] Jain, S.C. and Roulston, D.J. 1991. A simple expression for bandgap narrowing in heavily doped Si, Ge, GaAs and Ge_xSi_{1-x} strained layers, *Solid-State Electron.*, Vol. 34, p. 453.
[60] Jain, S.C., McGregor, J.M., Roulston D.J. and Balk, P. 1992. Modified simple expression for bandgap narrowing in n-type GaAs, *Solid-State Electron.*, Vol. 35, p. 639.
[61] Klaassen, D.B.M., Slotboom, J.W and de Graaf, H.C. 1992. Unified apparent bandgap narrowing in n and p-type silicon, *Solid-State Electron.*, Vol. 35, p. 125.
[62] Lundstrom, M.S., Harmon, E.S. and Melloch, M.R. 1996. Effective bandgap narrowing in doped GaAs, in Brozel, M.R. and Stillman, G.E., Eds., *Properties of Gallium Arsenide* (3rd edition), IEE/INSPEC, Institution of Electrical Engineers, London, p. 186.
[63] Tsang, C. and Street, R.A. 1979. *Phys. Rev.*, Vol. B19, p. 3027.
[64] Fritzsche, H., Ed. 1989. *Amorphous Silicon and Related Materials*, University of Chicago.
[65] Puigdollers, J. and Voz, C., personal communication.
[66] Street, R.A. 1991. *Hydrogenated Amorphous Silicon*, Cambridge University Press.

IIa-3

Ideal Efficiencies

Peter T. Landsberg, Faculty of Mathematical Studies,
University of Southampton, UK
Tom Markvart, School of Engineering Sciences,
University of Southampton, UK

1 Introduction

In this chapter we deal with the simplest ideas which have been used in the past to attain an understanding of solar cell efficiencies from a theoretical point of view. The first and most obvious attack on this problem is to use thermodynamics, and we offer four such estimates in Section 2. Only the first of these is the famous Carnot efficiency. The other three demonstrate that one has more possibilities even within the framework of thermodynamics. To make progress, however, one has to introduce at least one solid-state characteristic and the obvious one is the energy gap E_g. That this represents an advance in the direction of a more realistic model is obvious, but it is also indicated by the fact that the efficiency now calculated is lower than the (unrealistically high) thermodynamic efficiencies (Section 3). In order to get closer to reality we introduce in Section 4 the fact that the radiation is effectively reduced from the normal black-body value (Equation (6)) owing to the finite size of the solar disc. This still leaves important special design features such as the number of series-connected tandem cells and higher order impact ionisation, and these are noted in Section 5.

2 Thermodynamic Efficiencies

The formulae for ideal efficiencies of solar cells are simplest when based on purely thermodynamic arguments. We here offer four of these; they involve only (absolute) temperatures:

- T_a, temperature of the surroundings (or the ambient),
- T_s, temperature of the pump (i.e. the sun),
- T_c, temperature of the actual cell which converts the incoming radiation into electricity.

From these temperatures we form the following efficiencies [1]:

$$\eta_C \equiv 1 - T_a/T_s, \text{ the Carnot efficiency} \tag{1}$$

$$\eta_{CA} \equiv 1 - (T_a/T_s)^{\frac{1}{2}}, \text{ the Curzon–Ahlborn efficiency} \tag{2}$$

$$\eta_L \equiv 1 - (4/3)(T_a/T_s) + (1/3)(T_a/T_s)^4, \text{ the Landsberg efficiency} \tag{3}$$

$$\eta_{PT} \equiv \left[1 - (T_c/T_s)^4\right] [1 - T_a/T_c], \text{ the photo-thermal efficiency due to Müser} \tag{4}$$

In the latter efficiency the cell temperature is determined by the quintic equation

$$4T_c^5 - 3T_aT_c^4 - T_aT_s^4 = 0 \tag{5}$$

The names associated with these efficiencies are not historically strictly correct: for example, in Equations (2) and (3) other authors have played a significant part.

Figure1 [1] shows curves of the four efficiencies which all start at unity, when $T_a/T_s \equiv 0$, and they all end at zero, when $T_a = T_s$. No efficiency ever beats the Carnot efficiency, of course, in accordance with the rules of thermodynamics. Values near T_s = 5760–5770 K seem to give the best agreement with the observed solar spectrum and the total energy received on Earth but a less accurate but more convenient value of T_s = 6000 K is also commonly used. Using the latter value of T_s and T_a = 300 K as the temperature for the Earth, one finds

$$\eta_C = 95\%, \qquad \eta_{CA} = 77.6\%, \qquad \eta_L = 93.3\%, \qquad \eta_{PT} = 85\%$$

If $T_s = T_a = T_c$ one has in effect an equilibrium situation, so that the theoretical efficiencies are expected to vanish.

The above thermodynamic efficiencies utilise merely temperatures, and they lie well above experimental results. One needs an energy gap (E_g) as well to take us from pure thermodynamics to solid-state physics. Incident photons can excite electrons across this gap, thus enabling the solar cell to produce an electric current as the electrons drop back again. The thermodynamic results presented earlier, on the other hand, are obtained simply by considering energy and entropy fluxes.

3 Efficiencies in Terms of Energies

In order to proceed we need next an expression for the number of photons in black-body radiation with an energy in excess of the energy gap, E_g say, so that they can excite electrons across the gap. At black-body temperature T_s the number of photons incident on unit area in unit time is given by standard theory as an integral over the photon energy [2]:

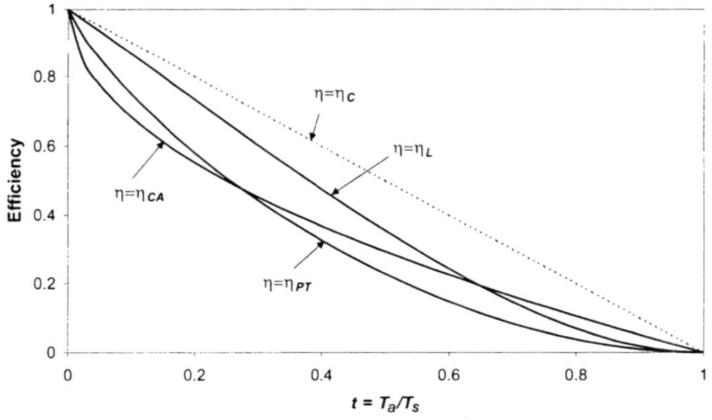

Figure 1 The efficiencies (1)–(4) as functions of T_a/T_s.

$$\Phi(E_g, T_s) = \frac{2\pi k^3}{h^3 c^2} T_s^3 \int_{E_g/kT_s}^{\infty} \frac{x^2 dx}{e^x - 1} \tag{6}$$

Now suppose that each of these photons contributes only an energy equal to the energy gap to the output of the device, i.e. a quantity proportional to

$$x_g \int_{x_g}^{\infty} \frac{x^2 dx}{e^x - 1} \quad (x_g \equiv E_g/kT_s) \tag{7}$$

To obtain the efficiency η of energy conversion we must divide this quantity by the whole energy which is in principle available from the radiation:

$$\eta = x_g \int_{x_G}^{\infty} \frac{x^2 dx}{e^x - 1} \Big/ \int_0^{\infty} \frac{x^3 dx}{e^x - 1} \tag{8}$$

Equation (8) gives the first of the Shockley–Queisser estimates for the limiting efficiency of a solar cell, the *ultimate efficiency* (see Figure 5). The argument neglects recombination in the semiconductor device, even radiative recombination which is always present (a substance which absorbs radiation can always emit it!). It is also based on the black-body photon flux (Equation (6)) rather than on a more realistic spectrum incident on the Earth.

We shall return to these points in Section 4, but first a brief discussion of Equation (8) is in order. There is a maximum value of η for some energy gap which may be seen by noting that $\eta = 0$ for both $x_g = 0$ and for x_g very large. So there is a maximum efficiency between these values. Differentiating η with respect to x_g and equating to zero, the condition for a maximum is

$$x_g = x_{g\,opt} = 2.17, \text{ corresponding to } \eta = 44\%.$$

This is still higher than most experimental efficiencies, but the beauty of it is that it is a rather general result which assumes merely properties of black-body radiation.

Let $f(x)$ be a generalised photon distribution function; then a generalised efficiency can be defined by

$$\eta = \frac{x_g \int_{x_g}^{\infty} f(x) dx}{\int_0^{\infty} x f(x) dx}. \tag{9}$$

The maximum efficiency with respect to x_g is then given by

$$x_{gopt} f(x_{gopt}) = \int_{x_{gopt}}^{\infty} f(x) dx \tag{10}$$

This is rather general and will serve also when the photon distribution departs from the black-body forms, and even for radiation in different numbers of dimensions.

4 Efficiencies using the Shockley Solar Cell Equation

A further step in finding the appropriate efficiency limits for single-junction solar cells can be made by estimating the relevant terms in the Shockley ideal solar cell equation (Equation (1) in Chapter IIa-1). To this end, further remarks must be made about the solar spectrum and solar energy incident on the Earth's surface. The ultimate efficiency, discussed in Section 3, was based on the black-body photon flux (Equation (6)), a rigorous thermodynamic quantity but not a very good estimate of the solar spectrum as seen on Earth. By virtue of the large distance between the Sun and the Earth, the radiative energy incident on the Earth's surface is less than that of Equation (6), by a factor f_ω which describes the size of the solar disk (of solid angle ω_s) as perceived from the Earth:

$$f_\omega = \left(\frac{R_s}{R_{SE}}\right)^2 = \frac{\omega_s}{\pi} \tag{11}$$

where R_s is the radius of the Sun (696×10^3 km), and R_{SE} is the mean distance between the Sun and the Earth (149.6×10^6 km), giving $\omega_s = 6.85 \times 10^{-5}$ sterad and $f_\omega = 2.18 \times 10^{-5}$. The resulting spectrum is shown in Figure 2 alongside the standard terrestrial AM1.5 spectrum (a further discussion of the spectra that are used for solar cell measurements in practice can be found in Chapter IV-1 which also shows the extraterrestrial spectrum AM0).

The maximum value of the photogenerated current I_{ph} now follows if we assume that one absorbed photon contributes exactly one electron to the current in the external circuit:

$$I_{ph} = A\, q\, f_\omega \Phi(E_g, T_s) \tag{12}$$

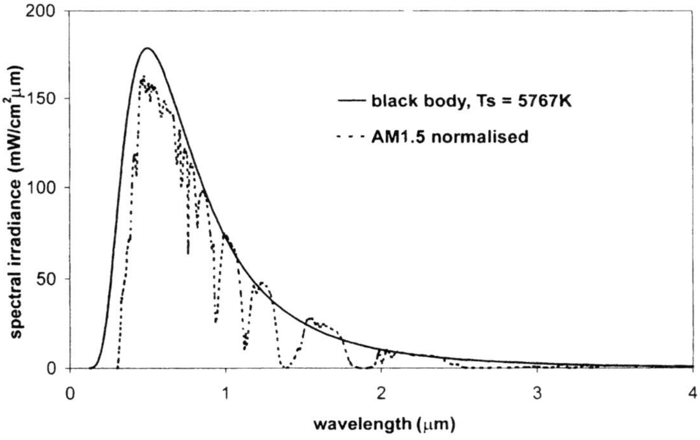

Figure 2 The black-body spectrum of solar radiation and the AM1.5 spectrum, normalised to total irradiance 1 kW/m^2, which is used for the calibration of terrestrial cells and modules.

where A is the illuminated area of the solar cell and q is the electron charge. The maximum photogenerated current density $J_{ph} = I_{ph}/A$ that can be produced by a solar cell with band gap E_g is shown in Figure 3. To allow a comparison with photocurrents measured in actual devices, Figure 3 is plotted for the AM1.5 solar spectrum which used for calibration of terrestrial solar cells, rather than for the black body spectrum used in Section 3.

The open circuit voltage V_{oc} can now be obtained using the photogenerated current I_{ph} (Equation (12)) and the (dark) saturation current I_o which appears in the ideal solar cell equation:

$$V_{oc} = \frac{kT}{q} \ln\left(1 + \frac{I_{ph}}{I_o}\right) \tag{13}$$

The current I_o can be obtained by a similar argument as the photogenerated current I_{ph} since, as argued by Shockley and Queisser, it can be equated to the black-body photon flux at the cell temperature T_c (in what follows, the cell temperature T_c will be assumed to be equal to the ambient temperature T_a):

$$I_0 = A\, q\, f_0 \Phi(E_g, T_a) \tag{14}$$

where the coefficient f_0 has been inserted to describe correctly the total area $f_0 A$ exposed to the ambient photon flux. Various values of f_0 (some dependent on the refractive index $\mathbf{n}$ of the cell material) can be found, appropriate for different device structures and geometries. The usual value is $f_0 = 2$, as suggested by Shockley and Queisser [2], since this radiation is incident through the two (front and rear) surfaces of the cell. A similar argument for a spherical solar cell yields an effective value $f_0 = 4$ [3]. Henry [4] gives $f_0 = 1 + \mathbf{n}^2$ for a cell grown on a semiconductor substrate but the value $f_0 = 1$ is also sometimes used (see, for

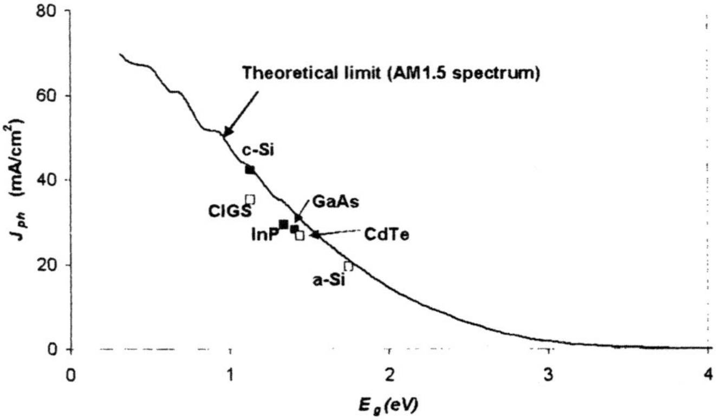

Figure 3. The theoretical limit on photogenerated current, compared with the best measured values. The curve is obtained by replacing the product $f_\omega\Phi(E_g,T_S)$ in Equation (12) by the appropriate AM1.5 photon flux. Full symbols correspond to crystalline materials, open symbols to thin films.

example, [5]). Green [6] gives a semi-empirical expression for the dark saturation current density $J_o = I_O/A$:

$$J_O \text{ (in Amps/cm}^2) = 1.5 \times 10^5 \, exp\left(-\frac{E_g}{kT_a}\right) \tag{15}$$

An approximate analytical method for estimating V_{oc} can also be useful, particularly as it stresses the thermodynamic origin of V_{oc}. Indeed, it can be shown [7] that, near the open circuit, the solar cell behaves as an ideal thermodynamic engine with Carnot efficiency $(1 - T_c/T_s)$. Ruppel and Würfel [3] and Araùjo [8] show that V_{oc} can be approximated to a reasonable accuracy by the expression

$$V_{oc} = \frac{E_g}{q}\left(1 - \frac{T_c}{T_s}\right) + \frac{kT}{q}\ln\frac{f_\omega}{f_0} + \frac{kT_c}{q}\ln\frac{T_s}{T_c} \tag{16}$$

which depicts the dependence of V_{oc} on the band gap E_g and on the cell temperature T_c. Figure 4 compares this theoretical values for the open circuit voltage with data for the best solar cells to-date from different materials.

Using now an expression for the fill factor (defined by Equation (3) in Chapter IIa-1), one readily obtains a theoretical estimate for the efficiency. Slightly different results may be encountered, principally by virtue of the different ways one can estimate the current and the voltage. Figure 5 shows the best-known result, the celebrated Shockley–Queisser ideal efficiency limit [2]. Shockley and Queisser call this limit the *nominal efficiency*, to be compared with the *ultimate efficiency* which is discussed in Section 3. Figure 5 shows two such curves: one labelled 'one-sun' corresponds to the AM0 solar intensity, as observed outside the Earth's atmosphere. A second curve, labelled 'maximum concentration'

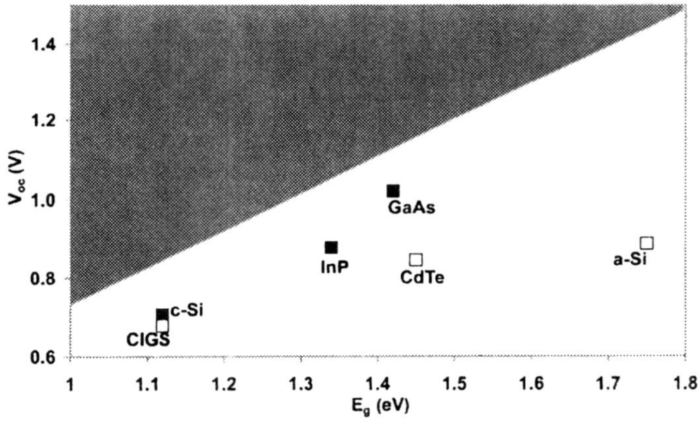

Figure 4 The theoretical Shockley–Queisser limit on open circuit voltage: values exceeding this limit lie in the shaded area of the graph. Line corresponding to Equation (16) appears as identical to within the accuracy of this graph. Full symbols correspond to crystalline materials, open symbols to thin films.

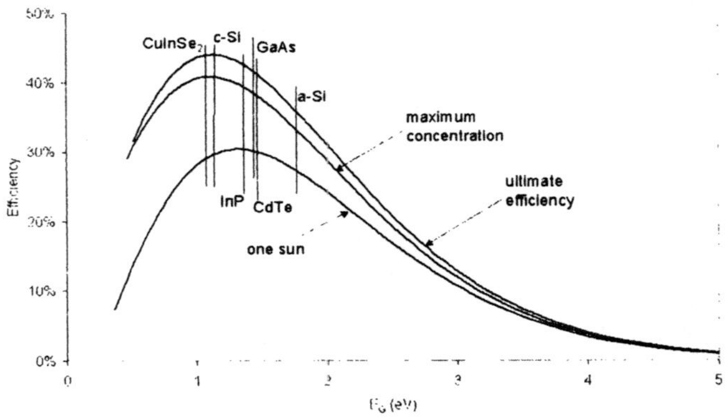

Figure 5 The 'ultimate' and two 'nominal' Shockley–Queisser efficiencies. Note that the black-body radiation with temperature $T_s = 6000$ *K has been used here, in keeping with the Shockley–Queisser work [2].*

corresponds to light focused on the cell, by a mirror or a lens, at the maximum concentration ratio of $1/f_\omega = 45{,}872$ [9].

The various unavoidable losses in photovoltaic energy conversion by single-junction solar cells can be depicted in a graph constructed by Henry [4] and analogous to Figure 6. There are two curves in this graph:

- The photogenerated current density J_{ph} from Equation (12) as a function of photon energy. J_{ph} is divided here by the total irradiance, making the area under this curve equal to unity by construction.
- The maximum voltage that can be extracted from the cell at the maximum power point. This curve is drawn in such a way that the ratio of lengths of the two arrows b/a is equal to V_m/E_g.

The three shaded areas then depict the three fundamental losses in a single junction solar cell (shown here for silicon with band gap E_g equal to 1.12 eV):

- Shaded area marked $h\nu < E_g$ is equal to the loss of current due to the inability of the semiconductor to absorb below-band-gap light.
- Shaded area marked $h\nu > E_g$ represents energy losses due to the thermalization of electron-hole pairs to the band gap energy.
- Hatched area marked $V < E_g$ corresponds to the combined thermodynamic losses due to V_{oc} being less than E_g, and losses represented by the fill factor FF.

The area of the blank rectangle then represents the maximum efficiency that can be obtained for a single junction cell made from a semiconductor with band gap E_g. The graph is drawn here for light with maximum possible concentration. A different 'voltage curve' would result if light with one-sun intensity were used.

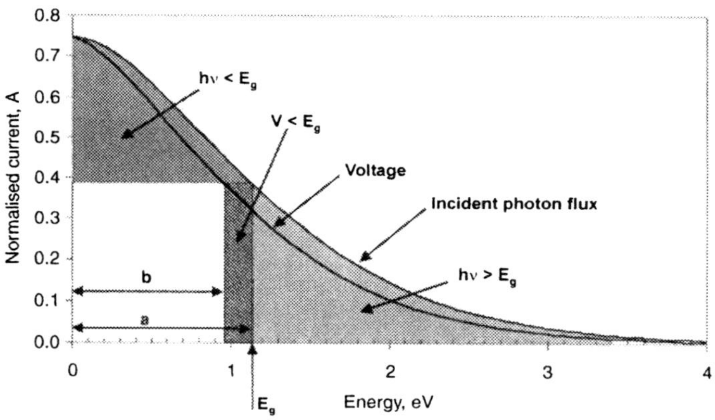

Figure 6 Henry's construction.

5 General Comments on Efficiencies

The ideal solar cell efficiencies discussed above refer to single-junction semiconductor devices. The limitations considered in the ultimate efficiency of Section 3 are due to the fact that the simplest semiconductor (i.e. one whose defects and impurities can be ignored) cannot absorb below band gap photons. Furthermore it is also due to the fact that the part of the energy of the absorbed photons in excess of the band gap is lost as heat. Radiative recombination at the necessary fundamental level was taken into account in the treatment of Section 4. It is sometimes argued that there are other 'unavoidable' losses, due to electronic energy transfer to other electrons by the Auger effect (electron–electron collisions) [10–12]. There is also the effect of band gap shrinkage, discussed in Chapter 2, and light trapping may also play a part [11]. None of these effects are discussed here, and the reader is referred to the relevant literature.

It is clear that it is most beneficial if one can improve the effect of a typical photon on the electron and hole density. This can be achieved, for example, if the photon is energetic enough to produce two or more electron–hole pairs. This is called impact ionisation and has been studied quite extensively. A very energetic photon can also project an electron high enough in to the conduction band so that it can, by collision, excite a second electron from the valence band. This also improves the performance of the cell. On the other hand, an electron can combine with a hole and the energy release can be used to excite a conduction band electron higher into the band. In this case energy is uselessly dissipated with a loss of useful carriers and hence of conversion efficiency. This is one type of Auger effect. For a survey of these and related effects, see [12]. These phenomena suggest a number of interesting design problems. For example, is there a way of limiting the deleterious results of Auger recombination [13]? One way is to try to 'tune' the split-off and the fundamental band-gaps appropriately. If one is dealing with parabolic bands, then the obvious way is to examine the

Table 1 The maximum efficiencies of tandem cells as a function of the number of cells in the stack, for different concentration ratios [17]. Note that de Vos [17] uses a slightly smaller value of f_ω than Shockley and Queisser, resulting in a marginally different maximum concentration ratio than used in Figure 5

Concentration ratio	Number of cells in the stack	Maximum efficiency (%)
1	1	31.0
	2	49.9
	3	49.3
	...	
	∞	68.2
46,300	1	40.8
	2	55.7
	3	63.9
	...	
	∞	86.8

Table 2 The currently best reported efficiencies of different types of solar cells [18]

	Efficiency (%)	J_{sc} (mA/cm^2)	V_{oc} (V)	FF (%)
Crystalline: single junction				
c-Si	24.7	42.2	0.706	82.8
GaAs	25.1	28.2	1.022	87.1
InP	21.9	29.3	0.878	85.4
Crystalline: multijunction				
GaInP/GaAs/Ge tandem	31.0	14.11	2.548	86.2
Thin-film: single junction				
CdTe	16.5	25.9	0.845	75.5
CIGS	18.9	34.8	0.696	78.0
Thin-film: multijunction				
a-Si/a-SiGe tandem	13.5	7.72	2.375	74.4
Photoelectrochemical				
Dye-sensitised TiO_2	11.0	19.4	0.795	71.0

threshold energies which an electron has to have in order to jump across the gap, and to make these large so as to make this jump difficult.

Then there is the possibility of placing impurities on the energy band scale in such a way as to help better use to be made of low-energy photons, so that they can now increase the density of electrons in the system. This is sometimes referred to as the impurity photovoltaic effect. So one can make use of it [14].

One can also utilise excitons to improve the efficiencies of solar cells. There may be as many as 10^{17} cm^{-3} excitons in silicon at room temperature. If they are split up in the field of a p-n junction, this will increase the concentration of current carriers and so increase the light generated current, which is of course beneficial.

We have here indicated some useful ideas for improving solar cells. There are of course many others, some of which are discussed in Chapters IIb-5, IId-1 and elsewhere [15]. Note, in particular, the idea of developing tandem cells in which photons hit a large band gap material first and then proceed gradually to smaller band gap materials. Tandem cells are now available with three of more stages. Solar cells with efficiency of order 20% are predicted to be produced on a large scale in the near future [16]. Table 2 shows the best laboratory efficiencies at the present time for different materials.

References

[1] Landsberg, P.T. and Badescu, V. 1998. Solar energy conversion: list of efficiencies and some theoretical considerations, *Prog. Quantum Electronics*, Vol. 22, pp. 211 and 231.

[2] Shockley, W. and Queisser, 1961. H. J. Detailed balance limit of efficiency of pn junction solar cells, *J. Appl. Phys.* Vol. 32, p. 510.

[3] Ruppel, W. and Würfel, P. 1980. Upper limit for the conversion of solar energy, *IEEE Trans. Electron Devices*, Vol. ED-27, p. 877.

[4] Henry, C.H. 1980. Limiting efficiencies of ideal single and multiple energy gap terrestrial solar cells, *J. Appl. Phys.* 51, p. 4494.

[5] Kiess, H, and Rehwald, W. 1995. On the ultimate efficiency of solar cells, *Solar Energy Materials and Solar Cells*, Vol. 38, pp. 45-55.

[6] Green, M.A. 1982 *Solar Cells*. Prentice Hall, New York.

[7] Baruch, P. and Parrott, J.E. 1990. A thermodynamic cycle for photovoltaic energy conversion, *J. Phys. D: Appl. Phys.* Vol. 23, p. 739.

[8] Araùjo, G.L. 1990. Limits to efficiency of single and multiple band gap solar cells, in Luque A. and Araùjo G.L., Eds., *Physical Limitations to Photovoltaic Energy Conversion*, Adam Hilger, Bristol, p. 106.

[9] Welford, W.T. and Winston, R. 1978. *The Physics of Non-imaging Concentrators*, Academic Press, New York, Chapter 1.

[10] Green, M.A. 1984. Limits on the open-circuit voltage and efficiency of silicon solar cells imposed by intrinsic Auger process, *IEEE Trans Electron Devices* Vol. ED-31, p. 671.

[11] Tiedje, T., Yablonovich, E., Cody, G.C. and Brooks, B.G. 1984. Limiting efficiency of silicon solar cells, *IEEE Trans Electron Devices*, Vol. ED-31, p.711.

[12] Landsberg, P.T. 1987. The band-band Auger effect in semiconductors, *Solid-State Electronics*, Vol. 30, p. 1107.

[13] Pidgeon, C.R., Ciesla, C.M. and Murdin, B.N. 1997. Suppression of non-radiative processes in semiconductor mid-infrared emitters and detectors, *Prog. Quantum Electron.* Vol. 21, p. 361.

[14] Kasai, H. and Matsumura, H. 1997. Study for improvement of solar cell efficiency by impurity photovoltaic effect, *Solar Energy Materials and Solar Cells*, Vol. 48, p. 93.

[15] Green, M.A. 2001. Third generation photovoltaics: Ultra high conversion efficiency at low cost, *Prog. Photovoltaics Res. Appl.*, Vol. 9, pp. 123–135.

[16] Wileke, G.P. 2002. The Frauenhoffer ISE roadmap for crystalline silicon solar cell technology, *Proc. 29th IEEE Photovoltaic Specialists Conf.*, New Orleans.

[17] deVos, A. 1980. Detailed balance limit of the efficiency of tandem solar cells, *J. Phys. D: Appl. Phys.* Vol. 13, p. 839. See also deVos, A. 1992. Endoreversible *Thermodynamics of Solar Energy Conversion*, Oxford University Press.

[18] Green, M.S., Emery, K.L., King, D.L., Igari, S. and Warta, W. 2002. Solar cell efficiency tables (version 20), *Prog. Photovoltaics Res. Appl.*, Vol. 10, pp. 355–360.

Part IIb

Crystalline Silicon Solar Cells

IIb-1

Crystalline Silicon: Manufacture and Properties

Francesca Ferrazza, Eurosolare S.p.A, Nettuno, Italy

1 Introduction

The majority of silicon wafers used for solar cells are Czochralski (CZ) single crystalline and directional solidification, or cast, multicrystalline (mc) material. The split between the two types of wafer is presently about 55% mc-Si and 45% CZ-Si. Until 1995 CZ wafers represented 60% of the substrates used by industry and mc-Si wafers around 25%. The fast scale up of commercially available multicrystalline wafers changed the picture rapidly. The remainder of the silicon substrates used by the industry are non-wafered sheets or ribbons that are of different types and have recently gained significant production figures, following long development phases. Non-wafer silicon accounted for about 4% of the market in 2001, up from 1–2% in the mid-1990s [1, 2].

2 Characteristics of Silicon Wafers for Use in PV Manufacturing

2.1 Geometrical Specifications

Most of the wafer substrates used in production facilities have dimensions relating to the diameters of monocrystalline silicon cylinders for the semiconductor industry (essentially 5 and 6 inch) that, in turn, have influenced standards for wafer carriers, automation, packaging etc. However, in order to maximise the power density of the modules, wafers are square, or pseudo-square in the case of monocrystalline silicon, that is cylinders are shaped as squares with rounded off corners. This reduces the surface area of the wafers by between 2% and 5% compared with a full square of same dimensions.

In the case of mc-Si, ingot sizes are designed to be compatible with multiple numbers of each of the standard wafer dimensions, in order to maximise geometrical yield. Yield considerations limit the possible wafer sizes achievable for any given ingot dimension, as much as expensive wafer cassettes, automation and packaging do later in the process. Table 1 reports the different sizes for commercially available wafers, including typical tolerances.

A SEMI® (Semiconductor Equipment and Materials International) standard, M61000, was developed with the purpose of covering the requirements for silicon wafers for use in solar cell manufacturing [3], including dimensional specifications, defects and electronic properties. Most commercial suppliers sell their wafer products using specifications that are close to those described by

Table 1 Commercially available wafer sizes.

Nominal size	Dimension (mm)	Diagonal (multi) (mm ± 1)	Diameter (mono) (inch)
103	103 ± 0.5	146	5
125	125 ± 0.5	177	6
150	150 ± 0.5	212	–

M61000. However, smaller wafers are usually 103 mm rather than 100 mm as specified, and there are some notable exceptions to the specifications, e.g. dimensions of wafers produced in-house by some of the early players who have developed their own standard and do not usually buy wafers on the market. Another obvious exception is provided by non-wafer substrates, the dimensions of which are in general determined by the growth equipment and technique. Some manufacturers use rectangular wafers. Other requirements, besides the geometrical definitions of the wafers, are thickness uniformity and reduced levels of cracks and saw marks that could adversely affect later processing.

Typical specifications for commercially available wafers are described in Table 2.

The absolute value of the wafer thickness has dropped by about 100 μm in the last decade, as a consistent cost reduction measure [4, 5], and is expected to decrease further in the next years [2] as automation and cell processing become more sophisticated and can allow effective handling of thin wafers. Some wafer and cell producers already have less than 300 μm wafers in their production lines, although in general wafer sizes in such cases are limited to the 100 cm^2 range in order to maintain mechanical yields in the high 90s. Similarly to the case of the area dimensions, thickness in non-wafer substrates is determined by the process, and is in general less homogeneous, providing one of the major differences between wafer and non-wafer cell technologies. A great deal of effort was put in the last decade in developing automated thickness measurement tools for manufacturing plants, to inspect wafer thickness variations in lots, an extremely difficult task in manually inspected wafer fabrication sequences.

2.2 Physical Specifications

Wafers are generally classified in terms of resistivity, type, and oxygen and carbon content. These data are generally present in all commercial specifications related to single and multicrystalline wafers, and refer to ASTM or equivalent standards. However, the PV community has had to face the unavoidable departure from standard test conditions of all parameters when measuring the properties of the inherently inhomogeneous nature of multicrystalline wafers, which led to agreement on relatively broad ranges for resistivity or upper thresholds for oxygen and carbon contents. Early concerns, for instance, of the influence of grain boundaries on the determination of resistivity using the

Table 2 Other dimensional specifications for typical PV wafers

Parameter	Value
Thickness of a batch	330 ± 40 μm
Total Thickness Variation (TTV) of a wafer	50 μm
Cracks	< 1 mm
Saw marks	< 10 μm
Bow	< 50 μm

four-point probe method are now somewhat more relaxed, after significant statistical feedback has provided comfort in the values proposed. Still, in strict terms, standards related to the measurement of resistivity in multicrystalline wafers do not exist, which is true of course for non-wafer silicon technologies as well. This is also true for other kinds of measurements, and the effort to develop meaningful characterisation tools for lower or inhomogeneous quality materials as compared to the semiconductor industry is a clear indication of such a need (see Chapter IIb-4). Furthermore, the increasing volumes of wafers in the growing PV market has forced a second, big departure from semiconductor wafer characterisation standards, imposing fast, non-destructive test methods to optimise costs and yields. In most cases, for instance, resistivity and type are measured at block rather than at wafer level. Table 3 shows the typical values for physical parameters of commercially available wafers for industrial processing. These are either multicrystalline or monocrystalline Czochralski. Float zone wafers for PV may become commercial products, and will be discussed in a later paragraph.

2.3 Physical Specifications

Minority carrier lifetime characterisation of commercial silicon is worth a paragraph on its own. This is by far the most complicated parameter to measure and to effectively relate to subsequent processing quality and yield. It is also most influenced by the inhomogeneity of multicrystalline silicon, as well as by thermal treatments. It became immediately evident to all PV manufacturers at the very beginning of the expansion of the multicrystalline silicon market that the identification of an appropriate tool for analysing and understanding the properties of mc-Si would have been one of the keys for the commercial success of the material. A generous number of attempts were made to adapt the existing lifetime measurements – rigorously valid for high quality polished single crystalline wafers – to provide meaningful values for mc-Si and even CZ-Si for the PV community. Also, as mentioned before, any acceptable test would need to be fast, cost effective, and obviously non-destructive as the number of samples to inspect was bound to be large. This focussed effort led to the development of a number of automated lifetime analysers, which at the end of the development process had relatively low resemblance with the semiconductor industry

Table 3 Physical specifications of commercial silicon wafers

Parameter	Value
Type	P – boron doped
Resistivity	0.5–3 ohm cm
Oxygen (mc-Si)	$< 8 \times 10^{17}$ at/cm^3
Oxygen (CZ-Si)	$< 1 \times 10^{18}$ at/cm^3
Carbon (mc-Si)	$< 1 \times 10^{18}$ at/cm^3
Carbon (CZ-Si)	$< 2 \times 10^{17}$ at/cm^3

counterparts. Probably the most successful commercial methods are the microwave photoconductance decay method (μ-PCD) performed directly on silicon blocks, and the photoconductance decay or quasi-steady state method developed by Ron Sinton which is discussed in Chapter IIb-4.

For the purposes of the present chapter, we will focus on the μ-PCD characterisation of mc-Si blocks. This measurement technique is commercially available and its use is widespread, although strong debates on the validity of the results occurred for many years. The measurement is based on the detection of the amplitude of the microwave field reflected by the sample surface. This amplitude variation depends on the conductivity, and thus also on the number of minority carriers generated by a short laser pulse [6]. The time in which the system recovers the initial state is associated with the quality of the semiconductor material and with the recombination mechanisms in the bulk and at the surface. It is generally rather complicated to separate different contributions, and PV silicon has peculiar characteristics which enhance difficulties, such as relatively high doping, rough surfaces and short diffusion lengths for minority carriers. Furthermore, as mentioned before, the industry requirements are for fast non-destructive techniques which enable prediction of later behaviour of the material in the processing line, in order to minimise the costs of processing low quality material as early as possible. For this reason, the industry has pushed towards the use of fast non-invasive block scanners since the early 1990s, despite the inability of the measurement systems to conform to any of the existing standards. A certain effort has been directed until relatively recently towards developing uniform measurement systems and procedures, which has proven once more the difficulty of the problem [7] and finally led to generally accepted principles subject to bilateral confirmation in the case of commercial relationships between wafer vendors and cell producers.

The main problems with block scanners, besides separation of bulk and surface components, lie in the fact that the measurement is actually performed on a very thin portion of the block. This is due to the absorption of the laser pulse in silicon (usually in the near infrared range) and to the high reflectivity of microwaves which only allow the field to penetrate a skin depth of the sample under examination. Other difficulties lie in the unpredictable behaviour of the material in three dimensions, lateral distribution of carriers, trapping effects, unknown injection levels, and the macroscopic saw damage affecting the control of the distance of the measurement head from the sample. Microwave block-scanners are unable to handle these problems, which are better taken into account by the technique described in Chapter IIb-4. The experience of crystal growers and the feedback from cell processing lines, however, led to the establishment of a method for the analysis of silicon blocks which is able to reject low quality material at block level, and which allows the identification of the correct cropping position for the rejection of tops and tails.

A typical map, performed with a commercially available automated system [8], of a standard block of mc-Si grown by the directional solidification method is shown in Figure 1. The silicon is boron doped to a resistivity of about 1×10^{16} at/cm^3, and the measurement is performed with a microwave field in

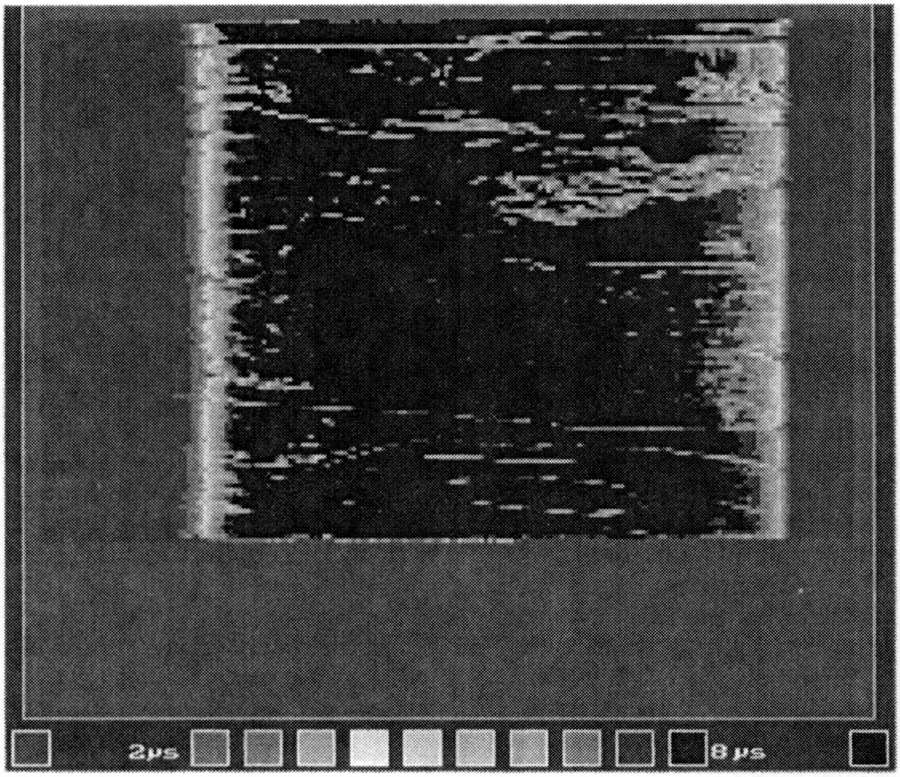

Figure 1 Lifetime map.

the GHz range coupled by an antenna for a sample irradiated by a laser diode pulse at 904 nm.

The absolute value of the minority carrier lifetime is surface limited, as the sample is measured 'as cut'. Normally, in fact, no impractical etching or passivation treatments are applied to the surface which therefore has a high recombination velocity. It is assumed that the surface is always in the same conditions, so any change in the relaxation time is associated with bulk properties. A map such as the one in Figure 1 takes a few minutes to be realised with a modern lifetime scanner such as the one in [8]. Early systems could take several hours to perform measurements with the same resolution. The low lifetime regions (in red) have different physical origins, and this is where the extensive material-to-cell correlation work performed over the years has been essential in comfortably introducing these instruments in the production environment [4, 5, 9]. The red zones at the top, in fact, are determined by the segregation of metals due to the refining process during solidification, and the wafers cannot effectively be used in cell processing, so they are rejected, and possibly remelted. The red zones at the bottom of the block are composed of a highly defected area – the initial crystal growth, highly dislocated and unusable for solar cells within about 1 cm from the start – and the 2–3 cm region of oxygen rich material which, despite its low initial lifetime, recovers after the thermal treatments used in cell processing and normally produces good quality cells. Care must be taken, therefore, to exactly determine the cut off (by any means possible!) between good and low quality material at the bottom of the block, and this is probably the most difficult part of the quality control procedure at block level.

Figure 2 shows a typical correlation between the lifetime at block level and cell performance, being evidence of a good performance of initially low quality material as detected by the block scanner [9]. The central part of the block is instead relatively uniform and produces good quality cells, in the range 12–15% depending on the particular process used, the higher value provided by silicon nitride-based sequences.

Typical values in the central part of the blocks are around 5–10 μs, depending on the specific measurement system used. A fundamental assumption of this method is a relative uniformity of the material in any given region of the block (i.e. all central regions behave similarly in same conditions), which is also a result of extensive correlation work [5, 9].

μ-PCD testers are also used to inspect incoming wafers in production lines, and in this case as well, there has been important correlation work to be able to confidently accept material for subsequent processing. This applies to CZ-Si wafers, which are inspected for uniformity as well as for the acceptance threshold value (which varies from case to case).

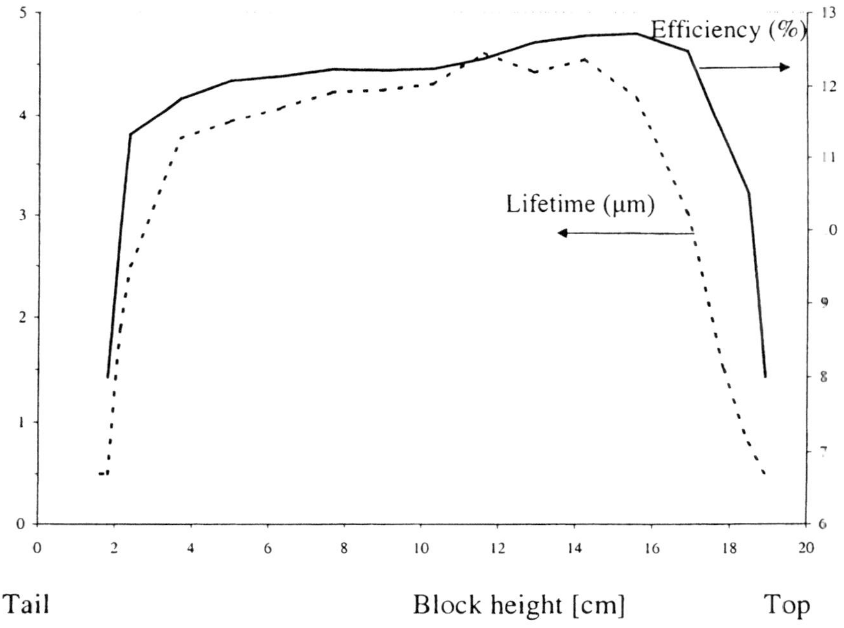

Figure 2 Correlation of lifetime with cell efficiency.

3 Feedstock Silicon

The commercial success of PV is driven critically by its cost. Silicon wafers account for about 50% of the total production cost of a module, a figure which has increased over the years, from the 33% of about 10 years ago, thanks to the constant improvements in technology which has allowed to identify the wafer as the ultimate cost limiting factor [10].

There is no source of silicon feedstock unique to the PV industry, so the issue of a possible feedstock shortage has been largely debated, and is still not concluded. About 10–15% of the silicon used by the microelectronics industry is available in various forms for PV use. This is in the range of 1800–2500 tons per year of higher quality scrap, and an extra 1500–2000 tons of lower quality material (e.g. pot scrap). Based on an effective usage rate of 10–15 tons per MWp produced, the amounts considered cannot feed the fast growing PV market for long. Whilst extensive research programmes have been conducted for many years to upgrade cheap metallurgical silicon to be an independent low cost silicon source for PV, none of the techniques proposed has reached commercial maturity [5]. The scare of a silicon shortage as early as the middle of the present decade has instead favoured several proposals for processes similar to those used for the production of polysilicon, but with looser specifications [11, 12].

From a practical point of view there is, in general, no constraint related to the geometrical specification of the starting material, so for the moment PV can enjoy low cost scrap such as silicon chips from the cutting processes of semiconductor manufacturing, popcorn silicon rods, tops and tails from crystal growth processes, etc. [13].

However, different crystallisation methods require different specifications. In general, monocrystalline and non-wafer technologies require high quality starting material, while the multicrystalline technology can allow a looser specification if some care is taken, due to its purifying characteristics – another point in its favour.

4 Crystal Preparation Methods

A number of techniques are available for the production of silicon wafers for the PV industry: CZ-Si and multicrystalline silicon (which have already been mentioned), magnetically confined multicrystalline silicon, float zone silicon and the non-wafer technologies (also already mentioned). In this paragraph we will briefly introduce the main features of each of them. The reader is encouraged to consult specific references for further details, as we will focus on the most relevant recent developments of the technologies under discussion.

4.1 Czochralski Silicon

The most common method for the growth of single crystalline ingots consists of pulling an oriented seed slowly out of the molten silicon contained in a pure

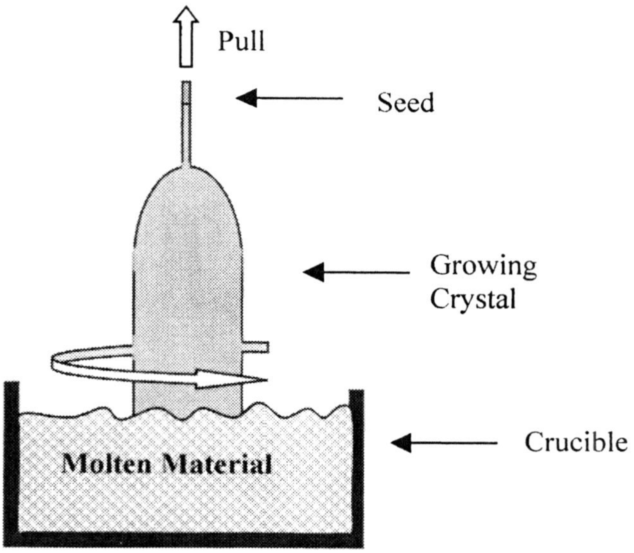

Figure 3 Schematic of CZ growth principle.

quartz crucible. The method is well known and extensively described in literature [10, 14].

What we will mention here is that a number of actions have been taken in the last 10 years to reduce the cost of CZ material, and regain competitiveness against multicrystalline silicon.

For instance, crystal growers now quite commonly use some kind of scrap silicon from the semiconductor industry as well as virgin poly as feedstock. Lower energy consumption, from the standard 100 kWh/kg figure to a promising 40 kWh/kg was recently reported due to improved furnace design, including heaters and gas distribution systems [15]. A crystallisation yield up to 70% from the standard 50% was also reported in the same study.

4.2 Multicrystalline Silicon

The realisation of multicrystalline silicon ingots is a relative simple process, and is based on controlling the extraction of heat from the melt in a quartz crucible in such a way that the interface between the growing solid and the ingot is as flat as possible. In this way, silicon grows in large columns of a few centimetres in section and as tall as 25 cm, and most detrimental impurities are segregated towards the top of the ingot. The critical steps to ensure a high quality and high yield process are in the design of the furnace for appropriate heat control, and in the quality of the quartz crucibles. A schematic of the general method is given in Figure 4.

Modern mc-Si furnaces are designed to minimise inhomogeneity, and maximise productivity [16, 17], and in the last few years a great deal of effort has

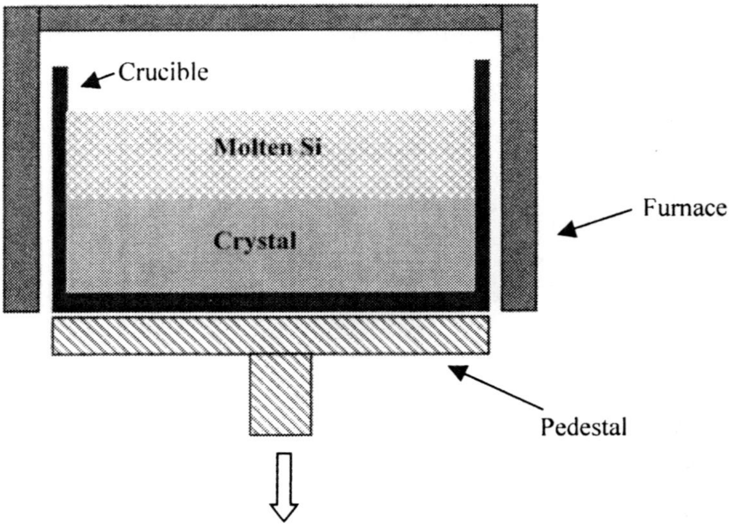

Figure 4 Schematic of mc-Si ingot growth.

been put in the study of appropriate models for growth control and optimisation [18]. Some of the distinctive features of mc-Si ingot growing are summarised in Tables 4 and 5.

As discussed in Section 2.3, the central part of the ingot enjoys a relatively uniform quality, and the purification ability of the process is witnessed by the following typical values for impurities other than O, C.

4.2.1 Charge Preparation

Ingots are normally doped at the level of about 1×10^{16} at/cm^3 of boron. This can be achieved with different mixtures of starting feedstock, and is a common practice in the PV industry which has to use feedstock form different sources and of different nature. For the range of resistivity values used in PV, no special requirements are needed for doping the ingot. If the feedstock is virgin poly or lowly doped silicon, highly doped silicon powder, available commercially with specifications of the B content, can be added to the charge. It is easy to control the final resistivity of an ingot given the starting characteristics of the material. A simple set of equations determines the amounts of each kind of feedstock to be added to the mix. The constraints are the weight of the ingot and the doping level, the latter given by the difference between donor and acceptor concentrations, assuming all impurities are ionised at room temperature. The relationship between resistivity and doping level is known from the literature [14] and a simple spreadsheet can be used to do the conversion.

4.2.2 Crucibles

Crucibles are one of the critical points of mc-Si technology. They are made of slip-cast silica, a technique known since the medieval age, which consists

Table 4 Typical features of DS mc-Si

Parameter	Typical value
Energy consumption	10 kWh/kg
Crystallisation yield	70–80%
Growth rate	5–10 mm/h
Ingot size	100–300 kg
Ingot base	Square, 66 cm × 66 cm
Ingot height	20–25 cm

Table 5 Impurity levels of a typical mc-Si ingot

Impurity	Typical value (ppma)
Fe	< 0.1
Al	0.5–2
Cu, Mn, Cr, Mg, Sr	< 0.1

of letting a plaster mould slowly absorb the quartz present in a water suspension. A layer of up to about 2 cm thick can be realised in this way, and the mould can have a double jacket to improve thickness uniformity. The crucible is then baked for mechanical resistance.

The crucibles currently used have been developed to withstand the high temperatures of a heavy silicon ingot growth process in order to avoid unwanted failures in the presence of liquid silicon. Crucibles are lined with a Si_3N_4-based coating to prevent liquid silicon sticking to the walls and subsequent cracking of the ingot due to the strong stress during solidification and cooling. However, only a limited number of companies manufacture crucibles worldwide, and the maximum size of a 'safe' crucible has probably already been reached (68 × 68 cm). This imposes a boundary condition on the design of future mc-Si furnaces [17].

4.3 Electromagnetic Continuous Casting

Electromagnetic continuous casting (EMC) uses an RF coil to induce currents in an appropriately designed circuit able to push the melt away from the walls, therefore making it unnecessary to use crucibles. A schematic of the furnace is shown in Figure 5 [19]. The process is carried out in an argon ambient at slight overpressure. The top end is open for the ingot to be pulled down while new feed material is added. The resulting ingot is a long bar of about 240 kg in weight. The idea behind this growth method is to completely avoid the use of any physical crucible by confining electromagnetically the charge. This gets rid, at one time, of two major issues of the DS techniques that have previously been described: expensive crucibles and related contamination. However, inhomogeneous nucleation occurs and grain size is also rather small, resulting in a low starting quality of the material [20], although a low oxygen content is reported. This kind

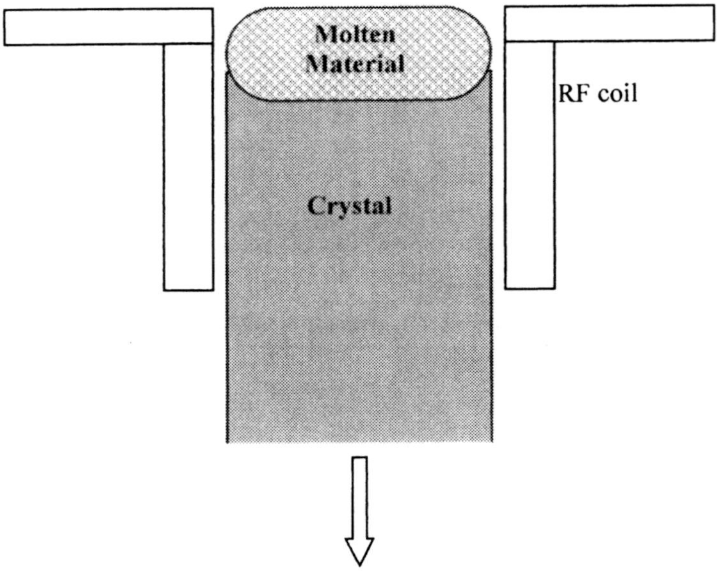

Figure 5 Schematic of EMC furnace.

of material is not commercially available yet, but there are announcements it could be shortly.

4.4 Float Zone Silicon

As float zone silicon typically is used for power electronic components and detectors, the advantage of using large diameter substrates has been limited and even today, the majority of all float zone crystals are only 100–125 mm in diameter, which is advantageous for the PV industry needs. Most product and process optimisation activities have been focused on achieving highly predictable yields when running many small series of different types, constantly varying with respect to crystalline orientation (or), diameter (25–150 mm), dopant type (n- or p-type), and resistivity range (0.01–100,000 Ω cm).

During the float zone growth method, a molten zone is passed along the silicon rod, eating up the raw polycrystalline silicon material and leaving behind a purified monocrystal. Modern FZ machines are now capable of accepting feedrods up to 2 m long with a weight between 60 and 100 kg. The bottom end of the feedrod is coned by a grinding operation, and during the process, the surface of this cone is heated to the melting temperature of silicon. This results in a thin layer of molten silicon continuously running down the feedrod bottom tip, and through the centre hole of the induction coil. The feedrod is heated by a skin current induced by an electromagnetic field. As the feedrod, the molten silicon and the finished crystal are freely suspended in the growth chamber; there is never direct physical contact between silicon and the surroundings, except for the ambient gas, typically argon. Contamination is therefore very low, and the process also allows purification of impurities which segregate in the melt.

As the feed rates of the feedrod and finished crystal can be controlled independently, there is no constrains on the diameter of the monocrystal and the diameter of the feedrod. Typical growth rates of the monocrystal are between 2 and 3 mm/min.

Beside the physical dimension of the monocrystal, the fundamental parameter that must be controlled is the shape of the phase boundaries, i.e. the free surface of the melt and the melt-solid interface where the crystallisation takes place. Factors affecting the phase boundaries are the induction coil current, the pull velocities of the feedrod and monocrystal, the rotation rates, the eccentricity of the rod and monocrystal with respect to the coil centre and additional heat sources.

The monocrystalline perfection of the finished crystal is very high, as neither volume defects (precipitates or voids), planar defects (twins grain boundaries or stacking faults) or line defects dislocations) are present. The purity of the finished monocrystal is very high, with oxygen and carbon as the two impurities of highest concentrations (upper limit at 1.0 and 2.0×10^{16} cm^{-3}, respectively). Also due to the purity of the feedrod material, the concentration of other impurities is very low, and the total concentration of all metal atoms typically lies below 10^{13} at/cm^3 [21, 22].

4.5 Non-wafer Technologies

The idea of lowering the wafer manufacturing costs by avoiding the wafer-cutting step with its silicon loss was the main motivation for the development of a number of silicon ribbon-growth technologies. The common feature to all

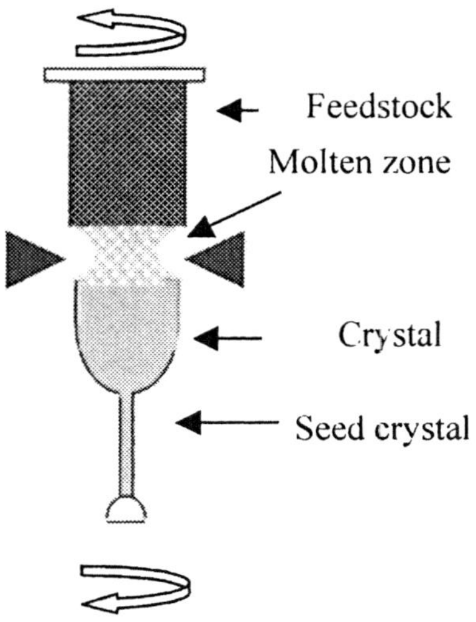

Figure 6 Schematic of FZ growth.

of them is the principle of a continuous production of a thin foil or sheet directly from the silicon melt, using different techniques to confine or stabilise the edges. From these different technologies, developed in R&D programmes such as the JPL Flat-plate solar module project [23], only few are used in commercial wafer production today. The most relevant to date are the edge-defined film fed growth (EFG) [24] (by far the most advanced in terms of industrial performance), the string ribbon (SR) [25] and the dendritic web [26] technology. A great deal of improvement was reported recently in all technologies, and in all cases industrial facilities are described or anticipated. In all cases tailored cell processing is needed to improve the starting quality of the material which is, in general, low. Also, in all cases the technologies appear to be capable of producing very thin sheets, in the 100 μm range. A promising technique from the point of view of productivity is the Ribbon-Growth on Substrate (RGS), originally developed by Bayer and now under development at ECN [27].

Other silicon ribbon technologies with the potential for high production rates by de-coupling ribbon production from crystal growth (such as the low angle silicon sheet (LASS) or the supporting web (S-Web)) are not yet developed to industrial production. Table 6 compares production speed and capacity of different silicon ribbon production technologies. The last column shows the number of furnaces for a 100 MWp production line. [27].

5 Shaping and Wafering

5.1 Shaping

The large, square-based mc-ingots are cut into smaller blocks using large blade or band machines. Blade machines are in general more robust and easy to use and maintain, but have the disadvantage of producing a relatively high kerf loss, up to 3–4 mm. Band saws, on the other hand, suffer from frequent band breakage and may produce waviness in the blocks, which will then need rectifying. However, modern band saws seem to have greatly improved from this point of view.

Monocrystalline silicon ingots instead are treated as the semiconductor counterpart for removing heads and tails, and are shaped to pseudo-square by removing parts of the rounded edges, a process which does not present particular problems, thanks to the relatively small dimension of the ingots.

Table 6 Comparison of different ribbon technologies

Material type	Pull rate (cm/min)	Throughput (m^2/h)	Furnaces per 100 MWp
EFG	1.7	1	100
SR	1–2	0.03–0.1	1200
RGS	600	45	2–3

In the case of mc-Si blocks, it is after the shaping step that blocks are inspected for minority carrier lifetime and resistivity, as described earlier, so finally tops and tails can be removed to leave the material ready to be wafered.

5.2 Wafering

Wafering of Si ingots for the PV industry is probably one of the only examples of technology successfully transferred to the semiconductor industry, which was originally developed for the PV industry. Cost constraints in PV in fact imposed the development of a slicing technique able to reduce kerf loss and increase productivity, as an alternative to slow, large kerf loss blade cutting techniques used until about 10 years ago [15]. On the other hand, the specifications for semiconductor grade wafers up to 300 mm diameter needed a totally new concept of machines for the control of taper, thickness variation and surface smoothness, so at the end both industries enjoy the development of wire saw technology.

Modern slicing technology is based on wire sawing, where a thin wire (160 μm diameter) web pushes an abrasive-based slurry into the silicon to be cut. In this way several wafers are cut at the same time, with high mechanical precision, and in a highly automated process.

The principle of wire sawing is shown in Figure 7. A spool of single bronze coated stainless steel wire up to several hundreds of kilometres long (!) is fed on high precision grooved wire guides. The feeding system, not detailed in the figure, is designed to allow high precision control of the wire tension, one of the critical parameters of the process, throughout the cut. The silicon block(s) is glued on a low cost glass support, which in turn is mounted on a motor-driven table which translates downwards through the web. The abrasive slurry is fed to the wire web through a nozzle, and allows the silicon to be cut. The abrasive is

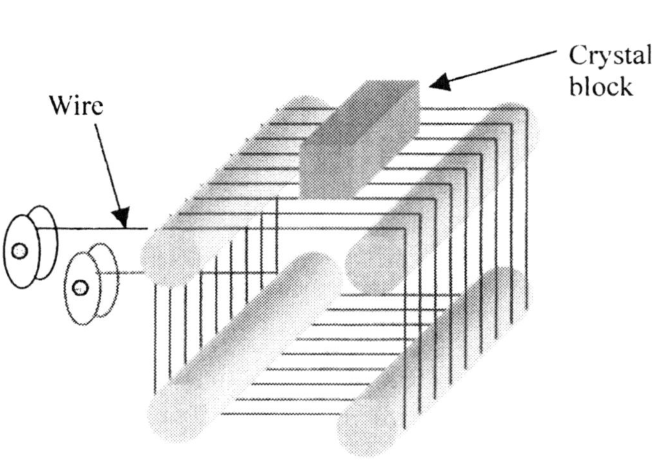

Figure 7 Principle of wire sawing.

Table 7 Typical wafering conditions for 350 μm thick, 125 mm square mc-Si wafers

Parameter	Value
Wire speed (m/s)	5–10
Table speed (μm/min)	300–400
Wire diameter (μm)	160–180
Wire guide pitch (μm)	550–570
SiC mesh	500–600
Slurry temperature (°C)	25 ± 5
Slurry mixture	1:1 (glycol based)
	5:8 (oil based)
Viscosity (g/l)	1600
Wire tension (N)	23–25

fine mesh silicon carbide powder. The process is completed when the wires reach the glass, thus allowing the wafers to be separated without any damage, as they are attached to the support through a very thin layer of glue. Wire speed and tension, table speed, slurry viscosity and temperature and abrasive characteristics are the main parameters to control in order to produce wafers that match the specifications indicated in an earlier paragraph. As the conditions tend to change during the process, e.g. because of the increase of the temperature of the slurry or because of the contamination of the slurry by the silicon dust produced by the process, it is critical to be able to adjust the parameters to avoid waves or thickness variations. The slurry can be based on oil or on other fluids which are water washable and can improve subsequent cleaning and handling steps, although the waste treatment system tends to be more sophisticated and needs a COD control device than with oil-based technology.

Typical parameters for cutting 125 mm square wafers over a length of about 30 cm are given in Table 7.

Large wafer manufacturers have developed automated washing and handling equipment for the steps following the cutting one.

References

[1] *Photon International*, March 2002.

[2] Bruton, T., 2002. General trends about photovoltaics based on crystalline silicon. *Proc. E-MRS 2001 Spring Meeting, Symposium E on Crystalline Silicon Solar Cells, Sol. Energy Mater. Sol. Cells*, Vol. 72, pp. 3–10.

[3] SEMI ™ Standard M61000 – SEMI ™ International Standards www.semi.org.

[4] Ferrazza, F., 1995. New developments and industrial perspectives of crystalline silicon technologies for PV, *Proc. 13th European Photovoltaic Solar Energy Conf.*, Nice, p. 3.

[5] Ferrazza, F. et al., 1998. The status of crystalline silicon modules, *World Renewable Energy Conf.*, Florence.

[6] Kunst, H. and Beck, G.1986. The study of charge carrier kinetics in semiconductors by microwave conductivity measurements, *J. Appl Phys*, Vol. 60(10), p. 3558.

[7] Schonecker, A. et al., 1997. Results of five solar silicon wafer minority carrier lifetime round robins organised by the SEMI M6 Solar Silicon Standardisation Task Force, *Proc. 14th European Photovoltaic Solar Energy Conf.*, Barcelona, p. 666.

[8] Semilab homepage www.semilab.hu.

[9] Ferrazza, F. et al., 1998. Cost effective solar silicon technology, *Proc. 2nd World Conference on Photovoltaic Solar Energy Conversion*, Vienna, p. 1220.

[10] Endroes, A., 2002. Mono- and tri-crystalline Si for PV application, *Proc. E-MRS 2001 Spring Meeting, Symposium E on Crystalline Silicon Solar Cells, Sol. Energy Mater. Sol. Cells*, Vol. 72, pp. 109–124.

[11] Maurits, J., 1998. Policrystalline Silicon-World Demand and Supply, Eighth NREL Workshop on Crystalline Silicon Solar Cell Materials and Processes, Colorado, 1998.

[12] Woditsch, F. and Koch, W., 2002. Solar grade silicon feedstock supply for PV industry, *Proc. E-MRS 2001 Spring Meeting, Symposium E on Crystalline Silicon Solar Cells, Sol. Energy Mater. Sol. Cells* , vol. 72, 2002, pp. 11-26.

[13] Aulich, H. and Schulze, F., 2002. Crystalline silicon feedstock for solar cells, *Prog. Photovolt: Res. Appl.*, Vol. 10, pp. 141–147.

[14] O'Mara, W., Herring, R., and Hunt, L, Eds., 1990. *Handbook of Semiconductor Silicon Technology*, Noyes Publication, p. 395.

[15] Jester, T., 2002. Crystalline silicon manufacturing progress, *Prog. Photovolt: Res. Appl.*, Vol. 10, pp. 99–106.

[16] Ferrazza, F., 1995. Growth and Post growth Processes of multicrystalline silicon for photovoltaic use, in: Polycrystalline Semiconductors IV – Physics, Chemistry and Technology, S. Pizzini, H. P. Strunk and J. H. Werner, Eds., in *Solid State Phenomena*, Transtec, Switzerland, Vols 51–52, pp. 449–460.

[17] Ferrazza, F., 2002. Large size multicrystalline silicon ingots. *Proc. E-MRS 2001 Spring Meeting, Symposium E on Crystalline Silicon Solar Cells. Sol. Energy Mater. Sol. Cells*, Vol. 72, pp. 77–81.

[18] Franke, D. et al., 2002. Silicon ingot casting: process development by numerical simulations, *Proc. E-MRS 2001 Spring Meeting, Symposium E on Crystalline Silicon Solar Cells, Sol. Energy Mater. Sol. Cells*, Vol. 72, pp. 83–92.

[19] Durand, F., 2002. Electromagnetic continuous pulling process compared to current casting processes with respect to solidification characteristics, *Proc. of the E-MRS 2001 Spring Meeting, Symposium E on Crystalline Silicon Solar Cells, Sol. Energy Mater. Sol. Cells*, Vol. 72, pp. 125–132.

[20] Perichaud, I., Martinuzzi, S. and Durand, F., 2002. Multicrystalline silicon prepared by electromagnetic continuous pulling: recent results and comparison to directional solidification material, *Sol. Energy Mater. Sol. Cells*, Vol. 72, pp. 101–107.

[21] Dietze, W., Keller W. and Muhlbauer, A., 1981. Float-zone grown silicon, in *Crystals, Growth, Properties, and Applications*, Vol. 5, Silicon, Springer-Verlag, p. 1.

[22] Luedge, A., Riemann, H., Hallmann, B., Wawra, H., Jensen, L., Larsen T. L. and Nielsen, A., 2002. High-speed growth of FZ silicon for photovoltaics, *Proc. High Purity Silicon VII*, Electrochemical. Society, Philadelphia.

[23] Flat plate solar array project: Vol. III, Silicon sheet: wafers and ribbons, Report *DOE/JPL-1012-125*, 1986.

[24] Kaleis, J., 2002. Silicon ribbons and foils – state of the art, *Proc. E-MRS 2001 Spring Meeting, Symposium E on Crystalline Silicon Solar Cells, Sol. Energy Mater. Sol. Cells*, Vol. 72, pp. 139–153.

[25] Hanoka, J., 2002. PVMat contribution towards Evergreen Solar's new factory, *Proc. 29th IEEE Photovoltaic Specialists Conference*, New Orleans, p.66.

[26] Meyer, D.L. et al, 2002. Production of thin (70–100 mm) crystalline silicon cells for conformable modules, *Proc. 29th IEEE Photovoltaic Specialists Conference*, New Orleans, p.110.

[27] Schonecker, A. et al., 2002. Ribbon growth-on-substrate: progress in high speed crystalline silicon wafer manufacturing, *Proc. 29th IEEE Photovoltaic Specialists Conf.*, New Orleans, p. 316.

IIb-2

Low Cost Industrial Technologies of Crystalline Silicon Solar Cells

Jozef Szlufcik[1], Guido Agostinelli[2], Filip Duerinckx[2], Emmanuel Van Kerschaver[2], Guy Beaucarne[2]

[1] Photovoltech, Industrial Area West-Grijpen, Grijpenlaan 18, 3300 Tienen, Belgium, tel.: +32 16 805 865, fax: +32 805 905
[2] Interuniversity Microelectronics Center, IMEC, Kapeldreef 75, B-3001 Leuven, Belgium, tel: +32 16 281080, fax: + 32 16 281501

1 Introduction

Although efficiency is important, the principal requirement for industry is low cost. Processing techniques and materials are selected for the maximal cost reduction while maintaining a relatively good efficiency. Industrial solar cells are fabricated in large volumes, on 5 inch or larger, Czochralski monocrystalline or multicrystalline silicon substrates.

Analysis indicates that the market price of the commercial PV modules lies in the range of 2.8–3.0 €/Wp. 40–50 % of the PV module cost is caused by the ingot growth (including the polysilicon feedstock material), single crystal ingot formation and wafering. The tendency here is to develop a cheap, good quality solar grade polysilicon feedstock material, to increase the substrate size, to reduce the kerf loss in slicing and to decrease the thickness of the substrates below 200μm. Cell fabrication and module assembly are each responsible for 25 to 30% of the final module cost.

Typical efficiency of commercially produced crystalline silicon solar cells lies in the range of 14%–17%. Because the efficiency of the cell influences the production cost at any production stage, considerable effort directed towards efficiency improvement. The required near future efficiency goals for industrial cells are 18–20% on monocrystalline and 16–18% on multicrystalline silicon. Based on laboratory scale achievements one can consider that production type cells able to fulfil the efficiency goal should possess most of the following features (providing that they can be introduced in a cost effective way):

- front surface texturing,
- optimised emitter surface concentration and doping profile,
- effective front surface passivation,
- fine line front electrode
- front electrode passivation:
  - point contact,
  - deep and highly doped emitter under the contact,
  - MIS contact,
- thin base i.e. much smaller than the minority carrier diffusion length,
- back surface passivation:
  - oxide and/or nitride passivation + local BSF(PERL)
  - floating junction structure
- or back surface field,
- back electrode passivation:
  - point contact,
  - deep back-surface diffusion under the contact,
- back reflector,
- back surface texture.
- antireflection coating optimised for encapsulation

The current status in the development of industrial type processing steps leading to an improved cell efficiency will be described in detail in the sections below.

2 Cell Processing

2.1 Substrates

Although until a few years ago, the standard substrate size was 10×10 cm^2, most cell manufacturers nowadays base their production lines on 12.5×12.5 cm^2 wafers. A further increase of the standard size is taking place, several (mostly Japanese) companies already producing 15×15 cm^2. The trend might continue further, the next size being 20×20 cm^2. The driving force towards these larger cell sizes results from the fact that the cell manufacturing and module assembly costs show little area dependence and that therefore the cost per Wp decreases with increasing cell size. On the other hand the optimum cell size is limited by series resistance and by limitation of module size due to handling, wind loads, module transportation and system assembly. Therefore cell sizes larger than 20×20 cm^2 are probably excluded. Due to the success of developments in multiwire sawing, wafer thickness of 150 μm or smaller are becoming feasible corresponding to final cell thickness of 120 μm or lower [1]. This allows an important material saving; at the same time, thinner cells correspond to the optimum thickness if efficient light trapping and surface passivation are possible. In spite of significant progress in slicing techniques; around 200 μm of high quality silicon per wafer is still lost in kerf waste.

Kerf loss can be completely avoided in the ribbon and sheet silicon technologies. Several technologies have been tried on the laboratory or pilot scale [2, 3, 4, 5, 6, 7]. At present, EFG (Edge-Defined Film-Fed Growth) is the most commercialised of the developed ribbon materials. The technology is based on growth of hollow octagon tubes of silicon after which the wafers are cut out by laser [4]. Another ribbon technology that has entered industrial production is the String Ribbon technology [6]. Here, two parallel strings are pulled through a silicon melt. The liquid film that forms between the strings solidifies to form a continuous polycrystalline ribbon.

2.2 Etching, Texturing and Optical Confinement

Silicon substrates used in commercial solar cell processes contain a near-surface saw-damaged layer which has to be removed at the beginning of the process. The thickness of the damaged layer depends on the technique used in wafering of the ingot. A layer with thickness of about 10 μm has to be etched from both sides of wafers cut by wire saw. The damage removal etch is typically based on 20–30 wt.% aqueous solution of NaOH or KOH heated to 80–90°C. The etching process has to be slightly modified when applied to multicrystalline substrates. Too fast or prolonged etching can produce steps at grain boundaries. This can lead to problems of interruptions of metal contacts.

The silicon surface after saw damage etching is shiny and reflects more than 35% of incident light. An important step in solar cell processing therefore consists of texturing the front surface, to create a structure causing the reflected ray to get a second chance to be coupled into the cell. Such surface texturing can reduce the optical reflection from more than 35 to less than 10%. In addition to the reduced reflection, front surface texturing ensures that light rays are coupled into the solar cell under an oblique angle, making it less probable to escape from the front surface after reflection at the rear. This light trapping effect results in an improvement in internal quantum efficiency in the wavelength range of 750–1000 nm. The effect is very important when thin silicon substrates (<200 μm) are used for material saving.

The reflection losses in commercial solar cells are reduced mainly by random chemical texturing [8, 9]. Monocrystalline silicon substrates with a surface orientation <100> can be textured by anisotropic etching at temperature of 70–80°C in a weak, usually 2wt.%, solution of NaOH or KOH with addition of isopropanol. This etch produces randomly distributed upside pyramids [9]. However, this process brings often production problems of repeatability, lack of pyramid size control and the presence of untextured regions [8]. The important parameters are: adequate surface preparation, temperature control, mixing rate and isopropanol concentration [9]. The solution to this problem requires the use of appropriate additives which enhance the pyramid nucleation process [8]. When the process is under control, uniformly distributed pyramids with height of 3–5 μm are optimal for low reflection losses and later metallization process. Figure 1 shows the SEM micrograph of a randomly textured <100> oriented silicon surface.

The random texturization process is not effective on multicrystalline substrates due to its anisotropic nature. Isotropic texturing methods based on photolithography and wet etching are not cost-effective. Many techniques such as defect etching, reactive ion etching, mechanical texturing or laser scribing have been tried by many groups [10, 11, 12, 13, 14, 15, 16, 17]. The best results, from the optical point of view, have been obtained by mechanical texturing and by reactive ion etching [14]. A very elegant technique of

Figure 1 SEM micrograph of a random textured <100> oriented silicon surface.

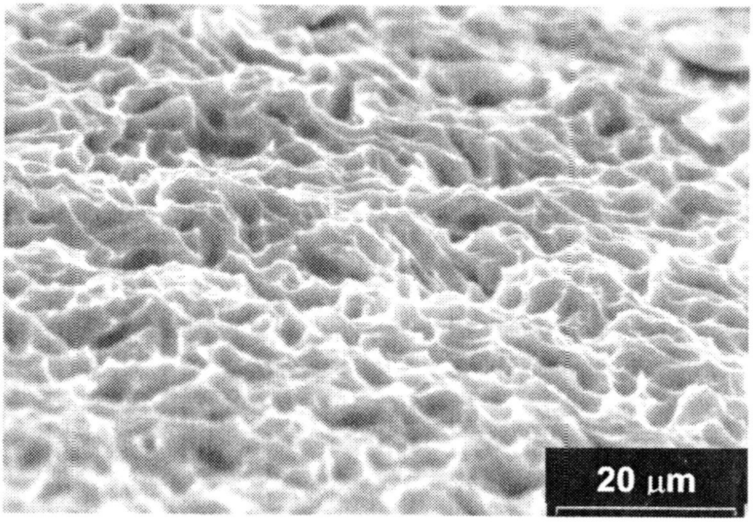

Figure 2 SEM micrograph of the surface of an acidic isotextured multicrystalline silicon wafer

texturing multicrystalline silicon is to etch the wafers in an acid mixture based on HF and HNO_3 [18, 19]. When saw damage is present, this etching process structures the surface in a way that is independent of the crystal orientation. This acidic isotexturing results in lower reflection than the traditional anisotropic etching and in a better conversion efficiency [19, 20]. An SEM picture of the surface of an acidic isotextured wafer is shown in Figure 2.

2.3 Cleaning

In an industrial high efficiency silicon solar cell process, wafers are typically cleaned after texturing and before surface passivation. Traditionally RCA clean [21], originally developed for use in microelectronics, is the most widespread cleaning recipe in solar cell processing. Although not often discussed in the solar cell technical literature, cleaning is very important for solar cell performance. Long diffusion lengths of minority carriers, necessary for high efficiency cells, require low levels of metal contamination at the silicon surface before a high temperature treatment. Moreover, a growing concern in the photovoltaic community is the chemical waste produced during cell processing. The conventional RCA cleaning consists of two steps normally referred to as SC1 and SC2. The SC1 step consisting of a $NH_4OH/H_2O_2/H_2O$ mixture, aims at organic particle removal, whereas the SC2 step (an $HCl/H_2O_2/H_2O$ mixture) is used to remove metal contaminants. A more detailed analysis reveals that for industrial solar cells – because of their large feature size and the absence of photolithographic processes – organic particles are not an important issue and therefore the SC1 step is not essential. On the contrary, it is known that the metal contamination resulting from the SC1 step may be

Table 1 Metal contamination removal (in 10^{10} at/cm² after IMEC vs. RCA+HF clean) [22]

Cleaning treatment	Ca	Fe	Cu	Zn
Starting level	15	1	2	1
IMEC	0.7	0.3	<0.1	0.7
RCA + HF	2.5	0.8	0.5	4.9
5 × (IMEC) +HF	<0.1	<0.1	<0.1	0.1
5 × (RCA) + HF	4.5	0.3	<0.1	<0.1

high and a one-to-one correlation between the metal concentration of the SC1 bath and the metal contamination of the silicon surface was found. This requires the use of sub ppb metal contamination specification of the chemicals.

A potential replacement for the standard RCA clean is the "IMEC-clean" [22]. This cleaning procedure, consisting of a H_2SO_4/H_2O_2 step followed by a 1% diluted HF step, reveals a perfect removal of metallic particles. Table 1 presents the metal contamination removal for RCA-clean and IMEC-clean.

If removal of metallic contamination is the only issue a single cleaning step in 1% diluted HCl yields excellent results. This is especially important for the cleaning step performed after texturing. A further advantage is the low consumption of chemicals. This results not only in important cost savings but also in a considerable reduction of chemical waste products.

2.4 Junction Formation

As has been proven by many researchers [23, 24, 25, 26], the optimum emitter doping profile should be relatively deep and moderately doped, or a shallow emitter with a high surface concentration. Both profiles combined with surface passivation by high quality thermal oxides or PECVD nitrides show reduced surface recombination losses and increased emitter collection efficiency. However, both industrial techniques used for front contact fabrication i.e. screen printing of silver pastes and electroless plating of Ni require a highly doped (P surface concentration above 10^{20} cm^{-3}) and a deep junction to obtain acceptable contact resistance and to avoid metallic impurity penetration towards the junction region. The typical emitter sheet resistance used in a screen printing metallization process is between 40 and 60 Ohm/sq. This can be achieved by diffusion from liquid $POCl_3$ or solid P_2O_5 sources in open tube furnaces or from screen printed, spray-on or spin-on P-sources followed by a conveyor belt furnace diffusion. A deep emitter and poor surface passivation lead to a voltage loss and collection losses at the short wavelengths of light.

One solution consists of a selective emitter: the heavy and deep diffusion under the contact fingers not only assures low contact resistance giving a good fill factor, but also reduces the contact contribution to front surface recombination losses. The emitter between the fingers is optimised for high spectral response and high open-circuit voltage thanks to a relatively low

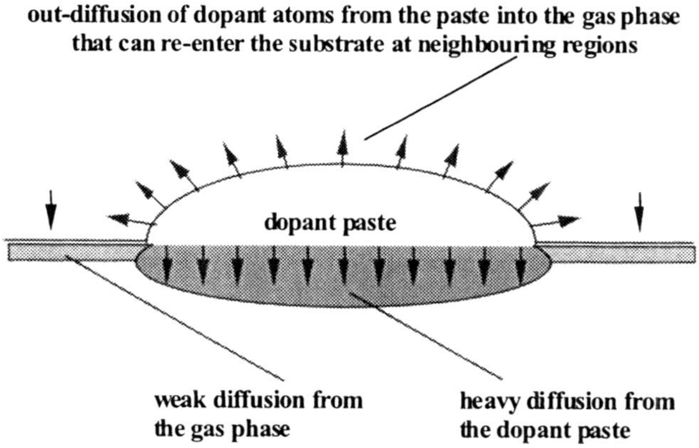

Figure 3 Schematic representation of a one-step selective emitter fabricated by out-diffusion via the gas phase of a locally printed rich phosphorus source.

surface concentration. The selective emitter structure can be realised by double diffusion [27] or by a single diffusion and etching back the active emitter region to the desired sheet resistance [28]. Industrial applicability is doubtful due to the complicated nature of the process based on etching back the deep emitter completely or partially. Another selective emitter process which circumvents this problem relies on P-paste printing on the contact region to create a deep emitter. Diffusion from this region via the gas phase creates a shallower diffusion in between the contacts (Figure 3). This one-step selective emitter has led to high efficiencies [29]. Other selective emitter schemes include self-aligned plasma etch-back emitters [30] and selective emitters by diffusion barriers [31].

While not able to achieve the ultra-high efficiencies of a selective emitter process, a shallow homogeneous emitter process could be the ideal compromise between avoiding contact resistance problems and improving a cell's blue response [32]. Research on classic front contact Ag-pastes and optimised process parameters have enabled good contacts on emitters with sheet resistance around 70 Ω/sq. Because of its simplicity this method is more industrially feasible and is currently applied in production lines.

An entirely different approach is presented by MIS inversion layer solar cell technology. Here the emitter diffusion process is completely eliminated by inducing an n+ inversion layer in the p-type silicon substrate. The inversion layer is induced by positive charge in the top silicon nitride antireflection coating layer. The high positive charge density is achieved by the incorporation of cesium [33].

2.5 Front Surface Passivation and Antireflection Coating

Surface recombination can be effectively decreased by many techniques. The most common one for the laboratory cells is to grow a thin thermal oxide and deposit a double antireflection coating by evaporating ZnS and MgF_2 layers [34], or by growing a thick thermal oxide up to 110 nm which serves at the same time as a passivating and anti-reflection coating (ARC) layer [35]. This gives a decreased surface recombination velocity at the Si-SiO_2 interface depending on the P surface concentration of the emitter and density of interface states. The interface quality is further improved by a low temperature anneal in forming gas. Such surface passivation is very effective especially with low emitter doping concentration and regularly leads to an enhancement in the photogenerated current and open circuit voltage. Both approaches are excluded in industrial processes due to the high cost and low throughput of vacuum processes and long time needed to grow a thick thermal oxide. The thickness of the passivating oxide in case of industrial cells would be 6–15 nm, which is thin enough not to disturb the optical system in combination with an antireflection coating, and thick enough to ensure an effective surface passivation.

Until the middle of the 1990s titanium dioxide (possibly in combination with a thin silicon dioxide as described above) was still the industrial choice for ARCs. Today, almost all the PV manufacturers make use of PECVD hydrogenated silicon nitride, due to its possibility of achieving optimal refraction index and excellent surface and bulk passivation properties at the same time. The increase in conversion efficiency observed in cast mc-Si by replacing TiO_2 with SiNx:H, with the same firing-through process, is remarkable: improvements well in excess of 10%, have been reported [36]. It is not possible to talk about a unique SiNx, since a real variety of different silicon nitride layers with different physical properties can be deposited with the same machine. Not only the refractive index, but the surface as well as bulk passivation properties are strongly dependent on the deposition parameters. Commercial systems for nitride deposition are all suitable for reaching surface passivation of typical industrial n^+ emitters equal, in practical terms, to that of any of the oxides, and a refraction index between roughly 1.9 and 2.4 with nominally zero extinction coefficient around 2.0–2.1 [37, 38, 39]. The real issue is whether *a particular nitride* is suitable for bulk hydrogenation, and whether a given bulk material is sensitive to nitride passivation [40] (see Section 2.8.3).

2.6 Front Contact Formation

The process of front contact formation is one of the most important solar cell processing steps. The applied metallization technique determines the shadowing and series resistance losses, determines the emitter diffusion profile and surface doping concentration, and dictates the choice of certain surface passivation techniques. High efficiency, large area solar cells require front electrodes with low series resistance and low area coverage. In order to meet these requirements, two basic techniques are implemented in mass solar cell production: laser grooved buried contact metallization and advanced screen printing processes.

In the first method, plated metal contacts are formed in deep grooves cut by laser or mechanically into a lightly diffused front surface protected by nitride or oxide. After etching and cleaning, the grooves are subjected to a second very heavy diffusion. The metallization is then obtained by a self-aligned plating process of nickel, copper and a thin layer of silver [41]. The advantages of these cells have been described in the literature [42] and include the very large height-to-width ratio of the finger metallization and the fine line width (typically 20–25 μm) (see Figure 4a). Based on this cell structure and using simplified processing, a cost effective production technology is now in operation, yielding average efficiencies between 16–17 % and occasionally up to 18% on 100 cm^2 CZ pseudo-square cells [43]. The main problem related to this technology is an environmental issue. The external costs of meeting environmental specifications of developed countries with processes producing an enormous amount of rinse water containing nickel and copper must be taken into account [44]. It can be a major issue for planning large volume plants above 100 MW.

In the screen printing method, a stainless steel or polyester mesh screen stretched on a metal frame is covered by a photo-emulsion layer. Openings – which define the front contact pattern – are photolitographically formed in the emulsion layer. Highly conductive silver paste is pushed by a squeegee through the openings in the screen onto substrates with well defined adjustable pressure.

Screen printing is a traditional industrial solar cell technology, existing since the beginning of the 1970s. The advantages of screen printed solar cells are the fact that the technology is well established and can be improved step by step without requiring large capital investment, the robustness of the production equipment and the low amount of chemical wastes. The first generation of photovoltaic devices made with this technology suffered from severe limitations (large shading losses, low fill factors and a deep emitter without surface passivation).

Due to all these factors the efficiency of screen printed cells was typically about 25% to 35% lower than the efficiency of buried-contact cells. These limitations, however, have in the last decade been overcome thanks to an important research effort mainly in Japan and Europe [14, 45, 46, 47]. Figure 4 depicts the cross-section of buried and screen printed contacts drawn in the same scale.

Further developments in the field of screen printable contact pastes [48] and adapted process parameters allow good contact formation on shallow homogeneous emitters. The printing of finger widths of 100 micron is a common practice on production scale. By optimising pastes and firing temperatures fill factors between 78 and 79% for 100 cm^2, screen printed-cells are obtained [49] in the lab as well as in production lines. Surface and bulk passivation is achieved during contact firing by the development of Ag-pastes capable of firing-through a hydrogen-rich silicon nitride layer. Screen printed metallization can also be combined with effective surface passivation in selective emitter structures. An interesting development in this field are the

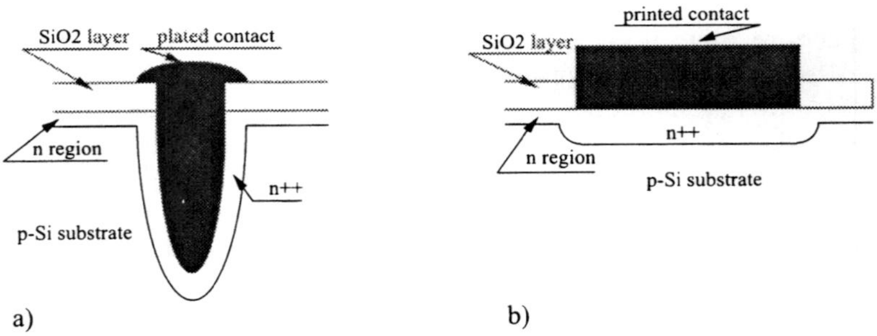

Figure 4 Cross section of a laser grooved buried plated contact (a) in comparison with a fine line screen printed contact (b).

self-doping pastes [50]. A self-aligned selective emitter process has been proposed based on contact pastes with incorporated dopant atoms. The dopant atoms can contribute to the lowering of the contact resistance beneath the electrode by an alloying procedure with silicon during contact firing [51].

Although screen printing is the most widely used contacting method for silicon solar cells, few studies were available to study the nature of the contact. Recent efforts by some European research groups lead to a convincing model of contact formation, and explain the important role of the glass frit [52, 53]. New paste developments are concentrated on environmentally benign pastes (Pb and Cd-free) as well as dedicated rear side Al-pastes that are capable of creating a strong Back-Surface-Field (BSF) on thin wafers without bending the wafer.

Among other techniques for a front contact formation there are reports of direct pen writing, offset printing [54], pad printing [55], hot melt ink technology [56], stencil printing [57, 58] and roller printing [59]. The direct pen writing with a very high aspect ratio has been applied for a front contact metallization of EFG silicon sheets solar cells [60].

2.7 Rear Structure

Currently in most of the industrial cell structures, rear electrode formation and surface passivation are achieved in a very convenient single step by alloying a screen printed aluminium paste with silicon. Aluminium forms a eutectic alloy with silicon at a temperature of 577°C. During the firing process, a liquid Al-Si phase is formed according to the Al-Si phase diagram. The molten Al-Si region acts, as well, as a sink for many impurities giving a gettering effect. During cooling down, the silicon recrystallizes and is doped with aluminium at its solubility limit creating a p^+ Back Surface Field (BSF) layer (Figure 5a) [61, 62]. A sufficient thickness of Al is required to achieve a significant contribution of Si in the formation of the liquid phase. Very low back surface recombination velocities (as low as 200 cm/s) have been reported for thick screen printed and evaporated aluminium layers (at least 20 and 10 μm thick, respectively) fired at a temperature above 800°C [62, 63, 64]. The

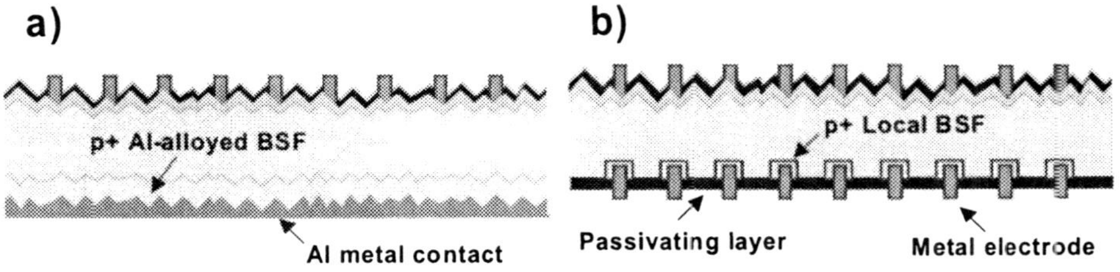

Figure 5 Schematic representations of different rear structures: a) Al-alloyed BSF and b) locally diffused BSF.

analysis of industrial type solar cells, however, has always led to determine the back surface recombination velocities of standard Al-alloys BSF being in the order of 10^3 cm/s [65, 66]. This process, in addition, can cause significant wafer warping due to the different thermal expansion coefficient of silicon and aluminium when thin substrates are used. Both these issues pose a significant technological challenge for the next generation of industrial silicon solar cells. For bulk silicon technology significant cost and energy payback time reduction pass necessarily through a more efficient use of the prime material- that is, thinner wafers. The leap to substrates with thickness well below 200 μm offers on one hand the possibility for higher cell efficiencies and, on the other hand, it imposes demanding requirements in terms of back surface passivation, optical confinement, and it presents process yield issues. In this context, local contact structures become commercially attractive and are making their way from lab into production. In this approach, most of the rear surface is coated with a dielectric, or other high-quality passivating layer, while the locally defined back selectrode (local BSF contact) covers between 1 and 4% of the total back surface area (Figure 5b). Excellent rear side passivation can be obtained in a variety of ways. Silicon oxide, silicon nitride in different stoichiometries [67], oxide/nitride stacks [68], amorphous silicon/silicon heterostructures [69] have all proven to be excellent termination of monocrystalline and multicrystalline silicon surfaces. The best results at the cell level so far, both at lab and pilot line scale, have always been obtained with thermally grown silicon oxide [66, 70] and, with slightly lower efficiencies, with silicon nitride. The latter has the advantage of a lower process temperature, which is less harmful to low-cost substrates. Amorphous silicon yields excellent values for surface passivation but its integration into the solar cell process is not immediate due to its low resistance to high temperature steps and stability issues when in contact with the metal electrode. Electrode formation has always been the process bottleneck in local contact structures since it required complicated structuring and lift-off steps. The ongoing development of firing-through screen-printing pastes and the very promising Laser Fired Contact (LFC) process [71] is likely to make Local BSF structures the standard in industrial solar cells in the coming years.

2.8 Substrate Material Quality Modification

A sufficiently high bulk minority-carrier lifetime is crucial to achieve high performance of crystalline silicon solar cells. Unlike the high quality and expensive float zone (FZ Si), as-prepared "solar-grade quality" Czochralski grown (CZ-Si) and multicrystalline (mc-Si) silicon substrates generally exhibit moderate bulk lifetimes depending on various factors that introduce generation-recombination (GR) centers within the bandgap. The major sources of these centers include oxygen and carbon content, metallic impurities, high densities of crystallographic defects, presence of grain and sub-grain boundaries, etc. Many of these factors are not completely avoidable given the low-cost substrate requirement for silicon photovoltaics. Several processing techniques can be applied that result in the removal of recombination centres and lead to a higher minority carrier lifetime at the end of the process than at the beginning. On the other hand, some mechanisms during and after the solar cell process lead to the generation of new recombination centres. These mechanisms should be understood and, if possible, avoided.

2.8.1 Gettering by Phosphorous Diffusion

Gettering is a technique that either reduces or eliminates metallic impurities in the substrates by localising and blocking them away from the device active regions, or by completely removing them from the substrate. The former is referred to as intrinsic gettering and the latter as extrinsic gettering. Gettering techniques are well-established in Si IC-technology for various purposes, for example, to reduce the leakage currents induced by GR centers in CCD image sensors, CMOS devices etc. In solar cell fabrication, it is desirable that the gettering treatment remains part of the cell processing and does not considerably increase the production cost. Phosphorus diffusion and aluminium diffusion are the most efficient gettering schemes used in Si solar cell processing.

Impurity migration towards gettering sites takes place as a consequence of a large emission of Si interstitials due to the formation of SiP particles by heavy P-diffusion [72, 73]. Enhanced solubility of metallic impurities in such heavily P-diffused regions, and impurity segregation at Si_3P_4 precipitates lead to efficient gettering [74]. In solar cell fabrication, P-diffusion can be performed prior to cell fabrication followed by the removal of the heavily diffused layers (pre-gettering), or as part of emitter formation depending on the optimal gettering conditions required for the material and on the costs involved. Certain selective emitter processes that involve two P-diffusions (see Section 0) benefit both from gettering and removal of heavily diffused layers with accumulated impurities. The gettering process is more complicated in mc-Si since, at high temperatures, a dissolution of impurities that are precipitated near bulk crystal defects is also initiated. Therefore in mc-Si, high temperature gettering is not desirable due to the competition between the gettering and dissolution rates of the precipitated impurities [74]. Gettering efficiency depends also on certain material properties, for example, on the interstitial oxygen $\{O_i\}$ content [75]. Some mc-Si wafers with lower $\{O_i\}$ reportedly showed better improvement after pre-gettering compared to those with higher

$\{O_i\}$[76, 77]. On a sensitive material of standard resistivity, the minority carrier lifetime increase can be spectacular and reach values up to 200 μs [78]. Even on low resistivity mc-Si wafers, lifetimes as high as 15 μs can be obtained after pre-gettering at low (<900°C) temperatures, enabling solar cells with diffusion length approaching the wafer thickness and V_{oc} values well above 630 mV to be fabricated [78, 79]. However, the incorporation of an additional pre-gettering step in a production line adds to the production cost. This additional cost can be limited by keeping the heavy P-diffused layer and etching back to the desired sheet resistance. This selective emitter process is, however, rather difficult to control in an industrial process.

2.8.2 Gettering by Aluminium Treatment

Formation of a p^+/p high/low junction at the rear side of the cell by re-growth from a "fast-alloyed" Al-Si melt is the most commonly used process for creating the back surface field (BSF) in industrial silicon cells (see Section 0). This treatment has an additional advantage of bulk gettering by prolonged firing or by a thermal anneal after the initial firing. At alloying times > 1 minute with screen printed Al, evidence of bulk impurity gettering has been noted at the grain boundaries of (100)/(111) CZ bi-crystals [80]. The effective grain boundary passivation obtained after Al diffusion has also been attributed to a much higher diffusion coefficient of Al along the grain boundaries [81]. Al-gettering effects have also been noted in single crystalline CZ-Si and FZ-Si materials. From the relatively low temperatures generally used in Al gettering, it can be said that the gettered species are the fast moving interstitial impurities such as Cu, Fe etc. Pilot line processes involving fast-alloying of screen printed Al-paste by firing, followed by the removal of excess aluminium and a subsequent thermal anneal for up to 1h, resulted in a considerable enhancement in the bulk diffusion length in large area mc-Si wafers [82]. This process of fast firing also prevents impurities contained in the Al source from diffusing into Si. High performance, screen printed CZ-Si cells with Al-gettering have also been reported [61]. The thickness of the Al-layer deposited on the rear side of the wafers is also of great importance in order to get ideal pp^+ junctions for low-enough surface recombination velocity [61, 83, 84, 85, 86, 87]. In ribbon materials such as EFG, efficiency enhancements of 1.4% absolute along with an increment of >60 μm in diffusion length have been reported after an 850°C gettering treatment with evaporated aluminium [88]. However, the on-going improvement of material and basic process steps make a separate Al-gettering step less and less necessary. In a recent publication, efficiencies up to 16.1 % and lifetime up to 100 μs have been reached in EFG ribbons with a simple process involving co-firing of screenprinted Ag and Al contacts [89]. Due to the large thickness requirement for the Al-source, and for the cost effectiveness of the process, it appears that screen printing is the most efficient way for depositing the Al-source.

2.8.3 Bulk Passivation from Silicon Nitride

The introduction of PECVD silicon nitride in industrial production of multicrystalline solar cells was justified not only by improved antireflection properties and surface passivation, but also by defect passivation in the bulk of the device. Several comparative experiments have clearly indicated a substantial performance increase when the solar cell is covered with silicon nitride and then briefly annealed at high temperature (>600°C), usually during the necessary firing step of screen-printed contacts [49, 90, 36]. It is widely accepted that the beneficial effect of silicon nitride comes from defect passivation by hydrogen atoms. During a short contact firing, hydrogen (present in large concentrations in PECVD silicon nitride) gets released, diffuses into the bulk of the material and passivates electrically active defects. This phenomenon has mostly been studied indirectly, through the performance improvement of solar cells. Only recently, it has been attempted to directly measure the concentration of hydrogen released into the silicon [91, 92]. The first studies indicate that the amount of incorporated hydrogen is surprisingly low ($<10^{14}$ cm^{-3}) considering the large effect on device performance. The phenomenon is far from being fully understood and appears to be rather complex. The passivation effect strongly depends on the material. Multicrystalline ribbon material is generally extremely sensitive to passivation from nitride (see e.g. [93]), whereas some types (but not all) of cast multicrystalline silicon show very little improvement. Moreover, the initial properties of the nitride layer before firing appear to determine the extent of passivation during subsequent firing. Next to the amount of hydrogen present in the layer, the density of the nitride layer appears to be a very important parameter [94, 40]. Furthermore, it is crucial to find the right process window for the firing process that delivers a good contact firing, H-release and alloy formation all at the same time.

2.8.4 Lifetime Degradation from B-O Complex Formation

While the processes described in the previous sections improve the electronic quality of the silicon material, other phenomena result in defect formation and lifetime decrease. A technologically important example is the lifetime degradation due B-O complex formation.

For a long time it has been observed that the performance of solar cells made from B-doped CZ-Si wafers degrades under illumination, to stabilize at an efficiency one to two absolute percent below the initial efficiency [95]. The reason is a substantial decrease of the minority carrier lifetime due to light-induced formation of defects. Interestingly, these defects are removed upon annealing in the temperature range of 150°C. This fact, in combination with the observation that the defect creation strongly depends on both the B content and the O concentration, led to the conclusion that the defect is a metastable complex involving both B and O [96, 97]. CZ-Si always contains a substantial amount of O ($>5 \times 10^{17}$ cm^{-3}) because of the manufacturing technique used. Recently, it has been shown that the defects are not strictly activated by light-induced generation but by the carrier recombination process [98, 99].

This phenomenon is particularly unwelcome as B-doped CZ-Si is one of the major materials for today's photovoltaic industry and limits the gain in performance one can obtain from using monocrystalline silicon instead of multicrystalline. Several solutions are being investigated [100]. One approach is to replace boron by gallium as p-type dopant, and the another one uses n-type wafers in place of p-type. Both approaches have shown potential for high and stable performance, but important manufacturing issues still need to be addressed.

3 Industrial Solar Cell Technologies

Most of the commercially fabricated crystalline solar cells are based on screen printing. The industrial production of more advanced solar cells technologies such as laser grooved buried contact, MIS and EFG solar cells contribute less than 10% to the total commercial production. The way in which the high efficiency solar cell features, described above, can be incorporated in the commercial processing sequence is outlined below.

3.1 Screen-Printed Solar Cells

Screen printing is a well established, simple, robust, continuous and easily adaptable process. Fully automated screen printing solar cell production lines are offered by many companies. From all high efficiency features described in Section 1, usually only the front surface texturing and Al-alloyed back surface field were included in the 1st generation screen printing solar cell process. This explains why the efficiency of some industrial crystalline solar cells stayed long in the range of 13% to 15%. In the last ten years an important effort has been made by several research groups and companies, mainly in Japan and Europe, in order to improve the screen printing process [101, 45, 46, 47]. The main progress has been made in the paste formulation, fabrication of new types of fine line screens and the development of modern screen printers.

The new type of silver pastes contains additives which permit, during the firing process, a selective dissolution of silicon dioxide and ARC layers of TiOx or SiNx while preventing deep penetration into bulk silicon. This gives the possibility to use a very simplified process with firing through passivating anti-reflection coatings. The improved screenprintability of new pastes and new types of metal screens [47] give the possibility to print narrower lines down to 50–60 μm with an improved aspect ratio. Therefore, the spacing of the screen printed contact lines can be reduced, leading to more lightly doped emitters with an improved short wavelength response. The problem of non-optimal surface passivation is tackled by the use of a selective emitter structure or a shallow homogeneous emitter.

A one-step selective emitter process based on selective printing of phosphorus paste has led to pilot line efficiencies of 17.9% on 10 Ωcm CZ-Si (100 cm^2) and 16.8% on multi-Si (100 cm^2) [29]. With a double diffusion and etch-back, 17.0% has been obtained on 156 cm^2 large multi-Si cells [32]. The high efficiency features of these new screen printed solar cells are [102]:

- Isotropic front surface texturing independent of grain orientation
- Reduced total front surface shadowing losses down to 6%
- Shallow emitter with optimised profile and front surface passivation
- Efficient bulk passivation based on thermal treatment of hydrogen-rich SiNx
- Back Surface Field (BSF)

The alignment of the front contacts on top of the highly doped n^{++} region is done automatically by the latest generation of screen printers equipped with digital cameras. The movement of the table and the squeegee is driven by linear motors controlled by a central computer.

A simplified shallow homogeneous process as presented in [103] retains many of the advanced features described above. An independently confirmed efficiency of 16.8% has been obtained on 200μm thick multi-Si substrates. The applied process sequence is shown in Table 2. Figure 6 presents a cross section of such a cell. All processing steps can be easily transferred to production lines.

A process that combines screen printing and grooving has demonstrated efficiencies above 17% on 100 cm^2 multicrystalline solar cells [45]. The screen printing process is also well suited for larger cell sizes: 15 × 15 cm^2 multicrystalline cells with efficiencies above 16% have also been demonstrated [104].

3.2 Buried Contact Solar Cells (BCSC)

The buried contact solar cell (BCSC) process has been developed at the University of New South Wales [41, 105]. Many aspects of the BCSC structure and its processing have been extensively described in the technical literature [105, 106, 107, 108, 109, 110, 111, 112]. A laboratory efficiency as high as 21.3 % on small area FZ material has been reported [113]. The buried contact solar cell is discussed in more detail in Chapter II-b5. A conventional commercial BCSC processing sequence licensed to many solar cell manufactures is presented in Table 3.

Table 2 Processing sequence for a thin screen-printed multicrystalline silicon solar cell based on a shallow homogeneous emitter, SiNx-passivation and Al-BSF without wafer warping [103]

Step no:	Process description
1.	Saw damage etching + isotropic texturisation in one single step
2.	Shallow n+ diffusion over the whole front surface (70 Ohm/sq.)
3.	Glass removal
4.	Edge junction isolation
5.	PECVD SiNx as front ARC and hydrogen source
6.	Printing + drying of adapted front and rear contacts
7.	Co-firing of both contacts: • Creation of contact to highly resistive emitter with low contact resistance by firing through SiNx • Surface and bulk passivation by hydrogen release from SiNx • Creation of Al-BSF for rear surface passivation without wafer bending

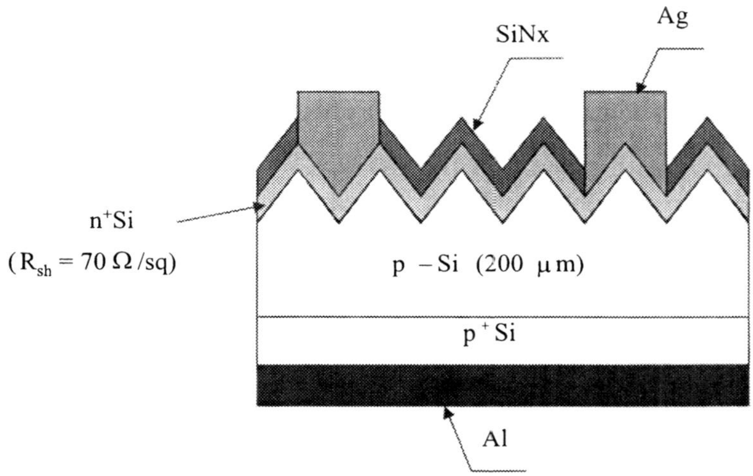

Figure 6 Cross section of an advanced thin screen printed solar cell with shallow homogeneous emitter and Al-BSF.

The buried contact solar cell structure embodies almost all characteristic features of high efficiency laboratory cells described earlier: shallow emitter diffusion with a very good surface passivation by a thick thermal oxide, a very fine metallization line width, front-contact passivation by heavy diffusion in the contact area, and Back Surface Field. One of the important processing steps is the growth of a very thick thermal oxide on the top surface which simultaneously acts as a diffusion mask, a plating mask and a surface passivation layer.

This process, however, has its disadvantages when commercial applications are considered: a large number of lengthy processing steps at high temperature(above 950°C for a longer total time up to 16 hours), expensive equipment, many careful pre-cleaning steps making the process complex and labour intensive [114]. Although the buried contact solar cell process has been licensed to many leading solar cell manufacturers, only one has succeeded

Table 3 Conventional commercial process sequence of buried contact solar cell [111]

Step no:	Process description
1.	Saw damage etching and random texturing
2.	Light n+ diffusion over the whole surface
3.	Growing of thick thermal oxide
4.	Mechanical or laser groove formation
5.	Groove damage etching and cleaning
6.	Second heavy diffusion in grooved areas only
7.	Aluminium evaporation on a back side
8.	High temperature Al alloying
9.	Electroless plating of nickel, sintering and etching
10.	Electroless plating of copper and silver
11.	Edge junction isolation by laser scribing

introducing it into a large volume production by simplification of many processing steps [115]. Production efficiencies close to 17% are obtained on CZ-Si. Application of a lab scale buried contact process on lower quality multicrystalline wafers resulted in top efficiencies of 17.6% (144 cm^2 cell) and 17.9% (25 cm^2 cell) [116].

A simplified buried contact solar cell process has been proposed [112]. The aim of this process simplification is to suit infrastructure and equipment existing already in many solar cell production plants based on screen printed contacts. The simplified process relies on a homogeneous emitter diffusion instead of the selective emitter scheme [117]. For Czochralski and multicrystalline materials, the advantage of the selective emitter structure is rather limited due to the dependence on the base quality. The thermal oxidation, which is a long process with a high thermal budget, can be replaced by TiO_x or SiN_x-coatings. Novel uses of TiO_2 are under study [118]. Cell results are scarce for these new simplified process steps. Other efforts in the field of buried–contact cells are focussed on double-sided structures, thinner wafers, new rear surface passivation (e.g. B-diffusion) [119] and environmentally friendly processing. More data on the buried contact approach can be found in [120] and Chapter II-b5.

3.3 Solar Cells on Silicon Ribbons

Ribbons offer a significant cost advantage over traditional crystalline silicon technology like CZ pulling or casting multicrystalline blocks. The cost saving arises from the elimination of the slicing process which is a significant cost contributor to CZ and multicrystalline wafers. At present, EFG (Edge-Defined Film-Fed Growth) is the most commercialised of the developed ribbon materials. The technology is based on growth of hollow octagon tubes of silicon after which the wafers are cut out by laser [4]. The total silicon material loss is less than 10%. The EFG material is characterised by a high crystal defect density such as grain boundaries, twins and dislocations [121] and is therefore very susceptible to hydrogen passivation.

High mechanical stress and the uneven surface of EFG sheets make the application of standard screen printing processes difficult because of high breakage. Several patented processing steps have been developed to passivate the highly defected EFG bulk material and to tackle the problem of contacting the uneven surface. A detailed description of the EFG solar cell process has not been published. Some information can be found in reference [122]. The process comprises several patented processing steps: spray-on of liquid P-source and diffusion in an IR-belt furnace, Plasma Enhanced Chemical Vapour Deposition (PECVD) of silicon nitride antireflection coating preceded by ammonia plasma treatment in order to produce hydrogen implantation, "pad printing" of solderable silver contacts on the back and drying, "pad printing" of aluminium paste on the back and drying, direct writing of silver paste on the front, drying and co-firing of all contacts in an IR-furnace.

Dedicated research has led to top lab efficiencies of 18.2% [123]. Hydrogenation from silicon nitride is of paramount importance for bulk improvement. Bulk lifetimes of 100 μsec have been achieved after a short high-temperature firing (1 sec at around 750°C).

A new interesting development in EFG growth is the demonstration of large diameter EFG cylinders [4]. The main driving force is the forecast of producing thinner and cheaper ribbons. Wafers with a thickness around 100 μm are foreseen while the standard octagonal EFG growth produces 280 μm thick wafers.

Besides EFG, the String Ribbon is the second ribbon-based technology available in a MW-production. As for all ribbon technologies, much attention is being paid to edge stabilisation, stress control, and to an increased throughput [124]. The string ribbon technology uses foreign high-temperature string materials as edge stabilisers when the silicon is pulled from the melt – not unlike forming a soap bubble [127]. Efficiencies up to 17.8% have been achieved at the R&D level [126]. Current fields of interest are dual and quad ribbon growth from a single crucible melt [127] with the aim of a further decrease in the costs of the ribbon growth. The latter is still in laboratory phase but the former is entering the production stage.

3.4 Emerging Industrial High Efficiency Technologies

Recently, there has been a renewed interest in the use of monocrystalline silicon, which might eventually reverse the historical trend of increasing use of multicrystalline silicon in PV manufacturing. One reason for this is the better understanding of the degradation phenomenon of B-doped CZ and the solutions that have been found. The other reason is the development of new high-efficiency processes that allow extracting the maximum performance out of the silicon material while being applicable to large production volumes. These emerging industrial high efficiency technologies are discussed in Chapters II-b5 and II-d2. Apart from the buried contact solar cell (also discussed in Section 3.2), the new approaches in industrial production or close to that stage are Heterojunction with Intrinsic Thin a-Si (HIT) solar cell [128], point contact solar cell (also discussed in the next section), and the Obliquely Evaporated Contact (OECO) solar cell [129]. The latter concept uses a Metal-Insulator-Semiconductor (MIS) contact to achieve low emitter saturation current in spite of direct contacting of a shallow emitter. By a combination of mechanical structuring of the wafer and oblique evaporation, contacts are created that provide high conductance and cause negligible shadowing losses.

3.5 Back-Contacted Solar Cells

The interdigitated back contact (IBC) and the point contact solar cells are among the most efficient solar cells both on laboratory and production scales. Of the IBC design, first introduced in 1975 by Schwartz et al. [130, 131], several variants have been researched, mainly as high efficiency solar cells for concentrator applications (see Chapter II-d2). They all have in common that the shadowing losses have been eliminated by removing all front surface metallisation, and the active junction is located at the rear surface of the cell. Next to the performance advantage, the elimination of the shadowing losses has an important aesthetic advantage as the visual aspect of the module becomes very uniform. This triggered the interest of developing low cost one-sun variants of this cell concept. To this end, the many steps involving lithography within the process flow needed to be replaced with low cost technologies. For this cell structure, it has managed to design a process flow completely based on screen printing [132]. Several research institutes are currently adapting industrial techniques to manufacture back junction cells [133, 134].

As the bulk recombination in most of the substrates used in industrial cell production is not compatible with the requirements imposed by the IBC structure, new concepts were required to manufacture rear contacted solar cells. In the emitter wrap through (EWT) structure [135], the active junction is located at the front surface of the cell. The connection between the emitter on the front surface and the emitter contact on the rear surface is realized by wrapping the emitter through the cell. This cell concept has been extensively studied and in all cases, this connection is achieved by drilling numerous holes through the wafer and extending the front surface junction with an additional junction region formed within the walls of these holes [136, 137, 138]. The potential of the concept was demonstrated by manufacturing highly efficient samples.

The Spheral Solar Technology is an alternative cell concept that combines a front surface junction with the absence of front-surface metallisation [139, 140]. In this technology, a junction is formed in the surface of silicon spheres. These spheres are positioned within a pre-formed aluminium foil that acts as the negative contact to the emitter region of the cell. A second aluminium foil acting as the positive contact to the core of the sphere is connected to the rear of this structure.

The metallisation wrap through (MWT) concept was introduced [141] as a back-contact cell structure with the active junction close to the front surface, but requiring only a limited number of holes through the wafer. This is achieved by maintaining a front surface grid that picks up the current on the cell surface and provides a low resistance path to the holes. During the metallisation sequence, the holes are filled with metal and a connection is established between the front surface grid and the external emitter contact formed adjacent to the base contact on the rear of the cell. Both within the screen printing [142] and buried contact technology [143], the process flow is very similar to the process flow of conventional solar cells and this has allowed the introduction of the cell structure into industrial production early in 2004.

While the development of the MWT structures has been focused on the individual cells maintaining compatibility with conventional tabbing and stringing for interconnection, the Pin-Up Module (PUM) concept has evolved into an integrated module manufacturing concept based on a similar cell structure [144].

In the Metallisation Wrap Around (MWA) concept the drilling of the holes is avoided by connecting the front grid to the external contacts at the rear by the deposition of metal around the edges of the wafer [145].

4 Cost of Commercial Photovoltaic Modules

A typical crystalline Si solar cell produces a voltage of 0.5 V around the maximum power point. In a module, the individual cells are usually connected in series to produce a voltage useful for practical application. Most industrial modules today comprise 36, 54, 72 cells connected in series. Other configurations of cells in a module are also being implemented, mainly in BIPV modules. A cross section of a typical PV module is shown on Figure 7. The cells are placed in a sandwich structure comprising glass/encapsulant sheet/cells/encapsulant sheet/back substrate. The cover glass has normally a low iron content to minimise light absorption and is generally either chemically or thermally tempered to increase the mechanical strength, specially against hailstorms. The encapsulant sheet is a transparent polymer that can be laminated onto glass. The most commonly used encapsulants are: polyvinyl butyral (PVB) or ethylene vinylacetate (EVA). For the back substrate, several materials are used: anodised aluminium, glass or polymers such as Mylar or Tedlar.

The price of commercially available photovoltaic modules is 2,8–3,2 €/Wp. The silicon wafer is responsible for 46%, cell fabrication 24% and module fabrication 30 % of the total module cost. A sensitivity analysis of different cost elements, performed for a base line screen printing process, shows that cell efficiency is the most significant factor in the cost reduction [146, 147]. Cost reduction is a driving force for introducing efficiency enhancement techniques into commercial cell production. However, care must be taken that novel techniques do not bring significant costs from increased capital investment and a lower yield associated with the higher complexity process [147]. The cost effectiveness of several efficiency enhancement techniques has been analysed [146] and has revealed that all techniques described in this chapter are commercially viable. Processing sequences such as evaporated

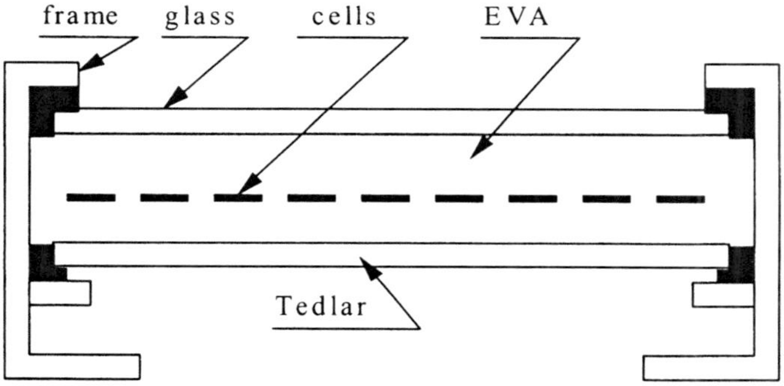

Figure 7 Cross section of a photovoltaic module.

Table 4 Manufacturing cost in $/Wp for different crystalline manufacturing scenario's at 500 MWp/year production. Case1: EFG Si sheets, printed contacts η= 15 %; Case2: direct solidification multi-Si, screen printed contacts, η= 15%; Case3: CZ mono-Si, LGBC, η= 18%; Case4: CZ mono-Si, screen printed contacts, η= 16%; Case5: mono-Si, MIS contacted, η= 17%. Wafer size in all cases is 125 mm × 125 mm and thickness 250 μm [148]

Fabrication step	Case1	Case 2	Case 3	Case 4	Case 5
Ingot growing	0.37	0.37	0.73	0.82	0.78
Wafering	0.00	0.29	0.24	0.28	0.28
Solar cell fab.	0.15	0.15	0.19	0.16	0.24
Module fab.	0.43	0.40	0.37	0.41	0.43
Factory Cost	0.95	1.21	1.53	1.67	1.73

front contacts and a selective emitter involving photolithography are not commercially applicable in their present form.

A study completed in 1994 in the framework of the European project APAS RENA CT94 0008 [148] demonstrated that practical implementation of new industrial processing schemes and increasing the market size above 500MWp/year should lead to drastic crystalline silicon photovoltaic module price reductions to 0.95–1.73 $/Wp depending on the wafer type and cell process. The different commercial technologies of mono- and multi-crystalline solar cells that have been studied include screen printing, laser grooved buried contact, MIS contacted diffused emitter, and a proprietary cell process on EFG silicon sheets. The study assumptions and cost breakdown are presented in Table 4. These results demonstrate that cell fabrication based on EFG silicon sheets gives the lowest module cost mainly due to the avoidance of the costly wafering step. The well known and established screen printed cells on multicrystalline wafers fabricated by directional solidification shows the next lowest cost of 1.21$/Wp. It is worth mentioning that top efficiencies above 16 %, and average efficiencies of 15.5% assumed by the model, have already been achieved on large-area screen-printed industrial multicrystalline cells. All the processes based on CZ wafers show higher costs because of the costly mono-Si wafers.

References

[1] Mueller, N.C.A. and Nasch, P.M. The challenge to implement thin wafer potential with wire saw cutting technology, in *Proc. 3rd World Conf. on Photovoltaic Energy Conversion, Osaka*, 2003.

[2] Eyer, A., Schilinger, N., Rauber, A. and Grabmaier, J.G., 1989. Continuous Processing of Silicon Sheet Material by SSP-Method in *Proc. 9th. European. Photovoltaic Solar Energy Conf.*, pp.17–18.

[3] Schonecker, A., Geerligs, L.J. and Muller, A., 2004. Casting technologies for solar silicon wafers: Block casting and ribbon-growth-on substrate, *Solid State Phenomena*, Vol. 95–96, pp. 149–158.

[4] Kalejs, J.P. and Schmidt, W., 1998. High Productivity Methods of Preparation of EFG Ribbon Silicon Wafers, in *Proc. 2nd World Conf. on Photovoltaic Energy Conversion*, pp. 1822–1825.

[5] Culik, J.S., Goncharovsky, I.S., Rand, J.A. and Barnett, A.M., 2002. Progress on 15-MW single-thread Silicon-Film (TM) solar cell manufacturing systems, *Prog. Photovoltaics*, Vol. 10, pp. 119–128.
[6] Hanoka, J.I., An overview of silicon ribbon growth technology, 2001. *Sol. Energy Mat. Sol. Cells*, Vol. 65, pp. 231–237.
[7] Meier, D.L., Davis, H.P., Garcia, R.A., Salami, J., Rohatgi, A., Ebong, A. and Doshi, P., 2001. Aluminum alloy back p-n junction dendritic web silicon solar cell, *Sol. Energy Mat. Sol. Cells*, Vol. 65, pp. 621–627.
[8] Gee, J. and Wenham, S., 1993. Advanced Processing of Silicon Solar Cells, *Tutorial Notebook, 23rd. IEEE Photovoltaic Specialists Conf.*
[9] King, D.L. and Buck, M.E., 1991. Experimental Optimization of an Anisotropic Etching Process for Random Texturization of Silicon Solar Cells, 1991. *Proc. 22nd IEEE Photovoltaic Specialists Conf.*, pp. 303–308.
[10] Grauvogl, M., Aberle, A. and Hezel, R., 1996. 17.1 % Efficient Truncated-Pyramid Inversion-Layer Silicon Solar Cells, *Proc. 25th IEEE Photovoltaic Specialists Conf.*, pp. 433–436.
[11] Kaiser, U., Kaiser, M. and Schindler, R. Texture Etching of Multicrystalline Silicon 1990. *Proc. 10th European Photovoltaic Solar Energy Conf.*, pp. 293–294.
[12] Narayanan, S., Zolper, J., Yung, F., Wenham, S., Sproul, A., Chong, C. and Green, M., 1990. 18% Efficient Polycrystalline Silicon Solar Cells, *Proc. 21st IEEE Photovoltaic Specialists Conf.*, pp 678–680.
[13] Pirozzi, L., Garozzo, M., Salza, E., Ginocchietti, G. and Margadona, D., 1994. The Laser Texturization in a Full Screen Printing Fabrication Process of Large Area Poy silicon Solar Cells, *Proc. 12th European Photovoltaic Solar Energy Conf.*, pp. 1025–1028.
[14] Inomata, Y., Fukui, K. and Shirasawa, K., 1996. Surface texturing of large area multicrystalline silicon solar cells using reactive ion etching method, *Technical Digest of 9th PVSEC*, pp. 109–110.
[15] Fujii, S., Fukawa, Y., Takahashi, H., Inomata, Y., Okada, K., Fukui, K. and Shirasawa, K., 2001. Production technology of large-area multicrystalline silicon solar cells, *Sol. En. Mat. Sol. Cells*, Vol. 65, pp. 269–275.
[16] Willeke, G., Nussbaumer, H., Bender, H. and Bucher, E., 1992. A simple and Effective Light Trapping Technique for Polycrystalline Silicon Solar Cells, *Sol. Energy Mat. and Sol. Cells*, Vol. 26, pp. 345–356.
[17] Bender, H., Szlufcik, J., Nussbaumer, H., Nijs, J., Mertens, R., Willeke, G. and Bucher, E., 1993. Polycrystalline Silicon Solar Cells with a Mechanically Formed Texturization, *Appl. Phys. Lett.*, Vol. 62, pp. 2941–2943.
[18] Sarti, D., Le, Q.N., Bastide, S., Goaer, G. and Ferry, D., 1995. Thin Industrial Multicrystalline Solar Cells and Improved Optical Absorption, *Proc 13th European Photovoltaic Solar Energy Conf.*, pp. 25–28.
[19] Einhaus, R., Van Kerschaver, E., Szlufcik, J., Nijs, J. and Mertens, R., 1997. Isotropic texturing of multicrystalline silicon wafers with acidic

texturing solutions, *Proc. 26th IEEE Photovoltaic Specialists Conf.*, pp. 167–170.
[20] De Wolf, S., Choulat, P., Vazsonyi, E., Einhaus, R., Van Kerschaver, E., De Clercq, K. and Szlufcik, J., 2000. Towards industrial application of isotropic texturing for multicrystalline silicon solar cells, *Proc. 16th European Photovoltaic Solar Energy Conf.*, pp. 1521–1523.
[21] Kern, W. and Poutinen, D.A., June 1970. Cleaning Solution Based on Hydrogen Peroxide for Use in Silicon Semiconductor Technology, *RCA Review*, pp. 187–206.
[22] Meuris, M., Mertens, P.W., Opdebeeck, A., Schmidt, H.F., Depas, M., Vereecke, G., Heyns, M.M. and Phillipossian, A., 1995. The IMEC Clean: A New Concept for Particle and Metal Removal on Si Surfaces, in *Solid State Technology*, Vol. 38, pp. 109–113.
[23] King, R.R., Sinton, R.A. and Swanson, R.M., 1990. Studies of Diffused Phosphorus Emitters: Saturation Current, Surface Recombination Velocity, and Quantum Efficiency, *IEEE Trans. on Electron. Devices*, Vol. ED-37, pp. 365–371.
[24] Morales-Acevado, A., 1986. Optimization of Surface Impurity Concentration of Passivated Emitter Solar Cells, *J. Appl. Phys.* Vol. 60., pp. 815–819.
[25] Wolf, M., 1986. The Influence of Heavy Doping Effects on Silicon Solar Cells Performance, *Solar Cells*, Vol. 17, pp. 53–63.
[26] Morales-Acevado, A., 1991. Theoretical Study of Thin and Thick Emitter Silicon Solar Cells, *J. Appl. Phys.* Vol. 70, pp. 3345–3347.
[27] Wenham, S., 1993. Burried-Contact Solar Cells, *Progress in Photovoltaics*, Vol. 1, pp. 3–10.
[28] Szlufcik, J., Elgamel, H., Ghannam, M., Nijs, J. and Mertens, R., 1991. Simple Integral Screen Printing Process for Selective Emitter Polycrystalline Silicon Solar Cells, *Appl. Phys. Lett.* Vol. 59, pp. 1583–1584.
[29] Horzel, J., Szlufcik, J. and Nijs, J., 2000. High Efficiency Industrial Screen Printed Selective Emitter Solar Cells, *Proc. 16th European Photovoltaic Solar Energy Conf.*, pp. 1112–1115.
[30] Ruby, D.S., Yang, P., Zaidi, S., Brueck, S., Roy, M. and Narayanan, S., 1998. Improved performance of Self-Aligned, Selective-Emitter Silicon Solar Cells, *Proc. 2nd World Conf. on Photovoltaic Energy Conversion*, pp. 1460–1463.
[31] Bultman, J.H., Kinderman, R., Hoornstra, J. and Koppes, M., 2000. Single step selective emitters using diffusion barriers, *Proc. 16th European Photovoltaic Solar Energy Conf.*, pp. 1424–1426.
[32] Duerinckx, F., Frisson, L., Michiels, P.P., Choulat, P. and Szlufcik, J., 2001. Towards highly efficient industrial cells and modules from multicrystalline wafers, in *Proc. 17th European Photovoltaic Solar Energy Conference*, pp. 1375–1378.
[33] Hezel, R., 1996. A Review of Recent Advances in MIS Solar Cells, *Proc. 6th. Workshop on the Role of Impurities and Defects in Silicon Device Processing*, pp. 139–153.

[34] Zhao, J. and Green, M., 1991. Optimized Antireflection Coatings for High-Efficiency Silicon Solar Cells, *IEEE Trans. on Electron. Devices*, Vol. 38, pp. 1925–1934.
[35] Aberle, A., Glunz, S., Warta, W., Knopp, J. and Knobloch, J., 1991. SiO_2-Passivated High Efficiency Silicon Solar cells: Process Dependence of Si-SiO_2 Interface Recombination, *Proc. 10th. European Photovoltaic Solar Energy Conf.*, pp. 631–635.
[36] Duerinckx, F. and Szlufcik, J., 2002. Defect passivation of industrial multicrystalline silicon solar cells based on PECVD silicon nitride, *Sol Energy Mat. Sol. Cells*, Vol. 72, pp. 231–246.
[37] Kerr, M., 2002. Surface, emitter and bulk recombination in silicon and development of silicon nitride passivated solar cells, Ph.D. thesis, The Australian National University.
[38] Kintzel, W., Bail, M., Auer, R. and Brendel, R., 2002. Efficient Surface Passivation by Silicon Nitride Using a Large Area Deposition System, Proc. *29th IEEE Photovoltaic Specialists Conf.*, New Orleans, p. 262.
[39] Duerinckx, F., 1999. Bulk and surface passivation of screen printed multicrystalline silicon solar cells based on plasma enhanced CVD of silicon nitride, *Ph.D. Thesis*, Katholieke Universiteit Leuven.
[40] Dekkers, H.F.W., De Wolf, S., Agostinelli, G., Duerinckx, F. and Beaucarne, G., 2004. Requirements of PECVD SiNx:H layers for bulk passivation of mc-Si, to be published in *Sol. Energy Mat Sol Cells*.
[41] Wenham, S.R. Green, M. A., 1988. Buried Contact Solar Cell, *US Patent 4,726,850.*
[42] Wenham, S., 1993. Buried-Contact Silicon Solar Cells, *Progress in Photovoltaics*, Vol 1. pp. 3–10.
[43] Mason, N.B., Jordan, D. and Summers, J.G., 1991. A High Efficiency Silicon Solar Cell Production Technology, *Proc. 10th European Photovoltaic Solar Energy Conf.*, pp. 280–283, 1991.
[44] Munzer, A., 1996. In *MONOCEPT: 1st. Progress Report of EC project JOR 3-CT-95-0011.*
[45] Nakaya, H., Nishida, M., Takeda, Y., Moriuchi, S., Tonegawa, T., Machida, T. and Nunoi, T., 1994. Polycrystalline Silicon Solar Cells with V-Grooved Surface *Sol. Energy Mat. Sol. Cells*, Vol. 34, pp. 219–225.
[46] Takayama, M., Yamashita, H., Fukui, K., Masuri, K., Shirasawa, K. and Watanabe, H., 1990. Large Area High Efficiency Multicrystalline Silicon Solar Cells, *Techn. Dig. of Int. PVSEC-5*, pp. 319–322.
[47] Nijs, J., Demesmaeker, E., Szlufcik, J., Poortmans, J., Frisson, L., De Clercq, K., Ghannam, M., Mertens, R. and Van Overstraeten, R., 1994. Latest Efficiency Results with the Screenprinting Technology and Comparison with the Buried Contact Structure, *Proc. First. World Conf. on Photovoltaic Energy Conversion*. pp. 1242–1249.
[48] Young, R.J.S. and Carroll, A.F., 2000. Advances in Front-Side Thick Film Metallisations for Silicon Solar Cells, *Proc 16th European Photovoltaic Solar Energy Conf.*, pp. 1731–1734.
[49] Szlufcik, J., De Clercq, K., De Schepper, P., Poortmans, J., Buczkowski, A., Nijs, J.

and Mertens, R., 1994. Improvement in Multicrystalline Silicon Solar Cells After Thermal Treatment of PECVD Silicon Nitride AR Coating, *Proc. 12th European Photovoltaic Solar Energy Conf.*, pp. 1018–102.

[50] Meier, D.L., Davis, H.P., Shibata, A., Abe, T., Kinoshita, K., Bishop, C., Mahajan, S., Rohatgi, A., Doshi, P. and Finnegan, M. 1998. Self-doping contacts and associated silicon solar cell structures, *Proc. 2nd World Conf. on Photovoltaic Solar Energy Conversion*, pp. 1491–1494.

[51] Rohatgi, A., Hilali, M., Meier, D.L., Ebong, A., Honsberg, C., Carroll, A.F. and Hacke, P., 2001. Self-Aligned self-doping selective emitter for Screen-Printed Silicon Solar Cells, *Proc 17th European Photovoltaic Solar Energy Conf.*, pp. 1307–1310.

[52] Schubert, G., Fischer, B., Fath, P., 2002. Formation and Nature of Ag thick Film Front Contacts on Crystalline Silicon Solar Cells, *Proc. Int. PV in Europe Conference*, pp. 343–346.

[53] Ballif, C., Huljin, D.M., Willeke, G. and Hessler-Wyser, A., 2003. Silver Thick-Film Contacts on Highly Doped n-type Silicon Emitters: Structural and Electronic Properties of the Interface, Applied Physics Letters, Vol. 82, No. 12, pp. 1878–1880.

[54] Dziedzic, A., Nijs, J. and Szlufcik, J., 1993. Thick-Film Fine-Line Fabrication Techniques – Application to Front Metallisation of Solar Cells, *Hybrid Circuits*, No. 30, pp. 18–22.

[55] Huljic, D., Thormann, S., Preu, R., Lüdemann, R. and Willeke, G., 2002. Pad Printed Front Contacts for c-Si Solar Cells – A Technological and Economical Evaluation, *Proc. 29th IEEE Photovoltaic Specialists Conf.*, pp. 126–129.

[56] Williams, T., McVicker, K., Shaikh, A., Koval, T., Shea, S., Kinsey, B. and Hetzer, D., 2002. Hot Melt Ink Technology for Crystalline Silicon Solar Cells, *Proc. 29th IEEE Photovoltaic Specialists Conf.*, pp. 352–355.

[57] Van Kerschaver, E., Choulat, P., Szlufcik, J. and Weldon, T., 2000. Stable and reliable metallisation through the application of electro-formed stencils, *Proc. 16th European Photovoltaic Solar Energy Conf.*, pp. 1528–1531.

[58] Hoornstra, J., Roberts, S., de Moor, H.H.C. and Bruton, T.M., 1998. First Experiences with Double Layer Stencil Printing for Low Cost Production Solar Cells, *Proc. 2nd World Conf. on Photovoltaic Solar Energy Conversion*, pp. 1527–1530.

[59] Fath, P., Marckmann, C., Bucher, E. and Willeke, G., 1995. Multicrystalline Silicon Solar Cells Using a New High Throughput Mechanical Texturization Technology and a Roller Printing Metallization Technique, *Proc. 13th European Photovoltaic Solar Energy Conf.*, pp. 29–32.

[60] Hanoka, J.I., Danielson, S.E., 1992. Method for Forming Contacts, *US Patent 5,151,377.*

[61] Cheeck, G.C., Mertens, R.P., Van Overstraeten, R. and Frisson, L., 1984. Thick Film Metallization for Silicon Solar Cells, *IEEE Trans. on Electron Deices.* Vol. ED-31, pp. 602–609.

[62] Lolgen, P., Leguit, C., Eikelboom, J.A., Steeman, R.A., Sinke, W.C., Verhoef, L.A., Alkemande, P.F.A. and Algra, E., 1993. Aluminium back-surface field doping profiles with surface recombination velocities below 200 cm/sec, *Proc 23rd IEEE Photovoltaic Specialists Conf.*, p. 231.
[63] Amick, J.A., Battari, F.J. and Hanoka, J.I., 1994. The Effect of Aluminium Thickness on Solar Cell Performance, *J. Electrochem. Soc.*, Vol. 141, pp. 1577–1585.
[64] Narashima, S., Rohatgi, A. and Weber, A.W., 1999. An optimized rapid aluminum back surface field technique for silicon solar cells, *IEEE Trans. Electron Devices*, Vol. 46 (7), pp. 1363–1370.
[65] Tool, C.J.J., Manshanden, P., Burgers, A.R. and Weeber, A.W., 2004. Wafer thickness and Performance of Multicrystalline Silicon Solar Cells, to be published in *Sol Energy Mat Sol Cells.*
[66] Glunz, S.W., 2004. New concepts for high-efficiency silicon solar cells, to be published in *Sol Energy Mat Sol Cells.*
[67] Cuevas, A., Kerr, M. and Schmidt, J., 2004. Passivation of crystalline silicon using silicon nitride, *Proc. 3rd World Conf. on Photovoltaic Energy Conversion.*
[68] Lee, J.Y., Dicker, J., Rein, S. and Glunz, S.W., 2004. Investigation of various surface passivation layers using oxide/nitride stacks of silicon solar cells, *Proc. 3rd World Conference on Photovoltaic Energy Conversion.*
[69] Dauwe, S., Schmidt J. and Hezel, R., 2002. Very low surface recombination velocities on p- and n-type silicon wafers passivated with hydrogenated amorphous silicon films, *Proc. 29th IEEE Photovoltaic Specialists Conf.*, pp. 1246–1249.
[70] Dauwe, S., Mitteldstat, L., Metz, A., Schmidt, J. and Hezel, R., 2004. Low temperature rear surface passivation schemes for >20% efficient silicon solar cells, *Proc. 3rd World Conference on Photovoltaic Energy Conversion.*
[71] Schneiderlöchner, E., Preu, R., Lüdeman, R. and Glunz, S.W., 2002. Laser fired rear contacts for crystalline silicon solar cells, *Progr. Photovoltaics, Vol. 10,* pp. 29–34.
[72] Ourmazd, A. and Schroter, W., 1984. Phosphorous gettering and intrinsic gettering of nickel in silicon, *Appl. Phys. Lett.*, Vol. 45, p. 781.
[73] Myers, S.M., Seibt, M. and Schroeter, W., 2000. Mechanisms of transition-metal gettering in silicon *J. Appl. Phys.* Vol. 88, pp. 3795–3819.
[74] Sopori, B., Jastrzebski, L. and Tan, T., 1996. A comparison of gettering in single and multicrystalline silicon for solar cells, *Proc. 25th IEEE Photovoltaic Specialists Conf.*, pp. 625–628.
[75] Gee, J., 1991. Phosphorous diffusions for gettering-induced improvement of lifetime in various silicon materials. *Proc. 22nd IEEE Photovoltaic Specialists Conf.*, pp. 118–23.
[76] Perichaud, I., Floret, F. and Martinuzzi, S., 1993. Limiting factors in phosphorous external gettering efficiency in multicrystalline silicon, *Proc. 23rd IEEE Photovoltaic Specialists Conf.*, pp. 243–247.

[77] Périchaud, I., Gettering of impurities in solar silicon, 2002. *Sol. Energy Mat. Sol. Cel.*, Vol. 72, pp. 315–326.

[78] Cuevas, A., Stocks, M., Armand, S., Stuckings, M., Blakers, A. and Ferrazza, F., 1997. High minority carrier lifetime in phosphorus-gettered multicrystalline silicon, *Appl. Phys. Let.* Vol. 24, pp. 1017–1019.

[79] Sivoththaman, S., Rodot, M., Nam, L., Sarti, D., Ghannam, M. and Nijs, J., 1993. Spectral response and dark I-V characterization of polycrystalline silicon solar cells with conventional and selective emitters, *Proc. IEEE Photovoltaic Specialists Conf.*, pp. 335–339.

[80] Orr, W. and Arienzo, M., 1982. Investigation of polycrystalline silicon BSF solar cells, *IEEE Trans. Eletron Devices*, Vol. 29, p. 1151.

[81] Kazmerski, L., 1985. Polycrystalline silicon: Impurity incorporation and passivation, *Proc. 6th European Photovoltaic Solar Energy Conf.*, pp. 83–89.

[82] Verhoef, L., Roorda, S., Van Zolingen, R. and Sinke, W., 1988. Improved bulk and emitter quality by backside aluminum doping and annealing of polycrystalline silicon solar cells, *Proc. 20th IEEE Photovoltaic Specialists Conf.*, pp. 1551–1556.

[83] Mandelkorn, J. and Lamneck, J.H., 1990. Simplified Fabrication of Back Surface Electric Field Silicon Cells and Novel Characteristics of Such Cells, Vol. 29, *Solar Cells*, pp. 121–130.

[84] Del. Alamo, J., Euguren, J., Luque, A., 1981. Operating Limits of Al-Alloyed High-Low Junctions for BSF Solar Cells, *Solid State Electronics*, Vol. 24, pp. 415–420.

[85] Lolgen, P., Sinke, W.C. and Verhoef, L.A., 1990. Bulk and Surface Contribution to Enhanced Solar-Cell Performance Induced by Aluminium Alloying, *Techn. Dig. of Int. PVSEC-5*, pp. 239–243.

[86] Amick, J.A., Battari, F.J. and Hanoka, J.I., 1994. The Effect of Aluminium Thickness on Solar Cell Performance, *J. Electrochem. Soc.*, Vol. 141, pp. 1577–1585.

[87] Lolgen, P., Leguit, C., Eikelboom, J.A., Steeman, R.A., Sinke, W.C., Verhoef, L.A., Alkemande, P.F.A. and Algra, E., 1993. Aluminium Back-Surface Field Doping Profiles with Surface Recombinbation Velocities Below 200 cm/sec. in *Proc. 23rd IEEE Photovoltaic Specialists Conf.*, p. 231.

[88] Sana, P., Rohatgi, A., Kalejs, J. and Bell, R., 1994. Gettering and hydrogen passivation of EFG multicrystalline silicon solar cells by aluminum diffusion and forming gas anneal, *Appl. Phys. Lett.*, Vol. 64, pp. 97–99.

[89] Rohatgi, A., Hilali, M.M. and Nakayashiki, K., 2004. High-efficiency screen-printed solar cell on edge-defined film-fed grown ribbon silicon through optimized rapid belt co-firing of contacts and high-sheet-resistance emitter. *Appl. Phys. Let.*, Vol. 84, pp. 3409–3411.

[90] Soppe, W.J., Schiermeier, S.E.A., Weeber, A.W., Steiner, A. and Schuurmans, F.M., 2000. A high throughput PECVD reactor for

deposition of passivating SiN layers, *Proc. 16th European Photovoltaic Solar Energy Conf.*, pp. 1420–1423.

[91] Jiang, F., Stavola, M., Rohatgi, A., Kim, D., Holt, J., Atwater, H. and Kalejs, J., 2003. Hydrogenation of Si from SiNx(H) films: Characterization of H introduced into Si, *Appl Phys Let.* Vol. 83, pp. 931–933.

[92] Dekkers, H.F.W., De Wolf, S., Agostinelli, G. and Szlufcik, J., 2004. Investigation on mc-Si bulk passivation using deuterated silicon-nitride, *Proc. 3rd World Conf. on Photovoltaic Energy Conversion*, Osaka.

[93] Rohatgi, A., Hilali, M.M. and Nakayashiki, K., 2004. High-efficiency screen-printed solar cell on edge-defined film-fed grown ribbon silicon through optimized rapid belt co-firing of contacts and high-sheet-resistance emitter, *Appl. Phys. Let.*, Vol. 84, pp. 3409–3411.

[94] Hong, J., Kessels, W.M.M., van Assche, F.J.H., Rieffe, H.C., Soppe, W.J., Weeber, A.W. and van de Sanden, M.C.M., 2003. Bulk passivation of multicrystalline silicon solar cells induced by high-rate-deposited (>1nm/s) silicon nitride films, *Prog. Photovoltaics*, Vol. 11, pp. 125–130.

[95] Fischer, H. and Pschunder, W., 1973. *Proc. 10th IEEE Photovoltaic Specialists Conf.*, Palo Alto; p. 404.

[96] Schmidt, J., Aberle, A.G. and Hezel, R., 1997. *Proc. 26th IEEE Photovoltaic Specialists Conf.*, Anaheim, p.13.

[97] Glunz, S.W., Rein, S., Warta, W., Knobloch, J. and Wettling, W., 1998. *Proc. 2nd World Conf. on Photovoltaic Energy Conversion*, Vienna, p. 1343.

[98] Glunz, S.W., Schaeffer, E. and Rein, S., 2004. Analysis of the defect activation in CZ-silicon by temperature-dependent bias-induced degradation of solar cells, *Proc. 3rd World Conf. on Photovoltaic Energy Conversion*, Osaka.

[99] Bothe, K., Hezel, R. and Schmidt, J., 2003. Recombination-enhanced formation of the metastable boron-oxygen complex in crystalline silicon *Appl. Phys. Let.*, Vol. 83, pp. 1125–1127.

[100] Saitoh, T., Hashigami, H., Rein, S. and Glunz, S., 2000. Overview of light degradation research on crystalline silicon solar cells, *Prog. Photovoltaics*, Vol. 8, pp. 537–547.

[101] Shirasawa, K., Takahasashi, H., Inomata, Y., Fukui, K., Okada, K., Takayaka, M. and Watanabe, H., 1994. Large Area High Efficiency Multicrystalline Silicon Solar Cells, *Proc. 12th European Photovoltaic Solar Energy Conf.*, pp. 757–760.

[102] Szlufcik, J., Duerinckx, F., van Kerschaver, E., Einhaus, R., Ziebakowski, A., Vazsonyi, E., De Clercq, K., Horzel, J., Frisson, L., Nijs, J. and Mertens, R., 1997. Simplified Industrial Type Processes for High Efficiency Crystalline Silicon Solar Cells, *Proc. 14th European Photovoltaic Solar Energy Conf.*, Barcelona, pp. 380–383.

[103] Duerinckx, F., Choulat, P., Allebé, C., Beaucarne, G., Young, R.J.S, Rose, M. and Raby, J.A., 2004. Improved Screen Printing Process for very thin multicrystalline silison solar cells, to be published in *Proc. 19th European Photovoltaic Solar Energy Conf.*

[104] Fujui, S., Fukawa, Y., Takahashi, H., Inomata, Y., Okada, K., Fukui, K.

and Shirasawa, K., 1999. Production Technology of Large Area Multicrystalline Silicon Solar Cells, *Techn. Digest of the Int. PVSEC-11*, pp. 123–126.

[105] Wenham, S.R. and Green, M.A. Laser grooved solar cell, *US Patent* 4,748,130.

[106] Chong, C.M., Wenham, S.R. and Green, M.A., 1988. High-Efficiency, Laser grooved buried contact silicon solar cells, *Appl. Phys. Lett.* Vol. 52, pp. 407–409.

[107] Wenham, S.R., Honsberg, C.B. and Green, M.A., 1994. Buried contact solar cells, *Sol. Energy Mat. Sol. Cells*, Vol. 34, pp. 101–110.

[108] Honsberg, C.B., Yun, F., Ebong, A., Tauk, M., Wenham, S.R. and Green, M.A., 1994. 685 mV Open-Circuit Voltage Laser Grooved Silicon Solar Cell, *Sol. Energy Mat. Sol. Cells*, Vol. 34. pp. 117–123.

[109] Green, M.A., Wenham, S.R., 1991. Present Status of Buried Contact Solar Cells, *Proc. 22nd IEEE Photovoltaic Specialists Conf.*, pp. 46–49.

[110] Green, M.A., Wenham, S.R., Honsberg, C.B. and Hogg, D., 1994. Transfer of Buried Contact Cell Laboratory Sequences into Commercial Production, *Sol. Energy Mat. Sol. Cells*, Vol. 34, pp. 83–89.

[111] Honsberg, C.B. and Wenham, S.R., 1995. New Insights Gained Through Pilot Production of High-Efficiency Silicon Solar Cells, *Prog. Photovoltaics* , Vol 3. pp. 79–87.

[112] Wenham, S.R., and Green, M.A., 1996. Silicon Solar Cells, *Prog. Photovoltaics*, Vol 4., pp. 3–33.

[113] Green, M.A., Wenham, S.R. and Zhao, J., 1993. Progress in High Efficiency Silicon Solar Cells and Module Research, *Proc. 23rd IEEE Photovoltaic Specialists Conf.* pp. 8–13.

[114] Narayanan, S., Wohlgemuth, J.H., Creager, J.B., Roncin, S.P. and Perry, J.M., 1993. Buried Contact Solar Cells, *Proc. 23rd IEEE Photovoltaic Specialists Conf.*, pp. 277–280.

[115] Jordan, D. and Nagle, J.P., 1994. New Generation of High-Efficiency Solar Cells: Development, Processing and Marketing, *Prog. Photovoltaics*, Vol. 2, pp. 171–176.

[116] McCann, M., Raabe, B., Jooss, W. and Fath, P., 2004. Buried Contact Solar Cells on Multicrystalline Silicon with Improved Surface Passivation, *Techn. Dig. of Int. PVSEC-14*, pp. 745–746.

[117] Cotter, J.E., Richards, B.S., Ferrazza, F., Honsberg, C.B., Leong, T.W., Mehrvarz, H.R., Naik, G.A. and Wenham, S.R., 1998. Design of a Simplified Emitter Structure for Buried Contact Solar Cells, in *Proc. 2nd World Conf. on Photovoltaic Solar Energy Conversion*, pp. 1511–1514.

[118] Richards, B., Cotter, J., Honsberg, C.B. and Wenham, S.R., 2000. Novel uses of TiO2 in Crystalline Silicon Solar Cells, *Proc. 28th IEEE Photovoltaic Specialists Conf.*, pp. 375–378.

[119] Slade, A.M., Honsberg, C.B. and Wenham, S.R., 2000. Processing and Characterisation of Boron Diffused Back Surface Fields in Mono- and Bifacial Buried Contact Solar Cells, *Proc 16th European Photovoltaic Solar Energy Conf.*, pp. 1763–1766.

[120] Honsberg, C.B., Cotter, J.E., McIntosh, K.R., Pritchard, S.C., Richards B.S. and Wenham, S.R., 1999. Design Strategies for Commercial Solar Cells Using the Buried Contact Technology, *IEEE Trans. Electron Devices*, Vol. 46, No. 10, pp. 1984–1992.
[121] Kardauskas, M., 1996. Processing of Large-Area Silicon Substrates with High Defect Densities into Higher Efficiency Solar Cells, *Proc. 6th. Workshop on the Role of Impurities and Defects in Silicon Device Processing*, pp. 172–176.
[122] Amick, J., Bottari, F.J., Hanoka, J.I., 1994. Solar Cell and Method of Making Same, *US Patent* no: 5,320,684.
[123] Rohatgi, A., Kim, D.S., Nakayashiki, K., Yelundur, V. and Rounsaville, B., 2004. High-Efficiency Solar Cells on Edge-Defined Film-Fed Grown (18.2%) and String Ribbon (17.8%) Silicon by Rapid Thermal Annealing, *Appl. Phys. Letters*, Vol. 84, No. 1, pp. 145–147.
[124] Hanoka, J.I., 1999. An Overview of Silicon Ribbon – Growth Technology, *Techn. Digest of the Int. PVSEC-11*, pp. 533–534.
[125] Wallace, R.L., Hanoka, J.I., Narasimha, S., Kamra, S. and Rohatgi, A., 1997. Thin Silicon String Ribbon for High Efficiency Polycrystalline Solar Cells, *Proc. 26th IEEE Photovoltaic Specialists Conf.*, pp. 99–102.
[126] Kim, D.S., Gabor, A.M., Yelundur, V., Upadhyaya, A.D., Meemongkolkiat, V. and Rohatgi, A., 2004. String Ribbon Silicon Solar Cells with 17.8% Efficiency, in *Proc. 3rd World Conf. on Photovoltaic Solar Energy Conversion.*
[127] Wallace, R.L., Sachs, E. and Hanoka, J.I., 2004. Multiple Ribbon Growth Using the String Ribbon Method, *Proc. 3rd World Conf. on Photovoltaic Solar Energy Conversion.*
[128] Sakata, H., Nakai, T., Baba, T., Taguchi, M., Tsuge, S., Uchihashi, K. and Kiyama, S., 2000. 20.7% highest efficiency large area (100.5 cm^2) HIT cell, *Proc. 28th IEEE Photovoltaic Specialists Conf.*, Anchorage, pp. 7–12.
[129] Hezel, R., Schmiga, C. and Metz, A., 2000. Next generation of industrial silicon solar cells with efficiencies above 20%, *Proc. 28th IEEE Photovoltaic Specialists Conf.*, Anchorage, pp. 184–187.
[130] Schwartz, R.J. and Lammert, M.D., 1975. *IEEE International Electron Devices Meeting*, Washington DC, p. 350.
[131] Lammert, M.D. and Schwartz, R.J., 1977. The Interdigitated Back Contact Solar Cell: A Silicon Solar Cell for Use in Concentrated Sunlight, *IEEE Trans. Electron Devices*, Vol. Ed.-24(4), pp. 337–342.
[132] McIntosh, K.R., Cudzinovic, M., Smith, D., Mulligan W. and Swanson, R., 2004. The Choice Of Silicon Wafer For The Production Of Low-Cost Rear-Contact Solar Cells, *Proc. 3rd World Conf on Photovoltaic Energy Conversion*, Osaka.
[133] Hezel, R., 2002. Novel Back Contact Silicon Solar Cells Designed For Very High Efficiencies and Low-Cost Mass Production, *Proc. 29th IEEE Photovoltaic Specialists Conf.*, New Orleans, p. 114.
[134] Guo, J. and Cotter, J., 2004. Interdigitated Backside Buried Contact Solar Cells, *Proc. 3rd World Conf. on Photovoltaic Energy Conversion*, Osaka.

[135] Gee, J., Schubert, W. and Basore, P., 1993. Emitter Wrap Through Solar Cell, *Proc. 23rd IEEE Photovoltaic Specialists Conf.*, Louisville, p. 265.

[136] Schönecker, A. et al., 2002. ACE Designs: The Beauty Of Rear Contact Solar Cells, *Proc. 29th IEEE Photovoltaic Specialists Conf.*, New Orleans, p. 107.

[137] Jooss, W., Neu, W., Faika, K., Knauss, H., Kress, A., Keller, S., Fath P. and Bucher, E., 2002. Process And Technology Development for Back Contact Silicon Solar Cells, *Proc. 29th IEEE Photovoltaic Specialists Conf.*, New Orleans, p. 122.

[138] Kray, D., Dicker, J., Osswald, D., Leimenstoll, A., Glunz, S., Zimmerman, W., Tentscher, K. and Strobl, G., 2004. Progress in High-Efficiency Emitter-Wrap-Through Cells on Medium Quality Substrates, *Proc. 3rd World Conf. on Photovoltaic Energy Conversion*, Osaka.

[139] Drewes, P., 2002. Spheral Solar – A Completely Different Photovoltaic Technology, *Proc. PV in Europe – From PV Technology to Energy Solutions*, Rome, p. 280.

[140] Nakata, J., 2003. *Asia Electronics Industry*, Dempa Publications Inc., p. 45

[141] Van Kerschaver, E., Einhaus, R., Szlufcik, J., Nijs, J. and Mertens, R., 1998. A Novel Silicon Solar Cell Structure With Both External Polarity Contacts on The Back Surface, *Proc. 2nd World Conf. on Photovoltaic Energy Conversion*, Vienna, p. 1479.

[142] Van Kerschaver, E. et al., 2004. AFRODITE: Power And Aesthetics for the Built Environment, to be published in *Proc. 19th European Photovoltaic Solar Energy Conf.*, Paris.

[143] Knauss, H., Terheiden B. and Fath, P., 2004. Large Area Metallisation Wrap Through Solar Cells Using Electroless Plating, *Tech. Digest Int. PVSEC-14*, Bangkok, p. 54.

[144] Bultman, J., Eikelboom, D., Kinderman, R., Tip, A., Tool, C., van den Nieuwenhof, M., Schoofs, C., Schuurmans, F. and Weeber, A., 2004. Fast And Easy Single Step Module Assembly for Back-Contacted C-Si Slar Cells with Conductive Adhesives, *Proc. 3rd World Conf. on Photovoltaic Energy Conversion*, Osaka.

[145] Jooss, W., Knauss, H., Huster, F., Fath, P., Bucher, E., Tölle, R. and Bruton, T., 2000. Back Contact Buried Contact Solar Cells With Metallisation Wrap Around Electrodes, *Proc. 28th IEEE Photovoltaic Specialists Conf.*, Anchorage, p. 176.

[146] Narayanan, S. and Wohlgemuth, J., 1994. Cost-benefit analysis of high efficiency cast polycrystalline silicon solar cell sequences, *Prog. Photovoltaics*, Vol. 2, pp. 121–128.

[147] Hogan, S., Darkazalli, D. and Wolfson, R., 1991. An Analysis of High Efficiency Si Cells Processing, *Proc. 10th. European Photovoltaic Solar Energy Conf.*, pp. 276–279.

[148] Final Report of the EC project Multi-Megawatt Upscaling of Silicon and Thin Film Solar Cell and Module Manufacturing *Music FM – APAS RENA CT94*.

IIb-3

Thin Silicon Solar Cells

Michael Mauk, Paul Sims, James Rand, and Allen Barnett[1],
AstroPower Inc., Solar Park, Newark, Delaware, USA

[1] For current address, see List of Contributors on page

1 Introduction, Background and Scope of Review

Thin silicon solar cells are an important class of photovoltaics that are currently the subject of intense research, development, and commercialisation efforts. The potential cost reductions realised by manufacturing solar cells in a thin device configuration are highly compelling and have long been appreciated. However, most work on thin film approaches to solar cells has centred on materials other than crystalline silicon because it was believed that the optical properties of crystalline silicon would limit its usefulness as a thin-film solar cell, and for various reasons, crystalline silicon did not readily lend itself to the common thin-film deposition technologies. Several early experimental efforts on making thin film crystalline silicon solar cells seemed to confirm at least some of these perceived difficulties. Still, there were proponents of thin crystalline silicon solar cells including Redfield [1], Spitzer et al. [2] and Barnett [3], who advocated their potential advantages and articulated design principles of light trapping and high-open-circuit voltage needed to achieve high efficiencies. Starting in the late 1980s and gaining momentum in the 1990s, thin-film crystalline silicon solar cells emerged (or perhaps more aptly 're-emerged') as a promising approach. This was partly due to several device design and materials processing innovations proposed to overcome the difficulties and limitations in using crystalline silicon as a thin film solar cell material, and also partly due to the stubborn lack of progress in many competing photovoltaic technologies. Progress on thin crystalline silicon solar cells has now reached a level where they are positioned to capture a significant share of the photovoltaics market, and a thin silicon solar cell may well become the successor to the conventional (thick) silicon wafer solar cells that are presently the mainstay of the photovoltaics industry. This chapter reviews the issues and technical achievements that motivate the current interest in thin silicon solar cells, and surveys developments and technology options.

Thin silicon solar cells is actually an umbrella term describing a wide variety of silicon photovoltaic device structures utilising various forms of silicon (monocrystalline, multicrystalline, polycrystalline, microcrystalline, amorphous, and porous), and made with an almost incredibly diverse selection of deposition or crystal growth processes and fabrication techniques. Thin silicon solar cells are distinguished from traditional silicon solar cells that are comprised of $\sim$0.3 mm thick wafers or sheets of silicon. The common defining feature of a thin silicon solar cell is a relatively thin (<0.1 mm) 'active' layer or film of silicon formed on, or attached to, a passive supporting substrate. Nevertheless, even this very general description may not subsume all the different variations of 'thin silicon solar cell' designs currently under investigation. One purpose of a review such as this is to provide some perspective and objective criteria with which to assess the merits and prospects of different approaches. However, such comparisons must be tempered with the realisation that the technology is still in a state of flux and relatively early development, and there are many disparate solar cell applications, each with a different emphasis on cost and performance. Thus, it is possible – if not probable – that no single solar cell technology will satisfy or otherwise be the best choice for every present or future application. For example, a thin-film solar cell design for large-scale (~megawatt) utility grid-connected power applications may not be the best choice for a small (~1 watt) minimodule battery-charger for walk lights. Further, the projected economics of solar cells is based on complicated and sometimes speculative assumptions of materials costs, manufacturing throughput and scale-up issues, as well as balance of systems constraints and costs, and it would be imprudent at this stage to 'downselect' the most promising technical path for the solar industry. From this vantage, it is fortunate that there is such a diverse choice of technology options under development that should lead to thin silicon solar cells with a wide range of cost and performance characteristics.

Such considerations not withstanding, Bergmann [4] has categorised thin silicon solar cells into three groups, and this delineation serves as rational and useful framework to discuss and review the subject.

1. Thin-film solar cells based on small-grained (< 1 micron) nanocrystalline or microcrystalline silicon films (2–3 microns in thickness) deposited on glass substrates typically using technologies adapted from thin-film amorphous silicon solar cells. These types of cells are usually made as p–i–n structures, sometimes in combination with amorphous silicon layers in heterojunction or tandem cell designs.
2. Thick (~30 micron) silicon layers deposited on substrates that are compatible with recrystallisation of silicon to obtain millimeter size grains. A diffused p–n junction is the preferred device design. Also included in this category are epitaxial silicon solar cells on upgraded metallurgical silicon substrates.
3. Transfer techniques for films of silicon wherein an epitaxial monocrystalline silicon film is separated from the silicon substrate upon which it seeded and bonded to a glass superstrate. The anticipated cost reductions in this technology are based on reuse of the seeding substrate.

In this chapter we will concentrate on the second and third types of solar cells listed above. The first type is more properly considered as an extension or outgrowth of amorphous silicon technology, although much of the technology and design considerations are also relevant to other kinds of thin silicon solar cells. We will briefly survey device designs deposition technologies for making nanocrystalline and microcrystalline thin-film silicon solar cells, but not discuss materials and device physics of this type of solar cell. Amorphous silicon and related thin-film solar cells are considered in a separate chapter of this book. Bergman [4] has provided several penetrating reviews of thin-film microcrystalline and nanocrystalline silicon solar cells. Our emphasis will be on polycrystalline silicon solar cells formed on ceramics, because most of our experience is with this type of solar cell. The third type of solar cell listed above is of general interest as it provides the shortest route to making thin silicon devices in which device issues can be analysed independent of material properties issues. For instance, light trapping and surface passivation effects can be studied in thin device structures with and without grain boundaries.

The plan of the chapter is as follows. We first provide a general discussion of effects and features that are common to most thin silicon solar cells such as light trapping designs, modelling, and methods of analysis. We next discuss voltage enhancement issues including surface passivation, followed by a review of grain boundary effects in thin silicon solar cells. A survey of technologies for producing thin silicon solar cells is then presented. This will include descriptions of various silicon deposition techniques such as chemical vapour deposition, melt and solution growth, as well as substrate issues and post-deposition recrystallisation to enhance grain size and texture. Low-temperature CVD methods for microcrystalline silicon solar cells is also covered, as are techniques for wafer thinning and transfer of thin layers to surrogate substrates. Finally, we end with a tabulation summarising some of the prominent experimental results for thin silicon solar cells.

2 Light Trapping in Thin Silicon Solar Cells

Thin silicon solar cells can greatly benefit from light trapping effects, which can offset the relatively weak absorption near-bandgap energy photons by increasing the optical path length of light within the solar cell structure. The basic idea of light trapping in a thin solar cell is shown in Figure 1. The back side of the solar cell with a grooved, blazed, textured or otherwise roughened surface is made reflective by, for example, a change in refractive index or by coating with a reflective material such as a metal. In a free-standing solar cell, this surface is the backside of the solar cell itself, but in thin solar cell formed on a substrate, the reflective backside 'mirror' is situated at the interface between the silicon film and the substrate. This latter design for a solar cell made on a substrate is obviously more difficult to implement than with a solar cell formed as a free-standing wafer or sheet. As suggested by the schematic ray tracing of Figure 1, it appears in this case the light makes at least several passes of the silicon layers. Thus, the effective optical thickness is several times the actual silicon physical thickness, in which case a thin silicon solar cell with light trapping could reap the same absorption and carrier generation as a conventional solar cell (without light trapping) of much greater thickness.

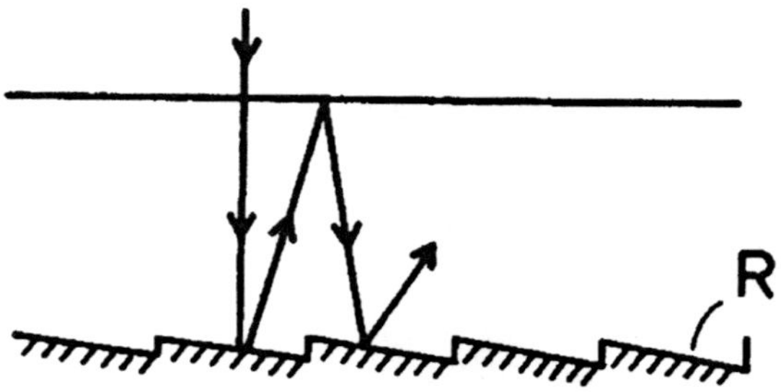

Figure 1 Illustration of the light-trapping concept [5].

Further, the generation of minority carriers would be relatively close to the p–n junction formed near the top surface of the solar cell, thus providing a high collection efficiency. To effect light trapping, it is necessary that at least one interface (front or back) deviates from planarity. If both front and back surfaces are smooth and parallel, a simple analysis shows that the first internal reflection at the front side will result in a large loss of the light due to transmission. This is especially true since the front surface of the solar cell will use an optical coating to reduce reflection (maximise transmission) of incident light. Since the transmission characteristics are symmetrical, the internal reflection from the front surface would be small. Instead, if the back surface is grooved, textured, or roughened as indicated in Figure 1, then light reflected from the back surface will in general be obliquely incident to the front surface, and if the angle of incidence exceeds a critical angle of about 16 degrees, then the light will be totally internally reflected as shown.

Nevertheless, in a real device with imperfect interfaces and diffuse components of light, a certain fraction of internally trapped light will eventually fall within the near-normal incidence angle for transmission, so some light leakage is inevitable and perfect confinement is impossible. An alternative approach is to texture or groove both the top and front surface, in which case, external reflection at the top surface is also reduced [5]. In theory, this is the best approach. In such cases, Yablonovitch and Cody [6] have predicted that for weakly absorbed light, the effective optical thickness a silicon solar cell with both surfaces textured to form Lambertian diffuse reflectors can be about 50 times greater than its actual thickness.

Enhancing the performance of thin crystalline silicon layers with light-trapping has been actively discussed in the literature since the 1970s [1, 2, 7]. In general, texturing one or both surfaces, and maximising the reflection at the back surface obtains optical path lengths greater than the thickness of the device. Texturing results in oblique paths for internally confined light and maximises total internal reflection at the illuminated device surface.

To quantify the effects of light trapping on device performance and optimisation, we define a parameter Z that indicates the ratio of effective optical thickness to actual thickness for weakly absorbed light. Z can thus be interpreted as the number of passes trapped light makes in the solar cell. Z can vary from 1 (no light trapping) to about 50, and is regarded as an adjustable parameter in the optimisation. Although this is an overly simplistic way to describe light trapping in a solar cell, it does not appear that the main conclusions of such and similar modelling depend on the details of the optical absorption and carrier generation due to light trapping phenomena. The main effect on solar cell efficiency is the total level of enhanced absorption and generation in the thin silicon active layers, rather than the microscopic details. However, such detailed modelling is necessary to optimise light trapping designs.

In Figure 2, an example result of the determination of optimal solar cell thickness as a function of light trapping (Z) is illustrated. In this analysis, the doping level was adjusted to yield a minority carrier diffusion length twice the layer thickness, thus insuring good collection efficiency. (We regarded this as

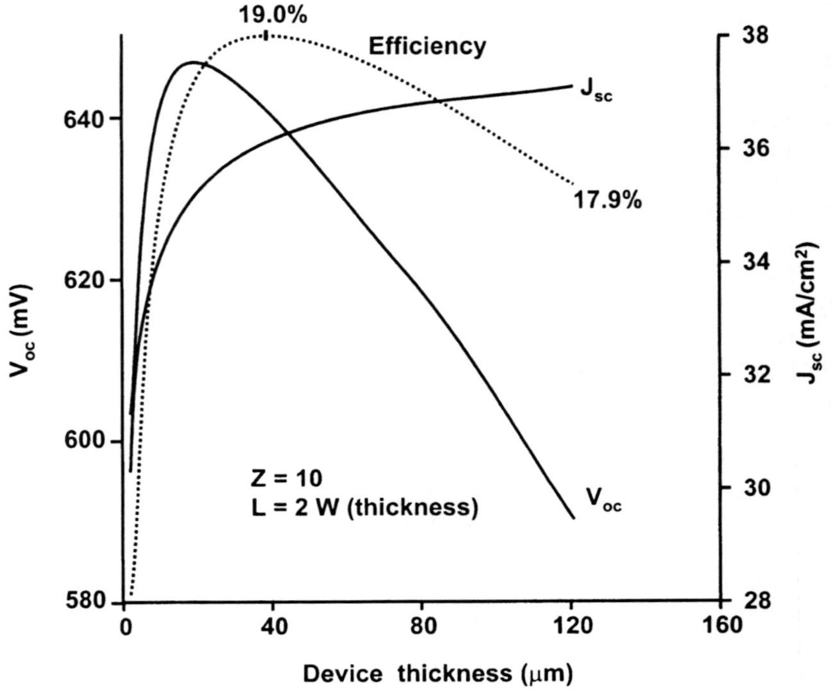

Figure 2 Thin silicon device performance predictions for the case where the optical thickness due to light trapping (Z) is ten times the device thickness and the diffusion length is twice the actual device thickness. The optimum silicon thickness using these assumptions is in the range of 30–40 microns [8].

a general design principle of thin silicon solar cells.) As a conservative estimate, the diffusion length for a given doping concentration was degraded by a factor of 5 from typical single-crystal values to account for the relatively inferior material quality generally expected for silicon deposited on a substrate. Modest surface passivation corresponding to front and back minority carrier surface recombination velocities of 1000 cm/s, a front surface solar-spectrum averaged reflection of 5%, and a series resistance of 0.1 ohm-cm were also assumed. The results of this analysis are shown in Figure 2 for a Z factor of 10. For this set of assumptions, the optimum efficiency occurs at silicon thicknesses between 30 and 40 microns.

2.1 Methods of Implementing Light Trapping

As might be imagined, there is ample opportunity for creative designs to effect light trapping in silicon solar cells, and light-trapping structures have been realised by many different methods. For purposes of review, three types of reflective surfaces can be distinguished – random texture, geometric or regular structuring, and the use of optical elements external to the silicon solar cell structure.

2.1.1 Random Texturing

Random texturing holds promise for two reasons: modelling predicts such random texturing can provide very reflective light trapping, and the perceived ease at which random texture can be experimentally realised. For Lambertian diffuse reflectors made by random surface or interface texturing, and where the angular distribution of reflected light follows a cosine law, Goetzberger [5] has shown that the fraction of light reflected by total internal reflection at the front surface is equal to $1 - 1/n^2$, where n is the refractive index of silicon (approximately 3.4). In this case, when weakly absorbed light is reflected from a diffuse back reflector, 92% will be internally reflected from the front surface. The 50-fold increase in optical path length suggested by Yablonovitch and Cody [6] is based on front and back surface texturing to effect Lambertian reflection. Unfortunately, the experimental realisation of a Lambertian reflector [9] has proved to be more difficult.

Random textures of varying effectiveness have been experimentally realised by the following methods:

- reactive ion etching [10–12]
- sand blasting [13]
- photolithography [14]
- natural lithography [15]
- porous etching[16]
- rapid thermal processing of an aluminium–silicon interface [17]
- random-textured ancillary dielectric layers such as ZnO or SnO_2 [18-21]

2.1.2 Geometrical or Regular Structuring

Surface structuring is relatively easy to realise in single-crystal silicon wafer surfaces by taking advantage of the anisotropic (crystal orientation-dependent) etch rates of alkaline solutions. This approach has been used to fabricate the majority of light-trapping structures demonstrated to date, a few examples of which are shown in Figure 3. Common structures include 54-degree pyramids, inverted pyramids, slats, and perpendicular slats. Regular patterns are achieved with photolithography in combination with anisotropic etching [22–24]. Random pyramids are formed with the use of anisotropic etchants on non-patterned surfaces [25, 26]. A similar result can be achieved in multicrystalline substrates (with much more effort) with the use of mechanical scribing, abrading or grinding or with laser ablation [27–32]. An option for very thin layers is conformal growth on a textured substrate [33] (Figure 3).

Modelling of geometrical textures is often carried out by computer-aided ray tracing techniques [35–38]. Ray tracing analysis makes no assumptions about the distribution of light within the absorber layer. Instead, the path of representative incident rays of light are plotted using geometric optics to follow the light path through as it is reflected and refracted at surfaces and interfaces. Attenuation of light due to optical absorption in bulk silicon is also incorporated in the model. This approach is useful for the analysis of regularly structured

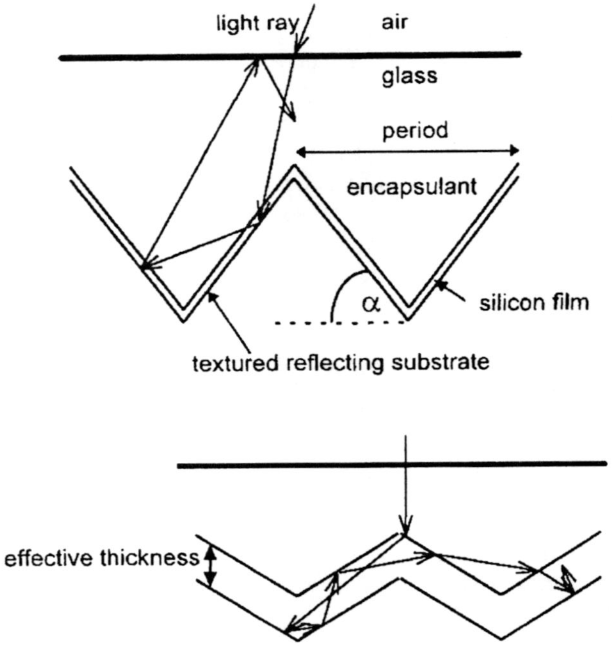

Figure 3 Examples of structured surfaces to effect light trapping in silicon solar cells [34].

surfaces [39] with complex geometries as well as unusual solar cell shapes such as the Spheral Solar™ Cell [40].

2.1.3 External Optical Elements

The solar cell can be overlayed with refractive optics elements such as prismatic cover slips (Figure 4) to effect a degree of light trapping. These optical elements were originally developed to ameliorate shading by the front contact grid metallisation, but they can also be used to redirect light back into the solar cell after it has escaped. External reflectors and optical cavities are two other examples of the use of external optical elements to implement light trapping. Miñiano et al. [41] have analysed light-confining cavities for concentrator solar cells.

2.2 Assessment of Light-Trapping Effects

Light trapping has been incorporated in structures with thickness ranging from less than 1 micron to 400 microns with varying degrees of success. The short-circuit current is the ultimate figure of merit for comparisons, as the objective of light trapping is to enhance absorption and contribute to minority carrier generation. Using short-current current to assess the effectiveness of light-trapping schemes is complicated by other losses, such as bulk recombination and front surface reflection. In laboratory settings spectral

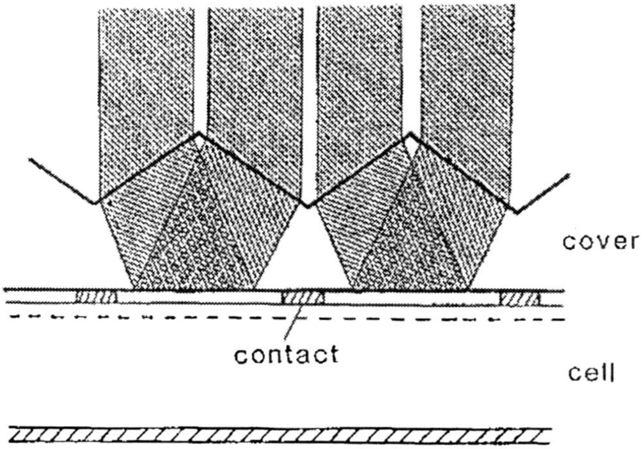

Figure 4 A prismatic cover slip can be used to effect light trapping in solar cells [42].

response and reflection data can be analysed to extract detailed information about light trapping; however, these techniques require assumptions that limit their usefulness in predicting final cell performance. Each analysis method is reviewed below.

2.2.1 Short-Circuit Current Analysis of Light Trapping

Modelling the current generated by a known thickness of silicon is straightforward when all absorbed photons contribute a charge carrier. A 'no light-trapping' scenario can be generated that has perfect anti-reflection properties (no front surface reflection), and only one pass of light through the device thickness (back surface is 100% absorbing with no contribution to current). To the extent an experimental result exceeds this model light trapping features are indicated. In real devices shading, imperfect AR, and parasitic absorption make the 'no light-trapping' model impossible to achieve [43]. Calibration of the light source is another source of error in this method.

Figure 5 shows modelling results and experimental data for short circuit current as a function of thickness. The effect of light trapping becomes more pronounced as the device thickness is reduced and the light not absorbed on the first pass through the device grows to a significant level. Theoretical analysis of Green [44] and Tiedje [45] are included in Figure 5 and serve as the 'best case' limit for perfect light trapping. The analysis of Green appears to predict perfect collection (J_{SC} = 44 mA/cm^2) independent of thickness. The method of Tiedje assumes random texture, Lambertian-type reflection, and more realistic modelling using a loss-cone analysis for front surface internal reflections. The baseline case with no light trapping was computed with PC-1D [46].

The experimental results shown in Figure 5 indicate that most laboratory and commercial devices have not produced a level of photogenerated current exceeding that which would be possible in a device of comparable thickness

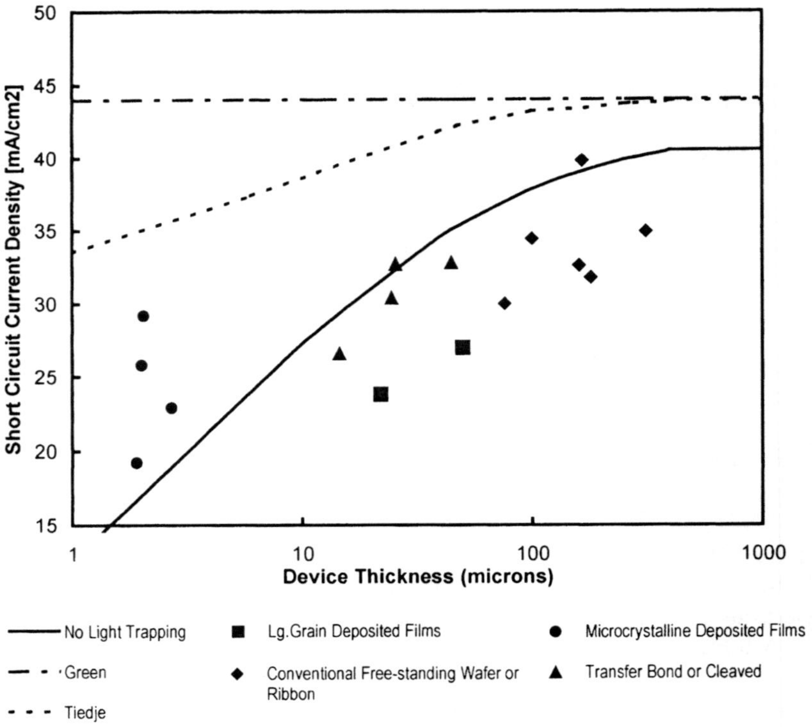

Figure 5 Summary of the last five years of published data on silicon layer thickness and short-circuit current. Theoretical analyses are shown as lines. Laboratory results are shown as solid symbols [47–68].

without light trapping. Thus, based solely on total current, it cannot be concluded that in these particular solar cells light trapping makes a significant contribution to solar cell performance. These devices may include some light-trapping enhancement of current, but additional losses due to imperfect anti-reflection coatings, bulk and surface recombination, possibly exacerbated by adding light trapping features to the solar cell, may have offset any gains from light trapping. This has prompted researchers to investigate other methods based on reflection and spectral response measurement to analyse light-trapping effects in ways that are not obscured by various losses [69].

2.2.2 'Sub-bandgap' Reflection Analysis of Light Trapping

The effectiveness of backside reflectors can be evaluated at long wavelengths ('sub-bandgap' energy photons), i.e., longer than about 1100 nm. The silicon layer is approximately transparent to light in this wavelength range. As shown in the inset of Figure 6, the measured total reflection for non-absorbed light at these wavelengths of a thin solar cell with a backside mirror is due to contributions of multiple internal reflections from the front and back surface. If absorption is negligible, the ratio between the measured reflection and the reflection expected of an infinitely thick slab of silicon gives an estimate of the effective light trapping in a device. This technique does not measure light trapping *per se*, but instead provides an indicator of the effectiveness of the backside mirror and the level of optical confinement for near-bandgap, weakly absorbed photons. This analysis can be especially decisive if test structures, both *with* and *without* backside reflectors, are compared. In thin silicon structures, the effect of the buried reflector is shown as an enhancement of sub-bandgap (>1050 nm wavelength) reflection, which is attributed to unabsorbed light reflected at the backside mirror and transmitted through the front surface. This escaping light boosts the measured front surface reflection (Figure 6), and the increased reflectance permits an estimate of the internal reflections in the silicon layer.

2.2.3 Extended Spectral Response Analysis of Light Trapping

The extended spectral response method was developed by Basore [71] as a means to estimate an effective optical path length for near-bandgap light in a thin silicon solar cell, and is based on an analysis of the internal quantum efficiency of a solar cell as a function of wavelength.

The method utilises a plot of $1/IQE$ vs. $1/\alpha$ where α is the optical absorption coefficient for silicon and *IQE* is the internal quantum efficiency. Two representative examples of such plots are shown in Figure 7, for a thick silicon wafer-based solar cell and a thin (4 micron) silicon solar cell, both of which have textured surfaces to effect light trapping. The $1/IQE(\lambda)$ vs. $1/\alpha$ curves show two linear regions, which appears to be a general feature of solar cells with light trapping.

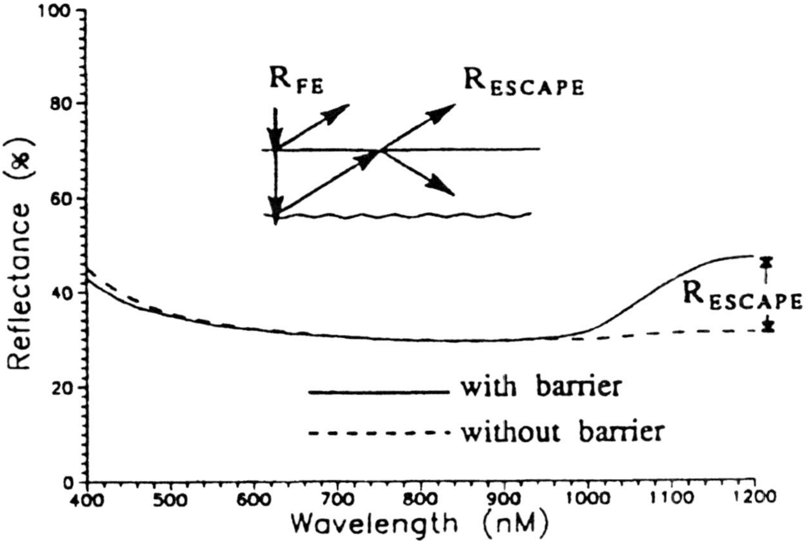

Figure 6 External reflection measurements in thin silicon structures with reflecting barrier layer between silicon layer and substrate [70].

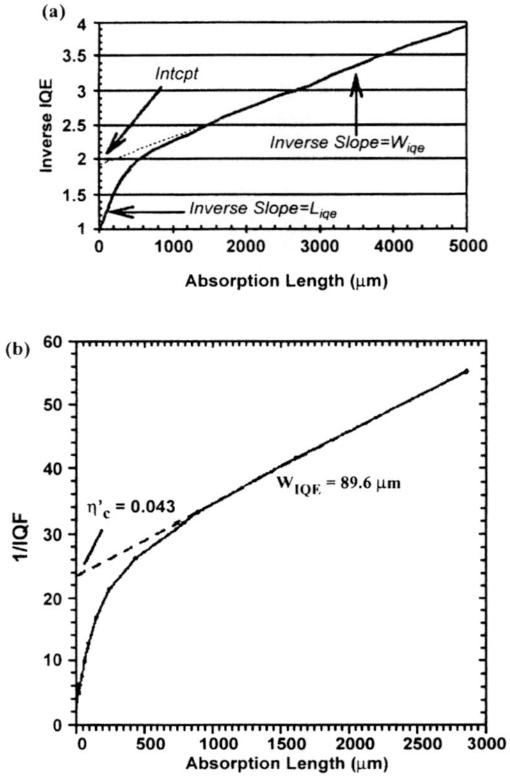

Figure 7 (a) Plot of $1/\alpha$ vs. $1/IQE$ for a textured (thick) wafer-based silicon solar cell [71]; and (b) textured 4 micron thick silicon solar cell [72].

We will not discuss the details of the somewhat involved analysis in which Basore shows that the slopes and intercepts of the linearised parts of these curves can be used to estimate the collection efficiency, the back reflectance, the back surface recombination velocity, and the effective minority carrier diffusion length. This extended spectral response analysis, although more complicated than other techniques, appears to be the most useful method for assessing light trapping in solar cells.

3 Voltage Enhancements in Thin Silicon Solar Cells

The factors that influence the open-circuit voltage of a silicon solar cell are the same whether the device is thin or thick. These include doping levels, the various bulk recombination mechanisms (defect-mediated Shockley–Read–Hall, radiative, and Auger), and surface recombination. The open-circuit voltage V_{OC} depends logarithmically on the dark diode current density J_0. The diode is analysed in terms of current components from the silicon layer (base) and a thin emitter layer formed on the surface of the base, usually by impurity diffusion. J_0 can vary by orders of magnitude depending on device parameters and silicon properties. The dark current can be calculated as

$$J_0 = \frac{q \cdot n_i^2}{N_D} \frac{D}{L} \left[\frac{(SL/D)\cosh(W_b/L) + \sinh(W_b/L)}{\cosh(W_b/L) + (SL/D)\sinh(W_b/L)} \right]$$

where q is the electronic charge, n_i is the intrinsic carrier concentration of silicon, N_D is the base doping concentration, S is the back surface recombination velocity which characterises the extent of minority recombination at the silicon/substrate interface, L and D are the minority carrier diffusion length and diffusivity in the base, respectively, and are sensitive to doping levels and material quality, and W_b is the thickness of the base. Actually, the above equation represents only the base layer contribution to J_0. There is an analogous equation for emitter contribution to J_0, but in well-designed solar cells, normally the base component is the dominant contribution to J_0. At any rate, our purpose here is to simply highlight the factors that contribute to J_0 and indicate the design principles to reduce J_0. In a thin silicon solar cell, the diffusion length will normally be longer than the layer thickness, i.e., $W_b/L < 1$, in which case the above equation simplifies to

$$J_0 = \frac{q \cdot n_i^2 \cdot S}{N_D}$$

In such cases, the dark current does not depend on diffusion length, but is directly proportional to the surface recombination velocity, thus underscoring the importance of surface passivation in thin silicon devices. Higher doping concentrations N_D will decrease J_0, so long as the doping does not degrade L such that $L < W$ and bulk recombination becomes significant. Another consideration of high doping but which is not evident from the above equations are bandgap narrowing effects. As doping levels exceed about 10^{19} atoms/cm^3, the effective bandgap of silicon is reduced, leading to increased intrinsic carrier concentrations n_i, and correspondingly increased J_0. Thus increasing doping to increase V_{OC} becomes self-defeating after a certain point.

3.1 Minority Carrier Recombination Issues in Thin Silicon Solar Cells

As already pointed out, an important potential, although not completely experimentally verified, advantage of thin silicon solar cells is their decreased sensitivity to minority carrier recombination. This permits higher doping levels to enhance open-circuit voltage, and leads to better tolerance of impurities and defects. A sensitivity analysis of solar cell efficiency to device thickness and minority carrier lifetime is shown in Figure 8. Minority carrier lifetime τ and diffusion length L are related as

$$L = \sqrt{D \cdot \tau}$$

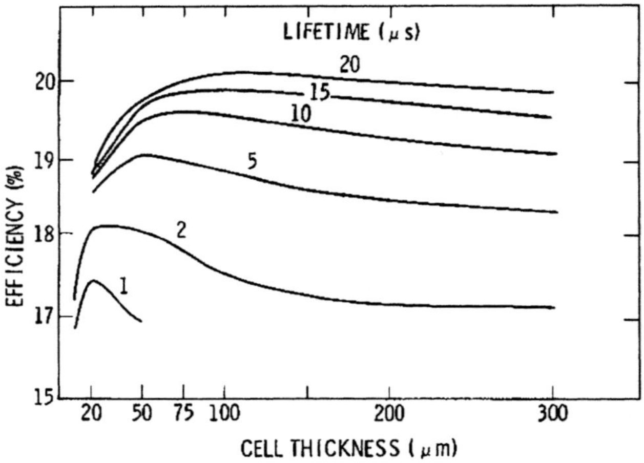

Figure 8 Sensitivity of solar cell efficiency to device thickness and minority carrier lifetime [74].

where D is the minority carrier diffusivity which is proportional to the minority carrier mobility. While D is not highly sensitive to impurity and defects, τ can easily vary by more than an order of magnitude in silicon of various purity and quality. The analysis summarised in Figure 11 shows the interesting result that for a given minority carrier lifetime there is an optimum thickness, and that for material with relative low minority carrier lifetimes the optimum thickness is less than 50 microns.

Films of silicon deposited on substrates will usually be polycrystalline, although not all thin silicon solar cells are necessarily polycrystalline; epitaxial films removed from a monocrystalline silicon substrate and bonded to a superstrate, as well as solar cells made by thinning monocrystalline silicon wafers need not have grain boundaries. In fact, these types of silicon solar cells provide an interesting control for exploring the effects of grain boundaries. Nevertheless, many low-cost approaches to thin silicon solar cells will produce material with varying grain sizes and textures. The effects of grain boundaries are complex and depend on the microstructure, film thickness, grain size distribution, junction depth, doping, etc. Diffused junctions can spike down grain boundaries, which may improve collection efficiency but also make the solar cell more prone to shunting effects (Figure 9). These issues are not unique to thin silicon solar cells, and in fact, polycrystalline cast silicon solar cells have been an established line of solar cells for many years. Grain boundaries act as surfaces for minority carrier recombination, and can be depleted or accumulated and exhibit space charge regions much as junctions and free surfaces. For single-crystal silicon solar cells, space-charge recombination is usually so small that it can be neglected, and the performance of the device, particularly open-circuit voltage, is controlled by bulk, surface, and shunt losses, but this may not be the case with multicrystalline solar cells. It was initially thought that polycrystalline silicon solar cells would suffer major short-circuit current and open-circuit voltage losses from grain-boundary recombination, which

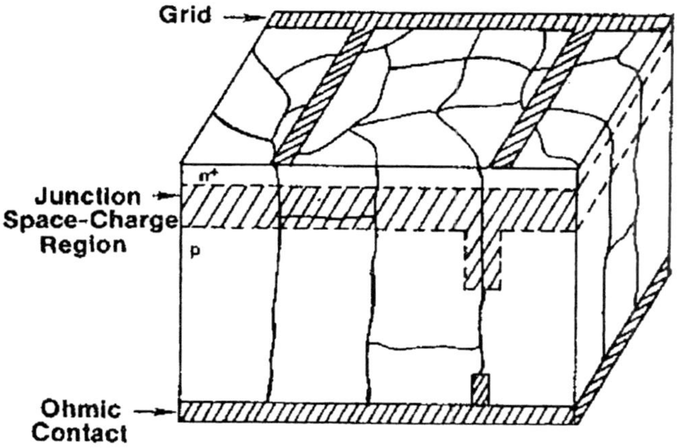

Figure 9 Geometry of grain boundaries in silicon solar cells [75].

would severely restrict the maximum light-generated current [73]. However, it has been shown that when the grain diameter is several times larger than the intra-grain (bulk) minority-carrier diffusion length, the short-circuit current is controlled not by grain-boundary recombination, but by the intra-grain diffusion length.

Other factors that may play a substantial role in determining the electrical performance of polycrystalline silicon solar cells are the presence of inclusions and tunnel junctions, both of which act as resistive shunts and degrade the open-circuit voltage and fill-factor locally. Although their impact on performance is fairly straightforward, it is not clear that either of these two possible defects is intrinsic to any polycrystalline silicon solar cell material or process. Accordingly, the analysis of the thin polycrystalline silicon solar cell is based on the relaxation of single crystal material and device properties due to the polycrystalline characteristics of the semiconductor.

As expected, increasing grain size results in better solar cell performance (Figure 10), but note the discontinuity between trends for p-i-n microcrystalline silicon solar cells and p-n multicrystalline silicon solar cells. The electric field in thin p-i-n cells is probably aiding collection efficiency and mediating the effect of grain boundary recombination. The grain structure in these thin microcrystalline silicon solar cells may also have a texture resulting in grain boundaries with less electrical activity (e.g., minority carrier recombination). Grain boundary effects are one of the most active areas for research in silicon solar cells and passivation techniques can be very effective in all types of multicrystalline silicon solar cells (Figure 11).

4 Silicon Deposition and Crystal Growth for Thin Solar Cells

In this section, we review some of the more important technologies used to realise thin silicon solar cells on supporting substrates (Figure 12). There is a

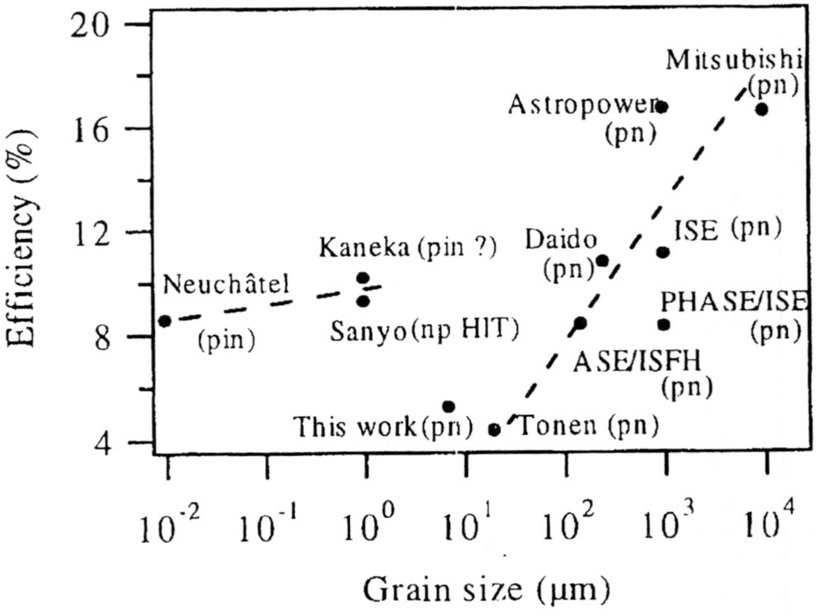

Figure 10 Effect of grain size on solar cell efficiency for pin microcrystalline silicon devices, and p–n multicrystalline thin silicon devices [76].

wide range of methods used to deposit semiconductor materials, virtually all of which have been applied to some extent or degree to the production of thin silicon solar cells. This section is offered as a survey of the diverse approaches; space limitations do not permit an in-depth review. Where possible, common issues and criteria and unifying design principles are noted.

4.1 Substrate Considerations

One of the key technological challenges to achieving a commercially viable thin-layer polycrystalline silicon solar cell technology is the development of a low-cost supporting substrate. The requirements for the substrate material are severe. Mechanical strength and thermal coefficient of expansion (TCE) matching are needed to prevent the film from breaking or deforming during handling and high-temperature processing. There are several good candidates for thermal expansion matched substrates including mullite, a compound of alumina and silica. The substrate must also provide good wetting and nucleation during the film growth process without contaminating the film. The substrate can be conducting or insulating, depending on device requirements. Finally, the substrate-silicon interface must provide a high degree of diffuse reflectivity and surface passivation.

The following substrates have been utilised to fabricate thin silicon solar cells:

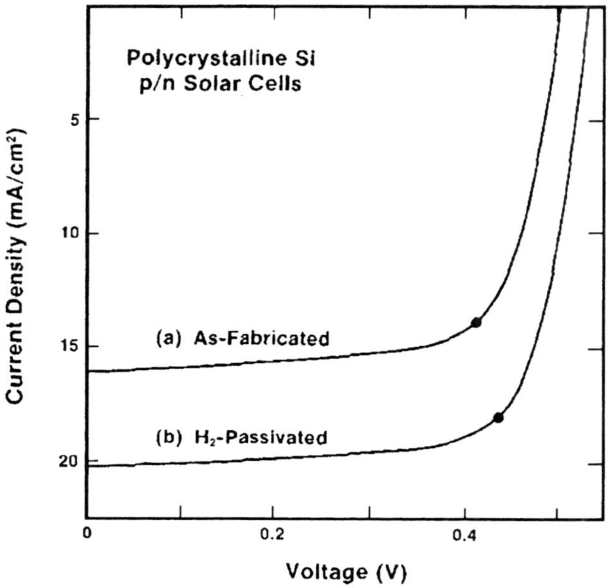

Figure 11 *Hydrogen passivation of multicrystalline silicon solar cells showing improvement in current and voltage [77].*

- glass [80–82]
- ceramics [83–90]
- steel [91–93]
- graphite [94–99]
- upgraded metallurgical silicon sheet or wafers [100–103]

4.2 High-Temperature Silicon Deposition Methods

It is useful to distinguish silicon deposition methods that employ high temperatures (>1000°C) from low-temperature deposition techniques. The high temperatures impose significant constraints on and limit the choice of substrates, especially if a post-deposition recrystallisation step is desired. High-temperature deposition methods are probably the only way to achieve high silicon deposition rates (e.g., 1–20 microns/min), and therefore if silicon layers of 10–50 microns thickness are required, such high-temperature steps may be the only viable option. High temperature deposition methods include:

- Melt growth or melt coating techniques where elemental silicon is melted and then deposited as a film or layer on a substrate.
- Chemical vapour deposition (CVD) where a silicon-containing gaseous precursor is thermally decomposed on a substrate.
- Liquid phase epitaxy (LPE) where silicon is precipitated from a molten metal solution.

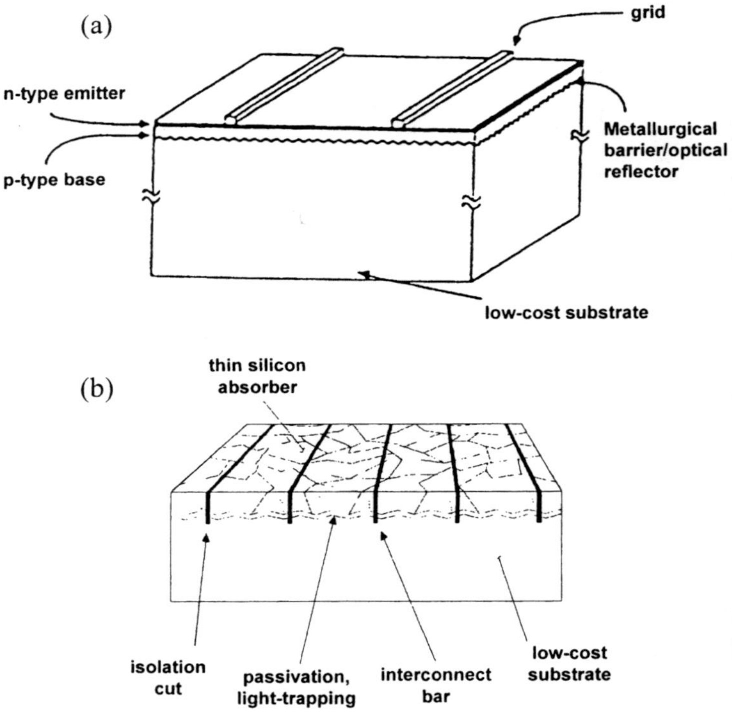

Figure 12 Generic polycrystalline thin-film silicon device structures: (a) the electrically active part of the device consists of a thin silicon layer on top on a passive mechanically supporting substrate [78]; (b) similar structure with interconnection achieved monolithically, a benefit unavailable to conventional wafer based devices [79].

4.2.1 Melt Growth Techniques

Melt growth processes characteristically have both high growth rates and good material quality [83, 84, 104]. An example melt coating process is shown in Figure 13. In these examples, a substrate is contacted with molten silicon, which wets the substrate and then solidifies as a silicon layer. In many cases, the substrate is drawn through a bath of molten silicon. Generally, such processes cannot produce layers much less than 100 microns in thickness.

4.2.2 Recrystallisation of Silicon

Related to melt growth are various recrystallisation techniques. These are generally not deposition processes per se, but instead are used to melt already-deposited silicon layers and recrystallise them in order to achieve a more favourable grain structure. In this case, the grain-structure of the as-deposited silicon layer is not critical, and the deposition process can be optimised for high-growth rates, large-areas and purity specifications. For instance, a plasma-enhanced CVD process such as shown in Figure 14 can be used to plate a silicon layer of desired thickness on a suitable substrate, such as a ceramic, which is compatible with a recrystallisation step.

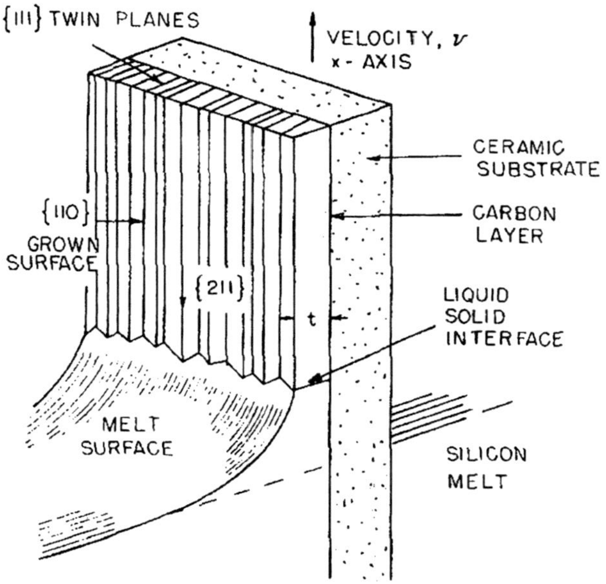

Figure 13 Honeywell silicon-on-ceramic dip coating process [83].

In the preferred techniques of recrystallisation, the deposited silicon layer is not usually simultaneously melted in its entirety. Instead, a zone-melting recrystallisation (ZMR) process is effected by localised heating to create a melted zone that moves or scans across the deposited silicon layer, melting the silicon at the leading edge and resolidifying a silicon layer at the trailing edge. Such ZMR techniques can yield millimetre to centimetre-sized grain structures. There are several ways to induce localised or zone melting of layers including moving point- or line-focused infrared lamps, travelling resistively heated strip heaters, and laser and electron beams. Figure 15 shows several of these zone melting recrystallisation techniques commonly used for silicon solar cell applications.

Much work has been done on optimising the quality of silicon layers produced by ZMR. For example, Figure 16 shows the effects of silicon layer thickness and zone melting scan speed on defect density of recrystallised silicon layers.

4.2.3 High-Temperature Silicon Chemical Vapour Deposition

Chemical vapour deposition or CVD is defined as the formation of a solid film on a substrate by reacting vapour-phase chemicals, or 'precursors', which contain the desired constituents [110]. For example, substrates can be coated with silicon layers by decomposition of gaseous silane (SiH_4) or trichlorosilane ($SiHCl_3$). In fact, many precursors are possible for silicon CVD, and silane or the chlorosilanes are probably the most commonly used, although for example, iodine and bromine compounds are also sometimes considered as silicon precursors. In general, Silicon CVD is a well-developed technology commonly used in integrated circuit fabrication, in which case it is often used for epitaxial

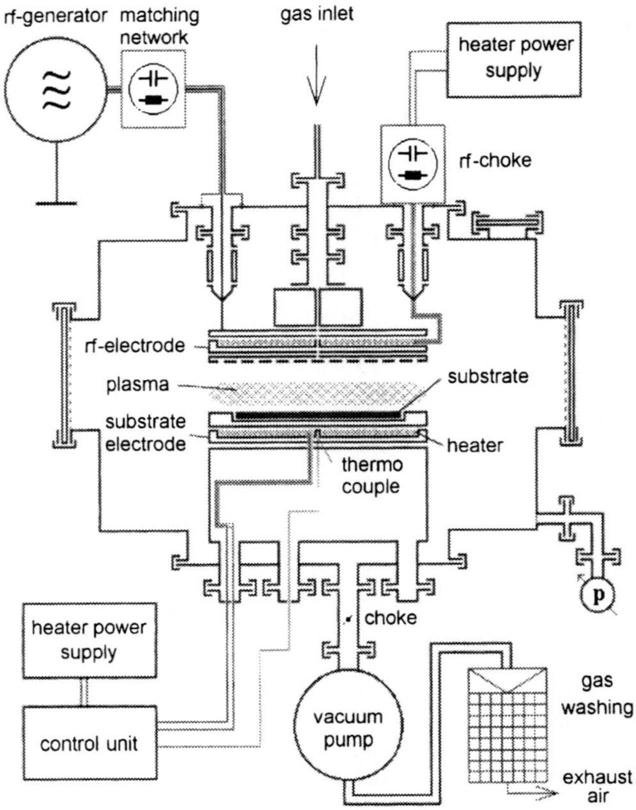

Figure 14 PECVD silicon deposition process [105].

growth of silicon layers on monocrystalline silicon substrates. For solar cell applications where CVD is used to deposit a 10–50 micron thick silicon layer for subsequent, post-deposition ZMR, the CVD is optimised for high precursor utilisation (i.e., the fraction of precursor converted to silicon), deposition efficiency (i.e., the fraction of deposited silicon that ends up on the substrate rather than the walls of the reactor chamber or the substrate susceptor), the deposition rate, the purity of the deposited silicon, areal uniformity, the potential to recover unreacted precursors or reaction product, and various safety and environmental issues.

For solar cell applications, three types of CVD are most used:

- atmospheric pressure chemical vapour deposition (APCVD)
- rapid thermal CVD (RTCVD)
- low-pressure CVD (LPCVD)

In all these types of CVD, a gaseous silicon precursor (e.g., SiH_4 or $SiHCl_3$), generally mixed with a dilution carrier gas such as hydrogen or nitrogen, is

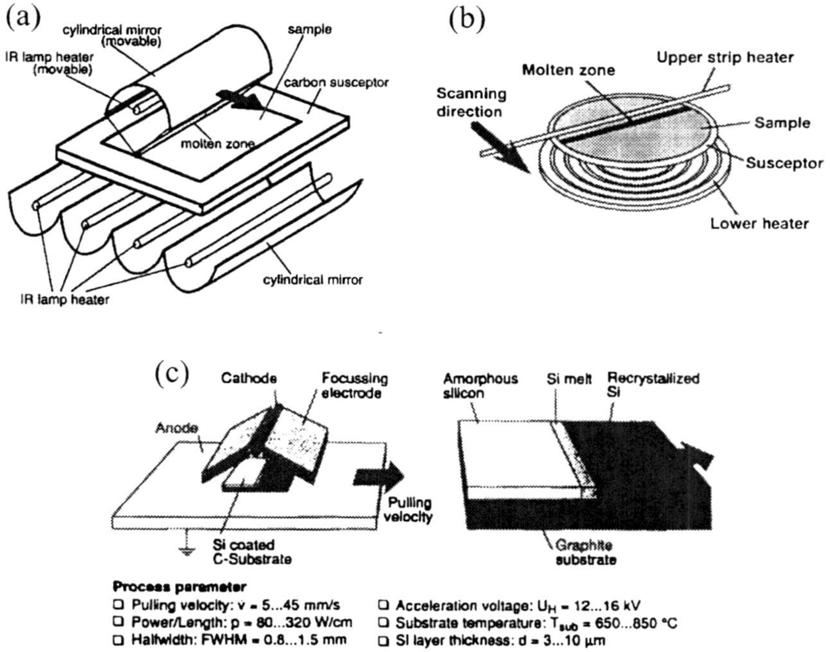

Figure 15 (a) Halogen IR lamp heating for ZMR [106]. (b) Travelling strip heaters for ZMR [107]. (c) Electron beam heating [108].

delivered into a reaction chamber. These gases move from the inlet to the outlet in a continuous stream to form the main gas flow region which bring the precursor into close proximity of the heated substrate. Some further description and details of the three main types of CVD processes used for thin silicon solar cells follows.

Atmospheric pressure CVD. APCVD systems were historically the first used for applications in the microelectronics industry [111, 112]. These systems are simple in design and are generally composed of three subsystems, a gas delivery system, a reactor, and an effluent abatement system. For the most part, APCVD is carried out at relatively high temperatures for silicon containing precursors (1100–1250°C). This allows for high deposition rates to be achieved. At these high temperatures, APCVD is in the mass-transport limited regime. This requires that a very uniform gas flow be achieved within the reactor to ensure that all areas of the heated substrate are exposed to equal amounts of precursor. This requirement is the primary concern during APCVD reactor geometry design and, to date, has limited this process to batch type processes with respect to CVD silicon.

Rapid thermal CVD. RTCVD is a variation of APCVD. It is based on the energy transfer between a radiant heat source and an object with very short processing times, of the order of seconds or minutes. This is typically from an optical heating system such as tungsten halogen lamps. The obvious benefit of this process is the

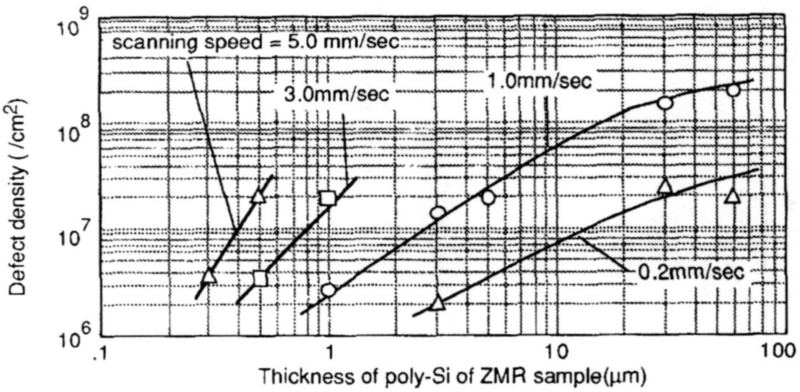

Figure 16 Relation between deposited silicon layer thickness, scan speed, and silicon defect density in ZMR [109].

fast cycle times for heating substrates to their required deposition temperature. A detailed explanation of this process along with its advantages and disadvantages can be found in Faller et al. [113].

Low-pressure CVD. LPCVD systems are inherently more complex than APCVD systems since they require robust vacuum systems that are capable of handling the often toxic corrosive precursor effluents. With respect to silicon deposition, LPCVD is generally conducted using vacuum pressures of 0.25–2.0 torr and temperatures of 550–700°C. At these pressures and temperatures LPCVD is in the surface rate-limited regime. It is important to note that at reduced pressures the diffusivity of the precursor is greatly enhanced. This allows for multiple wafers to be stacked very closely together, on the order of a few millimetres, and still achieve a highly uniform deposition. However in order to ensure this, very precise temperature control is necessary across the entire reactor, within 0.5–1°C is not uncommon. Since LPCVD is in the surface rate-limited regime it has the constraint of very low deposition rates. These low growth rates have limited its application in silicon solar cell fabrication.

A summary of silicon CVD growth rates for various precursors and deposition temperatures is given in Table 1.

4.2.4 Low-Temperature Chemical Vapour Deposition

Some low-temperature chemical vapour techniques can be distinguished from the high-temperature (>1000°C) CVD processes discussed above. These methods are employed with substrates such as glass which are not compatible with either a high-temperature deposition step nor a post-deposition recrystallisation step. The relatively slow growth rates inherent in a low-temperature deposition process necessitates thin device structures on the order of several microns thickness or less. The as-deposited silicon layers have average grain sizes of 1 micron or less and are characterised as microcrystalline. Hydrogenated microcrystalline silicon solar cells using a p–i–n structure which is similar to the amorphous silicon solar cell structure, can achieve very respectable conversion efficiencies in excess of 10%. The benefits of such microcrystalline silicon solar cells over amorphous silicon solar cells are a greater stability to light-induced degradation processes. The deposition processes for microcrystalline silicon solar cells are typically adaptations of those used for amorphous silicon solar cells.

Liquid-phase epitaxy (LPE). Liquid-phase epitaxy is a metallic solution growth technique that can be used to grow semiconductor layers on substrates. Silicon can be precipitated from solutions of a number of molten metals in the temperature range 600–1200°C. This method has been used to grow thin silicon solar cells on low-cost metallurgical grade (MG) silicon substrates. In this case, the MG silicon substrate is too impure for direct use in photovoltaics. Instead, the substrate is used as a substrate for the growth of high-purity layers of silicon by either LPE or CVD. The MG silicon provides a thermal-expansion matched substrate for the thin silicon solar cell. The grain size of MG substrates is relatively large (several millimetres to centimetres in lateral dimension), and as the epitaxial layer will replicate the grain structure of the silicon substrate, this approach will yield thin silicon solar cells with large grain sizes. One issue with using a MG silicon substrate is contamination of the solar cell device by outdiffusing substrate impurities. Other favourable features of LPE are the high mobility of adatoms in the liquid phase (as compared to surface diffusion upon which vapour-phase techniques depend), and the near-equilibrium growth conditions which reduce point defects and dislocations originating from the substrate. A conventional slideboat LPE system, similar to that used for making compound semiconductor optoelectronics devices, and suitable for R&D of LPE thin silicon solar cells is shown in Figure 17. This type of LPE system employs a programmed transient cooling mode, and is essentially a batch process. Steady-state LPE processes (Figure 18) using an imposed temperature difference across the growth solution in conjunction with a solid silicon source to replenish the solution have been proposed and developed for high-throughput production.

5 Thin Silicon Solar Cells Based on Substrate Thinning

Figure 19 shows a solar cell made by thinning and grooving the backside of a silicon wafer. Such solar cells obviously have little cost advantage in that they

Table 1 Common silicon precursors

Silicon precursor	Deposition temperature (°C)	Growth rates (μm/min)
Silicon tetrachloride ($SiCl_4$)	1150–1250	0.4–1.5
Trichlorosilane ($SiCl_3H$)	1000–1150	0.4–4.0
Dichlorosilane (SiH_2Cl_2)	1020–1120	0.4–3.0
Silane (SiH_4)	650–900	0.2–0.3
Disilane (Si_2H_6)	400–600	<0.1

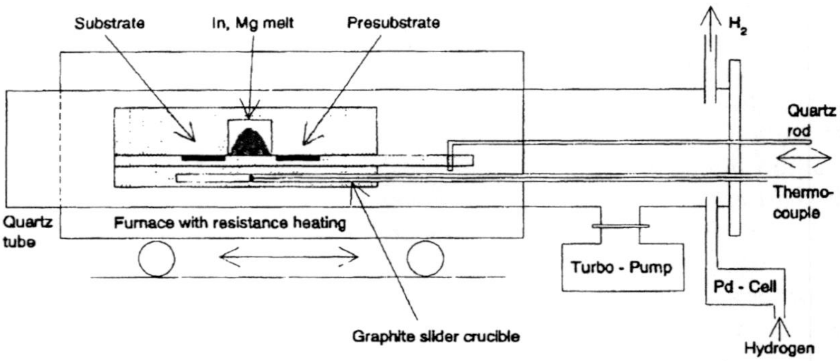

Figure 17 Small-scale LPE system [114].

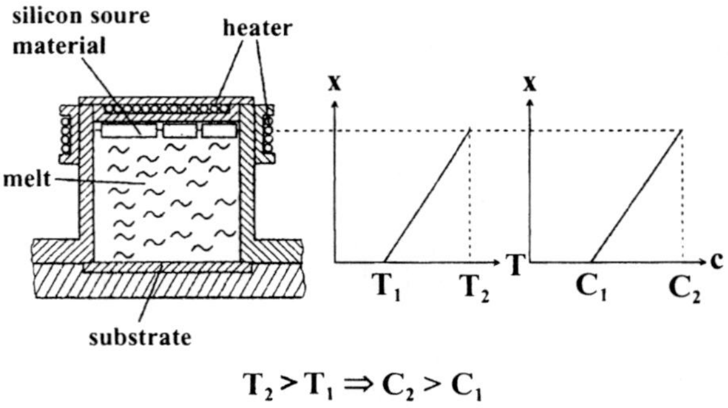

Figure 18 Principle of the temperature difference method [115].

utilise a high-quality silicon wafer and add considerable processing complexity. Because of the decreased sensitivity of performance to lifetime in such thin solar cells, they have application to space solar cells due to their potential radiation hardness. Further, these cells provide a means of studying basic effects such as light trapping and surface passivation in thin solar cell structures, without the complicating issues of material quality and grain boundaries. A similar type of thin solar cell is shown in Figure 20. This device structure permits a planar back mirror to be effected close to the front surface (Figure 20(a)). The dependence of short-circuit current on effective device thickness can be readily studied with this type of thin solar cell (Figure 20(b)).

Even thinner silicon solar cells can be made with silicon wafers using epitaxial growth, provided a superstrate is used for mechanical support. For instance, Figure 21 shows a thin silicon solar cell that made by layer transfer and wafer bonding techniques. An epitaxial silicon solar cell structure is grown on a monocrystalline silicon substrate. The structure is then bonded to a glass superstrate. The silicon substrate is then removed. Most simply, the removal of the silicon substrate can be effected by controlled etching, in which case the substrate is dissolved away. Various schemes have been proposed to separate the substrate from the epitaxial layer after bonding the solar cell structure to a

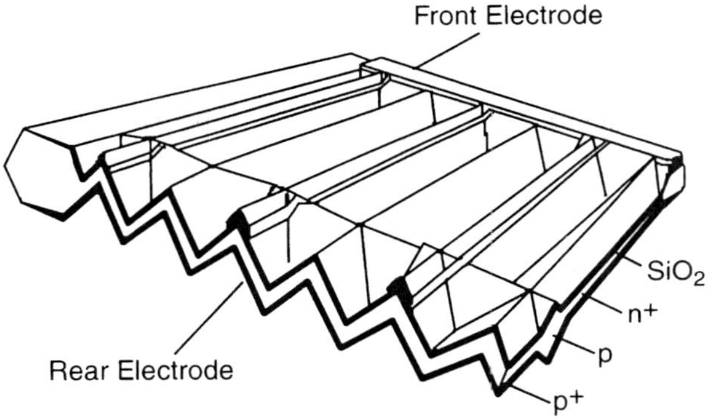

Figure 19 Thinned and grooved wafer-based silicon solar cell [116].

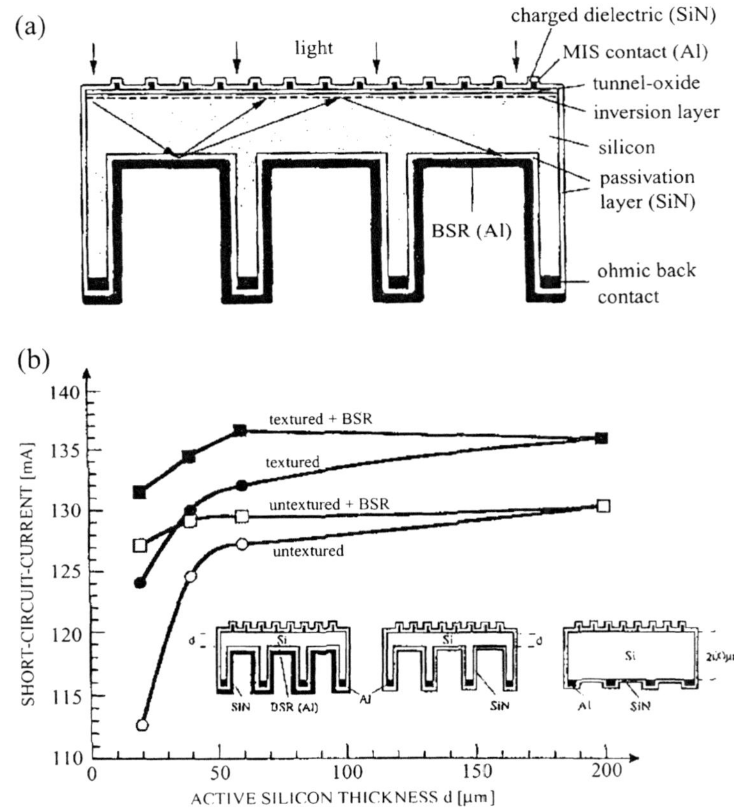

Figure 20 (a) Cross-section of ultra-thin, self-supporting MIS solar cell made by structuring a silicon wafer [117]. (b) Short-circuit current as a function of active layer silicon thickness for solar cell structures shown in inset [117].

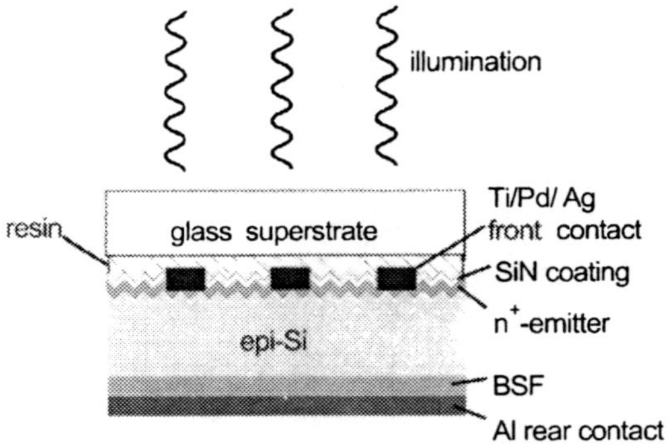

Figure 21 Monocrystalline silicon substrate made by layer transfer and bonding to a glass superstrate [119].

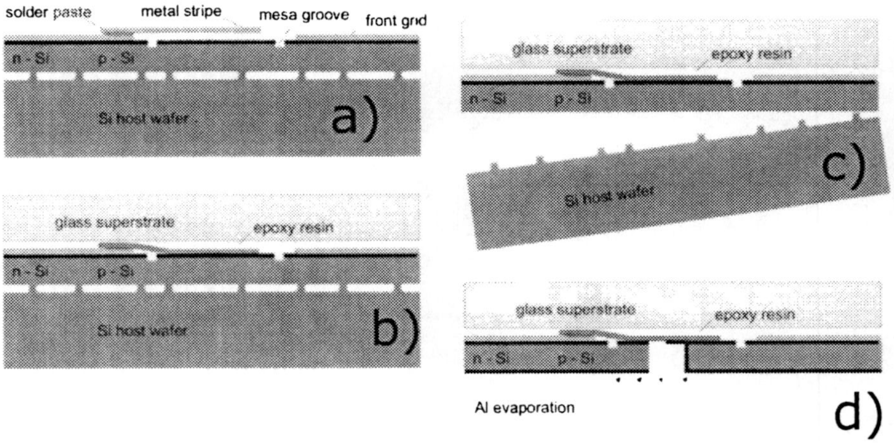

Figure 22 Schematic representation of series connection of thin-film Si transfer solar cells. (a) Two epitaxial thin-film silicon solar cells connected to the host wafer with the separation layer (columns). A metal stripe (Ag) is soldered to the front side grid of the left solar cell. Mesa grooves provide electrically isolation of the emitter in the interconnection area. (b) Epoxy resin fixes the superstrate glass to the surface of the cell. (c) Mechanical force removes the host wafer from the cell. (d) A groove structured via chip dicing sawing separates the two solar cells. Oblique deposition of aluminium creates the back contact of the solar cells and electrically connects the metal stripe that is in contact with the front side grid of the left solar cell with the back side contact of the right solar cell [120].

superstrate. One method of achieving this shown in detail in Figure 22. Some creative variations on this approach have been reported, as for example shown in Figure 23.

6 Summary of Device Results

We end this review by summarising results for thin silicon solar cells. Table 2 lists reports for thin silicon layers made by high-temperature growth methods, often

Table 2 High-temperature growth on foreign substrates (after Catchpole et al. [122])

Institution	Substrate	Deposition method	Grain size (μm)	Electrical	Reference
PHASE, ECN and CNRS-LMPM	Alumina	RTCVD	10	$\tau = 0.3\ \mu s$, $L = 10\ \mu m$	[123, 124]
PHASE, LPM and GEMPPM	Alumina	LPE with RTCVD seed	10		[125]
IMEC and PHASE	Alumina	RTCVD	5–10	$\tau = 0.5\ \mu s$	[126, 127]
ETL	Alumina	ECR-PCVD with EB-ZMR, diff. Barrier	10 × 200		[128]
ETL	Alumina	CVD with laser recryst., diff. Barrier		$\eta = 6.5\%$	[129]
TU-Berlin	Alumina	CVD	10	$L = 8\ \mu m$	[130]
PHASE and IMEC	Mullite	RTCVD	10–15	$\tau = 0.5\ \mu s$	[131]
MPI-F, MPI-M and LSG	Glassy carbon	LPE with RF plasma seed	10		[132]
MPI-F, Siemens and TU-Hamburg	Graphite	PECVD or sputtering + ZMR + CVD	100 × 1000		[133]
Fraunhofer ISE	Graphite	LPCVD, diff. Barrier, ZMR	mm × cm	$L = 30\ \mu m$, $\eta = 11.0\%$	[134]
Daido Hoxan	Graphite	LPE (temp diff) with diff. Barrier	100–300	$L = 30\ \mu m$	[135, 136]
PHASE, IMEC and Fraunhofer	Graphite	RTCVD with diff. Barrier	0.1–6	$L = 3\ \mu m$	[137]
ASE GmbH	Graphite	CVD with ZMR, diff. Barrier	100	$L = 10–15\ \mu m$	[138]
TU-Hamburg	Graphite	LPCVD with EB-ZMR	100 μm × cm		[139]
ECN	Si/SiAlON	LPE with plasma-spray	10–100		[140]
UNSW	High-temperature glass	LPE with a-Si seed	50		[141]
UNSW	High-temperature glass	LPE	100		[142]
MPI-F, U Erlangen and U Stuttgart NTT	High-temperature glass	CVD with LPCVD and SPC seed	2	$L = 2\ \mu m$, $\eta = 2\%$	[143, 144]
AstroPower	Tape cast ceramic	APCVD + ZMR	mm × cm	9.18%, 543 mV	[145]

Table 3 Summary of thin silicon-on-silicon device results (after McCann et al. [146]).

Institution	Substrate	Deposition method	Electrical	Reference
Single-crystal substrates				
ANU	p-type sc-Si	LPE and substrate thinning	18%, 666 mV	[147]
UNSW	p^+ sc-S	CVD	17.6%, 664 mV	[148]
Fraunhofer ISE	p^+ sc-Si	RTCVD	17.6%	[149]
MPI-F	sc-Si	CVD	17.3%, 661 mV	[150]
ANU	p^+ sc-Si	LPE and substrate thinning	17%, 651 mV	[151]
UNSW	p^+ sc-Si	LPE	16.4%, 645 mV	[152]
ASE GmbH	p^+ sc-Si	CVD	15.4%, 623 mV	[153]
Imec	p^+ sc-Si	CVD	14.9%, 635 mV	[154]
MPI-F	p^+ sc-Si	LPE	14.7%, 659 mV	[155]
Beijing SERI	p^+ sc-Si	RTCVD	12.1%, 626 mV	[156]
Mc-Si, ribbon and MG-Si substrates				
ANU	p-type mc-Si	LPE	15.4%, 639 mV	[157]
ANU	p^+ mc-Si	LPE	15.2%, 639 mV	[157]
IMEC and KU	p^+ mc-Si	APCVD	13.3%, 615 mV	[158]
Fraunhofer ISE	p^+ mc-Si	RTCVD	13.2%, 614 mV	[149]
IMEC	p^+ mc-Si	CVD and industrial cell Process	12.1%	[159]
IMEC, KU Leuven and Bayer	RGS-ribbons	CVD	10.4%, 558 mV	[160]
Fraunhofer ISE	SSP ribbons	RTCVD	8.0%, 553 mV	[161]
IMEC and KU	SSP pre-ribbon	CVD	7.6%	[162]
Leuven 1 Kristallzuchtung	mc-Si	Temp. diff. LPE	$\tau = 5$–$10\ \mu m$	[163]
NREL	MG-Si	LPE	$L = 42\ \mu m$	[164, 165]
Substrates with diffusion barriers				
Mitsubishi Electric	SiO_2 on Si	ZMR and CVD	16.4%, 608 mV	[166]
Fraunhofer ISE	SiO_2 (perforated) on SSP-Si	RTCVD and large area heating	11.5%, 562 mV	[167]
IMEC and Fraunhofer ISE	SiO_2 on Si	ZMR and CVD	9.3%, 529 mV	[168]
Fraunhofer ISE	SiO_2 on Si	LPCVD and ZMR	6.1%	[169]
Delft UT	SiO_2 on Si	CVD	Grain size 1–2 μm	[170]

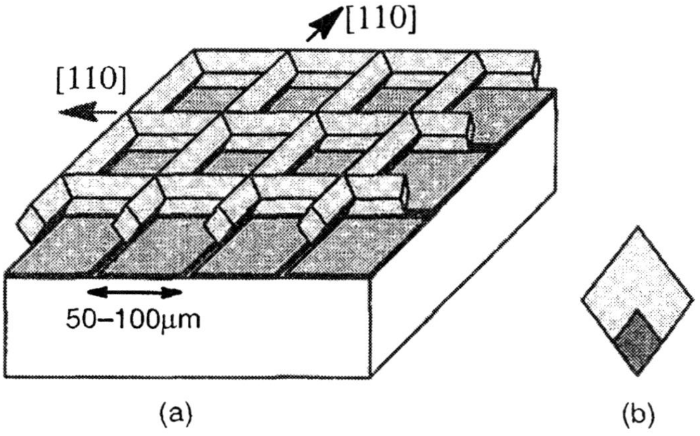

Figure 23 An example of the epitaxial lift-off process [121].

including a post-deposition recrystallisation (e.g., ZMR) step, on non-silicon substrates such as ceramics, graphite, and high-temperature glasses. Samples are characterised either by diffusion length or minority carrier lifetime, or in cases where a solar cell was made, by efficiency. Table 3 shows a similar summary of results for thin silicon solar cell structures on silicon-based substrates, which include oxidised silicon, MG silicon, and various types of silicon sheet.

References

[1] Redfield, D., 1975. Enhanced Photovoltaic Performance of Thin Silicon Films by Multiple Light Passes. *Proc. 11th IEEE Photovoltaic Specialists Conf.*, Scottsdale, pp. 431–432.

[2] Spitzer, M., Shewchun, J., Vera, E.S. and Loferski, J.J., 1980. Ultra High Efficiency Thin Silicon p–n Junction Solar Cells Using Reflecting Surfaces. *Proc. 14th IEEE Photovoltaic Specialists Conf.*, San Diego, pp. 375–380.

[3] Barnett, A.M., 1980. Thin Film Solar Cell Comparison Methodology. *Proc. 14th IEEE Photovoltaic Specialists Conf.*, San Diego, pp. 273–280.

[4] Werner, J.H. and Bergmann, R.B., 2001. Crystalline Silicon Thin Film Solar Cells. *Tech. Digest Int. PVSEC-12*, Jeju, pp. 69–72.

[5] Goetzberger, A., 1981. Optical Confinement in Thin Si-Solar Cells by Diffuse Back Reflectors. *Proc. 15th IEEE Photovoltaic Specialists Conf.*, Orlando, pp. 867–870.

[6] Yablonovitch, E. and Cody, G.D., 1982. Intensity Enhancement in Textured Optical Sheets for Solar Cells. *IEEE Trans. Electron Devices*, Vol. ED-29, pp. 300–305.

[7] Redfield, D., Multiple-Pass Thin-Film Silicon Solar Cell. *Applied Physics Letters*, Vol. 25(11), pp. 647–648.

[8] Barnett, A.M., Rand, J.A., Domian, F.A., Ford, D.H., Kendall, C.L., Rock, M.L. and Hall, R.B., 1988. Efficient Thin Silicon-Film Solar Cells on Low

Cost Substrates. *Proc. 8th European Photovoltaic Solar Energy Conf.*, Florence, Italy, pp. 149–155.

[9] Green, M.A. and Campbell, P., 1987. Light Trapping Properties of Pyramidally Textured and Grooved Surfaces. *Proc. 19th IEEE Photovoltaic Specialists Conf.*, New Orleans, pp. 912–917.

[10] Ruby, D.S., Yang, P., Zaidi, S. et al., 1998. Improved Performance of Self-aligned, Selective-emitter Silicon Solar Cells. *Proc. 2nd World Conf. on Photovoltaic Solar Energy Conversion*, Vienna, pp. 1460–1463.

[11] Schnell, M., Lüdemann, R. and Schaefer, S., 2000. Plasma Surface Texturization for Multicrystalline Silicon Solar Cells. *Proc. 28th IEEE Photovoltaic Specialists Conf.*, Anchorage, pp. 367–370.

[12] Wells, T., El-Gomati, M.M. and Wood, J., 1997. Low Temperature Reactive Ion Etching of Silicon with SF_6/O_2 Plasmas. *J. Vac. Sci. Technol. B*, Vol. 15, p. 397.

[13] Deckman, H.W., Roxlo, C.B., Wronski, C.R. and Yablonovitch, E., 1984. Optical Enhancement of Solar Cells. *Proc. 17th IEEE Photovoltaic Specialists Conf.*, pp. 955–960.

[14] Stocks, M.J., Carr, A.J. and Blakers, A.W., 1994. Texturing of Polycrystalline Silicon. *Proc. 24th IEEE Photovoltaic Specialists Conf.*, Hawaii, pp. 1551–1554.

[15] Deckman, H.W., Wronski, C.R., Witzke, H. and Yablonovitch, E., 1983. Optically Enhanced Amorphous Solar Cells. *Appl. Phys. Lett.*, Vol. 42(11), pp. 968–970.

[16] Tsuo, Y.S., Xiao, Y., Heben, M.J., Wu, X., Pern, F.J. and Deb, S.K., 1993. Potential Applications of Porous Silicon in Photovoltaics. *Proc. 23rd IEEE Photovoltaic Specialists Conf.*, Louisville, pp. 287–293.

[17] Cudzinovic, M. and Sopori, B., 1996. Control of Back Surface Reflectance from Aluminum Alloyed Contacts on Silicon Solar Cells. *Proc. 25th IEEE Photovoltaic Specialists Conf.*, Washington DC, pp. 501–503.

[18] Rothwarf, A., 1985. Enhanced Solar Cell Performance by Front Surface Light Scattering. *Proc. 18th IEEE Photovoltaic Specialists Conf.*, Las Vegas, pp. 809–812.

[19] Gee, J,M., King, R.R. and Mitchell, K.W., 1996. High-Efficiency Cell Structures and Processes Applied to Photovoltaic-Grade Czochralski Silicon. *Proc. 25th IEEE Photovoltaic Specialists Conf.*, Washington DC, pp. 409–412.

[20] Gee, J.M., Gordon, R. and Liang, H., 1996. Optimization of Textured-Dielectric Coatings for Crystalline-Silicon Solar Cells. *Proc. 25th IEEE Photovoltaic Specialists Conf.*, Washington DC, pp. 733–736.

[21] Hegedus, S.S. and Deng, X., 1996. Analysis of Optical Enhancement in a-Si n–i–p Solar Cells Using a Detachable Back Reflector. *Proc. 25th IEEE Photovoltaic Specialists Conf.*, Washington DC, pp. 1061–1064.

[22] Tobin, S.P., Keavney, C.J., Geoffroy, L.M. and Sanfacon, M.M., 1988. Experimental Comparison of Light-Trapping Structures for Silicon Solar Cells. *Proc. 20th IEEE Photovoltaic Specialists Conf.*, Las Vegas, pp. 545–548.

[23] Rand, J.A., Hall, R.B. and Barnett, A.M., 1990. Light Trapping in Thin Crystalline Silicon Solar Cells. *Proc. 21st IEEE Photovoltaic Specialists Conf.*, Orlando, pp. 263–268.
[24] Restrepo, F. and Backus, C.E., 1976. *IEEE Trans. Electron Devices*, Vol. 23, pp. 1195–1197.
[25] Campbell, P., Wenham, S.R. and Green, M.A., 1988. Light-Trapping and Reflection Control with Tilted Pyramids and Grooves. *Proc. 20th IEEE Photovoltaic Specialists Conf.*, Las Vegas, pp. 713–716.
[26] King, D.L. and Buck, M.E., 1991. Experimental Optimization of an Anisotropic Etching Process for Random Texturization of Silicon Solar Cells. Proc. *22nd IEEE Photovoltaic Specialists Conf.*, Las Vegas, pp. 303–308.
[27] Terheiden, B., Fath, P. and Bucher, E., 2000. The MECOR (Mechanically Corrugated) Silicon Solar Cell Concept. *Proc. 28th IEEE Photovoltaic Specialists Conf.*, Anchorage, pp. 399–402.
[28] Fath, P. et al., 1995. Multicrystalline Silicon Solar Cells Using a New High-Throughput Mechanical Texturization Technology and a Roller Printing Metallization Technique. *Proc. 13th European Photovoltaic Solar Energy Conf.*, Nice, pp. 29–32.
[29] Narayanan, S., Wenham, S.R. and Green, M.A., 1989, *Tech. Digest. Int. PVSEC-4*, Sydney, p. 1111.
[30] Kaiser, U., Kaiser, M. and Schindler, R., 1991. Texture Etching of Multicrystalline Silicon. *Proc. 10th European Photovoltaic Solar Energy Conf.*, Lisbon, pp. 293–294.
[31] Willeke, G., Nussbaumer, H., Bender, H. and Bucher, E., 1992. Mechanical Texturization of Multicrystalline Silicon Using a Conventional Dicing Saw and Bevelled Blades. *Proc. 11th European Photovoltaic Solar Energy Conf.*, Montreux, pp. 480–483.
[32] Stocks, M.J., Carr, J.J. and Blakers, A.W., 1994. Texturing of Polycrystalline Silicon. *Proc. First World Conf. on Photovoltaic Energy Conversion*, Hawaii, pp. 1551–1554.
[33] Campbell, P. and Keevers, M., 2000. Light Trapping and Reflection Control for Silicon Thin Films Deposited on Glass Substrates Textured by Embossing. *Proc. 28th IEEE Photovoltaic Specialists Conf.*, Anchorage, pp. 355–358.
[34] Thorp, D., Campbell, P. and Wenham, S.R., 1996. Absorption Enhancement in Conformally Textured Thin-Film Silicon Solar Cells. *Proc. 25th IEEE Photovoltaic Specialists Conf.*, Washington DC, pp. 705–708.
[35] Smith, A.W., Rohatgi, A. and Neel, S.C., 1990. Texture: A Ray-Tracing Program for the Photovoltaic Community. *Proc. 21st IEEE Photovoltaic Specialists Conf.*, Orlando, pp. 426–431.
[36] Sopori, B.L. and Marshall, T., 1993. Optical Confinement in Thin Silicon Films: A Comprehensive Ray Optical Theory. *Proc. 23rd IEEE Photovoltaic Specialists Conf.*, Louisville, pp.127–132.

[37] Rau, U., Meyer, T., Goldbach, M., Brendel, R. and Werner, J.H., 1996. Numerical Simulation of Innovative Device Structures for Silicon Thin-Film Solar Cells. *Proc. 25th IEEE Photovoltaic Specialists Conf.*, Washington DC, pp. 469–472.

[38] Abouelsaood, A.A., Ghannam, M.Y., Poortmans, J. and Mertens, R.P., 1997. Accurate Modeling of Light Trapping in Thin Film Silicon Solar Cells. *Proc. 26th IEEE Photovoltaic Specialists Conf.*, pp. 183–186.

[39] Campbell, P. and Green, M.A., 1987. Light Trapping Properties of Pyramidally Textured Surfaces. *J. Appl. Physics*, July, pp. 243–249.

[40] Bisconti, R. and Ossenbrink, H.A., 1995. Light Trapping in Spheral Solar^{TM} Cells. *Proc. 13th European Photovoltaic Solar Energy Conf.*, Nice, pp. 386–389.

[41] Minano, J.C., Luque, A. and Tobias, I., 1992. Light-Confining Cavities for Photovoltaic Applications Based on the Angular-Spatial Limitation of the Escaping Beam. *Applied Optics*, Vol. 31(16), pp. 3114–3122.

[42] Zhao, J., Wang, A., Blakers, A.W. and Green, M.A., 1988. High Efficiency Prismatic Cover Silicon Concentrator Solar Cells. *Proc. 20th IEEE Photovoltaic Specialists Conf.*, Las Vegas, pp. 529–531.

[43] Gee, J.M., 1988. The Effect of Parasitic Absorption Losses on Light Trapping in Thin Silicon Solar Cells. *Proc. 20th IEEE Photovoltaic Specialists Conf.*, Las Vegas, pp. 549–554.

[44] Green, M.A. et al., 1995. Enhanced Light-Trapping in 21.5% Efficient Thin Silicon Solar Cells. *Proc. 13th European Photovoltaic Solar Energy Conf.*, Nice, pp. 13–16.

[45] Tiedje, T., Yablonovitch, E., Cody, G.D. and Brooks, B.G., 1984. Limiting Efficiency of Silicon Solar Cells. *IEEE Trans. Electron Devices*, Vol. ED-3 1, pp. 711–716.

[46] Basore, P.A., 1990. Numerical Modeling of Textured Silicon Solar Cells Using PC-1D. *IEEE Trans. Electron Devices*, ED-37, p. 337.

[47] Zimmerman, W. and Eyer, A., 2000. Coarse-Grained Crystalline Silicon Thin Film Solar Cells On Laser Perforated SIO_2 Barrier Layers. *Proc. 28th IEEE Photovoltaic Specialists Conf.*, Anchorage, pp. 233–236.

[48] Bruton, T.M., Roberts, S., Heasman, K.C. and Russell, R., 2000. Prospects For High Efficiency Silicon Solar Cells In Thin Czochralski Wafers Using Industrial Processes. *Proc. 28th IEEE Photovoltaic Specialists Conf.*, Anchorage, pp. 180–183.

[49] Berge, C., Bergmann, R.B., Rinke, T.J. and Werner, J.H., 2001. Monocrystalline Silicon Thin Film Solar Cells By Layer Transfer. *17th European Photovoltaic Solar Energy Conf.*, Munich, pp. 1277–1281.

[50] Zimmerman, W., Bau, S., Eyer, A., Haas, F. and Oßwald, D., 2000. Crystalline Silicon Thin Film Solar Cells On Low Quality Silicon Substrates With And Without SiO_2 Intermediate Layer. *Proc. 16th European Photovoltaic Solar Energy Conf.*, Glasgow, pp. 1144–1147.

[51] Finck von Finckenstein, B., Horst, H., Spiegel, M., Fath, P. and Bucher, E., 2000. Thin MC SI Low Cost Solar Cells With 15% Efficiency. *Proc. 28th IEEE Photovoltaic Specialists Conf.*, Anchorage, pp. 198–200.

[52] Zahedi, C., Ferrazza, F., Eyer, A., Warta, W., Riemann, H., Abrosimov, N.V., Peter, K. and Hötzel, J., 2000. Thin Film Silicon Solar Cells On Low-Cost Metallurgical Silicon Substrates By Liquid Phase Epitaxy. *Proc. 16th European Photovoltaic Solar Energy Conf.*, Glasgow, pp. 1381–1384.

[53] Tanda, M., Wada, T., Yamamoto, H., Isomura, M., Kondo, M. and Matsuda, A., 1999. Key Technology for μc-Si Thin-Film Solar Cells Prepared at a High Deposition Rate. *Tech. Digest Int. PVSEC-11*, Sapporo, pp. 237–238.

[54] Hanoka, J.I., 1999. An Overview of Silicon Ribbon-Growth Technology. *Tech. Digest Int. PVSEC-11*, Sapporo, pp. 533–534.

[55] Brendel, R., Auer, R., Feldrapp, K., Scholten, D., Steinof, M., Hezel, R. and Schultz, M., 2002. Crystalline Thin-Film Si Cells From Layer Transfer Using Porous Si (PSI-Process). *Proc. 29th IEEE Photovoltaic Specialists Conf.*, New Orleans, pp. 86–89.

[56] Meier, D.L., Jessup, J.A., Hacke, P. and Granata, Jr., S.J., 2002. Production of Thin (70–100 μm) Crystalline Silicon Cells for Conformable Modules. *Proc. 29th IEEE Photovoltaic Specialists Conf.*, New Orleans, pp. 110–113.

[57] Cudzinovic, M.J. and McIntosh, K.R., 2002. Process Simplifications to the Pegasus Solar Cell – Sunpower's High-Efficiency Bifacial Silicon Solar Cell. *Proc. 29th IEEE Photovoltaic Specialists Conf.*, New Orleans, pp. 70–73.

[58] Schmidt, J., Oberbeck, L., Rinke, T.J., Berge, C. and Bergmann, R.B., 2001. Application of Plasma Silicon Nitride To Crystalline Thin-Film Silicon Solar Cells. *Proc. 17th European Photovoltaic Solar Energy Conf.*, Munich, pp. 1351–1354.

[59] Münzer, K.A., Eisenrith, K.H., Schlosser, R.E. and Winstel, M.G., 2001. 18% PEBSCO – Silicon Solar Cells For Manufacturing. *Proc. 17th European Photovoltaic Solar Energy Conf.*, Munich, pp. 1363–1366.

[60] Tayanaka, H., Nagasawa, A., Hiroshimaya, N., Sato, K., Haraguchi, Y. and Matsushita, T., 2001. Effects of Crystal Defects in Single-Crystalline Silicon Thin-Film Solar Cell. *Proc. 17th European Photovoltaic Solar Energy Conf.*, Munich, pp. 1400–1403.

[61] Terheiden, B., Fischer, B., Fath, P. and Bucher, E., 2001. Highly Efficient Mechanically V-Textured Silicon Solar Cells Applying A Novel Shallow Angle Contacting Scheme. *Proc. 17th European Photovoltaic Solar Energy Conf.*, Munich, pp. 1331–1334.

[62] Schneiderlöchner, E., Preu, R., Lüdemann, R., Glunz, S.W. and Willeke, G., 2001. Laser-Fired Contacts. *Proc. 17th European Photovoltaic Solar Energy Conf.*, Munich, pp. 1303–1306.

[63] Glunz, S.W., Dicker, J., Kray, D., Lee, J.Y., Preu, R., Rein, S., Schneiderlöchner, E., Sölter, J., Warta, W. and Willeke, G., 2001. High Efficiency Cell Structures For Medium-Quality Silicon. *Proc. 17th European Photovoltaic Solar Energy Conf.*, Munich, pp. 1286–1292.

[64] Yamamoto, K., Yoshimi, M., Suzuki, T., Okamoto, Y., Tawada, Y. and Nakajima, A., 1997. Thin Film Poly-Si Solar Cell With Star Structure on

Glass Substrate Fabricated at Low Temperature. *Proc. 26th IEEE Photovoltaic Specialists Conf.*, Anaheim, 1997, pp. 575–580.

[65] Yamamoto, K., Yoshimi, M., Tawada, Y., Okamoto, Y. and Nakajima, A., 1999. Cost Effective and High Performance Thin Film Si Solar Cell Towards the 21st Century. *Tech Digest Int. PVSEC-11*, Sapporo, pp. 225–228.

[66] Golay, S., Meier, J., Dubail, S., Faÿ, S., Kroll, U. and Shah, A., 2000. First pin/pin Micromorph Modules By Laser Patterning. *Proc. 28th IEEE Photovoltaic Specialists Conf.*, Anchorage, pp. 1456–1459.

[67] Meier, J., Vallat-Sauvain, E., Dubail, S., Kroll, U., Dubail, J., Golay, S., Feitknecht, L., Torres, P., Fischer, D. and Shah, A., 1999. Microcrystalline Silicon Thin-Film Solar Cells by the VHF-GD Technique. *Tech. Digest Int. PVSEC-11*, Sapporo, pp. 221–223.

[68] Shah, A., Meier, J., Torres, P., Kroll, U., Fischer, D., Beck, N., Wyrsch, N. and Keppner, H., 1997. Recent Progress On Microcrystalline Solar Cells. *Proc. 26th IEEE Photovoltaic Specialists Conf.*, Anaheim, pp. 569–574.

[69] Rand, J.A. and Basore, P.A., 1991. Light-Trapping Silicon Solar Cells Experimental Results and Analysis. *Proc. 22nd IEEE Photovoltaic Specialists Conf.*, Las Vegas, pp. 192–197.

[70] Rand, J.A., Ford, D.H., Bacon, C., Ingram, A.E., Ruffins, T.R., Hall, R.B. and Barnett, A.M., 1991. Silicon-Film Product II: Initial Light Trapping Results. *Proc. 10th European Photovoltaic Solar Energy Conf.*, Lisbon, pp. 306–309.

[71] Basore, P.A., 1993. Extended Spectral Analysis of Internal Quantum Efficiency. *Proc. 23rd IEEE Photovoltaic Specialists Conf.*, Louisville, pp. 147–152.

[72] Yamamoto, K., Suzuki, T., Yoshimi, M., and Nakajima, A., 1996. Low Temperature Fabrication of Thin Film Polycrystalline Si Solar Cell on the Glass Substrate and Its Application to the a-Si:H/Polycrystalline Si Tandem Solar Cell. *Proc. 25th IEEE Photovoltaic Specialists Conf.*, Washington DC, pp. 661–664.

[73] Rothwarf, A., 1976. Crystallite Size Considerations in Polycrystalline Solar Cells. *Proc. 12th IEEE Photovoltaic Specialists Conf.*, Baton Rouge, pp. 488–495.

[74] Mokashi, A.R., Daud, T. and Kachare, A.H., 1985. Simulation Analysis of a Novel High Efficiency Silicon Solar Cell. *Proc. 18th IEEE Photovoltaic Specialists Conf.*, Las Vegas, pp. 573–577.

[75] Milstein, J.B., Tsuo, Y.S., Hardy, R.W. and Surek, T., 1981. The Influence of Grain Boundaries on Solar Cell Performance. *Proc. 15th IEEE Photovoltaic Specialists Conf.*, Orlando, pp. 1399–1404.

[76] Beaucarne, G., Bourdais, S., Slaoui, A. and Poortmans, J., 2000. Carrier Collection in Fine-Grained p–n Junction Polysilicon Solar Cells. *Proc. 28th IEEE Photovoltaic Specialists Conf.*, Anchorage, pp. 128–133.

[77] Kazmerski, L.L., 1984. Silicon Grain Boundaries: Correlated Chemical and Electro-Optical Characterization. *Proc. 17th IEEE Photovoltaic Specialists Conf.*, Orlando, pp. 379–385.

[78] Barnett, A.M., Rand, J.A., Domian, F.A., Ford, D.H., Kendall, C.L., Rock, M.L. and Hall, R.B., 1988. Efficient Thin Silicon-Film Solar Cells on Low-Cost Substrate. *Proc. 8th European Photovoltaic Energy Conf.*, Florence, pp. 149–155.

[79] Ford, D.H., Rand, J.A., Barnett, A.M., DelleDonne, E.J., Ingram, A.E. and Hall, R.B., 1997. Development of Light-Trapped, Interconnected, Silicon-Film Modules. *Proc. 26th IEEE Photovoltaic Specialists Conf.*, Anaheim, pp. 631–634.

[80] Silier, M., Konuma, A., Gutjahr, E., Bauser, F., Banhart, C., Zizler, V., Schöllkopf, H. and Frey, H., 1996. High-Quality Polycrystalline Silicon Layers Grown on Dissimilar Substrates from Metallic Solution, I. *Proc. 25th IEEE Photovoltaic Specialists Conf.*, Washington, DC, pp. 681–684.

[81] Bergmann, R., Kühnle, J., Werner, J.H., Oelting, S., Albrecht, M., Strunk, H.P., Herz, K. and Powalla, M., 1994. Polycrystalline Silicon for Thin Film Solar Cells. *Proc. 24th IEEE Photovoltaic Specialists Conf.*, Hawaii, pp. 1398–1401.

[82] Andrä, G., Bergmann, J., Ose, E., Schmidt, M., Sinh, N.D. and Falk, F., 2002. Multicrystalline LLC-Silicon Thin Film Cells on Glass. *Proc. 29th IEEE Photovoltaic Specialists Conf.*, New Orleans, pp. 1306–1309.

[83] Zook, J.D., Shuldt, S.B., Maciolek, R.B. and Heaps, J.D., 1978. Growth, Evaluation and Modeling of Silicon-on-Ceramic Solar Cells. *Proc. 13th IEEE Photovoltaic Specialists Conf.*, Washington DC, pp. 472–478.

[84] Heaps, J.D., Schuldt, S.B., Grung, B.L., Zook, J.D. and Butter, C.D., 1980. Continuous Coating of Silicon-on-Ceramic. *Proc. 14th IEEE Photovoltaic Specialists Conf.*, San Diego, pp. 39–48.

[85] Minagawa, S., Saitoh, T., Warabisako, T., Nakamura, N., Itoh, H. and Tokuyama, T., 1976. Fabrication and Characterization of Solar Cells Using Dendritic Silicon Thin Films Grown on Alumina Ceramic. *Proc. 12th IEEE Photovoltaic Specialists Conf.*, Baton Rouge, pp. 77–81.

[86] Barnett, A.M., Fardig, D.A., Hall, R.B., Rand, J.A. and Ford, D.H., 1987. Development of Thin Silicon-Film Solar Cells on Low-Cost Substrates. *Proc. 19th IEEE Photovoltaics Specialists Conf.*, New Orleans, pp. 1266–1270.

[87] van Roosmalen, J.A.M., Tool, C.J.J., Huiberts, R.C., Beenen, R.J.G., Huijsmans, J.P.P., Sinke, W.C., 1996. Ceramic Substrates for Thin-Film Crystalline Silicon Solar Cells. *Proc. 25th IEEE Photovoltaic Specialists Conf.*, Washington DC, pp. 657–660.

[88] Shuldt, S.B., Heaps, J.D., Schmit, F.M., Zook, J.D. and Grung, B.L., 1981. Large Area Silicon-on-Ceramic Substrates for Low Cost Solar Cells. *Proc. 15th IEEE Photovoltaic Specialists Conf.*, Orlando, pp. 934–940.

[89] Slaoui, A., Rusu, M., Fosca, A., Torrecillas, R., Alvarez, E. and Gutjar, A., 2002. Investigation of Barrier Layers on Ceramics for Silicon Thin Film Solar Cells. *Proc. 29th IEEE Photovoltaic Specialists Conf.*, New Orleans, pp. 90–93.

[90] DelleDonne, E., Ingram, A., Jonczyk, R., Yaskoff, J., Sims, P., Rand, J. and Barnett, A., 2002. Thin Silicon-on-Ceramic Solar Cells. *Proc. 29th IEEE Photovoltaic Specialists Conf.*, New Orleans, pp. 82–85.

[91] Barnett, A.M., Mauk, M.G., Zolper, J.C., Hall, R.B. and McNeely, J.B., 1984. Thin-Film Silicon and GaAs Solar Cells on Metal and Glass Substrates. *Tech. Digest Int. PVSEC-1*, Kobe, pp. 241–244.

[92] Barnett, A.M., Mauk, M.G., Hall, R.B., Fardig, D.A. and McNeely, J.B., 1985. Design and Development of Efficient Thin-Film Crystalline Silicon Solar Cells on Steel Substrates. *Proc. 6th European Photovoltaic Solar Energy Conf.*, London, pp. 866–870.

[93] Barnett, A.M., Hall, R.B., Fardig, D.A. and Culik, J.S., 1985. Silicon-Film Solar Cells on Steel Substrates. *Proc. 18th IEEE Photovoltaics Specialists Conf.*, Las Vegas, pp. 1094–1099.

[94] Kunze, T., Hauttmann, S., Seekamp, J. and Müller, J., 1997. Recrystallized and Epitaxially Thickened Poly-Silicon Layers on Graphite Substrates. *Proc. 26th IEEE Photovoltaic Specialists Conf.*, Anaheim, pp. 735–738.

[95] Chu, T.L., Mollenkopf, H.C., Singh, K.N., Chu, S.S. and Wu, I.C., 1975. Polycrystalline Silicon Solar Cells for Terrestrial Applications. *Proc. 11th IEEE Photovoltaic Specialists Conf.*, Scottsdale, pp. 303–305.

[96] Merber, M., Bettini, M. and Gornik, E., 1984. Large Grain Polycrystalline Silicon Films on Graphite for Solar Cell Applications. *Proc. 17th IEEE Photovoltaic Specialists Conf.*, Orlando, pp. 275–280.

[97] Pauli, M., Reindl, T., Krühler, W., Homberg, F. and Müller, J., 1994. A New Fabrication Method for Multicrystalline Silicon Layers on Graphite Substrates Suited for Low-Cost Thin Film Solar Cells. *Proc. 24th IEEE Photovoltaic Specialists Conf.*, Hawaii, pp. 1387–1390

[98] Lin, A.Z., Fan, Z.Q., Sheng, H.Y. and Zhao, X.W., 1982. Thin-Film Polycrystalline Silicon Solar Cell. *Proc. 16th IEEE Photovoltaic Specialists Conf.*, San Diego, pp. 140–145.

[99] Lüdemann, R., Schaefer, S., Schüle, C. and Hebling, C., 1997. Dry Processing of mc-Silicon Thin-Film Solar Cells on Foreign Substrates Leading to 11% Efficiency. *Proc. 26th IEEE Photovoltaic Specialists Conf.*, Anaheim, p. 159

[100] Chu, T.L., Chu, S.S., Duh, K.Y. and Yoo, H.I., 1976. Silicon Solar Cells on Metallurgical Silicon Substrates. *Proc. 12th IEEE Photovoltaic Specialists Conf.*, Baton Rouge, pp. 74–78.

[101] Chu, T.L., Chu, S.S., Stokes, E.D., Lin, C.L. and Abderrassoul, R., 1978. Thin Film Polycrystalline Silicon Solar Cells. *Proc. 13th IEEE Photovoltaic Specialists Conf.*, Washington DC, pp. 1106–1110.

[102] Hötzel, J., Peter, K., Kopecek, R., Fath, P., Bucher, E. and Zahedi, C., 2000. Characterization of LPE Thin Film Silicon on Low Cost Silicon Substrates. *Proc. 28th IEEE Photovoltaic Specialists Conf.*, Anchorage, p. 225.

[103] Chu, T.L., Stokes, E.D., Chu, S.S. and Abderrassoul, R., 1980. Chemical and Structural Defects in Thin Film Polycrystalline Silicon Solar Cells. *Proc. 14th IEEE Photovoltaic Specialists Conf.*, San Diego, pp. 224–227.

[104] Belouet, C., Hervo, C., Mautref, M., Pages, C. and Hervo, J., 1982. Achievement and Properties of Self-Supporting Polysilicon Solar Cells Made From RAD Ribbons. *Proc. 16th IEEE Photovoltaic Specialists Conf.*, San Diego, pp. 80–85.

[105] Heemeier, J., Rostalsky, M., Gromball, F., Linke, N. and Müller, J., 2002. Thin Film Technology for Electron Beam Crystallized Silicon Solar Cells on Low Cost Substrates. *Proc. 29th IEEE Photovoltaic Specialists Conf.*, New Orleans, pp. 1310–1313.

[106] Deguchi, M., Morikawa, H., Itagaki, T., Ishihara, T., Namizaki, H., 1991. Large Grain Thin Film Polycrystalline Silicon Solar Cells Using Zone Melting Recrystallization. *Proc. 22nd IEEE Photovoltaic Specialists Conf.*, Las Vegas, pp. 986–991.

[107] Takami, A., Arimoto, S., Naomoto, H., Hamamoto, S., Ishihara, T., Kumabe, H. and Murotani, T., 1994. Thickness Dependence of Defect Density in Thin Film Polycrystalline Silicon Formed on Insulator By Zone-Melting Recrystallization. *Proc. 24th IEEE Photovoltaic Specialists Conf.*, Hawaii, pp. 1394–1397.

[108] Reindl, T., Krühler, W., Pauli, M. and Müller, J., 1994. Electrical and Structural Properties of the Si/C Interface in Poly-Si Thin Films on Graphite Substrates. *Proc. 24th IEEE Photovoltaic Specialists Conf.*, Hawaii, pp. 1406–1409.

[109] Kawama, Y., Takami, A., Naomoto, H., Hamamoto, S. and Ishihara, T., 1996. In-Situ Control in Zone-Melting Recrystallization Process for Formation of High-Quality Thin Film Polycrystalline Si. *Proc. 25th IEEE Photovoltaic Specialists Conf.*, Washington DC, pp. 481–484.

[110] Tauber, R.N. and Wolf, S., 2000. *Silicon Processing for the VLSI Era, Volume 1 – Process Technology* (2nd Edition), Lattice Press, Sunset Beach.

[111] Kern, W. and Ban, V., 1978. Chemical Vapor Deposition of Inorganic Thin Films. In: Vossen, J.L. and Kern, W., Eds., *Thin Film Processes*, Academic Press, New York, pp. 257–331.

[112] Hammond, M., 1979. Introduction to Chemical Vapor Deposition. *Solid State Technology*, December, p. 61.

[113] Faller, F.R., Henninger, V., Hurrle, A. and Schillinger, N., 1998. Optimization of the CVD Process for Low Cost Crystalline Silicon Thin Film Solar Cells. *Proc. 2nd World Conf. on Photovoltaic Solar Energy Conversion*, Vienna, pp. 1278–1283.

[114] Wagner, B.F. Schetter, Ch., Sulima, O.V. and Bett, A., 1993. 15.9% Efficiency for Si Thin Film Concentrator Solar Cell Grown by LPE. *Proc. 23rd IEEE Photovoltaic Specialists Conf.*, Louisville, pp. 356–359.

[115] Thomas, B., Müller, G., Wilde, P.-M. and Wawra, H., 1997. Properties of Silicon Thin Films Grown by the Temperature Difference Method (TDM). *Proc. 26th IEEE Photovoltaic Specialists Conf.*, Anaheim, pp. 771–774.

[116] Ida, M., Hane, K., Uematsu, T., Saitoh, T. and Hayashi, Y., 1989. A Novel Design for Very-Thin, High Efficiency Silicon Solar Cells with a New Light Trapping Structure. *Tech. Digest PVSEC-4*, Sydney, pp. 827–831.

[117] Hezel, R. and Ziegler, R., 1993. Ultrathin Self-Supporting Crystalline Silicon Solar Cells with Light Trapping. *Proc. 23rd IEEE Photovoltaic Specialists Conf.*, Louisville, pp. 260–264.
[118] Markvart, T., 2000. *Solar Electricity (2nd edition)*, John Wiley & Sons, Chichester.
[119] Schmidt, J., Oberbeck, L., Rinke, T.J., Berge, C. and Bergmann, R.B., 2001. Application of Plasma Silicon Nitride to Crystalline Thin-Film Silicon Solar Cells. *Proc. 17th European Photovoltaic Solar Energy Conf.*, Munich, p. 1351.
[120] Rinke, T.J., Hanna, G., Orgassa, K., Schock, H.W. and Werner, J.H., 2001. Novel Self-Aligning Series-Interconnection Technology for Thin Film Solar Modules. *Proc. 17th European Solar Energy Conf.*, Munich, p. 474.
[121] Weber, K.J., Catchpole, K., Stocks, M. and Blakers, A.W., 1997. Lift-Off of Silicon Epitaxial Layers for Solar Cell Applications. *Proc. 26th IEEE Photovoltaic Specialists Conf.*, Anaheim, p. 474.
[122] Catchpole, K.R., McCann, M.J., Weber, K.J. and Blakers, A.W., 2001. A Review of Thin-Film Crystalline Silicon for Solar Cell Applications. Part 2: Foreign Substrates. *Sol. Energy Mater. Sol. Cells*, Vol. 68, pp. 173–215.
[123] Angermeier, D., Monna, R., Slaoui, A., Muller, J.C,. Tool, C.J., Roosmalen, J.A., Acosta, S. and Ayral, A., 1997. Analysis of Silicon Thin Films on Dissimilar Substrates Deposited By RTCVD For Photovoltaic Application. *Proc. 14th European Photovoltaic Solar Energy Conf.*, Barcelona, p. 1452.
[124] Slaoui, A., Monna, R., Angermeier, D., Bourdias, S. and Muller, J.C., 1997. Polycrystalline Silicon Films on Foreign Substrates By a Rapid Thermal-CVD Technique. *Proc. 26th IEEE Photovoltaic Specialist Conf.*, Anaheim, p. 627.
[125] Bourdais, S., Monna, R., Angermeier, D., Slaoui, A., Rauf, N., Laugier, A., Mazel, F., Jorand, Y. and Fantozzi, G., 1998. Combination of RT-CVD and LPE for Thin Silicon-Film Formation on Alumina Substrates. *Proc. 2nd World Conf. on Photovoltaic Solar Energy Conversion*, Vienna, pp. 1774–1777.
[126] Beaucarne, G., Hebling, C., Scheer, R. and Poortmans, J., 1998. Thin Silicon Solar Cells Based on Re-Crystallized Layers on Insulating Substrates. *Proc. 2nd World Conf. on Photovoltaic Solar Energy Conversion*, Vienna, p. 1794.
[127] Beaucarne, G., Poortmans, J., Caymax, M., Nijs, J. and Mertens, R., 1997. CVD-Growth of Crystalline Si on Amorphous or Microcrystalline Substrates. *Proc. 14th European Photovoltaic Solar Energy Conf.*, Barcelona, p. 1007.
[128] Takahashi, T., Shimokawa, R., Matsumoto, Y., Ishii, K. and Sekigawa, T., 1997. *Sol. Energy Mater. Sol. Cells*, Vol. 48, p. 327.
[129] Shimokawa, R., Ishii, K., Nishikawa, H., Takahashi, T., Hayashi, Y., Saito, I., Nagamine, F. and Igari, S., 1994. *Sol. Energy Mater. Sol. Cells*, Vol. 34, p. 277.
[130] Nell, M.E., Braun, A., von Ehrenwell, B., Schmidt, C. and Elstner, L., 1999. Solar Cells From Thin Silicon Layers on Al_2O_3. *Tech. Digest Int. PVSEC-11*, Sapporo, p. 749–750.

[131] Angermeier, D., Monna, R., Bourdais, S., Slaoui, A., Muller, J.C., Beaucarne, G. and Poortmans, J., 1998. Thin Polysilicon Films on Mullite Substrates For Photovoltaic Cell Application. *Proc. 2nd World Conf. on Photovoltaic Solar Energy Conversion*, Vienna, p. 1778.
[132] Gutjahr, A., Silier, I., Cristiani, G., Konuma, M., Banhart, F., Schöllkopf, V. and Frey, H., 1997. Silicon Solar Cell Structure Grown By Liquid Epitaxy on Glass Carbon. *Proc. 14th European Photovoltaic Solar Energy Conf.*, Barcelona, p. 1460.
[133] Pauli, M., Reindl, T., Krühler, W., Homberg, F. and Müller, J., 1996. *Sol. Energy Mater. Sol. Cells*, Vol. 41/42, p. 119.
[134] Ludemann, R., Schaefer, S., Schule, C. and Hebling, C., 1997. Dry processing of mc-Silicon Thin-Film Solar Cells on Foreign Substrates Leading to 11% Efficiency. *Proc. 26th IEEE Photovoltaic Specialists Conf.*, Anaheim, p. 159.
[135] Mishima, T., Kitagawa, Y., Ito, S. and Yokoyama, T., 1998. Polycrystalline Silicon Films For Solar Cells By Liquid Phase Epitaxy. *Proc. 2nd World Conf. on Photovoltaic Solar Energy Conversion*, Vienna, p. 1724.
[136] Ito, S., Kitagawa, Y., Mishima, T. and Yokoyama, T., 1999. Direct-Grown Polycrystalline Si Film On Carbon Substrate By LPE. *Tech. Digest Int. PVSEC-11*, Sapporo, pp. 539–540.
[137] Monna, R., Angermeier, D., Slaoui, A., Muller, J.C., Beaucarne, G., Poortmans, J. and Hebling, C., 1997. Poly-Si Thin Films on Graphite Substrates By Rapid Thermal Chemical Vapor Deposition For Photovoltaic Application. *Proc. 14th European Photovoltaic Solar Energy Conf.*, Barcelona, p. 1456.
[138] Campe, H.V., Nikl, D., Schmidt, W. and Schomann, F., 1995. Crystalline Silicon Thin Film Solar Cells. *Proc. 13th European Photovoltaic Solar Energy Conf.*, Nice, p. 1489.
[139] Kunze, T., Hauttmann, S., Kramp, S. and Muller, J., 1997. Thin Recrystallized Silicon Seed Layers on Graphite Substrates. *Proc. 14th European Photovoltaic Solar Energy Conf.*, Barcelona, p. 1407.
[140] Schiermeier, S.E., Tool, C.J., van Roosmalen, J.A., Laas, L.J., von Keitz, A. and Sinke, W.C., 1998. LPE-Growth of Crystalline Silicon Layers on Ceramic Substrates. *Proc. 2nd World Conf. on Photovoltaic Solar Energy Conversion*, Vienna, p. 1673.
[141] Shi, A., Young, T.L., Zheng, G.F. and Green, M.A., 1993. *Sol. Energy Mater. Sol. Cells*, Vol. 31, p. 51.
[142] Shi, Z.,Young, T.L. and Green, M.A.., 1994. Solution Growth of Polycrystalline Silicon on Glass at Low Temperatures. *Proc. 1st World Conf. on Photovoltaic Energy Conversion*, Hawaii, p. 1579.
[143] Bergmann, R.B., Brendel, B., Wolf, M., Lölgen, P. and Werner, J.H., 1998. High Rate, Low Temperature Deposition of Crystalline Silicon Film Solar Cells on Glass. *Proc. 2nd World Conf. on Photovoltaic Solar Energy Conversion*, Vienna, pp. 1260–1265.
[144] Brendel, R., Bergmann, R.B., Fischer, B., Krinke, J., Plieninger, R., Rau, U., Reib, J., Strunk, H.P., Wanka, H. and Werner, J.H., 1997. Transport

Analysis For Polycrystalline Silicon Solar Cells on Glass Substrates. *Proc. 26th Photovoltaic Solar Conf.*, Anaheim, p. 635.

[145] DelleDonne, E., Ingram, A., Jonczyk, R., Yaskoff, J., Sims, P., Rand, J. and Barnett, A., 2002. Thin Silicon-on-Ceramic Solar Cells. *Proc. 29th IEEE Photovoltaic Specialists Conf.*, New Orleans, p. 82.

[146] McCann, M.J., Catchpole, K.R., Weber, K.J. and Blakers, A.W., 2001. A Review of Thin-Film Crystalline Silicon for Solar Cell Applications. Part 1: Native Substrates. *Sol. Energy Mater. Sol. Cells*, Vol. 68, pp. 135–171.

[147] Blakers, A.W., Weber, K.J., Stuckings, M.F., Armand, S., Matlakowski, G., Stocks, M.J. and Cuevas, A., 1995. 18% Efficient Thin Silicon Solar Cell By Liquid Phase Epitaxy. *Proc. 13th European Photovoltaic Solar Energy Conf.*, Nice, p. 33.

[148] Zheng, G.F., Wenham, S.R. and Green, M.A., 1996. *Prog. Photovoltaics*, Vol. 4, p. 369.

[149] Faller, F.R., Henninger, V., Hurrle, A. and Schillinger, N., 1998. Optimization of the CVD Process For Low-Cost Crystalline-Silicon Thin-Film Solar Cells. *Proc. 2nd World Conf. on Photovoltaic Solar Energy Conversion*, Vienna, p. 1278.

[150] Werner, J.H., Arch, J.K., Brendel, R., Langguth, G., Konuma, M., Bauser, E., Wagner, G., Steiner, B. and Appel, W., 1994. Crystalline Thin Film Silicon Solar Cells. *Proc. 12th European Photovoltaic Solar Energy Conf.*, Amsterdam, pp. 1823–1826.

[151] Blakers, A.W., Weber, K.J., Stuckings, M.F., Armand, S., Matlakowski, G., Carr, A.J., Stocks, M.J., Cuevas, A. and Brammer, T., 1995. *Prog. Photovoltaics*, Vol. 3, p. 193.

[152] Zheng, G.F., Zhang, W., Shi, Z., Gross, M., Sproul, A.B., Wenham, S.R. and Green, M.A., 1996. *Sol. Energy Mater. Sol. Cells*, Vol. 40, p. 231.

[153] Campe, H.V., Nikl, D., Schmidt, W. and Schomann, F., 1995. Crystalline Silicon Thin Film Solar Cells. *Proc. 13th European Photovoltaic Solar Energy Conf.*, Nice, p. 1489.

[154] Evrard, O., Demesmaeker, E., Vermeulen, T., Zagrebnov, M., Caymax, M., Laureys, W., Poortmans, J., Nijs, J. and Mertens, R., 1995. The Analysis of the Limiting Recombination Mechanisms on High Efficiency Thin Film Cells Grown With CVD Epitaxy. *Proc. 13th European Photovoltaic Solar Energy Conf.*, Nice, p. 440.

[155] Werner, J.H., Kolodinski, S., Rau, U., Arch, J.K. and Bauser, E., 1993. *Appl. Phys. Lett.*, Vol. 62, p. 2998.

[156] Wang, W., Zhao, Y., Xu, Y., Luo, X., Yu, M. and Yu, Y., 1998. The Polycrystalline Silicon Thin Film Solar Cells Deposited on SiO_2 and Si_3N_4 by RTCVD. *Proc. 2nd World Conf. on Photovoltaic Solar Energy Conversion*, Vienna, p. 1740.

[157] Ballhorn, G., Weber, K.J., Armand, S., Stocks, M.J. and Blakers, A.W., 1997. High Efficiency Thin Multicrystalline Silicon Solar Cells By Liquid Phase Epitaxy. *Proc. 14th European Photovoltaic Solar Energy Conf.*, Barcelona, p. 1011.

[158] Vermeulen, T., Poortmans, J., Caymax, M., Nijs, J., Mertens, R. and Vinckier, C., 1997. The Role of Hydrogen Passivation in 20 μm Thin-Film Solar Cells on p^+ Multicrystalline-Si Substrates. *Proc. 14th European Photovoltaic Solar Energy Conf.*, Barcelona, p. 728.

[159] Vermeulen, T., Deurinckx, F., DeClercq, K., Szlufcik, J., Poortmans, J., Laermans, P., Caymax, M., Nijs, J. and Mertens, R., 1997. Cost-Effective Thin Film Solar Cell Processing on Multicrystalline Silicon. *Proc. 26th IEEE Photovoltaic Specialists Conf.*, Anaheim, p. 267.

[160] Vermeulen, T., Evrard, O., Laureys, W., Poortmans, J., Caymax, M., Nijs, J., Mertens, R., Vinckier, C. and Hofs, H.-U., 1995. Realization of Thin Film Solar Cells in Epitaxial Layers Grown on Highly Doped RGS-Ribbons. *Proc. 13th European Photovoltaic Solar Energy Conf.*, Nice, p. 1501.

[161] Faller, F.R., Schillinger, N., Hurrle, A. and Schetter, C., 1997. Improvement and Characterization of Mi-Si Thin-Film Solar Cells on Low Cost SSP Ribbons. *Proc. 14th European Photovoltaic Solar Energy Conf.*, Barcelona, p. 784.

[162] Vermeulen, T., Poortmans, J., Said, K., Evrard, O., Laureys, W., Caymax, M., Nijs, J., Mertens, R. and Vinckier, C., 1996. Interaction between Bulk and Surface Passivation Mechanisms in Thin Film Solar Cells on Defected Silicon Substrates. *Proc. 25th IEEE Photovoltaic Specialists Conf.*, Washington, DC, p. 653.

[163] Thomas, B., Muller, G., Heidborn, P. and Wartra, H., 1997. Growth of Polycrystalline Silicon Thin Films Using the Temperature Difference Method. *Proc. 14th European Photovoltaic Solar Energy Conf.*, Barcelona, p. 1483.

[164] Wang, T.H., Ciszek, T.F., Schwerdtfeger, C.R., Moutinho, H. and Matson, R., 1996. *Sol. Energy Mater. Sol. Cells*, Vol. 41–42, p. 19.

[165] Ciszek, T.F. and Gee, J.M., 1997. Crystalline Silicon R&D at the US National Center for Photovoltaics. *Proc. 14th European Photovoltaic Solar Energy Conf.*, Barcelona, p. 53.

[166] Ishihara, T., Arimoto, S., Kumabe, H., Murotani, T., 1995. *Progr. Photovoltaics*, Vol. 3, p. 105.

[167] Zimmerman, W., Bau, S., Haas, F., Schmidt, K. and Eyer, A., 1998. Silicon Sheets From Powder as Low Cost Substrates For Crystalline Silicon Thin Film Solar Cells. *Proc. 2nd World Conf. on Photovoltaic Solar Energy Conversion*, Vienna, p. 1790.

[168] Beaucarne, G., Hebling, C., Scheer, R. and Poortmans, J., 1998. Thin Silicon Solar Cells Based on Recrystallized Layers on Insulating Substrates. *Proc. 2nd World Conf. on Photovoltaic Solar Energy Conversion*, Vienna, p. 1794.

[169] Hebling, C., Gaffke, R., Lanyi, P., Lautenschlager, H., Schetter, C., Wagner, B. and Lutz, F., 1996. Recrystallized Silicon on SiO_2-Layers for Thin-Film Solar Cells. *Proc. 25th IEEE Photovoltaic Specialists Conf.*, Washington, DC, p. 649.

[170] van Zutphen, A.J., Zeman, M., Tichelaar, F.D. and Metselaar, J.W., 1997. Deposition of Thin Film Silicon By Thermal CVD Processes For

IIb-4

Characterisation and Diagnosis of Silicon Wafers and Devices

Andrés Cuevas, Faculty of Engineering and IT,
Australian National University, Canberra, Australia
Ronald A. Sinton, Sinton Consulting Inc.,
Boulder, Colorado, USA

1 Introduction

Monitoring the fabrication process of a solar cell entails a number of different measurements, including mechanical, optical and electronic properties. This chapter focuses on the latter, the most important of which are described both for the silicon wafers and for solar cell precursors and finished devices. Emphasis is placed on measurements of the minority carrier lifetime and their interpretation to extract relevant information about the properties of the silicon material, the surface passivation and the emitter diffusions. The correlation between the effective lifetime measured at the various stages of fabrication and the final device characteristics is discussed. Direct measurements of the open-circuit voltage under variable illumination conditions are shown to provide most of the parameters needed by the process engineer to assess the ultimate potential efficiency of the solar cells, including diode saturation currents and ideality factors, shunting resistance, intrinsic fill factor and pseudo conversion efficiency (that is, excluding series resistance losses). These Illumination–V_{oc} techniques can be applied before the final cell metallisation, at steps back into the process as far as the junction formation, which makes them very attractive for process control and optimisation. The electrical characterisation of the finished devices is a necessary requirement in solar cell production, and the standard methods to measure the one-sun I–V curves are covered in Chapter IV-1.

2 Measurement of the Bulk Lifetime and Surface Passivation of Silicon Wafers

Once created within a semiconductor, photogenerated electrons and holes last, on average, a finite time called the *lifetime*. When carriers are continuously generated, as in a solar cell, the value of the lifetime determines the stable population of electrons and holes. This population should desirably be as high as possible because it determines the voltage produced by the device. A second, equally important aspect of the lifetime is that it is directly related to the *diffusion length*, which is the average distance that carriers can travel from the point of generation to the point of collection (the p–n junction). The relationship between L and τ is $L_n = \sqrt{D_n \tau_n}$, with a value for the diffusion coefficient of $D_n = 27\ \text{cm}^2/\text{s}$, for electrons (minority carriers) in 1 Ω cm p-type silicon. The diffusion length should be greater than the wafer thickness or the longest generation depth to ensure a high photogenerated current. Since the lifetime determines both the voltage and the current of the device, its characterization is of the utmost importance.

2.1 Lifetime Testing Methods

There are three basic approaches to measure the lifetime, depending on the way that an excess of carriers is created in the semiconductor. In the *transient decay method*, a photogeneration is terminated abruptly and the rate at which carriers disappear, dn/dt, is measured, together with the excess electron concentration,

Δn, itself (note that every photon generates one electron-hole pair, so that $\Delta n = \Delta p$). If no current is flowing from the device, then the rate of carrier density change is equal to the recombination rate:

$$\frac{d(\Delta n)}{dt} = -\frac{\Delta n}{\tau_{eff}} \tag{1}$$

This equation implies an exponential decay of the carrier density with time, which means that 37% of electrons are still present after one lifetime, decreasing to 5% after 3 lifetimes. This method, classically known as *Photoconductance decay*, or *PCD* [1], is quite robust, because it is based on the measurement of the relative change of Δn with time. It is, nevertheless, advisable to measure the absolute value of Δn at which the lifetime has been determined, since the lifetime is, in general, strongly dependent on the carrier injection level.

In the *steady-state method* [2] a constant generation rate of known value is maintained, and the effective lifetime (to be discussed more fully below and in Section 2.4) is determined from the balance between generation and recombination:

$$G = \frac{\Delta n}{\tau_{eff}} \tag{2}$$

This simple expression assumes a uniform generation rate across the thickness of the sample and also a uniform excess carrier density, Δn. Nevertheless, in cases where they may be non-uniform, Equation (2) is still applicable, with Δn representing an average value.

If the illumination varies slowly, quasi-steady-state conditions prevail within the semiconductor, leading to the *quasi-steady-state photoconductance method*, or *QSSPC* [3]:

$$\tau_{eff} = \frac{\Delta n(t)}{G(t) - \frac{\partial \Delta n}{\partial t}} \tag{3}$$

An absolute measurement of the excess carrier density, Δn, is required to determine the lifetime in the *steady-state* and *QSSPC* methods. In addition, the generation rate needs to be determined accurately as well. The latter is measured using a photodetector (for example a calibrated solar cell). The detector gives the total photon flux incident on the surface of the wafer. For the standard solar spectrum the number of photons per second and cm^2 with energy greater than the bandgap of silicon, 1.12 eV is $N_{ph} = 2.7 \times 10^{17}$ cm^{-2} s^{-1}. This flux is commonly referred to as *one sun* intensity and it gives, multiplied by the electronic charge, the upper limit of the current density for a silicon solar cell, 43.25 mA cm^{-2}. Silicon wafers absorb only a fraction of these photons, depending on the reflectivity of the front and back surfaces, possible faceting of those surfaces, and the thickness of the wafer. The value of the absorption fraction for a polished, bare silicon wafer is $f_{abs} \approx 0.6$. If the wafer has an optimised antireflection coating, such as a 70 nm thick silicon nitride or titanium oxide layer, $f_{abs} \approx 0.9$, while a textured wafer with antireflection coating can approach $f_{abs} \approx 1$. The uncertainties associated with the determination of f_{abs} can be kept very small by using tables or graphs that can be calculated using optical models. The generation rate per unit volume G can then be evaluated from the incident photon flux and the wafer thickness:

$$G = \frac{N_{ph} f_{abs}}{W} \tag{4}$$

One of the strengths of the *QSSPC* method is the ease with which the carrier injection level can be scanned to give a full picture of recombination processes within the semiconductor. A steady-state condition can be maintained as long as the time constant for the changing illumination is long compared to the effective carrier lifetime being measured. Using flashlamps and attenuating filters, a range of illumination intensities from 10^{-5} to 1000 suns can be investigated without significant heating of the wafers. Some PCD methods, such as microwave PCD (μ-PCD), use a small signal excitation. For these methods, a scan of carrier injection level can be implemented by taking measurements at several different levels of steady-state bias light.

Note that the lifetime in Equations (1)–(3) has been labelled τ_{eff} to indicate that it is an effective parameter that may encompass several different recombination and transport mechanisms. It is, nevertheless, possible to discriminate between these various mechanisms, as discussed in the following sections. For all three testing methods, the excess electron concentration, Δn, needs to be measured. There are several techniques to do this based on different properties of semiconductor materials. The simplest and most common of them is to measure the conductance of the wafer and the way it changes with illumination and time. The excess photoconductance for a wafer of thickness W is given by

$$\Delta\sigma_{\mathrm{L}} = qW(\mu_n + \mu_p)\Delta n \tag{5}$$

Typical mobility values for 1 Ω cm silicon in low injection are $\mu_n = 1100$ cm^2 V^{-1} s^{-1} for electrons and $\mu_p = 400$ cm^2 V^{-1} s^{-1} for holes. The electron and hole mobilities are functions of the dopant density and injection level.

The most straightforward apparatus to measure the photoconductance is based on applying electrical contacts to the silicon (typically an ingot or a wafer), forcing a flow of current through it and measuring the resulting voltage drop [4]. This can also be done in a contactless fashion by using an inductor that forms part of a radio-frequency circuit; this circuit produces a voltage that is proportional to the conductivity of the wafer, with the added advantage that the relationship between conductance and voltage is practically linear over a broad range [5]. The microwave-detected photoconductance decay systems are based on directing a microwave beam to the silicon wafer and measuring the reflected microwave power, which is proportional to the conductance of the wafer [6, 7]. The excess carrier density can also be directly determined by probing the sample with infrared light and measuring the amount of free-carrier absorption [8, 9], or using an infrared camera to visualize the free-carrier infrared (IR) absorption for the entire wafer with a fine resolution [10]. Schroder [11] gives a comprehensive description of these methods, their theoretical background and practical implementation. Additional methods, also described in [11], such as the surface photovoltage (SPV) and the short-circuit current response, measure the minority carrier diffusion length. Both rely on the formation of a surface space charge region to collect minority carriers. The SPV approach achieves this by chemically treating the surface so that a surface charge is created that forms the space charge region. A transparent contact is used to measure the voltage associated with the collection of carriers by the space charge region. The latter is usually in the mV range to maintain linearity and simplify the analysis. When a real junction exists, as in a finished solar cell, it is preferable to measure the spectral response of the short-circuit current, an analysis of which can give information on the bulk diffusion length, surface recombination velocity and optical light trapping [12]. The formation of a liquid electrolyte-semiconductor junction that allows the extraction of the short-circuit current is also possible, and constitutes the basis for the Elymat technique [11]. Different applications of the above techniques are also described in a collection of papers published by the ASTM [13].

Depending on the specific details of their practical implementation, in particular the magnitude of the excitation used, lifetime testing methods can be broken down into two categories, large-signal and small signal methods. Many commercial systems, such as the microwave-PCD, use a modulated laser beam to create a relatively small number of carriers in the semiconductor. The resulting small signal is then separated from a background illumination provided by a bias light (and background noise) using electronic amplification and lock-in techniques. It should be kept in mind that the result of such measurement is, in general, different from the true recombination lifetime. Converting the small-signal lifetime into the actual lifetime requires measuring the former as a function of injection level, followed by integration [14, 15].

2.2 Variability of the Carrier Lifetime

The lifetime should not be assumed to be a constant, single value. It can vary considerably depending on the process history of the sample (including possible contamination and thermal degradation), measurement conditions (injection level and temperature), and it can degrade or recover when the sample is exposed to light or annealed at certain temperatures. The measured, apparent lifetime can also be affected by non-recombination mechanisms, such as carrier trapping. Figure 1 shows, as an example of variability, the lifetime measured over a broad range of carrier density injection levels for two FZ silicon wafers, one of which was purposely contaminated with iron. The behaviour of the lifetime can be understood in most cases with the assistance of the theoretical models that describe the most important physical mechanisms. It should be kept in mind that the measured lifetime, τ_{eff}, may include several of them simultaneously.

In practical silicon wafers, the most important recombination losses occur through crystallographic defects and impurities that create energy levels within the band gap. The effect of such recombination centres can be described by the Shockley–Read–Hall model, which predicts that the injection-level dependence of the lifetime, τ_{SRH}, is a function of the dopant density, N_A, recombination centre density N_{SRH}, defect energy level E_T and capture cross-sections:

$$\frac{1}{\tau_{SRH}} = \frac{N_A + \Delta n}{\tau_{p0}(n_1 + \Delta n) + \tau_{n0}(N_A + p_1 + \Delta n)} \tag{6}$$

In this expression, which applies to p-Si, $\Delta n = \Delta p$ is the excess carrier density (assuming negligible trapping), and τ_{n0} and τ_{p0} are the fundamental electron and hole lifetimes, which are related to the recombination centre density, the thermal velocity $v_{th} = 1.1 \times 10^7$ cm s^{-1}, and the capture cross-sections via $\tau_{n0} = 1/(v_{th}\sigma_n N_{SRH})$ and $\tau_{p0} = 1/(v_{th}\sigma_p N_{SRH})$. The magnitudes n_1 and p_1 are given by:

$$n_1 = \mathcal{N}_c \exp\left(\frac{E_T - E_C}{k_B T}\right) \qquad p_1 = \mathcal{N}_v \exp\left(\frac{E_C - E_G - E_T}{k_B T}\right) \tag{7}$$

The values for the effective densities of states at the conduction and valence band edges are $\mathcal{N}_c = 2.86 \times 10^{19}$ and $\mathcal{N}_v = 3.10 \times 10^{19}$ cm^{-3}. For recombination centres located near the middle of the energy gap, Equation (6), predicts an increased lifetime between very low and very high injection level from τ_{no} to $\tau_{no} + \tau_{po}$. Nevertheless, every recombination centre is characterised by a distinctive set of parameters E_T, τ_{n0} and τ_{p0}. This leads to different injection and temperature dependences of the lifetime, and can be used to identify the characteristic signature of each centre. The diversity of possible lifetime curves and the determination of E_T, τ_{n0} and τ_{p0} based on the injection level or temperature dependence of the lifetime are reviewed in [16]. Figure 1, which is part an injection-level study for the case of iron–boron pairs in silicon [17], shows a good fit of Equation (6) to the experimental data. The light-soaked results show a widely varying lifetime in the low-injection range of interest for solar cells.

The upper limit to lifetime that may be measured for an otherwise perfect silicon sample is determined by two intrinsic mechanisms, Coulomb-enhanced Auger and band to band recombination, with Auger being the most important of the two. Simple empirical expressions for the intrinsic lifetime for p-type and n-type silicon have recently been proposed [18]:

$$\frac{1}{\tau_{intrinsic}} = (\Delta n + N_A)\left(6 \times 10^{-25} N_A^{0.65} + 3 \times 10^{-27} \Delta n^{0.8} + 9.5 \times 10^{-15}\right) \tag{8}$$

$$\frac{1}{\tau_{intrinsic}} = (\Delta p + N_D)\left(1.8 \times 10^{-24} N_D^{0.65} + 3 \times 10^{-27} \Delta p^{0.8} + 9.5 \times 10^{-15}\right) \tag{9}$$

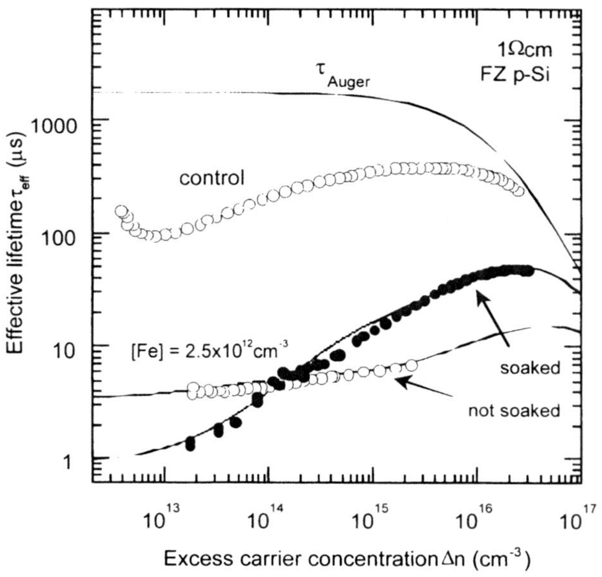

Figure 1 Measured effective carrier lifetime τ_{eff} as a function of the excess carrier concentration Δn for a 1 Ω cm boron-doped FZ silicon wafer contaminated with iron before and after light soaking. An uncontaminated control wafer is also shown. Both had the surfaces passivated with PECVD SiN.

For perspective, Equation (8) is plotted in Figure 1 for a 1 Ω cm boron-doped wafer. The intrinsic (Auger) lifetime is quite high, and will only impact very high efficiency or concentrator solar cells. However, it has a strong dependence on carrier density at high injection and should therefore be taken into account when using injection level dependence techniques to characterize and separate recombination mechanisms, as described in this section and those that follow.

2.3 Lifetime Instabilities

For p-type silicon solar cells, an important type of instability of the lifetime is a degradation induced by light. Two cases of such degradation have been documented extensively and are given here as examples: one is due to the presence of iron, the other to oxygen, both forming complexes or pairs with boron. While iron is avoidable, oxygen is innate to CZ grown silicon (except advanced CZ methods using magnetic confinement of the melt), which typically has an oxygen concentration in the vicinity of 7×10^{17} cm^{-3}. The effect only occurs when both boron and oxygen are present, and is more severe the higher the concentration of both is [19]. The conversion efficiency of 1 Ωcm CZ silicon cells has been reported to degrade by 4% relative after about 50 hours one sun illumination, while for 0.45 Ω cm the degradation is greater (about 7.5%) and for 10 Ω cm it is lower (1% or less). The lifetime of typical 1 Ω cm boron doped silicon degrades by more than a factor of ten after 5 hours exposure to one sun illumination, approaching a final lifetime of 10–20 μs [20].

Similar levels of efficiency degradation of 3–4% relative (about 0.5% absolute) after 30 minutes one-sun illumination have been attributed to the presence of iron [21]. In effect, iron is a well known lifetime killer and the change in lifetime upon exposure is an established method to determine its concentration in silicon. There are significant differences between Fe and Fe–B pairs and B–O complexes. (1) In the case of Fe, an anneal at 210°C degrades the lifetime (in much the same way as illumination) [22]. In the case of B–O complexes, the lifetime recovers after annealing at moderate temperatures. (2) The shape of the injection level dependent lifetime curve changes differently: the curves before and after degradation cross over in the case of Fe and Fe–B (see Figure 1), while they remain approximately parallel in the case of B–O. (3) An additional differentiating feature is that the lifetime of Fe contaminated silicon recovers after long-term storage in the dark.

2.4 Surface Component of the Effective Lifetime

The ability to separate out the different recombination mechanisms and identify where the major losses occur within the solar cell device is an essential part of the characterisation and diagnosis process. It is useful to express mathematically that the *effective lifetime*, τ_{eff}, is the net result of summing up all the recombination losses that occur within the different regions that constitute a given silicon wafer or a solar cell. The surfaces frequently have a significant impact on the measured effective lifetime. This is represented by the *surface recombination velocities*, S_{front} and S_{back}, at the front and back sides of the wafer as:

$$\frac{1}{\tau_{\text{eff}}} - \frac{1}{\tau_{\text{intrinsic}}} = \frac{1}{\tau_{\text{SRH}}} + \frac{S_{front} + S_{back}}{W} \tag{10}$$

where W, as in Section 2.3, is the thickness of the wafer. Surface recombination is commonly attributed to Shockley–Read–Hall processes and can, therefore be expected to vary with the carrier injection level, in much the same way as the bulk lifetime [23]. In addition, when a dielectric layer is used to passivate the surface, a space charge region may form. This effect has been documented for silicon dioxide [24] and SiN [25]. Since similar injection level dependencies can be due to either surface or bulk recombination, their separation is not trivial. The preferred method, applicable to both steady-state and transient methods, involves the preparation of several wafers with different thicknesses and identical bulk and surface properties [11]. Alternatively, by using wafers with a very high bulk lifetime, an upper bound for S can be determined by assuming $\tau_{SRH} = \infty$ in Equation (10).

Equation (10) is a simplified expression, and would predict a zero effective lifetime when the surface recombination velocity is very high. In reality there is a limit on how low the effective lifetime can be because electrons and holes have to travel towards the surfaces by diffusion in order to recombine, which is a relatively slow mechanism. It can be calculated that, for uniform steady-state photogeneration (produced by infrared light) or for a transient decay measurement, the minimum effective lifetime is, respectively, given by the following expressions:

$$\tau_{eff\,(S=\infty)QSSPC} = \tau_n \left(1 - \frac{2L_n}{W}\tanh(\frac{W}{2L_n})\right) \qquad \tau_{eff(S=\infty)PCD} = \frac{W^2}{\pi^2 D_n} \tag{11}$$

For a typical p-type 0.03cm thick wafer and an electron diffusion coefficient $D_n = 27\text{cm}^2\text{s}^{-1}$, the lifetime that can be expected for a transient decay measurement is $\tau_{eff} = 3.4\mu s$. The steady-state surface-limited lifetime is a function of the bulk lifetime τ_n and diffusion length L_n, and also of the wavelength spectrum of the light source. In the limit $L_n >> W$, Equation (11) simplifies to $\tau_{eff(QSSPC)} = W^2/(12D_n)$.

The dependence of the measured effective lifetime on the surface recombination velocity is shown in Figure 2 for the example of a 300 μm thick, 1 Ω cm p-type wafer and four different values of the bulk lifetime. The graph indicates the lifetime that would be measured for this wafer with either 400 nm or 1000 nm light using a steady-state photoconductance method. The curves were calculated for minority-carrier densities in low injection conditions using PC1D computer simulation [26]. Transient methods, such as μ-PCD, will determine the lifetime approximately as shown by the 1000 nm curves, except that in the limit of high surface recombination the difference between the measured effective lifetime and the actual bulk lifetime is reduced by 22%, as follows from Equation (11). Note that the diffusion limitation on the measurable lifetime varies quadratically with the wafer thickness. Several main features can be identified on this plot.

1 The measured effective lifetime is equivalent to the actual bulk lifetime if the surfaces are well passivated, independent of the wavelength of illumination (lower part of Figure 2). The demands on surface passivation can, nevertheless, be relaxed when determining relatively low bulk lifetimes. For example, for 0.1 μs bulk lifetime surface recombination velocities up to 1000 cm/s can be tolerated. On the other hand, to accurately determining high bulk lifetimes requires very good passivation. For example, the surface recombination velocity should be less than 10 cm/s to determine $\tau_{bulk} > 100\ \mu s$.

2. If the surface recombination velocity is greater than 10^5 cm/s, typical of unpassivated silicon wafers, infrared light (for example 1000 nm) measurements are more indicative of the bulk lifetime. The sensitivity to τ_{bulk} is excellent if the latter is lower than 2 μs, and a one-to-one correspondence can be made between the IR measured effective lifetime and the actual bulk lifetime. The sensitivity is still reasonable to resolve bulk lifetimes up to 10 μs, but is poor for higher lifetimes. Importantly, however, a pass-fail test for τ_{bulk} up to 10 μs can be established for bare, unpassivated wafers by using infrared light. For example, a measured lifetime of 2 μs indicates an actual bulk lifetime greater than 10 μs.

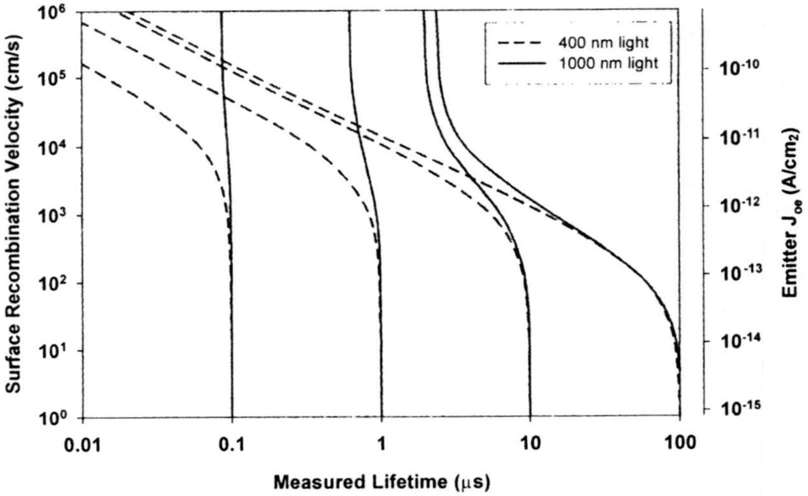

Figure 2 Effective lifetime for a 300 μm, 1 Ω cm p-type wafer as a function of the surface recombination velocity. The curves, modelled with PC1D for steady-state conditions, correspond to wafers with 0.1, 1, 10 and 100 μs bulk lifetimes. Two sets of curves are shown, for 400 nm and 1000 nm, monochromatic illumination, respectively.

3. When the surface recombination velocity is greater than 1000 cm/s (upper portion of Figure 2), the effective lifetime is significantly different for blue and IR light, and this allows the bulk lifetime and the surface recombination velocity to be uniquely determined by measuring with both wavelengths [27]. For example, from Figure 2, a wafer that measures 0.65 μs with 1000 nm light and 0.05 μs with 400 nm light has a bulk lifetime of 1 μs and a surface recombination velocity of 1×10^5 cm/s. This method exploits the fact that the photogeneration from blue light is very sensitive to the front surface recombination velocity, while the IR light penetrates deep into the bulk of the wafer. Note that the higher lifetime range (above 10 μs) cannot be discriminated even using this method, because the relatively high surface recombination velocity (SRV) above 1000 cm/s completely masks bulk recombination.

2.5 Emitter Component of the Effective Lifetime

Dopant-diffused, or *emitter* regions are commonly characterised through a *saturation current density*, J_{oe}. This parameter encompasses recombination within the bulk of the thin diffused region, which normally occurs through the Auger process, as well as the recombination at the heavily doped surface. If both sides of a p-type wafer are diffused, then the effective lifetime can be expressed as:

$$\frac{1}{\tau_{\text{eff}}} - \frac{1}{\tau_{\text{intrinsic}}} = \frac{1}{\tau_{\text{SRH}}} + \left[J_{\text{oe}(front)} + J_{\text{oe}(back)}\right] \frac{(N_A + \Delta n)}{q n_i^2 W} \qquad (12)$$

In this expression, N_A is the dopant density of the wafer and $J_{oe(front)}$ and $J_{oe(back)}$ are the *saturation current densities* that characterise the front and back emitter regions, respectively. At 25°C, $qn_i^2 = 12$ C cm^{-6}. The fact that the emitter recombination term has a different dependence on the carrier injection level than the bulk and surface recombination terms allows determination of the J_{oe} by examining the injection level dependence of the lifetime, particularly in the high injection regime.

It is straightforward to consider the case when one of the surfaces is diffused (hence characterised by an emitter saturation current) and the other is not (hence characterised by a surface recombination velocity) by combining Equations (10) and (12). In fact, both concepts are equivalent in low injection, $S_{eff} \approx J_o N_A / q n_i^2$. The scale on the right of Figure 2 shows the emitter-saturation current densities that correspond to the surface recombination velocities on the left axis, for a 1 Ω cm wafer. Equations (10) and (12) assume an approximately uniform carrier density across the wafer. The region of applicability of this assumption is visually displayed in Figure 2 as the region where the 400 nm curves and the 1000 nm curves overlap.

Figure 3 shows a family of curves for different values of emitter saturation current density and a bulk doping of 1.5×10^{16} cm^{-3} (1 Ω cm). The data is plotted as inverse measured lifetime, following the form of Equation (12). This equation indicates that for measurements performed at minority carrier densities above the doping density, the emitter saturation current density will be given by the slope of the line for each curve in Figure 3. The variations in lifetime due to the SRH recombination in the $1/\tau_{bulk}$ term occur primarily at carrier densities

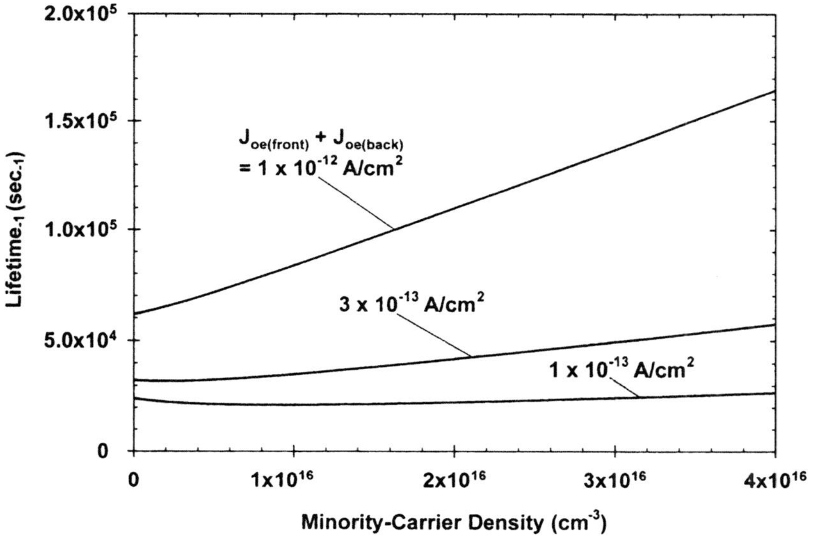

Figure 3 The inverse of the measured lifetime for a 1 Ω cm p-type wafer with three different emitters. Calculation from Equation (12) assuming a low injection lifetime of 50 μs and subtracting out the Auger contribution to the recombination.

less than the doping level, so that they do not significantly affect the slope of the data above the doping level.

There is a limited range for which the use of Equation (12) for extracting J_{oe} is valid. For very high-injection levels, the near-surface recombination due to the doped emitter becomes very high, and the effective lifetime decreases correspondingly. The limit imposed by the finite diffusivity of carriers towards the surface discussed above in Equation (11) applies also to the case of very high J_{oe} (in high injection conditions the ambipolar diffusivity, $D_A \approx 18$ cm^2 s^{-1}, substitutes the minority carrier one, D_n). Because of this, the highest J_{oe} that can be discriminated is, assuming equal front and back diffusions:

$$J_{\mathrm{oe}} \ll \frac{6qn_i^2 D_A}{W[N_A + \Delta n]} \approx \frac{1300}{W[N_A + \Delta n]} \tag{13}$$

This indicates that lower substrate doping, thinner wafers, or both can be used to extend the range of J_{oe} values that can be determined and optimise the range of data that can be used. The thinner and more lightly doped the wafer, the wider the range of minority-carrier densities that can be fit to Equation (12).

For optimal measurements of silicon wafers after emitter diffusion, it is advisable to use wavelengths of light longer than 700 nm in order to have a relatively uniform generation within the wafer and minimize the fraction of light absorbed within the emitter. For a detailed discussion of the qualifications for this data analysis for various cell geometries, lifetimes, and J_{oe} values, see references [28, 29]. For wafers with high recombination on one or both surfaces, a measurement strategy can be devised using front illumination, back illumination, IR and blue light to determine the values for recombination at the surfaces and in the bulk [27, 30].

In addition to this analysis for lifetime and emitter saturation current density, an analysis comparing measurement with red and blue monochromatic light can be used in order to specifically characterize the internal quantum efficiency of the heavily-doped emitter region [31].

2.6 Carrier Trapping Effects

Some materials, in particular multicrystalline silicon, show abnormally high photoconductance and apparent lifetime at very low carrier densities. The phenomenon affects both transient-decay and steady-state photoconductance measurements. The conventional analysis of the data results in an apparent *increase* of the effective lifetime τ_{eff} as the carrier injection level *decreases*. The physical reason for this behaviour is, in most cases, the trapping of minority carriers (electrons in p-type material). The semiconductor under illumination has a certain number of excess free electrons Δn and of trapped electrons n_t. Charge neutrality requires that the number of excess holes equals the sum of trapped and free electrons, $\Delta p = \Delta n + n_t$. This results in an anomalously high excess photoconductance, with an extra term due to trapping compared to Equation (5):

$$\sigma_L = q(\mu_n + \mu_p)\Delta n + q\mu_p n_t \tag{14}$$

Therefore, the photoconductance is a poor indicator of the recombination of electron-hole pairs except at high light intensities, when $\Delta n \gg n_t$. This does not affect other lifetime testing methods such as those based on device voltage or current measurements. If the effect of trapping on the photoconductance is properly accounted for, it is frequently possible to accurately determine the lifetime from a photoconductance measurement at all but the lowest minority-carrier densities.

Figure 4 [32] shows data from multicrystalline silicon where the transition from the SRH lifetime, in a range 1–60 μs, shifts up by an order of magnitude or more at the lowest carrier densities. This high apparent lifetime that is approached at low carrier density corresponds to the lifetime for trapped electrons before being released back into the conduction band. This trapping effect has been studied with application of a more detailed model than Equation (14), the Hornbeck–Haynes model [33], to multicrystalline and contaminated FZ silicon [34, 35]. The trapping levels and lifetimes can be separated from the SRH effects in many cases as long as data is taken at very low photoconductance values, to establish the the behavior of the second term in Equation (14). Then the data at moderate carrier densities can be corrected, enabling accurate characterization of the SRH levels. Simple corrections using

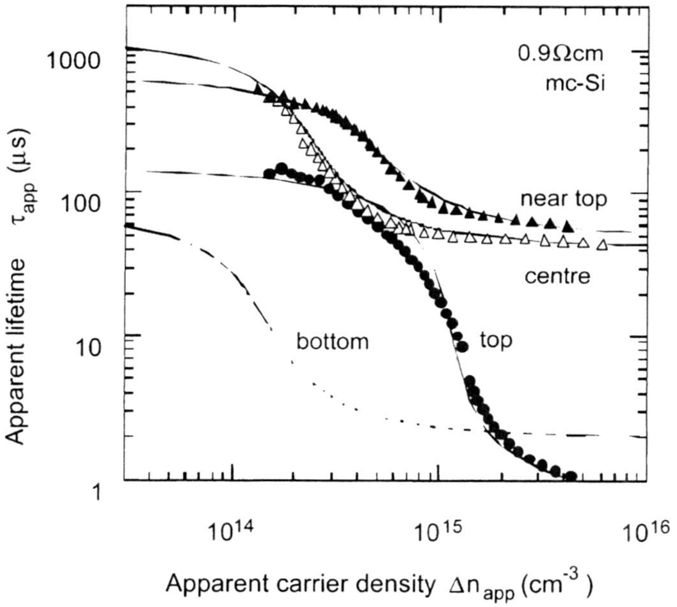

Figure 4 Multicrystalline silicon wafers from different regions of a cast ingot showing trapping effects.

bias light have been compared to the complete model and have been found to be adequate for many practical situations [36]. For some materials, particularly multicrystalline silicon, the trapping can obscure photoconductance data in the carrier-density range corresponding to the operating point for solar cells, generally between 1×10^{13} and 1×10^{15} cm^{-3}. In such cases the voltage characterisation techniques, described in the following section, can give less uncertainty in the lifetime determination.

3 Relationship Between Device Voltage and Carrier Lifetime

The output voltage of a solar cell is determined by the excess minority carrier density at the junction edge of the quasi-neutral base region, defined to be at the plane $x = 0$ in the following expression:

$$\Delta n(0)[N_A + \Delta n(0)] = n_i^2 \exp\left(\frac{V}{k_B T/q}\right) \tag{16}$$

In open-circuit conditions, this minority carrier density is the result of a balance between photogeneration and recombination. The conditions are identical to those of the steady-state method to measure the lifetime, described in Section 2. It is therefore possible to obtain Δn from a photoconductance measurement and determine an implicit, or expected voltage. Complete I–V characteristic curves can be obtained by plotting the implicit voltage as a function of the illumination, expressed in unit of *suns* (1 sun = 1 kW/m^2).

Direct measurements of the voltage are possible as soon as a junction exists in the wafer. By taking data at a range of illumination intensities, *Illumination–V_{oc}* curves can be constructed and analysed to obtain diode saturation currents, ideality factors, shunts and ideal efficiencies [37]. Applications of this are shown in Section 4. A generalised expression to obtain the I–V curves from voltage measurements under quasi-steady-state illumination or during a transient open-circuit decay (OCVD) has also been developed [38]. This formulation extends the validity for accurate measurement to the regime when the light varies at a rate comparable to the minority-carrier lifetime or is extinguished entirely (as in OCVD measurements).

In addition, solving for Δn as a function of voltage and substituting this value in Equations (1), (2) or (3) converts any voltage measurement into a lifetime measurement. Note that these equations for the lifetime use an average carrier density, which in some cases can be substantially lower than the value of $\Delta n(0)$ obtained from the voltage through Equation (16) [30]. Trapping effects have a negligible effect on the measured device voltage, and *Illumination–V_{oc}* data can be used to determine lifetimes at minority carrier densities less than 10^9 cm^{-3}. Note that temperature control, knowledge of the intrinsic carrier density n_i, and the dopant density are necessary to obtain the lifetime from a voltage measurement, and vice-versa. In contrast, photoconductance lifetime measurements require knowledge of carrier mobilities and are relatively insensitive to temperature.

4 Applications to Process Monitoring and Control of Silicon Solar Cells

Although in-process monitoring of the electronic properties of the silicon wafer using minority-carrier measurements has been extensively used for high-efficiency solar cell optimisation at research laboratories, this has not been true in typical large-scale production of terrestrial cells. Several possible industrial process-control applications are detailed here for single- and multi-crystalline solar cells and illustrated with examples. The main electronic properties that can be measured during the fabrication process of typical commercial silicon solar cells include the resistance, carrier lifetime and device voltage. It is possible to measure the resistance of the substrate, doped layers, and metallisation at each stage in the process. The effective lifetime of minority carriers in the silicon can also be monitored at each stage using photoconductance measurements (these may require metal etching after the metallisation is applied to the cell). Finally, after phosphorus diffusion, measurements can be made of the junction voltage, which is a clear indicator of device performance.

4.1 Resistance Measurements

The resistivity of the silicon bulk material can be measured at the ingot or wafer levels using contacting or non-contacting arrangements [11]. Knowledge of the resistivity and, subsequently, the net doping density, is very important in the optimisation of the fabrication process and the interpretation of solar cell performance, since both the voltage and the current depend on it. The lifetime and diffusion length are also correlated to the resistivity of the substrate, with lower resistivities usually being accompanied by lower lifetimes. A higher resistivity, on the other hand, makes the device more dependent on the quality of the surfaces.

The sheet resistance of the emitter (usually phosphorus) diffusion is another essential control parameter. It is correlated to the dopant density profile, which determines the emitter saturation current density, the emitter quantum efficiency, the contact resistance properties, and the tolerance to shunt formation during firing. The sheet resistance can be measured by the 4-point-probe method [11], or by contactless inductive coupling [5] if the wafer bulk resistivity is subtracted. Resistance measurements can also be used to monitor the thickness of metal deposited at the rear of the wafers (an alternative is to measure the weight differential) and the resulting alloying of the aluminium to form a highly doped p-type back-surface field (BSF) layer.

Measuring the contact resistance between the metal and the semiconductor usually requires separate experiments and special test structures. The recently developed Corescan instrument [39] provides detailed information about the contact resistance and the emitter sheet resistance on the finished solar cell. The technique, which is destructive, is based on scanning a probe directly on the silicon and the metal fingers to map the voltage drop vs. position across the illuminated solar cell. Such voltage maps can be used to optimise the front grid metallisation and diagnose metal contact problems [40].

The global series resistance of the device can be measured by a number of methods. It is important to emphasise that not all the methods provide the relevant value of series resistance. Methods based on measurements of the device under illumination are more realistic than methods in the dark [41]. The most robust method is based on a comparison between the open-circuit I–V curve and the photovoltaic I–V curve, as illustrated in Section 4.3.

4.2 Minority-Carrier Lifetime Measurements

In R&D laboratories, most of the fabrication steps are characterized in detail using lifetime measurements. Frequently, these detailed studies are done under ideal conditions to achieve unambiguous results. For example, studies of wafers vs. position in the original cast block have often tested the bulk lifetime after each process step by etching off the surface and subsequently applying a high-quality surface passivation, in order to determine the effect of each step on bulk lifetime without confounding effects of the surface [42, 43]. Alternatively, the surface can be etched back and passivated using a liquid, such as hydrofluoric acid [44] or iodine in ethanol [43, 45]. Satisfactory low-temperature passivation has also been obtained with corona-charged photoresist [46] and polymer films [47]. Many studies have also been performed on wafers with the surfaces as they exist after each step in the process [48, 49]. Such measurements do not require special wafer preparation and are ideal for industrial process control in production lines. Especially on multicrystalline material, high-resolution mapping of the wafer has been used to map the response of different grains and grain boundaries to gettering and hydrogenation [50]. These spatially resolved methods include microwave-PCD, modulated free carrier absorption, surface photovoltage, IR carrier density imaging, microwave phase-shift techniques and light beam induced current (LBIC) [51].

4.2.1 Measurements of Lifetime in a Boule or Block

At the level of the block or boule, the lifetime is often mapped to determine the region of potentially good wafers. The top and tail of a boule have low lifetimes and are discarded or recycled. Cast ingots have low lifetime at the top and bottom of the blocks, as well as near the edges of the crucible. Detailed two-dimensional high-resolution μ-PCD measurements are often used to characterise these blocks or boules using commercially available instruments. The initial lifetimes for as-grown material are, however, not perfectly correlated with final solar cell efficiency. It has been shown, for example, that some regions of cast blocks that might be discarded based on initial lifetime measurements recover lifetime during the high-temperature steps in cell processing due to gettering effects [49, 52]. Many of these low lifetime regions can also result in good solar cells after hydrogen bulk passivation procedures. Despite these complexities, a better understanding of the initial silicon lifetime and the effects of the process on the various regions of boules and blocks is expected to result in better strategies for optimising the silicon growth and choosing the regions of silicon to submit to the expensive sawing, wafer cleaning, and cell fabrication processes. Rejecting material at this early stage has great value. Tight tolerances for incoming wafers at the beginning of cell fabrication will also permit better optimisation and control of the fabrication process.

4.2.2 Lifetime Measurements on Bare Wafers

In principle, measurements on bare wafers should not be necessary if the boule or block has been measured prior to sawing. In practice, a measurement at the wafer level can indicate if a sufficient thickness of silicon was removed at the saw-damage etch step. In addition, many silicon solar cell manufacturers do not fabricate the wafers, but purchase them, and it is of great interest for both the vendor and the manufacturer to have the capability to measure the minority-carrier lifetime of the individual bare wafers. This is not simple, since the measured effective lifetime of a wafer without surface passivation can be very low and poorly correlated to the bulk lifetime of the material.

For typical commercial solar cells, the efficiency is relatively constant for bulk lifetimes greater than 5 μs, corresponding to diffusion lengths greater than 116 μm. Therefore, the main point of a lifetime measurement at the stage of bare wafers is to determine if the wafer has a minimum required lifetime of 2–5 μs. Wafers with greater lifetime than this clearly 'pass' and wafers with lower lifetime 'fail'. This type of pass/fail test is possible even in the absence of surface passivation by using infrared light, as discussed in Section 2.4.

As seen in Figure 2, measurements on bare wafers will generally result in effective lifetimes in the range 0.1–2 μs. The low levels of photoconductance corresponding to these lifetimes makes it critically important to determine the lifetime with a data analysis that removes the artifacts that come from trapping, as discussed in Section 2.6. Due to the dependence of the lifetime on injection level discussed in Section 2.2, all measurements of lifetime should be performed at the same minority-carrier injection level. This injection level should preferably be chosen to be relevant to the maximum power operating point of the finished solar cell.

4.2.3 Effective Lifetime After Emitter Diffusion

After the emitter diffusion is an ideal point to measure the lifetime, since its value at this stage is often very predictive of the final solar cell efficiency. The front phosphorus diffusion and oxidation acts as a surface passivation. In normal operating conditions of low injection, a diffused region is seen from the base of the cell as a surface recombination velocity, whose value is determined by the emitter-saturation current density J_{oe} of the diffused region (see Section 2.5). This correspondence between SRV and J_{oe} is shown in Figure 2 for a 1 Ω cm p-type wafer. It can be noted that J_{oe} values below 10^{-12} A cm^{-2} provide a reasonable passivation (SRV < 1000 cm/s). Yet such J_{oe} = 10^{-12} A cm^{-2}

would limit the measurable effective lifetime to approximately 12 μs in 1 Ω cm material, even if the bulk lifetime may be much higher. This surface-like role of the diffused regions applies to both aluminium alloyed BSF regions and to open-circuited phosphorus (or boron) diffused regions. Representative values for industrial phosphorus (n^+ region) and aluminium (p^+ region) diffusions are 8×10^{-13} A cm^{-2} and 5×10^{-13} A cm^{-2}, respectively. It should be noted that the measured lifetime limit imposed by a given value of J_{oe} is less restrictive for higher resistivity wafers.

The techniques discussed in Section 2.5 can be applied to determine both the bulk lifetime and the emitter saturation current density for the phosphorus diffusion. It is frequently advisable to use special test wafers with a lower substrate doping to optimise the range of carrier injection data that can be used for the analysis and extend the range of J_{oe} values that can be determined. Once J_{oe} is determined, its contribution to the low injection range can be subtracted. Therefore data taken in two minority-carrier density injection ranges allows both the bulk lifetime in the relevant range of cell operation and the emitter saturation current density to be uniquely determined. The exact analysis will depend on the technology used for the emitter diffusion, since this step may result on a diffused region at the front only, or at both surfaces. If emitter surface passivation is done in a subsequent, separate step, the same measurement methodology discussed above can be used to determine the bulk lifetime and the emitter saturation current density that results from that process step. This provides an ideal way to characterize the effectiveness of emitter passivation by, for example, oxidation or silicon nitride deposition.

4.3 Voltage Measurements for Characterisation of Process Steps After Junction Formation

Although measurements of device voltage have traditionally been reserved until the completion of its fabrication, an earlier voltage measurement is possible. This offers a powerful monitoring and process control tool complementary to the lifetime techniques described above. The most common, and frequently sole, characterisation used in industry consists of the final I–V curve testing of the solar cell under simulated one sun illumination. Whereas undeniably important and necessary, this test has limited usefulness as a diagnostic and control tool. Research laboratories have used I_{sc}–V_{oc} measurements to gain further insight into the device. Very early in the history of solar cell development, it was realised that the I_{sc}–V_{oc} curve contained information about the fundamental diode characteristic free from series resistance effects [53]. Comparison of I_{sc}–V_{oc} data with the final I–V curve of the solar cell can then determine the series resistance very precisely. In a recent variation of the I_{sc}–V_{oc} method, the *Quasi-Steady-State open-circuit voltage*, or *QSSV*$_{oc}$ [54], the device is kept in open-circuit at all times, but its short-circuit current is not actually measured at every light intensity. Instead, the incident light intensity is measured with a calibrated reference solar cell. This illumination intensity can be converted to a measure of current by using as a scaling factor, either the short-circuit current of the cell (if already known) or the modelled photogeneration in the test sample. It is often overlooked that such *Illumination–*V_{oc} data can be obtained and analysed as soon as a junction is formed in the solar cell fabrication. The *Illumination–*V_{oc} method only requires that the contacts to the p^+ and n^+ regions of the cell be better than the input impedance of the instrument used to measure the voltage. This very simple requirement means that, in many cases, the characteristic curve can be taken at any point in the process after the emitter diffusion by simply probing the appropriate areas of the silicon, giving valuable information in the middle stages of processing the wafers. Some care must be taken to ensure that the probed contacts are ohmic, rather than rectifying. The resistance between two probe contacts on the p-type region can be checked to verify that sufficiently good contact is made. Similarly, the n-type contacts can also be checked. This ensures that the subsequently measured voltage will be indicative of the junction voltage.

In the $QSSV_{oc}$ technique, the light intensity can be swept with a 1/e time constant sufficiently long (for example 4 ms), such that the solar cell voltage is in quasi-steady-state with the light intensity for typical effective carrier lifetimes encountered in most silicon solar cells (usually well below 4 ms after the device is completed). The short overall duration of the illumination (typically 12 ms), means that the sample does not heat significantly even at high light intensities, which allows data acquisition over a wide range of illumination intensities. The main methods of analysing these data are described below.

4.3.1 Standard Diode Analysis of Illumination–V_{oc} *Curves*

An example of *Illumination–*V_{oc} data is shown in Figure 5 in the form of the standard semi-logarithmic diode characteristic curve [32]. In this case *Illumination–*V_{oc} data was taken after cell completion, and the measured short-circuit current was used to convert the illumination scale from suns to current density. A classical interpretation of these characteristics leads to the determination of saturation current densities and ideality factors. The data in Figure 5 can be fit with a double exponential diode model using $J_{o1} = 4 \times 10^{-13}$ A cm^{-2} ($n = 1$ component), $J_{o2} = 3.7 \times 10^{-8}$ A cm^{-2} ($n = 2$ component), and a shunt resistance, $R_{shunt} = 2.4 \times 10^3\ \Omega$. It is important to realise that the saturation current densities obtained from this analysis are global parameters, encompassing recombination within the emitter and base regions, as well as at the surfaces and within any space charge regions. It is usually the case that J_{o1} is higher than the emitter saturation current J_{oe} discussed in preceding sections. Another important observation is that, whereas a satisfactory fit to the data is usually possible using the double exponential model, the transition from ideality 1 to ideality 2 (or greater) factors is often due to the variability of the bulk lifetime or the surface recombination velocity [55]. Process control using the analysis of Figure 5 should optimise the voltage at 0.05 to 0.1 suns, corresponding to the maximum power operating voltage of the solar cell.

Note that voltage can always be translated into lifetime, as described in Section 3. In some cases, a voltage measurement prior to metallisation is valuable and warranted, for example to check the p–n junction properties before

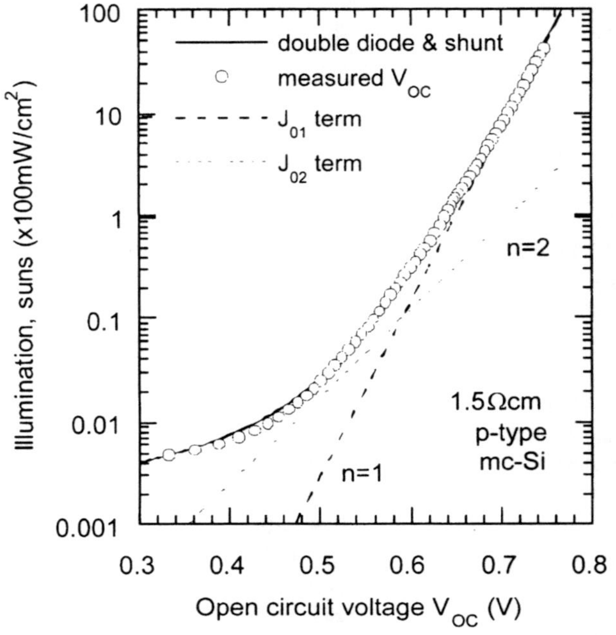

Figure 5 Illumination–V_{oc} *data plotted in a semi-logarithmic scale, with a double exponential diode curve fit. The experimental data was taken using the QSS*V_{oc} *technique for a 1.5 Ω cm multicrystalline silicon solar cell.*

metal sintering. Often though, the contactless lifetime measurements are easier and preferable at the stage of the fabrication process immediately after emitter diffusion. However, once the aluminium BSF is formed, voltage measurements become the preferred diagnostic technique, since the p-type contact can now be easily probed, while the alloyed Al layer with segregated metal at the surface presents a high conductance and must be etched before good photoconductance measurements can be made.

4.3.2 Voltage Monitoring of Contact Formation

Although not immediately obvious, data taken under open-circuit conditions can be valuable for monitoring the properties of the solar cell contacts. Metal contacts can be modelled as a Schottky potential barrier, which is formed by most metals on silicon, in parallel with some form of leakage current. The leakage might be from the metal locally doping or spiking the silicon or from thermally assisted tunnelling through the potential barrier. Under one-sun conditions, for a well-formed contact, the Schottky barrier is effectively shorted by the leakage. However, at a sufficiently high light intensity on a poorly formed contact, the Schottky diode will build up a voltage opposing the solar cell junction voltage by generating a current that the leakage is unable to fully shunt. An aluminium metallisation that has not sufficiently fired through a phosphorus diffusion on the back of the solar cell will also produce an opposing voltage at high illumination intensities. Note that Schottky-type contacts usually result in low fill factors and efficiencies. This poor contact effect is indicated as an ideality factor less than unity, or even a voltage that decreases at high illumination intensity. By monitoring the open-circuit voltage at light intensities significantly higher than the operating conditions, this effect can be used to anticipate problems in the contact formation. Departures from ideal behaviour can often be seen before they result in a yield loss due to low efficiency. This method is both a good diagnostic and a process-control technique.

4.3.3 Photovoltaic I–V curves from Illumination–V_{oc} data

An alternative presentation of *Illumination–V_{oc}* (or I_{sc}–V_{oc}) data is shown in Figure 6. By using the superposition principle, an implied photovoltaic I–V curve can be constructed from the open-circuit voltage measurements of Figure 5. At each open-circuit voltage, the implied terminal current is given by:

$$J_{terminal} = J_{sc}(1 - \text{suns}) \tag{17}$$

This yields the familiar photovoltaic I–V curve format, permitting the customary interpretation of fill factor, efficiency and shunt. The parameters that matter most to solar cell performance are now visually obvious. For example, it can immediately be seen if the shunt is having a major effect on the maximum

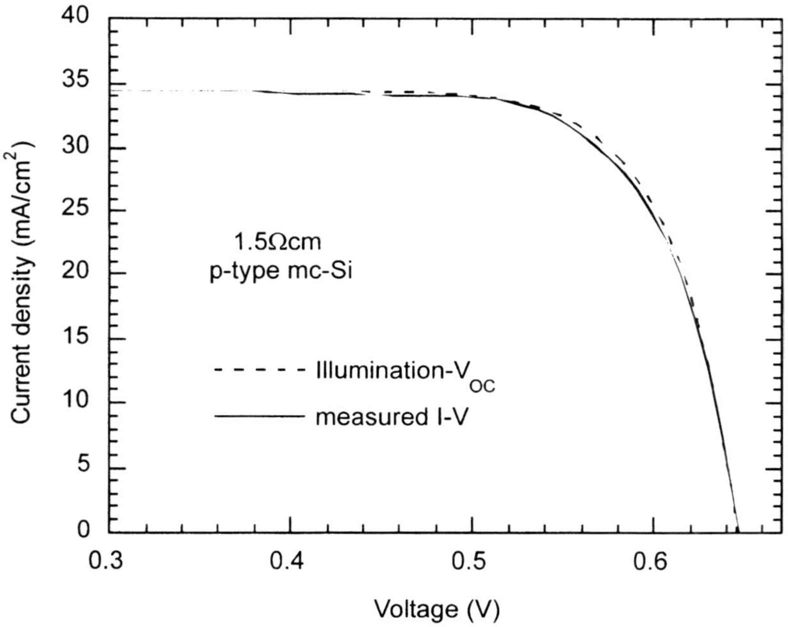

Figure 6 The same Illumination–V_{oc} *data as in Figure 5, plotted as a photovoltaic I–V curve and compared to the I–V curve taken on the finished cell.*

power point or not, and the upper bound on fill factor and efficiency (without series resistance effects) is clearly displayed. As discussed in the previous section, this 'pseudo' I–V curve can be measured very early in the process, by probing the silicon after junction formation. This allows the qualification of basic materials and device properties in terms of potential device performance, before the 'back end' processing may complicate the interpretation. The metallisation and sintering process steps can subsequently also be monitored for voltage loss and shunts using the $QssV_{oc}$ technique.

Eventually, by comparing the *Illumination–*V_{oc} curve with the actual output I–V curve of the finished solar cell, the series resistance can be determined with precision. Since the implied I–V curve from the open-circuit voltage has the shunt and ideality factors fully included, the differences between the two curves are clearly isolated to be due to series resistance. The latter is simply given by the voltage difference between the two curves at the knee, in the vicinity of the maximum power point, divided by the current. As an example, the same data in Figure 5 for a laboratory multicrystalline silicon cell is shown in Figure 6 [32]. The curves closely follow each other except near the maximum power point, where series resistance effects are greater. The measured output I-V curve gives an efficiency of 17.39%, with a fill factor of 0.785. The curve constructed from the *Illumination–*V_{oc} data indicates a pseudo-efficiency of 17.47%, with a fill factor of 0.790. This indicates that the series resistance is very small for this device and that the main limitation to voltage and fill factor is due to recombination losses within the solar cell.

Acknowledgement

We would like to thank M. Kerr and D. Macdonald for collaborations in developing expertise in this area of research and in discussions and preparation of this manuscript.

References

[1] Stevenson, D.T. and Keyes, R.J., 1995. Measurement of Carrier Lifetimes in Germanium and Silicon. *J. Appl. Phys.*, Vol. 26, pp. 190–195.

[2] Bube, R.H., 1960. *Photoconductivity of Solids*. Wiley, New York.

[3] Sinton, R.A. and Cuevas, A., 1996. Contactless determination of current–voltage characteristics and minority-carrier lifetimes in semiconductors from quasi-steady-state photoconductance data. *Appl. Phys. Lett.*, Vol. 69(17), pp. 2510–2512.

[4] ASTM, 1981. *Designation: F-28-75. Measuring the minority carrier lifetime in bulk germanium and silicon*. American Society for Testing of Materials.

[5] Miller, G.L., Robinson, D.A.H., and Wiley, J.D., 1976. Contactless measurement of semiconductor conductivity by radio frequency free carrier power absorption. *Rev. Sci. Instrum.*, Vol. 47(7), pp. 799–800.

[6] Deb, S. and Nag, B.R., 1962. Measurement of carriers in semiconductors through microwave reflection. *J. Appl. Phys.*, Vol. 33(4), p. 1604.

[7] Kunst, M. and Beck, G., 1986. The study of charge carrier kinetics in semiconductors by microwave conductivity measurements. *J. Appl. Phys.*, Vol 60(10), pp. 3558–3566.

[8] Harrick, N.J., 1956. Lifetime measurements of excess carrier kinetics in semiconductors. *J. Appl. Phys.*, Vol. 27(12), pp. 1439–1442.

[9] Glunz, S.W. and Warta, W., 1995. High resolution lifetime mapping using modulated free-carrier absorption. *J. Appl. Phys.*, Vol. 77(7), pp. 3243–3247.

[10] Brendel, R., Bail, M., and Bodman, B., 2002. Analysis of photoexcited charge carrier density profiles in Si wafers by using an infrared camera. *Appl. Phys. Lett.*, Vol. 80(3), pp. 437–439.

[11] Schroder, D.K., 1998. *Semiconductor Material and Device Characterization.* 2nd ed. John Wiley and Sons, New York.

[12] Basore, P.A., 1993. Extended spectral analysis of internal quantum efficiency. *Proc. 23rd IEEE Photovoltaic Specialists Conf.*, Louisville, pp. 147–152.

[13] ASTM, 1998. *Recombination lifetime measurements in silicon.* In: Gupta, D.C., Bacher, F.R. and Hughes, W.M. Eds., Vol. STP 1340, American Society for Testing of Materials.

[14] Brendel, R. and Wolf, M., 1995. Differential and actual surface recombination velocities. *Proc. 13th European Photovoltaic Solar Energy Conf.*, Nice, pp. 428–431.

[15] Schmidt, J., 1999. Measurement of differential and actual recombination parameters on crystalline silicon wafers. *IEEE Transactions on Electron Devices*, Vol. 46, 1999, pp. 2018–2025.

[16] Rein, S., Rehrl, T., Warta, W., and Glunz, S.W., 2002. Lifetime spectroscopy for defect characterization: Systematic analysis of the possibilities and restrictions. *J. Appl. Phys.*, Vol. 91(4), pp. 2059–2070.

[17] Macdonald, D., Cuevas, A., and Wong-Leung, J., 2001. Capture cross sections of the acceptor level of iron–boron pairs in p-type silicon by injection-level dependent lifetime measurements. *J. Appl. Phys.*, Vol 89(12), pp. 7932–7939.

[18] Kerr, M.J. and Cuevas, A., 2002. General parameterization of Auger recombination in crystalline silicon. *J. Appl. Phys.*, Vol. 91(4), pp. 2473–2480.

[19] Schmidt, J., Aberle, A.G., and Hezel, R., 1997. Investigation of carrier lifetime instabilities in Cz-grown silicon. *Proc. 26th IEEE Photovolt. Specialists Conf.*, Washington DC, pp. 13–17.

[20] Glunz, S.W., Rein, S., Warta, W., Knobloch, J. and Wettling, W., 1998. On the degradation of CZ-silicon solar cells. *Proc. 2nd World Conf. on Photovoltaic Energy Conversion*, Vienna, pp. 1343–1346.

[21] Reis, J.H., King, R.R. and Mitchell, K.W., 1996. Characterization of diffusion length degradation in Czochralski silicon solar cells. *Appl. Phys. Lett.*, Vol. 68(23), pp. 3302–3304.

[22] Zoth, G. and Bergholz, W., 1990. A fast, preparation-free method to detect iron in silicon. *J. Appl.. Phys.*, Vol. 67(11), pp. 6764–6771.
[23] Aberle, A.G., Robinson, S.J., Wang, A., Zhao, J., Wenham, S.R. and Green, M.A., 1993. High-efficiency silicon solar cells: fill factor limitations and non-ideal diode behaviour due to voltage-dependent rear surface recombination velocity. *Prog. Photovoltaics*, Vol. 1, pp. 133–143.
[24] Glunz, S.W., Biro, D., Rein, S. and Warta, W., 1999. Field-effect passivation of the SiO_2–Si interface. *J. Appl. Phys.*, Vol. 86, pp. 683–691.
[25] Kerr, M.J. and Cuevas, A., 2002. Comprehensive study of the doping and injection-level dependence of stoichiometric plasma silicon nitride passivation for silicon solar cells. *Proc. 29th IEEE Photovoltaic Specialists Conf.*, New Orleans, pp. 102–105.
[26] Basore, P.A. and Clugston, D.A., 1998. *PC1D V5.3*. University of New South Wales, Sydney, Australia.
[27] Bail, M. and Brendel, R., 2000. Separation of bulk and surface recombination by steady state photoconductance measurements. *Proc. 16th European Photovoltaic Solar Energy Conf.*, Glasgow, pp. 98–101.
[28] Kane, D.E. and Swanson, R.M., 1985. Measurement of the emitter saturation current by a contactless photoconductivity decay method. *Proc. 18th IEEE Photovoltaic Specialists Conf.*, Las Vegas, pp. 578–583.
[29] Cuevas, A., 1999. The effect of emitter recombination on the effective lifetime of silicon wafers. *Sol. Energy Mater. Sol. Cells*, Vol. 57, pp. 277–290.
[30] Cuevas, A. and Sinton, R.A., 1997. Prediction of the open-circuit voltage of solar cells from the steady-state photoconductance. *Prog. Photovoltaics*, Vol. 5, pp. 79–90.
[31] Cuevas, A., Sinton, R.A., Kerr, M., Macdonald, D. and Mackel, H., 2002. A contactless photoconductance technique to evaluate the quantum efficiency of solar cell emitters. *Sol. Energy Mater. Sol. Cells*, Vol. 71(3), pp. 295–312.
[32] Macdonald, D.H., 2001. *Recombination and Trapping in Multicrystalline Silicon Solar Cells*. PhD thesis, Australian National University, Canberra.
[33] Hornbeck, J.A. and Haynes, J.R., 1955. Trapping of minority carriers in silicon. I. p-type silicon. *Phys. Rev.*, Vol. 97(2), pp. 311–321.
[34] Macdonald, D. and Cuevas, A., 1999. Trapping of minority carriers in multicrystalline silicon. *Appl. Phys. Lett.*, Vol. 74(12), pp. 1710–1712.
[35] Macdonald, D. and Cuevas, A., 2000. Understanding carrier trapping in multicrystalline silicon. *Sol. Energy Mater. Sol. Cells*, Vol. 65(4), pp. 509–516.
[36] Macdonald, D., Sinton, R.A. and Cuevas, A., 2001. On the use of a bias-light correction for trapping effects in photoconductance-based lifetime measurements in silicon. *J. Appl. Phys.*, Vol. 89(5), pp. 2772–2778.
[37] Sinton, R.A. and Cuevas, A., 2000. A quasi-steady-state open-circuit voltage method for solar cell characterization. *Proc. 16th European Photovoltaic Solar Energy Conf.*, Glasgow, pp. 1152–1155.

[38] Kerr, M.J., Cuevas, A. and Sinton, R.A., 2001. Generalized analysis of quasi-steady-state and transient decay open circuit voltage measurements. *J. Appl. Phys.*, Vol. 91(1), pp. 399–404.
[39] van der Heide, A.S.H., Schonecker, A., Wyers, G.P. and Sinke, W.C., 2000. *Proc. 16th European Photovoltaic Solar Energy Conf.*, Glasgow, pp. 1438.
[40] van der Heide, A.S.H., Bultman, J.H., Hoornstra, J., Schonecker, A., Wyers, G.P. and Sinke, W.C., 2002. Optimizing the front side metallization process using the Corescan. *Proc. 29th IEEE Photovoltaic Specialists Conf.*, New Orleans, pp. 340–343.
[41] Araujo, G.L., Cuevas, A. and Ruiz, J.M., 1986. Effect of distributed series resistance on the dark and illuminated characteristics of solar cells. *IEEE Transactions on Electron Devices*, Vol. ED-33(3), pp. 391–401.
[42] Macdonald, D., Cuevas, A. and Ferrazza, F., 1999. Response to phosphorus gettering of different regions of cast multicrystalline silicon ingots. *Solid-State Electronics*, Vol. 43, pp. 575–581.
[43] Rohatgi, A., Yelundur, V., Jeong, J., Ebong, A., Meier, D., Gabor, A.M. and Rosenblum, M.D., 2000. Aluminium-enhanced PECVD SiN_x hydrogenation in silicon ribbons. *Proc. 16th European Photovoltaic Solar Energy Conf.*, Glasgow, pp. 1120–1123.
[44] Yablonovitch, E., Allara, D.L., Chang, C.C., Gmitter, T. and Bright, T.B., 1986. Unusually low surface-recombination velocity on silicon and germanium substrates. *Physical Review Letters*, Vol. 57, pp. 249–252.
[45] Horanyi, T.S., Pavelka, T. and Tutto, P., 1993. In situ bulk lifetime measurement on silicon with a chemically passivated surface. *Appl. Surf. Sci.*, Vol. 63, pp. 306–311.
[46] Schmidt, J. and Aberle, A.G., 1998. Easy-to-use surface passivation technique for bulk carrier lifetime measurements on silicon wafers. *Prog. Photovoltaics*, Vol. 6, pp. 259–263.
[47] Biro, D. and Warta, W., 2002. Low temperature passivation of silicon surfaces by polymer films. *Sol. Energy Mater. Sol. Cells*, Vol. 71, pp. 369–374.
[48] Stocks, M., Cuevas, A. and Blakers, A., 1997. Process monitoring of multicrystalline silicon solar cells with quasi-steady state photoconductance measurements. *Proc. 26th IEEE Photovoltaic Specialists Conf.*, Anaheim, pp. 123–126.
[49] Coletti, G., Iuliis, S.D. and Ferrazza, F., 2001. A new approach to measure multicrystalline silicon solar cells in a production process. *Proc. 17th European Photovoltaic Solar Energy Conf.*, Munich, pp. 1640–1642.
[50] Geiger, P., Kragler, G., Hahn, G., Fatch, P. and Bucher, E., 2002. Spatially resolved lifetimes in EFG and string ribbon silicon after gettering a hydrogenation steps. *Proc. 29th IEEE Photovoltaic Specialists Conf.*, New Orleans, pp. 186–189.
[51] Warta, W., 2002. Defect and impurity diagnostics and process monitoring. *Sol. Energy Mater. Sol. Cells*, Vol. 72, pp. 389–401.

[52] Cuevas, A., Macdonald, D., Kerr, M.J., Samundsett, C., Sloan, A., Leo, A., Mrcarica, M., Winderbaum, S. and Shea, S., 2000. Evidence of impurity gettering by industrial phosphorus diffusion. *Proc. 28th IEEE Photovoltaic Specialists Conf.*, Anchorage, pp. 108–111.

[53] Wolf, M. and Rauschenbach, H., 1963. *Advanced Energy Conversion*, Vol. 3, pp. 455–479.

[54] Sinton, R.A. and Cuevas, A., 2000. A quasi-steady open-circuit voltage method for solar cell characterisation. *Proc. 16th European Photovoltaic Solar Energy Conf.*, Glasgow, pp. 1152–1155.

[55] MacDonald, D. and Cuevas, A., 2000. Reduced fill factors in multicrystalline silicon solar cells due to injection-level dependent bulk recombination lifetimes. *Prog. Photovoltaics*, Vol. 8(4), pp. 363–375.

IIb-5

High-Efficiency Silicon Solar Cell Concepts

Martin A. Green, Centre for Advanced Silicon Photovoltaics and Photonics, University of New South Wales, Sydney, Australia

1 Introduction

The vast majority of solar cells sold up to 2003 used crystalline or multicrystalline silicon wafers in combination with a simple screen-printing approach to applying the metal contacts. This approach has the advantage of being well established with the ready availability of appropriate equipment, such as screen-printers and furnaces for drying and firing the screened metal patterns, since these already had been developed for the thick-film hybrid microelectronics area. This limited the capital requirements and risks involved in setting up cell manufacture in an era where the viability of the photovoltaics industry was marginal.

However, there are disadvantages with the simplicity of the screen-printing approach. One is the limited cell performance that results. Commercial solar cell modules based on this approach are limited to efficiencies in the 10–13% range [1], corresponding to cell efficiencies of 12–15%. Also, as wafers are thinned to reduce material costs, both performance and yield become more difficult to maintain, limiting the economic benefits of such thinning.

Now that the photovoltaic industry's future seems more assured, with rapidly growing markets and increasingly positive cash flows, it will become important to pay more attention to the full economics of cell manufacture. It is well known that the starting wafers account for about 50% of total module costs and encapsulants account for another large portion, with such material costs accounting for over 70% of finished module costs. The cost of processing wafers into cells is less than 20% of the total cost of the finished product.

Simple arithmetic based on these figures shows that doubling the cell processing costs would give a cheaper product if the resulting cells were 20% more efficient on a relative basis. The product additionally would have more value due to both the prestige associated with better performance and the reduced systems costs resulting from its use. The implications of this simple calculation often are not fully appreciated, with the notion that high efficiency automatically means high cost ingrained into the subconscious of many.

As the photovoltaic industry becomes more mature, it seems likely that such costing issues will be more closely examined than warranted in the 'hand-to-mouth' days of the past. It will become important to explore approaches that offer higher performance than possible with the screen-printing approach.

With this as the basic rationale, this chapter will explore approaches that offer the potential for higher silicon solar cell efficiency than seems possible from screen-printing. First, the recent history of high-efficiency laboratory cell development will be outlined, highlighting features improved in each successive generation. The features that limit screen-printed cells to the relatively modest performance levels previously mentioned will then be identified. Two commercial high-efficiency cell designs that overcome some of these limitations will then be described, followed by an exploration of other approaches that may have some potential.

2 High-Efficiency Laboratory Cells

2.1 Silicon Space Cell Development

The evolution of record silicon laboratory cell efficiency is shown in Figure 1, showing several stages in the evolution of cell design. After an initial period of rapid evolution in the 1950s, design stabilised for more than a decade on the conventional space cell of Figure 2(a). Key features include the use of 10 Ω cm p-type substrates to maximise radiation resistance, and the use of a nominally 40 Ω/square, 0.5 μm deep phosphorus diffusion. Although it was known that lighter diffusions gave better blue response, this value was chosen because it was found to be more resistant to shunting by the top contact metallisation during processing [2]. These cells remained the standard for space use for more than a decade and are even now still specified for some space missions. Energy conversion efficiency was 10–11% under space radiation, and 10–20% higher on a relative basis under terrestrial test conditions.

Towards the end of the 1960s, the benefits of a rear Al treatment became apparent, particularly for cells which were thinner than normal [3, 4]. The corresponding increase in space cell efficiency to 12.4% was attributed to the gettering action of the Al treatment [5].

More detailed work showed that it was the presence of a heavily doped region beneath the rear contact which gave rise to these beneficial effects [6]. These benefits were postulated to arise from spillage of majority carriers from the rear

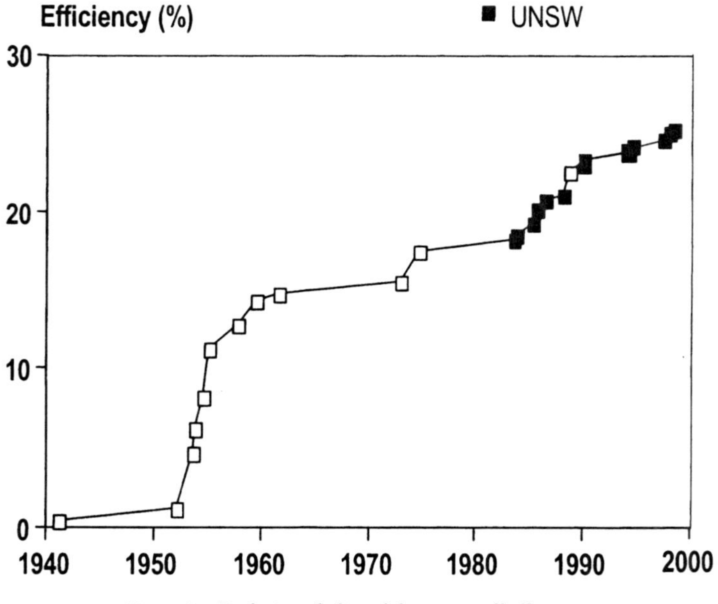

Figure 1 Evolution of silicon laboratory cell efficiency.

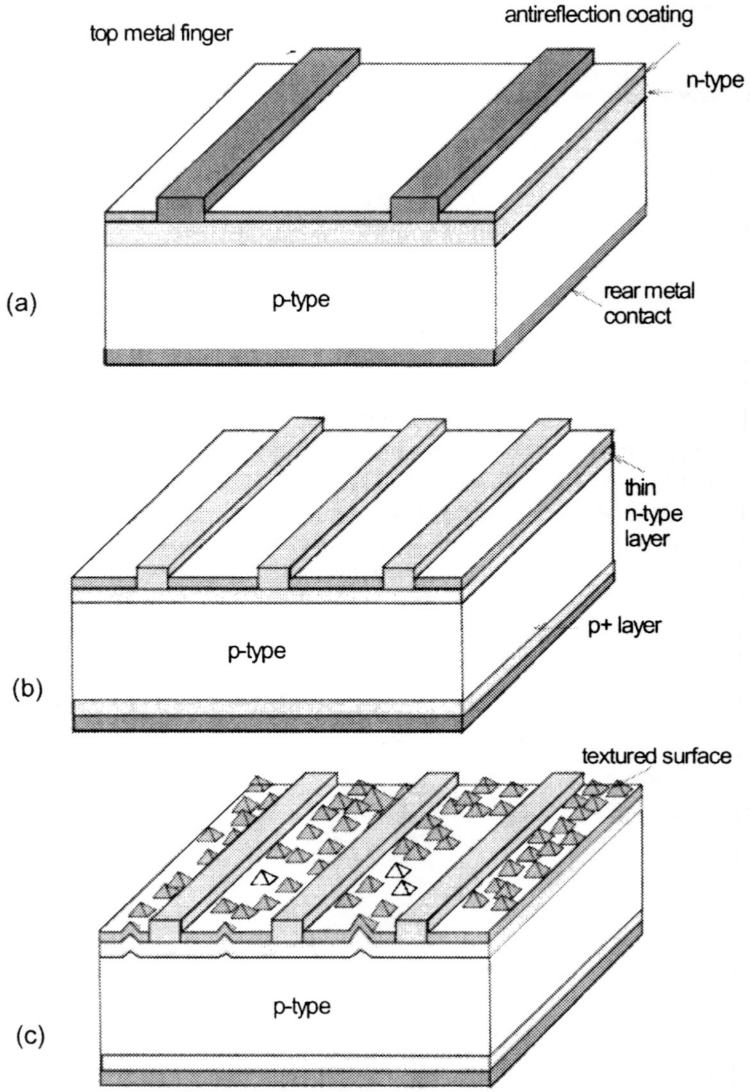

Figure 2 (a) Space silicon cell design developed in the early 1960s which became a standard design for over a decade; (b) shallow junction 'violet' cell; (c) chemically textured non-reflecting 'black' cell.

doped region into the bulk region of the cell, thus increasing the effective bulk concentration and hence the open-circuit voltage [6]. Although the correct explanation, in terms of a reduction in the effective rear surface recombination velocity, was soon forthcoming [7], the effect is still rather inappropriately, but almost universally, referred to as the back surface field (BSF) effect [6].

The conventional space cells had a relatively poor response to wavelengths which were shorter than 0.5 μm, due to the relatively heavy top junction diffusion, as previously noted, and the use of an SiO AR coating which absorbs light below this wavelength. A marked improvement in performance was demonstrated in the early 1970s by replacing such junctions with much shallower (0.25 μm), higher sheet resistivity junctions, and redesigning the entire cell to accommodate this change, as shown in Figure 2(b).

To accommodate the increased sheet resistivity of the diffused layer, much finer metal finger patterns were also incorporated using photolithography to define their geometry. As a consequence, the cell resistance was actually lower than in conventional cells. Improved antireflection coatings, based upon TiO_2 and Ta_2O_5, were also incorporated. These were less absorbing than SiO, as well as providing a better optical match between the cells and the cover glass to which space cells are normally bonded. The thickness of these coatings was also selected so they would be effective at shorter wavelengths than were traditional coatings, thus giving the cells their characteristic violet appearance. (A subsequent development, made possible by the higher refractive index of the new AR coating materials, was the use of double layer AR coatings.)

The final design change in these 'violet' cells was the use of lower resistivity 2 Ω cm substrates. Since the improved output of the cells at blue wavelengths was quite tolerant to radiation exposure, the overall radiation tolerance remained at least equal to that of the conventional devices, while giving an improved voltage output. The combination of improved open-circuit voltage (due to the change in substrate resistivity), improved current output (due to the removal of 'dead layers', better antireflection coatings, and lower top contact coverage), and improved fill factor (due to the improved open-circuit voltage and decreased cell resistance) resulted in a massive increase of 30% in performance as compared to an average space cell of conventional design. Efficiencies of 13.5% were obtained under space radiation, with terrestrial efficiencies close to 16% being demonstrated [8].

Not long after the superior performance of the violet cell had been established, a further boost in performance was obtained by texturing the top surface of the cell [9]. The idea of mechanically forming pyramids on the top surface of the cell to reduce reflection had been suggested some time earlier [10]. In the black cell, a simpler approach was used which relied upon the random nucleation of selective etches to expose (111) crystallographic planes in a substrate originally of (100) surface orientation. The intersecting (111) planes so exposed formed small, square-based pyramids of random size which were distributed randomly over the cell surface.

This has two advantages for cell performance. One is that light striking the sides of the pyramids is reflected downwards and hence has at least one more chance of being coupled into the cell. A second advantage is that light coupled into the cell enters obliquely. Most light will be coupled in at the first point of incidence on the pyramids. This light is refracted in at an angle of about 48° to the original surface, resulting in an increase in the path length of weakly absorbed light by a factor of 1.35 compared to a non-textured cell. The effect is similar to that of an increase in the silicon absorption coefficient or in the bulk diffusion length by the same factor. A third feature of the texturing approach is the high degree of trapping of light within the cell that is possible. For terrestrial cells, this is an advantage since it improves the long wavelength response of cells. However, for space cells, it is a disadvantage since it increases the absorption of sub-bandgap photons in the rear contact of the cell. This causes the cells to operate at a higher temperature in the space environment, largely negating the previous advantages. Combined with a greater potential for mechanical damage during array assembly by knocking peaks from pyramids, this has meant that textured cells have not been as widely used in space as their apparent performance advantage initially seemed to warrant.

These 'black' cells gave an energy conversion efficiency of 15.5% under AM0 radiation, corresponding to an energy conversion efficiency of about 17.2% under the current terrestrial standard (Global AM1.5, 100 mW/cm^2, 25°C). Such were the strengths of the texturing concept, when combined with the technological improvements incorporated into the violet cell, that it was almost a decade before any further significant improvement in cell performance was demonstrated. These improvements ultimately came about as a result of an increased open-circuit voltage due to the development of improved surface and contact passivation approaches.

2.2 High-Efficiency Terrestrial Cells

The simplicity of the surface passivation afforded by its thermal oxide is one of the key features of silicon technology which explains its present dominance in microelectronics. Unfortunately for photovoltaics, the refractive index of silicon dioxide is too low for its use as an effective antireflection coating in high-performance cells. In fact, if present in any reasonable thickness (greater than 20 nm) on the top surface of the cell, it will limit the ability to reduce reflection by the subsequent deposition of any number of compensating layers [11]. Hence, if oxide passivation is to be used on the cell surface exposed to light, the oxide layer has to be very thin.

The potential of such oxide passivation became clear around 1978 [12, 13]. All subsequent high efficiency silicon cells have taken advantage of the passivation provided by thin thermal oxide layers to maximise their open-circuit voltage and short-wavelength response.

Contacts made to the surface are generally regions of high recombination velocity. Best cell performance will be obtained when the electronic activity at such contacts is 'passivated'. The earliest approach was to passivate by isolating the contact from minority carriers by interposing a heavily doped region. As already discussed, rear-contact passivation via the 'back surface field' effect resulted in significant gains in cell performance. Heavily doped regions, localized to those areas where the top contact is made to the cell, are used in most recent high efficiency cell designs.

A second approach to reducing contact effects is to minimise the contact area [14,15]. The benefits were demonstrated by increased open-circuit voltages on low resistivity substrates [16]. Most recent high-efficiency cells employ this low contact area approach. A third approach is to employ a contacting scheme in which the contact itself has an inherently low recombination velocity. The MINP (metal-insulator-NP junction) cell of Figure 3 was the first successfully to exploit this approach [17]. The thin surface passivating oxide is continued under the metal, thus reducing its effective recombination velocity. Polysilicon [18] and semi-insulating polysilicon (SIPOS) [19] contact passivation have also been demonstrated. It appears that a thin interfacial oxide layer may play an important role in both of these schemes [19, 20]. More recently, excellent surface passivation has been demonstrated by a combination of a very thin layer of lightly doped amorphous silicon followed by a layer of doped amorphous silicon [21].

As seen in Figure 1, the performance levels established by 'violet' and 'black' cells in the mid-1970s remained unchallenged for close to a decade. Cells successfully incorporating the above mentioned oxide and contact passivation approaches, along the top surface, were the first to exceed these levels.

The MINP cell (Figure 3), the first silicon cell to demonstrate 18% efficiency, employs top contact passivation by the use of a thin oxide layer underlying this contact, as well as top surface passivation by a slightly thicker oxide layer. This difference in thickness complicates processing but was found necessary to achieve maximum device performance. The top contact metallisation is a Ti/Pd/Ag multilayer. The use of a low work function metal such as Ti as the contact layer is essential with this approach. This is to produce an electrostatically induced accumulation layer in the underlying silicon, an important factor in reducing contact recombination. The cells used alloyed aluminium to give a heavily doped region near the rear contact and were fabricated on polished (100)-oriented 0.2 Ω cm substrates. Surface passivation is easier to achieve on polished rather than textured or 'as-lapped' surfaces. To minimise reflection losses, a double layer antireflection coating was used which

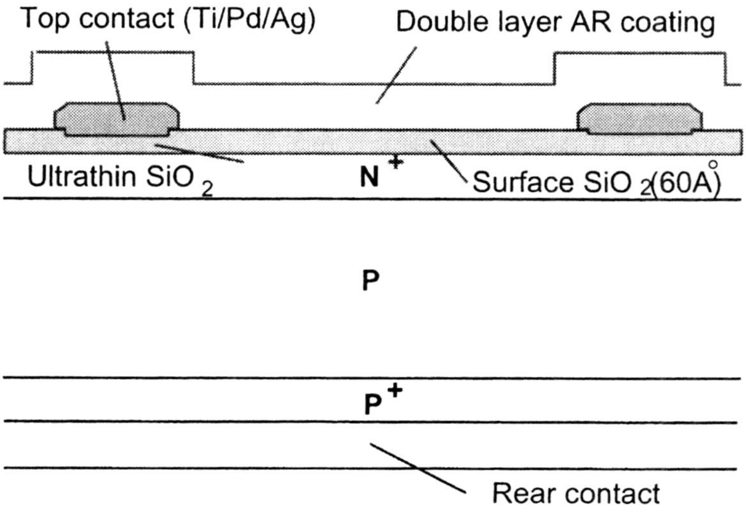

Figure 3 Metal–insulator–NP junction (MINP) solar cell.

consisted of approximately a quarter-wavelength of ZnS on top of the thin oxide, followed by a quarter-wavelength of MgF_2.

The PESC (Passivated Emitter Solar Cell) structure, shown in Figure 4, further improved cell efficiency. It is similar to the MINP cell structure, except that electrical contact is made directly through narrow slots in the thin oxide. In this case, contact passivation is obtained by minimising the contact area. In this case, surface texture is also used.

By combining the benefits of surface texture with the strengths of the PESC approach, the first non-concentrating silicon cells with an efficiency greater than 20% were fabricated in 1985 [22]. Rather than pyramidal texturing, the cells used microgrooving to achieve the same effect. The microgrooves were defined by using selective etches to expose crystal planes. Photolithographically designed oxide patterns were used to protect the surface against etching where this was not required, and so determine the final pattern of microgrooves. This approach was found to be easier to combine with fine line photolithography than was the normal pyramidal texturing.

The key characteristics of the PESC sequence are oxide surface passivation, self-aligned contacts through this oxide, high sheet resistivity top junction diffusion, alloyed aluminium rear surface passivation and antireflection control by texturing or double layer antireflection coating. The PESC sequence has proved to be very rugged and repeatable. Within one year of the initial 20% results, two groups had reported results approaching this figure (when referred to present calibrations) using almost identical structures [23, 24]. The sequence has since been reproduced in many laboratories, with commercial product now available for space and concentrator cells or for high value-added applications such as solar car racing.

The next major advance in cell performance came as a result of applying surface and contact passivation approaches to both top and rear surfaces.

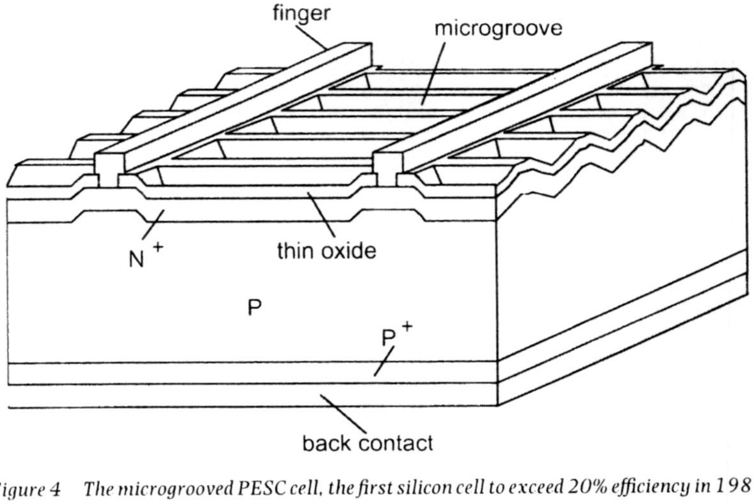

Figure 4 The microgrooved PESC cell, the first silicon cell to exceed 20% efficiency in 1985.

The rear point contact solar cell of Figure 5 achieved this landmark in cell design. Since both contacts are on the rear of the cell, the design places enormous pressure both upon the quality of surface passivation along both top and rear surfaces and upon post-processing carrier lifetimes. To achieve design objectives in these areas, full advantage had to be taken of 'state of the art' microelectronics technology.

Although originally developed for concentrator cells [25], the device design was modified for one-sun use by adding a phosphorus diffusion along the illuminated surface [26]. This produced the first one-sun silicon cells of efficiency above 22%.

Combining the earlier PESC sequence with similar double-sided surface passivation and chlorine based processing [27] produced an improved device, the PERL cell (Passivated Emitter, Rear Locally-diffused cell) shown in Figure 6. This took silicon cell efficiency to 23% by the end of the 1980s, an enormous improvement over the figure of 17%, the highest value only 7 years earlier. The PERL cell shares many features in common with the rear point contact cell, including almost complete enshroudment in a passivating oxide layer and small area contacts passivated by local heavy diffusions. However, it is a more robust design, being more tolerant of poor surface passivation and poor bulk lifetimes.

Since then, further improvements in PERL cells has taken their efficiency to close to 25%. Major improvements include the growth of much thinner oxide for top surface passivation which allows the direct application of a double layer antireflection coating to increase short-circuit current [11], the use of an annealing sequence for this top oxide and localised top contact points to increase

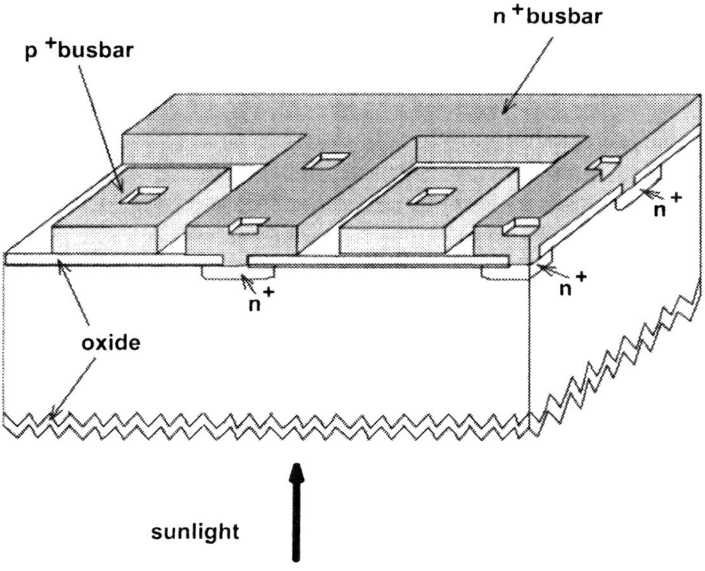

Figure 5 Rear point contact solar cell which demonstrated 22% efficiency in 1988 (cell rear shown uppermost).

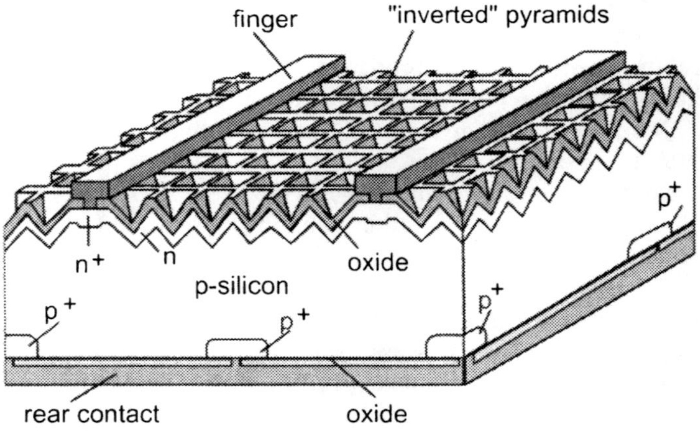

Figure 6 The PERL cell which took efficiency above 24% in the early 1990s.

open-circuit voltage and improved rear surface passivation and reduced metallisation resistance to improve fill factor.

2.3 PERL Cell Design Features

To maximise cell performance, as much light as possible of useful wavelengths should be coupled into and absorbed by the cell. Modern cell designs such as the PERL cell of Figure 6 incorporate several features of a primarily optical nature to achieve this result.

The inverted pyramids along the top surface serve primarily in such an optical role. Most light incident on this structure will hit one of the side walls of the pyramids at the first point of incidence with the majority of this light coupled into the cell. That reflected will be reflected downwards, ensuring that it has at least a second chance of entering into the cell. Some of the light incident near the bottom of the pyramids has three such chances. The pyramids are covered by an oxide layer of appropriate thickness to act as a quarter-wavelength antireflection coating. In more recent designs, this oxide is grown thin and a double layer antireflection coating applied [11].

Light coupled into the cell moves obliquely across the cell towards the rear surface with most absorbed on the way. Weakly absorbed light reaching the rear is reflected by the very efficient reflector formed by the combination of the rear oxide layer covered by an aluminium layer [28]. The reflectance from this combination depends upon the angle of incidence of the light and the thickness of the oxide layer, but is typically above 95% for angles of incidence close to the normal, decreasing to below 90% as the incidence angle approaches that for total internal reflection at the silicon/oxide interface (24.7°), and increasing to close to 100% once this angle is exceeded.

Light reflected from the rear then moves towards the top surface. Some reaching this surface strikes a face of a pyramid of opposite orientation to that which coupled it into the cell. Most of this immediately escapes from the cell. Light striking other faces of the pyramid is totally internally reflected. This results in about half the light striking the top surface internally at this stage being reflected back across the cell towards the rear contact. The amount of light escaping after the first double pass depends on the precise geometry involved. It can be reduced by destroying some of the symmetries involved, for example by using tilted inverted pyramids or by using a 'tiler's pattern' [28]. The latter approach is currently used in PERL cell designs.

The combination of the inverted pyramids and the rear reflector therefore forms a very efficient light-trapping scheme, increasing the pathlength of weakly absorbed light within the cell. Effective pathlength enhancement factors [28] above 40 are measured. The light trapping boosts the infrared response of the cell. The external responsivity (amps per watt of incident light) of PERL cells peaks at longer wavelengths at higher values than previous silicon cells with values of 0.75 A/W measured at 1.02 μm wavelength. Energy conversion efficiency under monochromatic light peaks at the same wavelength with values above 45% measured [29]. Further improvements could push this figure to above 50% at 1.06 μm.

Other optical losses are due to reflection from, and absorption in, the top metal fingers of the cell. This can be minimised by making these lines as fine as possible with, ideally, as large an aspect ratio (height to width ratio) as possible. Alternatively, optical approaches can be used to steer incoming light away from these lines or to ensure that light reflected from them eventually finds its way to the cell surface [30, 31].

Present PERL cells lose about 5% of incoming light due to absorption or reflection loss associated with these metal fingers, when combined with reflection from the unmetallised top surface of the cell. They also lose 1–2% in performance from the use of a less than optimum light-trapping scheme and from less than 100% reflection of light from the rear surface of the cell. There is therefore some scope for small to moderate gains in performance by further improving the optical properties of these cells.

Although such advanced cell designs have been used for spacecraft and high value terrestrial applications such as solar car racing, the multiple photolithographic steps required in their fabrication make them too expensive for low cost terrestrial applications [32]. They do, however, provide a reference point for the discussion of the compromises involved in lower cost designs, as discussed in the following sections.

3 Screen Printed Cell Limitations

3.1 Structure

Figure 7 shows the structure of a typical crystalline screen-printed cell. The normal cell processing sequence would consist of [33]: saw damage removal

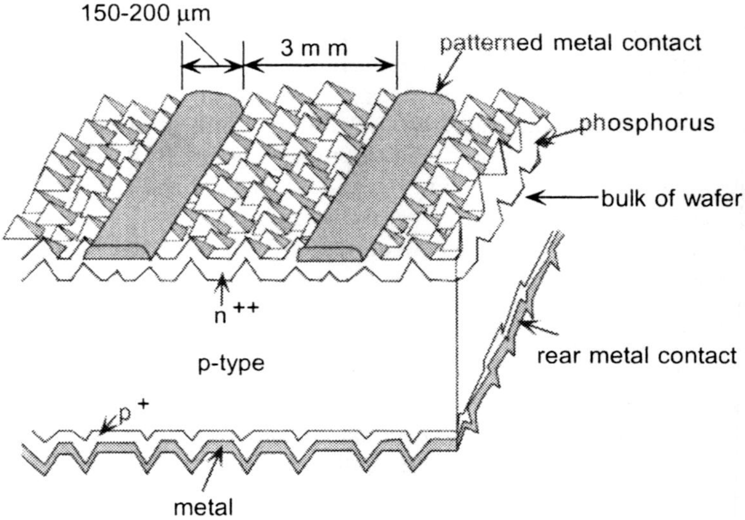

Figure 7 Screen printed crystalline silicon solar cell (not to scale).

from the starting wafer by etching; chemical texturing of the top surface, if a (100) crystalline wafer were the starting material; top surface diffusion to about 40 Ω/square; edge junction isolation normally by 'coin stacking' the cells and etching the edges in a plasma; etching to remove diffusion oxides; screening of front metal paste through a suitably patterned screen; drying and firing of the front surface metallisation; screening of the rear surface metal paste; drying and firing of rear metal contact; cell testing and sorting.

Using the normal boron-doped, 'solar grade' Czochralski silicon wafers of 0.5–5 Ω cm resistivity, the resulting cell efficiency is typically 12–15%. Application of a antireflection coating (usually TiO_2 or Si_3N_4) gives about 4% relative improvement.

For multicrystalline substrates, the use of an antireflection coating is mandatory due to the inability to reduce reflection substantially by texturing the multicrystalline wafer surface. This is due to the random orientations of the grains in the multicrystalline material which results in only a fraction being suitable for texturing. A recent development in this area has been the use of remote plasma enhanced chemical vapour deposited nitride coatings that are extremely beneficial for multicrystalline cells.

The major disadvantages of the screen printing approach relate to the cost of the metal pastes used in the process and the relatively low cell efficiency which results. The latter is due most fundamentally to the restricted line-width possible by screen printing. The relatively high contact resistance between the paste and the silicon is another constraint. The low aspect ratio (height/width) of the final lines due to paste thickness shrinkage during firing is another problem compounded by the low conductivity of the fired paste (about 3 times lower than that of pure silver).

There have been several investigations of the feasibility of screening pastes other than those based on silver. Nickel and copper have been investigated without success [34].

To reduce the linewidth, special fine screens can be used. However, the screens have been too frail for use in commercial production, although often used to produce a good research result [35].

The contact resistance between the paste and silicon can be a sensitive function of the precise firing environment and temperature. The glass frit (dispersed glass particles used in the paste as a binder) forms an oxide precipitate along the interface between the screened paste and the silicon surface. This contributes to the high contact resistance, although often, phosphorus is added to the paste to decrease the contact resistance to n-type material. Etching of the top surface with HF after firing is known to decrease the contact resistance, presumably by etching away some of the interfacial frit in this area. However, this is also considered to reduce the reliability of the cell in a moist environment.

The rear surface contact resistance is generally less of a problem. Even though contact is being made to more lightly doped material, a much larger contact area is available. Furthermore, the addition of aluminium to the silver paste or the use of an aluminium paste precursor can increase the doping level in the surface region by alloying. Under appropriate firing conditions, a significant 'back surface field' effect can be obtained by the use of such aluminium.

3.2 Typical Performance

Typically, the screen printing approach will produce cell open circuit voltages in the 580-620 mV range, depending on substrate resistivity, short circuit current densities in the 28–32 mA/cm^2 range, and fill factors for large area cells in the 70–75% range. For a large area cell, typically 10–15% of the top surface of the cell will be shaded by the screened metallization which, as indicated in Figure 7, typically consists of metal fingers of about 150–200 microns width, spaced about 2–3 mm apart. Due to the low conductivity of the paste, an interconnect busbar design, as indicated in Figure 8, is mandatory for large area cells. Although contributing to the large shading loss of this approach, this design also has the advantage of improving the cell tolerance to cracks. If scaling to large area cells, the number of busbar interconnections would have to be increased.

To maintain reasonable contact resistance, quite low sheet resistivities for the top surface diffusion are required. The 40 Ω/square typically used results in a significant loss in blue response of the cells due to dead layers along the surface. Higher sheet resistivities will improve the blue response but at the expense of cell fill factor. The heavy diffusion also limits the open-circuit voltage output of the cells. Oxide surface passivation is not of much benefit in improving performance due to this limitation. Using improved quality substrates such as floatzone silicon also generally does not result in any substantial performance improvement again due to this limitation.

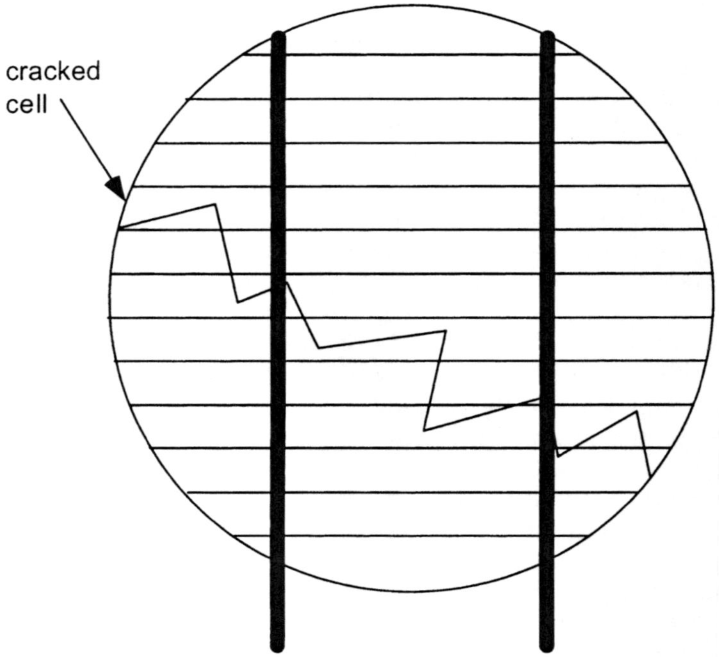

Figure 8 Cell design using the strip metal cell interconnects as busbars to reduce finger resistive losses and to improve tolerance to cracks.

3.3 Improved Technology

The limitations of the screen printing approach may not be fundamental to this approach but may be able to be overcome by introducing new ideas.

To reduce the linewidth, finer screens can be used although these require weaker material in their construction and have not been sufficiently durable for volume manufacturing. There are frequently alternative claims in this area [35], but commercial practice is a telling indicator of what is feasible here. Approaches for defining metal contacts, similar to ink jet printing approaches, have also been used in commercial production.

Some recent progress has been reported in improving the contact resistance obtainable by screen printing [35, 36]. Approaches similar to those shown in Figure 9 have also been suggested which rely on having different top layer sheet resistivities in contacted and non-contacted areas [37, 38]. A conductive antireflection coating such as formed by conducting metal oxides could be used in conjunction with this scheme to allow reasonable metal spacing [39].

However, it is doubtful that a high efficiency cell can ever be produced using the screen printed approach, with good screen printed laboratory cells only giving about 16.7% efficiency [11]. Even if both the linewidth and the contact resistance problems are successfully solved in production, there are still challenges due to conductivity and small aspect ratio.

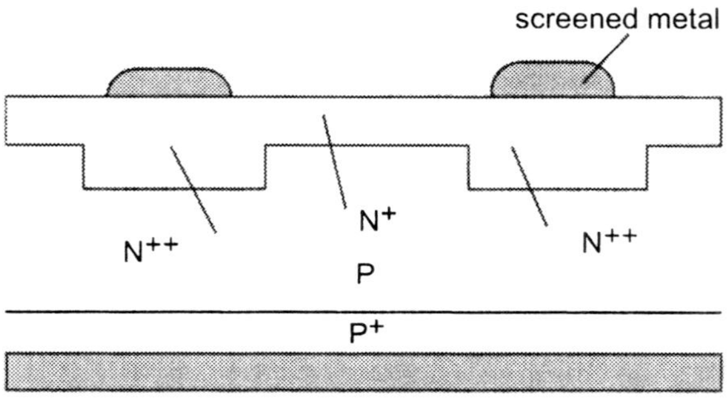

Figure 9 Ideal situation for fine linewidth screen printed metallisation with contact being made to selectively doped areas.

4 Buried Contact Cells

4.1 Structure

The buried contact solar cell of Figure 10 was developed to overcome the efficiency limitations of the screen printed cell approach previously described. The most distinctive feature of this approach is the use of grooves in the top surface to locate the cell metallisation. Although originally investigated using screen printed metallisation sequences (where the metal was forced into the groove during the screening operation), the most successful designs have used electrolessly plated metal contacts [40].

Cell processing bears some resemblance to the screen printing sequence previously described. After saw damage removal and texturing, the surface is lightly diffused and an oxide grown over the entire surface. This oxide serves multiple purposes during cell processing and is the key to the relative simplicity of the processing. Note that there is no need to remove the diffusion oxide as in the screen printed approach. Grooves are then cut into the top cell surface either using a laser scribing machine, mechanical cutting wheels, or other mechanical or chemical approaches.

After cleaning of the grooves by chemical etching, these are subject to a second phosphorus diffusion, much heavier than the first. This produces selective doping in the contact areas (previously discussed as desirable in connection with screen printed cells and as used in high efficiency silicon PERL cells). Aluminium is then deposited on the rear surface by evaporation, sputtering, screen printing or plasma deposition.

After firing of the aluminium and etching to remove oxides, cell metallisation is then completed by electroless plating of a nickel/copper/silver layer. A thin layer of nickel is first deposited. This is then fired and a substantial thickness of copper then deposited. Finally, the thin layer of silver is formed on the top surface by displacement plating. All these processes are electroless meaning that canisters containing the wafers are simply immersed in the plating solution.

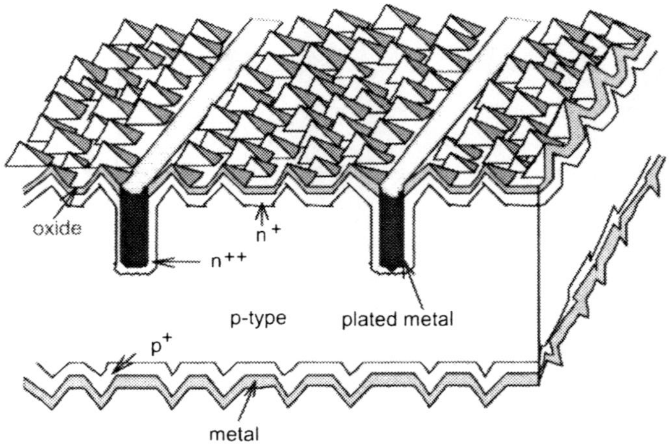

Figure 10 Buried contact solar cell.

An improved sequence uses Si_3N_4 in lieu of the oxide during the processing [41]. This layer withstands high temperature steps while giving better antireflection properties in the final cell. BP Solar have reported excellent results with this approach [41–43] as has the University of Konstanz where a record 17.5% efficiency has been demonstrated on large-area multicrystalline silicon substrates with this approach [44].

4.2 Performance Analysis

The buried contact cell gives substantial performance advantage over screen printed cells. Figure 11 shows the comparison reported by BP Solar, using identical starting substrate material. Approximately 30% performance advantage is reported in this case, although incremental improvements in screen printed cells have since reduced some of this gap. At the same time, BP Solar report that processing costs *per unit area* are virtually identical, within 4% of that of the screen printed cell [42]. This gives a much lower cost/watt, given the higher power output per unit area. Although capital costs are higher, the lower material costs compensate. A study involving representatives of several European manufacturers and research institutes reached essentially identical conclusions [32].

There are several reasons for this improved performance. The higher fill factor is due to the better conductance of the metallisation fingers and the lower contact resistance between these and the heavily doped grooved regions. The higher voltage is due to the high sheet resistivity; of the diffusion over the top surface, combined with the excellent surface passivation provided by the overlying oxide in these areas, and with the contact passivation provided by the heavy doping in the grooved areas. An open-circuit voltage approaching 700 mV has been demonstrated for this combination, close to the highest ever demonstrated for laboratory silicon cells [45].

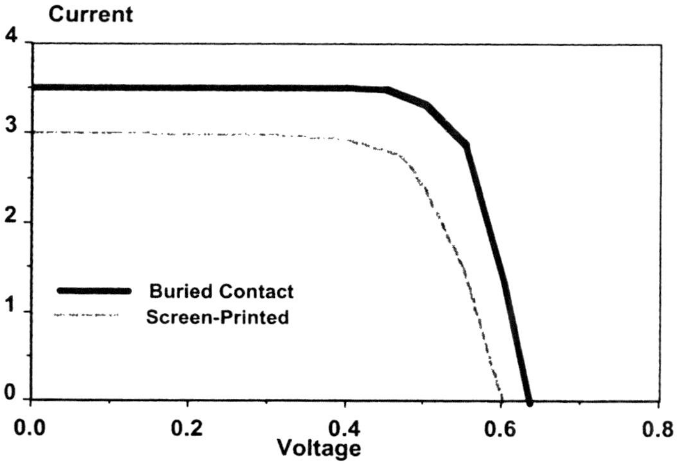

Figure 11 Output characteristics of buried contact cells compared to screen printed cells (after [43]).

The high current output is due to the relatively low top surface shading feasible with this approach even in large area cells. This is a result, most fundamentally, of the high aspect ratio possible with buried contacts. Aspect ratios as high as 5:1 have been demonstrated, although not essential in some metallisation design approaches. Low metallisation linewidths of 15–20 μm are feasible with laser grooving and linewidths of 40 μm are feasible with mechanical approaches. The low metallisation loss also makes the approach very well suited for the increasingly large size of commercial cells, an aspect hinted at, but not fully explored, in an earlier study [32]. The other contributor to the improved output current is the improved blue response resulting from the almost ideal surface properties in non-contacted areas.

The processing sequence has also shown itself capable of producing considerable gettering benefits during processing. The damage during the grooving process appears to be beneficial. Laser damage can produce effective gettering when applied to the rear of the wafers. Similarly, the laser grooves on the top surface might be expected to be effective gettering sites. One of the advantages of the buried contact sequence is that after grooving, the top surface is subjected to a heavy diffusion in the grooved areas. The phosphorus will diffuse preferentially in damaged areas and automatically passivate damage introduced by the grooves. Direct evidence for this effect has been observed [46]. Aluminium on the rear these cells provides another well established gettering strategy as does heavy phosphorus diffusion as used in the grooved areas [47].

4.3 Production Experience

Buried contact solar cell technology has been licensed to several major cell manufacturers, with pilot production experience reported by some. A high-efficiency array was fabricated by Telefunken [48] for the Swiss car 'Spirit of Biel', which convincingly won the 1990 World Solar Challenge, the solar car race from Darwin to Adelaide. Array efficiency was 17%, then, the highest ever for silicon. The array gave 25% more power than that of the second placed car, which used enhanced screen printing technology [49].

BP Solar has reported on both manufacturing yields and process economics [41]. Using the same 'solar grade' CZ substrates as in their screen printing process, BP reports substantial efficiency improvement for the technology (~30%) and cell efficiencies of 17.5–18%. Economic analysis shows that the approach, as developed by BP with nitride antireflection coating, is well suited for polycrystalline material [41]. BP Solar supplied modules for the first reasonably sized installation using buried contact technology, a 20 kW system powering the Marzili furnicular railway in Berne. This was the most efficient flat-plate system of this size installed up to this time [50]. BP Solar also supplied 550 kW of these modules for the 1 MW Union Fenosa system near Toledo in Spain [51]. When it began operation in 1994, this was Europe's largest PV installation. In 1994, BP Solar launched the Saturn 585 module, an 85 W module based on 123 mm square buried contact solar cells as its 'top-of-the-line' commercial product [52], with module size more recently doubled to 170 W [1]. Production capacity at BP Solar was reported to be 10 MW/year in1998, with an increase to 80 MW/year planned to be in place by 2004.

Solarex, since incorporated into BP Solar, has reported preliminary work using mechanical dicing wheels to form the grooves [53]. Experiments at UNSW with 35 ganged dicing blades have produced grooves of 2% depth uniformity over the wafer surface with a processing time of 3 s/cell. The attraction of this approach is lower equipment costs, although consumable costs are higher than with laser grooving. A small pilot production line operated by Unisearch Ltd in conjunction with UNSW has also given good yields of laser grooved cells with efficiencies of 19–20% demonstrated in production volumes of 10,000 cells per year [54].

The consensus of these pilot line and production studies is that the buried contact process, when transferred into production, can give efficiency margins of 20–30% over screen printing. Although more processing steps are involved than in the simplest screen printing approach, expensive silver pastes are eliminated so that processing costs per unit area are not greatly different, with costs per watt of product lower [32, 55]. Marketing experience has shown that higher selling prices are feasible for this product due to the lower balance of systems costs in installed systems and the perceived higher quality due to the superior performance.

5 HIT Cell

The HIT cell (heterojunction with intrinsic thin layer) combines both crystalline and amorphous silicon cell technology to produce cells with conversion efficiency in production similar to the buried contact cell and some of the highest

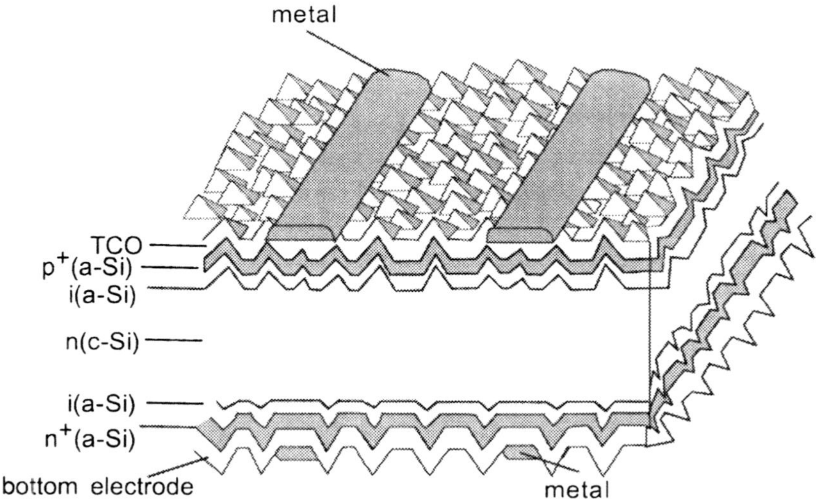

Figure 12 HIT cell structure using a textured n-type silicon wafer with amorphous silicon heterojunctions on both front and rear surfaces.

efficiency large area laboratory devices ever reported. The basic device structure is shown in Figure 12.

The starting substrate is an n-type silicon wafer, the opposite polarity from most other commercial product. This is a fortunate choice since such substrates are free from the light induced degradation effects that limit the performance of cells made on p-type, boron doped substrates, attributed to the formation of boron-oxygen complexes under illumination. After texturing, very thin layers of intrinsic hydrogenated and p-type amorphous silicon are deposited on the top surface and intrinsic and n-type on the rear surface. As is usual in amorphous silicon technology, these layers are contacted by transparent conducting oxide (TCO) layers, in turn contacted by screen printed metal contacts.

The low sheet resistivities possible with these TCO layers relax the constraints previously described on the screen printed metallisation parameters. However, these layers are quite absorbing and the underlying doped layers of amorphous silicon are inactive for photocurrent collection, resulting in poor blue response of the cell. About 10% of available current is lost by absorption in these layers. However, the bandgap of amorphous silicon is very much higher than in silicon and the quality of the interface between the amorphous and crystalline material is excellent. This good interface has produced some of the highest open-circuit voltages seen in silicon cells (710–720 mV).

This approach has produced record large-area laboratory cell performance of up to 21.0% [56, 57]. The differences between these laboratory devices and commercial devices are not clear. Commercial devices have more modest efficiencies but nonetheless result in a nominal module efficiency of 15.2% in Sanyo's 'top of the line' product, the highest value presently quoted [1]. Sanyo appears to have produced about 16 MW of HIT cell product during 2001 [58].

6 Nitride-Based Approaches

6.1 Remote Plasma Passivation

One limitation of the HIT cell structure arises from the absorption in the amorphous silicon layer and in the transparent conductor required on top of it to provide lateral conductivity. However, over the last decade, one development of note in silicon photovoltaics has been the use of remote plasma passivation, both of surfaces using silicon nitride [59] and of bulk regions using hydrogen [60].

The low temperatures associated with this passivation step give the potential for the relatively simple processing of high performance devices. It seems it should be possible to develop relatively simple high efficiency approaches based on this approach. Some of the work being conducted in this area is discussed in the following sections.

6.2 MIS-n^+p or MINP Approaches

The MINP cell structure of Figure 3 was important in the history of cell development since it produced the first 18% efficient cell, demonstrating the first improvement in silicon cell performance for close to a decade [17]. More recent work has sought to simplify the structure while incorporating the benefits of nitride surface passivation.

The challenge is to find a simple approach to obtaining the fine linewidths required for good performance. One approach uses a metal shadow masking process to produce the device structure of Figure 13 on multicrystalline wafers. The basic sequence is gettering of the starting wafer by a heavy phosphorus diffusion, removal of the gettering layer, diffusion of the n^+ emitter, evaporation of an Al grid on the cell rear, alloying at 850°C to form a local back surface field (LBSF), passivation of the rear by remote plasma nitride and evaporation of a rear Al contact/reflector over the entire rear surface, evaporation of the front Al contact onto a thin thermally grown tunnel oxide, also patterned using a shadow mask, followed by remote plasma nitride antireflection coating. An alternative sequence avoids the first patterned rear Al deposition step by instead patterning the nitride by mechanical abrasion.

A key result with this approach has been the demonstration of 18% cell efficiency on multicrystalline wafers [60], one of the highest values obtained for a cell larger than 1 cm^2 area, despite the relative simplicity of the approach. Problem areas are the use of shadow masks which are wasteful of deposited material, and rear shunting problems when the simpler technique above is used [60].

An alternative approach avoids the use of shadow masks by combining the mechanical abraded rear approach with obliquely evaporated contacts (OECO) onto a mechanically structured top surface. Using crystalline substrates with chemical texturing of the front surface (Figure 14) after larger scale mechanical structuring has resulted in large area cells of efficiency of 20%, using float-zoned substrates [61].

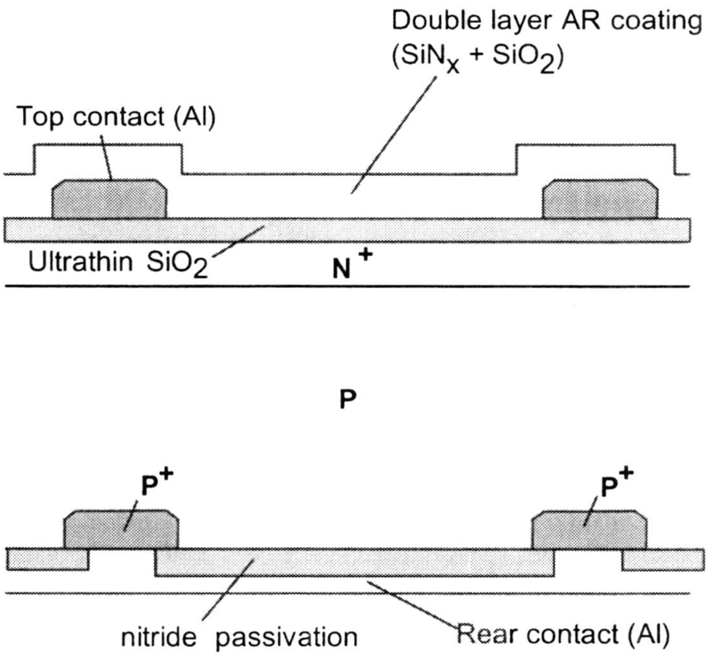

Figure 13 Simplified MINP cell fabricated using metal masks.

7 Conclusions

With the costs of present wafer-based silicon approaches dominated by material costs, particularly those of the wafers, encapsulants and low-iron tempered glass superstrates, increasing cell efficiency is an effective, if often counter-intuitive, approach to reducing the cost of the final product. Processing costs can be allowed to double if this results in a 20% improvement in cell performance. Work that stresses the trimming of processing costs at the expense of cell efficiency is not always as sensible as it might at first appear, in such a material-cost-dominated scenario.

Recent years have seen a diversification of manufacturing into higher efficiency approaches. Several manufacturers have realised the potential of plasma nitride based processing sequences to improve the performance of standard screen-printed cells on both multicrystalline and crystalline substrates, as first documented close to 20 years ago by Kyocera [62].

More revolutionary departures from the standard approach have been made by BP Solar, with its 'top of the line' Saturn processes, based on laser grooved, buried contacts, and Sanyo, with its crystalline/amorphous silicon hybrid HIT cells. These cells have demonstrated a clear advantage over standard cells in production, resulting in product with a clear performance margin.

In production, there is still the potential to capture more of the improvements in silicon cell performance that have been demonstrated in the laboratory. One

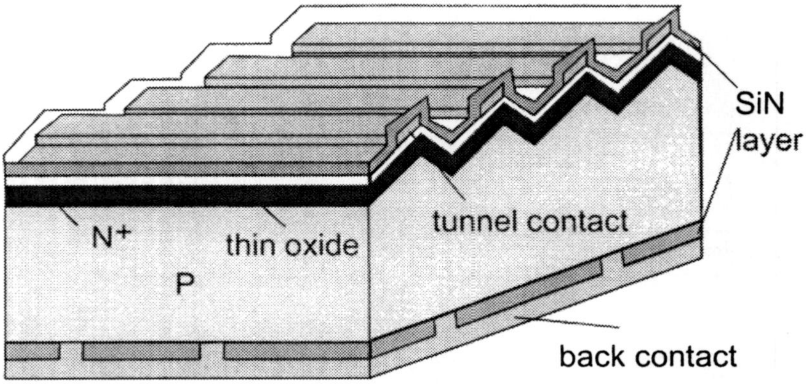

Figure 14 Possible implementation of a maskless MINP process on a mechanically textured wafer using oblique evaporation of top contact to eliminate this mask and mechanical abrasion to provide rear contact.

factor that has only quite recently been recognised as an impediment to this has been the tendency for boron-doped silicon to form an active boron-oxygen defect under illumination, restricting the quality of fielded material.

Several strategies, including the use of other dopants, have been suggested as a way of avoiding this problem. If successful, this gives rise to the possibility of material of much higher quality than currently used with only slight modification to the crystal growth economics. Such a development would increase the margin between the basic screen-printing and the more sophisticated approaches, giving real prospects for commercial cell efficiencies above 20%.

Acknowledgements

The author gratefully acknowledges an Australian Government Federation Fellowship. The Centre for Third Generation Photovoltaics is supported under the Australian Research Council Research Centre Scheme. The author thanks the Centre for Photovoltaic Engineering for its permission to reproduce material and figures from the text 'Silicon Solar Cells: Advanced Principles and Practice'.

References

[1] Schmela, M., 2003. Multicultural on the Roofs: Market Survey on Solar Modules, *Photon International*, February, pp. 32–41.

[2] Smith, K.D., Gummel, H.K., Bode, J.D., Cuttriss, D.B., Nielson, R.J. and Rosenzweig, W., 1963. The Solar Cells and their Mounting. *Bell Sys. Tech. J.*, Vol. 41, pp. 1765–1816.

[3] Iles, P.A., 1970. Increased Output from Silicon Solar Cells. *Proc. 8th IEEE Photovoltaic Specialists Conf.*, Seattle, pp. 345–352.

[4] Gereth, R., Fischer, H., Link, E., Mattes, S. and Pschunder, W., 1970. Silicon Solar Technology of the Seventies. *Proc. 8th IEEE Photovoltaic Specialists Conf.*, Seattle, p. 353.

[5] Gereth, R., Fischer, H., Link, E., Mattes, S. and Pschunder, W., 1972. *Solar Cell Technology Energy Conversion*, Vol. 12, pp. 103–107.

[6] Mandelkorn, J. and Lamneck, J.H., 1973. A New Electric Field Effect in Silicon Solar Cells. *J. Appl. Phys.*, Vol. 44, p. 4785.

[7] Godlewski, M.P., Baraona, C.R. and Brandhorst, H.W., 1973. Low-High Junction Theory Applied to Solar Cells. *Proc. 10th IEEE Photovoltaic Specialists Conf.*, Palo Alto, pp. 40–49.

[8] Lindmayer, J. and Allison, J., 1973. The Violet Cell: An Improved Silicon Solar Cell. *COMSAT Tech. Rev.*, Vol. 3, pp. 1–22.

[9] Haynos, J., Allison, J., Arndt, R. and Meulenberg, A., 1974. The Comsat Non-Reflective Silicon Solar Cell: A Second Generation Improved Cell. *Int. Conf. on Photovoltaic Power Generation*, Hamburg, p. 487.

[10] Rudenberg, H.G. and Dale, B., 1961. Radiant Energy Transducer, US Patent 3,150,999, filed 17 February 1961.

[11] Zhao, J. and Green, M.A., 1991. Optimized Antireflection Coatings for High Efficiency Silicon Solar Cells. *IEEE Trans. Electron Devices*, Vol. 38, pp. 1925–1934.

[12] Fossum, J.G. and Burgess, E.L., 1978. High Efficiency p^+ n n^+ Back-Surface-Field Silicon Solar Cells. *Applied Physics Letters*, Vol. 33, pp. 238–240.

[13] Godfrey, R.B. and Green, M.A., 1979. 655 mV Open Circuit Voltage, 17.6% Efficient Silicon MIS Solar Cells. *Applied Physics Letters*, Vol. 34, pp. 790–793.

[14] Green, M.A., 1975. Enhancement of Schottky Solar Cell Efficiency above its Semiempirical Limit. *Applied Physics Letters*, Vol. 28, pp. 287–268.

[15] Lindmayer, J. and Allison, J.F., 1976. Dotted Contact Fine Geometry Solar Cell. US Patent 3,982,964, September 1976.

[16] Arndt, R.A., Meulenberg, A. and Allison, J.F., 1981. Advances in High Output Voltage Silicon Solar Cells. *Proc. 15th IEEE Photovoltaic Specialists Conf.*, Orlando, pp. 92–96.

[17] Green, M.A., Blakers, A.W., Shi, J., Keller, E.M. and Wenham, S.R., 1984. High-Efficiency Silicon Solar Cells. *IEEE Trans. on Electron Devices*, Vol. ED-31, pp. 671–678.

[18] Lindholm, F.A., Neugroschel, A., Arienze M. and Iles, P.A., 1985. Heavily Doped Polysilicon Contact Solar Cells. *Electron Device Letters*, Vol. EDL-6, pp. 363–365.

[19] Yablonovitch, E., Gmitter, T., Swanson R.M. and Kwark, Y.H., 1985. A 720 mV Open Circuit Voltage, SiO_X:c-Si:SiO_X Double Heterostructure Solar Cell. *Applied Physics Letters*, Vol. 47, pp. 1211–1213.

[20] Van Halen, P. and Pulfrey, D.L., 1985. High-Gain Bipolar Transistors with Polysilicon Tunnel Junction Emitter Contacts. *IEEE Trans. on Electron Devices*, Vol. ED-32, p. 1307.

[21] Tanaka, M., Taguchi, M., Takahama, T., Sawada, T., Kuroda, S., Matsuyama, T., Tsuda, S., Takeoka, A., Nakano, S., Hanafusa, H. and Kuwano, Y., 1993. Development of a New Heterojunction Structure (ACJ-HIT) and its Application to Polycrystalline Silicon Solar Cells. *Progress in Photovoltaics*, Vol. 1, pp. 85–92.

[22] Blakers, A.W. and Green, M.A., 1986. 20% Efficiency Silicon Solar Cells. *Applied Physics Letters*, Vol. 48, pp. 215–217.

[23] Saitoh, T., Uematsu, T., Kida, T., Matsukuma, K. and Morita, K., 1987. Design and Fabrication of 20% Efficiency, Medium-Resistivity Silicon Solar Cells. *19th IEEE Photovoltaic Specialists Conf.*, New Orleans, pp. 1518–1519.

[24] Callaghan, W.T., 1986. Evening presentation on Jet Propulsion Lab. Photovoltaic activities. *Proc. 7th European. Photovoltaic Solar Energy Conf.*, Seville.

[25] Sinton, R.A., Kwark, Y., Gan, J.Y. and Swanson, R.M., 1986. 27.5% Si Concentrator Solar Cells. *Electron Device Letters*, Vol. EDL-7, p. 567.

[26] King, R.R., Sinton, R.A. and Swanson, R.M., 1988. Front and Back Surface Fields for Point-Contact Solar Cells. *Proc. 20th IEEE Photovoltaic Specialists Conf.*, Las Vegas, pp. 538–544.

[27] Blakers, A.W., Wang, A., Milne, A.M., Zhao, J., Dai, X. and Green, M.A., 1989. 22.6% Efficient Silicon Solar Cells. *Proc. 4th International Photovoltaic Science and Engineering Conf.*, Sydney, pp. 801–806.

[28] Green, M.A., 1995. *Silicon Solar Cells: Advanced Principles and Practice.* Bridge Printery, Sydney.

[29] Green, M.A., Zhao, J., Wang, A. and Wenham, S.R., 1992. 45% Efficient Silicon Photovoltaic Cell Under Monochromatic Light. *IEEE Electron Device Letters*, Vol. 13, pp. 317–318.

[30] Green, M.A., Zhao, J., Blakers, A.W., Taouk, M. and Narayanan, S., 1986. 25-Percent Efficient Low-Resistivity Silicon Concentrator Solar Cells. *IEEE Electron Device Letters*, Vol. EDL-7, pp. 583–585.

[31] Cuevas, A., Sinton, R.A. and Swanson, R.M., 1990. Point- and Planar-Junction P–I–N Silicon Solar Cells for Concentration Applications, Fabrication, Performance and Stability. *Proc. 21st IEEE Photovoltaic Specialists Conf.*, Kissimimee, pp. 327–332.

[32] Bruton, T., Luthardt, G., Rasch, K-D. Roy, K., Dorrity, I.A., Garrard, B., Teale, L., Alonso, J., Ugalde, U., Declerqu, K., Nijs, J., Szlufcik, J., Rauber, A., Wettling, W. and Vallera, A., 1997. A Study of the Manufacture at 500 MWp p.a. of Crystalline Silicon Photovoltaic Modules. *Proc. 14th European Photovoltaic Solar Energy Conf.*, Barcelona, pp. 11–26.

[33] Green, M.A. and Wenham, S.R., 1995. Silicon Cells: Single Junction, One Sun, Terrestrial, Single- & Multi-Crystalline. In: Partain, L., Ed., *Solar Cells and Their Applications*, Wiley, New York.

[34] Final Report, Flat Plate Solar Array Project, Vol. V, *Jet Propulsion Laboratory*, Publication 86-31, October 1986.

[35] Mertens, R., 1994. Silicon Solar Cells. *Proc. 12th European Photovoltaic Solar Energy Conf.*, Amsterdam, pp. 1–6.

[36] King, R.R., Mitchell, K.W. and Gee, J.M., 1994. Back Surface Cell Structures for Reducing Recombination in CZ Silicon Solar Cells. *Proc. First World Conf. on Photovoltaic Energy Conversion*, Hawaii.

[37] Wenham, S.R., Willison, M.R., Narayanan, S. and Green, M.A., 1985. Efficiency Improvement in Screen Printed Polycrystalline Silicon Solar Cells by Plasma Treatments. *Proc. 18th IEEE Photovoltaic Specialists Conf.*, Las Vegas, p. 1008.

[38] Szlufcik, J., Elgamel, H.E., Ghannam, M., Nijs, J. and Mertens, R., 1991. Simple Integral Screenprinting Process for Selective Emitter Polycrystalline Silicon Solar Cells. *Appl. Phys. Lett.*, Vol. 59, pp. 1583–1584.

[39] Mardesich, N., 1981. Solar Cell Efficiency Enhancement by Junction Etching and Conductive AR Coating Processes. *Proc. 15th IEEE Photovoltaic Specialists Conf.*, Kissimmee, pp. 446–449.

[40] Wenham, S.R., 1993. Buried-Contact Silicon Solar Cells. *Progress in Photovoltaics*, Vol. 1, pp. 3–10.

[41] Bruton, T.M., Mitchell, A. and Teale, L., 1991. Maximizing Minority Carrier Lifetime in High Efficiency Screen Printed Silicon BSF Cells. *Proc. 10th European Photovoltaic Solar Energy Conf.*, Lisbon, pp. 667–669.

[42] Bruton, T.M., Mason, N.B. and Summers, J.G., 1992. Towards Production of High Efficiency Terrestrial Solar Cells. *Proc. 6th International Photovoltaic Science and Engineering Conf.*, New Delhi, pp. 21–24.

[43] From data sheet, BP Saturn Solar Cells, 1991.

[44] Jooss, W., Fath, P., Bucher, E., Roberts, S. and Bruton, T., 2002. Large Area Multicrystalline Silicon Buried Contact Solar Cells with Bulk Passivation and Efficiency of 17.5%. *Proc. 29th IEEE Photovoltaic Specialists Conf.*, New Orleans, pp. 202–205.

[45] Honsberg, C.B., Yun, F., Ebong, A., Taouk, M., Wenham, S.R. and Green, M.A., 1994. 685 mV Open Circuit Voltage Laser Grooved Silicon Solar Cell. *Sol. Energy Mater. Sol. Cells*, Vol. 34, pp. 117–124.

[46] Chan, B.O., 1993. Defects in Silicon Solar Cell Materials. PhD Thesis, University of New South Wales.

[47] Sopori, B.L., Jastrzebski, L., Tan, T.Y. and Narayanan, S., 1994. Gettering Effects in Polycrystalline Silicon. *Proc. 12th European Photovoltaic Solar Energy Conf.*, Amsterdam, pp. 1003–1006.

[48] Boller, H.-W. and Ebner, W., 1989. Transfer of the BCSC-Concepts Into an Industrial Production Line. *Proc. 9th European Photovoltaic Solar Energy Conf.*, Freiburg, pp. 411–413.

[49] Kyle, C., 1991. Racing with the Sun: The 1990 World Solar Challenge. Engineering Society for Advancing Mobility: Land, Sea, Air and Space, SAE Order No. R-111.

[50] Knöpfel, H. and Nordmann, Th., 1992. 24 kW Photovoltaic Interconnected Network Installation to Power the Marzili Funicular Railway, Berne. *Proc. 11th European Photovoltaic Solar Energy Conf.*, Montreux, pp. 1475–1482.

[51] Alonso, M., Pottbrock, R., Voermans, R., Villa, J.J. and Yordi, B., 1994. 1 MW Photovoltaic Power Station Toledo/Spain – Plant Description and

Gained Experience during Construction – Toledo PV. *Proc. 12th European Photovoltaic Solar Energy Conf.*, Amsterdam, pp. 1163–1166.

[52] BP Saturn product sheet, April 1994.

[53] Wohlgemuth, J. and Narayanan, S., 1991. Buried Contact Concentrator Solar Cells. *Proc. 22nd IEEE Photovoltaic Specialists Conf.*, Las Vegas, pp. 273–277.

[54] Wenham, S.R., Wu, Y., Xiao, R.D., Taouk, M., Guelden, M., Green, M.A. and Hogg, D., 1992. Pilot Line Production of Laser Grooved Silicon Solar Cells. *Proc. 11th European. Photovoltaic Solar Energy Conf.*, Montreux, pp. 416–422.

[55] Bruton, T.M., 1994. Fabrication of Laser Grooved Buried Contact Si Solar Cells. *Proceedings 1st EU International Workshop on Crystalline Silicon Solar Cells*, Spain.

[56] Sakaata, H., Nakai, T., Baba, T., Taguchi, M., Tsuge, S., Uchihashi, K. and Kiyama, S., 2000. 20.7% Highest Efficiency Large Area (100.5 cm^2) HIT^{TM} Cell. *Proc. 28th IEEE Photovoltaic Specialists Conf.*, Anchorage, p. 7.

[57] Green, M.A., Emery, K., King, D.L., Igari, S. and Warta, W., 2002. Solar Cell Efficiency Tables (Version 20). *Progress in Photovoltaics*, Vol. 10, pp. 355–360.

[58] Schmela, M., 2003. A Bullish PV Year: Market Survey on World Cell Production in 2001. *Photon International*, March, pp. 42–48.

[59] Aberle, A.G., 1999. *Crystalline Silicon Solar Cells*. University of New South Wales.

[60] Mittelstadt, L., Dauwe, S., Metz, A., Hezel, R. and Hassler, C., 2002. Front and Rear Silicon-Nitride-Passivated Multicrystalline Silicon Solar Cells with an Efficiency of 18.1%. *Progress in Photovoltaics*, pp. 35–39.

[61] Metz, A. and Hezel, R., 2001. Easy-to-Fabricate 20% Efficiency Large-Area Silicon Solar Cells. *Sol. En. Matls. Sol. Cells*, Vol. 65, pp. 325–330.

[62] Kimura, K., 1984. Recent Developments in Polycrystalline Silicon Solar Cell. *Technical Digest, 1st International Photovoltaic Science and Engineering Conf.*, Kobe, pp. 37–42.

Part IIc

Thin Film Technologies

IIc-1

Amorphous Silicon Solar Cells

David E. Carlson, BP Solar, Linthicum, Maryland, USA
Christopher R. Wronski, Center for Thin Film Devices, Pennsylvania State University, USA

1 Introduction

Significant progress has been made over the last two decades in improving the performance of amorphous silicon (a-Si) based solar cells and in ramping up the commercial production of a-Si photovoltaic (PV) modules, which is currently more than 40 peak megawatts (MW_p) per year. The progress in a-Si solar cell technology can be attributed to concurrent advances in the areas of new and improved materials, novel cell designs and in the development of large-area deposition techniques suitable for mass production. There are currently more than 40 research, development and engineering organisations at universities, companies and government laboratories around the world that are actively investigating a-Si PV technology.

The first investigation of amorphous silicon deposited from a silane discharge was performed by Chittik et al. in 1969 [1]. Subsequent work carried out on this material showed that it had a much lower density of defects than evaporated or sputtered amorphous silicon. The ability to dope this material both n- and p-type was independently discovered by Carlson [2] and Spear and LeComber [3]. Carlson and Wronski [4] showed that a-Si had useful optoelectronic properties when they reported the first results on a-Si solar cells with a conversion efficiency of 2% in 1976, and the efficiency was increased to 5% shortly afterwards [5]. These device results sparked a worldwide interest not only in a-Si solar cells but also in other applications. This heightened interest led to a wide range of fundamental studies on the a-Si alloys materials as well as a large effort to improve the performance of a-Si solar cells.

It soon became clear that hydrogen was playing an important role in determining the optoelectronic properties of these materials [6], and that discharge-deposited a-Si is actually an alloy of hydrogen and silicon or hydrogenated amorphous silicon (a-Si:H). Subsequently, many organisations started investigating the intrinsic optoelectronic and photovoltaic properties of a-Si:H alloys. In 1977, Staebler and Wronski [7] observed large changes in the photoconductivity and dark conductivity of a-Si:H when the material was exposed to sunlight. The metastable changes, commonly known as the Staebler-Wronski Effect (SWE), are perfectly reversible upon annealing at or above 150°C for a few hours [7]. These light-induced changes manifest themselves in both thin film materials and solar cells. The early discovery of the SWE had a critical effect on the development of a-Si solar cell technology by impacting the design of the devices and the optimisation of the a-Si:H alloys. Approaches were developed to minimise the effects of the SWE on the light-soaked (or stabilised) cell efficiencies, which rely on engineering the cells to have active layers as thin as possible [8].

The development of high-performance a-Si based solar cells and their technology advanced along several fronts. Continuous improvements were made in a-Si:H materials by investigating a wide range of deposition conditions for plasma-induced decomposition of silane, a process that is commonly referred to as plasma-enhanced chemical vapour deposition (PECVD). While the early deposition work was performed using primarily DC and RF PECVD [2], subsequent studies showed that good quality a-Si alloys could be deposited using VHF [9] (~30–110 MHz) and microwave (~2.45 GHz) PECVD [10, 11]. These studies established the optimum decomposition conditions for a-Si:H and led to the development of commercial processes that allow a-Si solar cell structures to deposited on large-area substrates at relatively low substrate temperatures (~200–250°C) [12]. Other work established the highly beneficial effects of diluting the silane with hydrogen on the initial and stabilised properties of a-Si:H [13]. It is now recognised that hydrogen dilution leads to the growth of protocrystalline a-Si:H [14], which is now generally used in the fabrication of high-performance a-Si solar cells.

Another key advance in the development of a-Si:H alloys for solar cells was the discovery that the bandgap can be changed by varying the incorporation of hydrogen [15]. Moreover, as in the case of the crystalline materials, the bandgap of a-Si-based alloys can be varied by alloying with carbon or germanium. The development of wide bandgap p-type a-Si:C:H alloys has allowed the fabrication of heterojunction solar cells that exhibit very little absorption in the p-layer and also yield high built-in potentials [16]. The development of narrow bandgap a-Si:Ge:H alloys has led to the fabrication of both tandem and triple junction cells with relatively high efficiencies [17, 18]. While alloying with carbon or germanium does create additional defect states, relatively good optoelectronic properties can be obtained over bandgaps ranging from ~1.3 eV (~75 at.% Ge) to ~2.1 eV (~15 at.% C). This ability to tune the bandgap of a-Si based alloys has been an underlying factor in the improvement in the performance of a-Si based solar cells which has led to initial efficiencies as high as 15.2% [19]. The use of multijunction solar cell structures allows one to use relatively thin component cells, which in turn helps to reduce the degradation resulting from the SWE [8].

However, there are serious constraints imposed on the reduction in the thickness of the junctions since this leads to a decrease in the absorption of sunlight (and a corresponding decrease in short circuit currents) and also to an increase in shorts and shunts. The difficulty with low light absorption in thin cells was greatly reduced with the development of efficient optical enhancement obtained by introducing textured rather than smooth optical reflectors [20, 21]. Such optical enhancement, which was first successfully applied to a-Si based solar cells, is now extensively used in all types of thin film solar cells including thin film crystalline solar cells.

There has also been significant progress made over the last few decades in developing large-scale processing for a-Si based PV modules. All commercial a-Si based PV modules are fabricated using PECVD to deposit the a-Si alloy layers since excellent uniformity can be obtained over substrates more than 1 m^2 in area. In addition, early work at RCA Laboratories showed that most shorts and shunts could be cured by applying a reverse bias [22], and this curing process has allowed manufacturers to obtain relatively high yields in the production of large-area a-Si based PV modules [23]. Other work at RCA Laboratories led to the development of a laser scribing process that automatically patterned the a-Si on insulating substrates (such as glass) so that the resulting product was a large-area, monolithic PV module of a-Si solar cells connected in series [24].

2 Amorphous Silicon Alloys

The optoelectronic properties of amorphous silicon alloys vary over a wide range of parameters and are strongly influenced by the plasma deposition conditions used in PECVD reactors. In the this section, we will discuss the influence of the deposition conditions on the microstructure and optoelectronic properties, and also review the important aspects of doping and light-induced degradation.

2.1 Deposition Conditions and Microstructure

Although historically a variety of deposition methods has been used to deposit hydrogenated amorphous silicon, the most common method by far is PECVD, also known as glow-discharge deposition. The decomposition of the feedstock gases is carried out with plasmas that are generated over an extremely wide frequency range that includes DC, RF (13.56 MHz), VHF (30–110 MHz); and microwave (2.45 GHz) frequencies. A source gas such as silane (SiH_4) is decomposed by electron impact into a mixture of radical and ionic species that land on the substrate to produce the solid film. The PECVD process contains many variables that determine the quality of the materials produced: the substrate temperature, the pressure, the flow rate of the source gases, the plasma power, the frequency, the electrode spacing, and the source gases. In addition, the electronic properties of the a-Si alloys can be adversely affected by impurities such as oxygen, carbon and nitrogen [25, 26], which can be introduced by air leaks, residual water vapour and pump oil contamination.

The temperature of the substrate controls the reactions on the growing surface, and thus is a critical parameter in determining the quality of the materials. The gas pressure determines the mean free path for the collisions of the gas molecules and influences whether the reactions are at the growing surface or in the gas. The flow rate of the source gases is an important deposition parameter since it determines the residence time of the gas species in the plasma and hence affects the growth kinetics. The power controls the rate of dissociation of the gas and therefore also the film growth rate. The frequency used also affects the nature of the plasmas, and in particular the ion bombardment, which becomes less significant at VHF and microwave frequencies.

Finally, the nature of the plasmas and the growth processes also change with the introduction of the alloy forming gases, GeH_4 and CH_4, and with the addition of the n-type or p-type dopant gases. However, in all the types of depositions using PECVD, hydrogen plays a key role in reducing defects and improving the quality of the materials. The large number of variables are *interdependent* making PECVD a very complex process. This large number of variables, however, also makes the process and thus the material properties very flexible.

The a-Si alloy deposition process consists of four steps. First, the silane and hydrogen molecules are dissociated by electron impact into a reactive mixture of neutrals and radicals, which consists of a variety of radicals and ions of silane and hydrogen, unreacted silane, and higher silane species (such as disilane and trisilane). This mixture is then transported to the surface of the growing film via gas diffusion, a process during which further chemical reactions of the different species can occur. The molecules arriving on the surface can then react with and adsorb to the growing film. Finally, the reaction byproducts, mainly hydrogen and unreacted silane radicals, desorb from or are etched off the surface by incoming reactive species, and are then pumped away. It has been established that high quality films with low defect densities are generally obtained with pure silane (or hydrogen-diluted silane), low RF power, and a substrate temperature of 200–300°C [27].

The low temperature PECVD process offers a number of technological advantages. Not only can it be readily scaled up to produce photovoltaic modules with very large areas, but this process also allows an extremely high degree of uniformity to be obtained over large areas. PECVD is also a deposition process that allows controlled changes in composition to be carried out during growth with very high precision.

The optoelectronic properties of the a-Si:H based materials depend very strongly on their microstructure and its evolution during growth. Because the hydrogen bonding is very sensitive to the deposition conditions, the presence of hydrogen in the plasma is closely related to the quality of material since it not only passivates dangling bonds, but it is also believed to be responsible for a reconstruction of the network [28, 29].

The beneficial effect of hydrogen available from the feedstock gases (such as SiH_4, GeH_4, CH_4, etc.) has been successfully augmented by diluting the feedstock gases with hydrogen gas (H_2) [30, 31], and this is being extensively utilised in the deposition of high-performance solar cells. The dilution with hydrogen has a large effect on the growth beginning with the nucleation and coalescence of the thin films and then in controlling the bulk as well as the growing surface [32]. With relatively low hydrogen dilution, for example $R = [H_2]/[SiH_4] = 10$, the growth of a-Si:H films not only becomes dependent on the nature of the substrate but the growth also changes so as to make the microstructure thickness dependent. As a result, during growth, the materials, which are initially amorphous, eventually become microcrystalline as the films become thicker.

Using in-situ real time spectroscopic ellipsometry [32], deposition phase diagrams have been developed that describe how the microstructure and phase evolve during the growth of Si:H films [33–36]. Such diagrams identify the thickness regime within which a-Si:H grows, the thickness regime which involves a transition to a mixed phase of amorphous and microcrystalline Si:H (μc-Si:H) and then the thickness at which the film becomes single-phase μc-Si:H. The hydrogen-dilution gas flow ratio is used most often as the key parameter in these phase diagrams since it provides the most direct control over the phase. However, the microstructure and phase can also be controlled by other parameters such as the substrate temperature, power and total gas pressure [37]. An example phase diagram is shown in Figure 1 for a-Si:H deposited as a function of the dilution gas flow ratio $R = [H_2]/[SiH_4]$ on a $R = 0$ amorphous silicon substrate (i.e., a-Si:H prepared without H_2-dilution).

The striking and key characteristic of the phase diagram in Figure 1 is how the hydrogen dilution affects the film thickness at which the transition from an

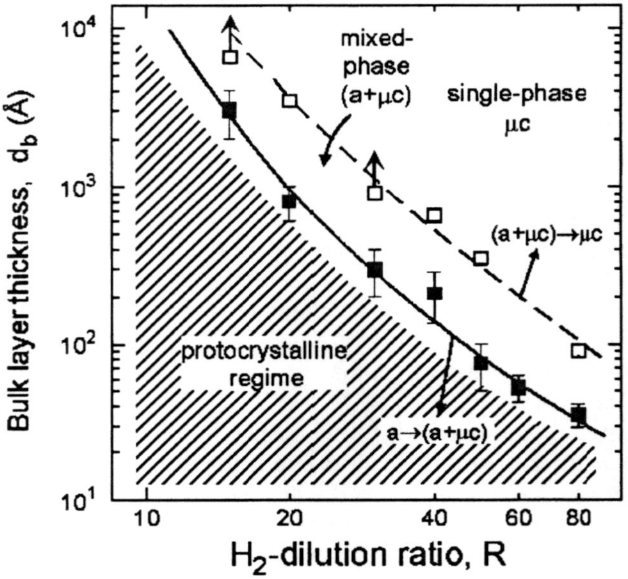

Figure 1 Film thickness, d_b, at which the different phase transitions occur during Si:H film growth plotted as a function of the hydrogen dilution ratio R.

amorphous to a mixed phase occurs. The rapid decrease in this thickness at high hydrogen dilutions has to be taken into account as it limits the thickness of the i-layers in cells that can be fabricated with purely a-Si:H. Such deposition phase diagrams have led to the concept of the *protocrystalline* Si:H growth regime, shaded in Figure 1 [14, 36].

Despite the evolutionary nature of the Si:H materials prepared with moderate hydrogen dilution, these materials exhibit uniform bulk properties over extended regions of thickness. In addition to the unique evolutionary growth behaviour exhibited under the protocrystalline Si:H growth conditions, the protocrystalline material itself exhibits unique optoelectronic properties. One of these is that the hydrogen incorporated into the protocrystalline Si:H with higher R increases its bandgap [38]. Consequently by taking into account the evolutionary nature it is possible to maximise the bandgap for different thickness layers while maintaining their excellent properties as protocrystalline a-Si:H [39, 40]. It also allows doped protocrystalline a-Si:H deposited with high R to be successfully applied in forming the p-contacts in n–i–p solar cells [41, 42]. Because these layers are very thin, the highest open-circuit voltages (V_{OC}) are obtained with protocrystalline p layers and not, *as has been extensively claimed* [43], with p-type microcrystalline Si:H, which actually results in low open-circuit voltages.

2.2 Optoelectronic Properties

The absence of long-range order broadens the distribution of states compared to crystalline Si, forming bandtails and a continuous distribution of localised states in the gap that reduces the carrier scattering length to atomic distances. The localised states that lie in the bandgap are a result of the disorder and structural defects such as broken bonds [7, 44]. These gap states determine many of the electronic properties since they cause carrier trapping and act as recombination centres [45]. Amorphous silicon based alloys have a very high absorption coefficient due to the random nature of the atomic ordering so they behave like a direct bandgap semiconductor [46]. The electron densities of states are still a function of energy, but there is now a continuous distribution of localised states. The equivalent of the bandgap in crystalline silicon becomes a region of extended states in amorphous silicon where electrons and holes can move as free carriers. Both the strong light absorption and the values of the mobility gaps, which are adjustable by alloying, make a-Si:H attractive for solar cell technology since it offers the possibility of very thin cells as compared to crystalline silicon cells.

The bandgaps and optical absorption, α, of the a-Si:H based materials can be changed by the extent to which hydrogen, germanium and carbon are incorporated to form a-Si:H, a-SiGe:H and a-SiC:H alloys [15, 47, 48]. Incorporation of hydrogen into the a-Si:H network not only removes defects and their states in the gap, but also widens the gap. Using hydrogen dilution of silane, it is possible to obtain bandgaps of around 2 eV without deteriorating the microstructure and the electronic properties [38, 40]. However the rapid transition in the microstructure with high hydrogen dilution limits the protocrystalline a-SiH regimes to very thin layers, which does not allow them to be used as intrinsic absorber layers. As in the case of a-Si:H, hydrogen dilution allows one to deposit a-SiGe:H and a-SiC:H alloys with better microstructure, and hence lower defect densities. However, the electronic properties of a-SiGe:H and a-SiC:H alloys are not as good as a-Si:H, and their deterioration with the increase of either germanium or with carbon limits the range of bandgaps that can be used in efficient cells [30, 34, 49].

Both the bandgap and the densities of gap states can be changed by controlling the hydrogen content and the microstructure of a-Si:H. The densities of dangling bond defects is less than 10^{16} cm^{-3} in PECVD a-Si:H materials with a hydrogen content of ~10 at.%, and the variation of the defect density with energy exhibits a rapid falloff in the vicinity of the tail states. The exponential regions of the optical absorption (α) between about 10^3 and 10 cm^{-1} arise from the absorption in valence band tail states, which are due to gap states introduced by the disorder [50]. The densities of the valence band tail states are significantly higher than those of the conduction band tails [51] even in the recently developed materials with a highly ordered network [52]. The absorption that is useful in creating free carriers in solar cells is at values of α greater than about 10^3 cm^{-1} which corresponds to photons with energies greater than the bandgap of the material. The phase transitions in protocrystalline a-Si:H set an upper limit of ~1.9 eV for the bandgap of a-Si:H while the increase in defects with the incorporation of germanium set the lower limit at ~1.3 eV for the bandgap of a-SiGe:H. These bandgaps, however, cover a sufficiently wide range to offer the flexibility required for constructing efficient single-junction, as well as multijunction solar cells.

A knowledge of the densities of states (DOS) and their energy distributions in the gap of a-Si:H is necessary for quantitative analysis that describes the role of both native and light-induced defects on the optoelectronic properties of the material. The determination of the complete density of states distribution for amorphous silicon is extremely difficult because there is no periodic structure and it is prepared under non-equilibrium conditions and is therefore metastable. Many models for the a-Si:H DOS have been proposed, all of which envisage conduction and valence band-tails. However, each model has distinctly different distributions of states near the middle of the gap. One distribution of densities of electronic states that is found useful in the self consistent interpretation of results of both a-Si:H films and cells is illustrated in Figure 2. The differences in the optical and electronic transport properties of amorphous silicon associated with such a DOS with those of its crystalline counterpart must be taken into account in the design of a-Si solar cells.

Deep lying gap states consist of neutral dangling-bond states (D^0) in the middle of the gap, negatively charged defect states (D^-) below the middle of the gap, and positively charged defect states (D^+) above the middle of the gap [53–55]. It is these deep lying states that are very important in determining the collection of photogenerated carriers in a-Si solar cells. Alloying with Ge has little effect on the tail states but a large effect on increasing the densities of the deep lying states.

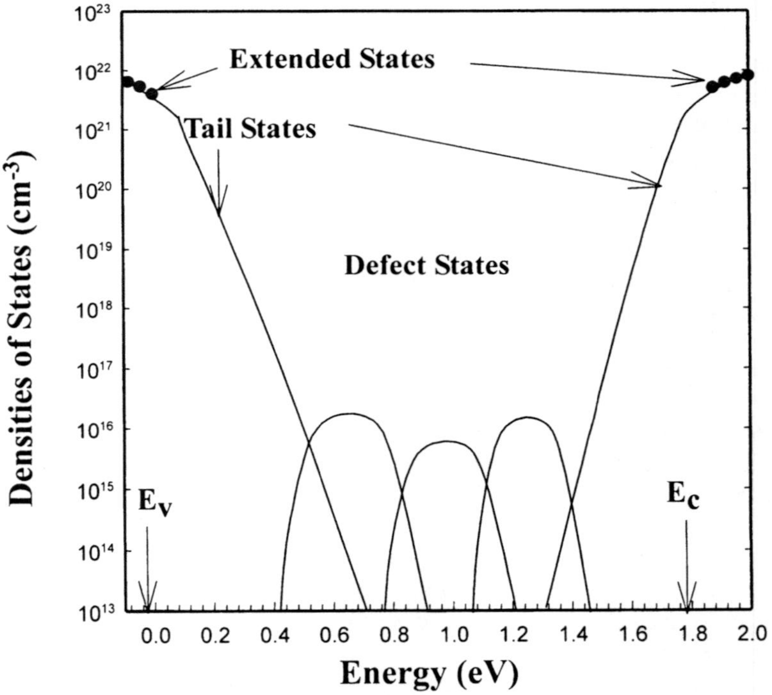

Figure 2 Schematic representation of the density of electronic state distribution in a-Si:H showing the conduction band edge (E_C), the valence band edge (E_V), the extended states, tail states, and defect states.

There is also little effect on the electronic properties of the materials with low levels of impurities such as O, N, C which are generally less than several times 10^{18} cm^{-3} [56]. The semiconductor properties and the stability of the a-Si:H based alloys, however, do depend on the growth processes, the incorporation of hydrogen and the resultant microstructure [13, 32, 57]. By optimising the growth conditions it has been possible to obtain a-Si:H materials with properties that are outstanding for an amorphous semiconductor. Intrinsic (undoped) materials have Fermi levels near the middle of the energy gap, and free carrier transport occurs in extended states with mobilities that are significantly lower than those in crystalline silicon, being only about 10 and 1 cm^2 V^{-1} s^{-1} for electrons and holes respectively [51].

The densities of the deep lying states, which are as low as 10^{15} to 10^{16} cm^{-3} in the a-Si:H based materials, are key to efficient solar cell operation. Consequently, the hole and electron lifetimes are in the range of 10^{-8} to 10^{-6} s, and the space charge densities in the junctions are low enough to allow the electric fields in the junctions to extend over the entire thickness of efficient solar cells [45]. The a-SiGe:H alloys retain good semiconductor properties with densities of midgap states around 10^{16} cm^{-3} even with about 60 at.% of germanium in the alloy. In the case of the a-SiC:H alloys, even a relatively low incorporation of carbon (< 10 at.%) has a large effect on the microstructure, and the resulting density of defects, especially after light soaking, is too high for use as an absorber layer in solar cells.

2.3 Doping

The undoped materials used in solar cells have the Fermi levels at or near midgap but are slightly n-type. Because the a-Si:H based materials have such low densities of midgap defects, they are the only amorphous materials that can be doped both n- and p-type. The introduction of donors or acceptors can readily move the Fermi level towards the conduction or the valence band, respectively. As in the case of crystalline silicon, n-type doping is achieved by incorporating phosphorous into the materials, and p-type doping by incorporating boron. The incorporation of high densities of these dopants however introduces defect states near midgap, which limit the doping efficiency and drastically reduce free carrier lifetimes [58]. As a result, doped materials cannot be used as active absorber layers in solar cells as is the case in p/n junction crystalline silicon solar cells.

Instead, thin films of either p-type a-SiC:H [59] or p-type protocrystalline Si:H [42] are used in p/i heterojunctions while n-type a-Si:H or n-type μc-Si:H layers are used as ohmic contacts. The p-type a-SiC:H and the p-type *protocrystalline*-Si:H materials result in excellent window contacts since their quasi Fermi levels are located about 0.4 eV and 50 meV from their respective valence bands. The n-type a-Si: H and n-type μc-Si:H provide excellent ohmic contacts to the i-layers since their Fermi levels are about 0.2 eV and 50 meV from their respective conduction bands. By incorporating such p-type and n-type materials into p-i–n and n–i–p cells, it is possible to obtain built-in voltages well over one volt [60]. The use of thin p-type and n-type contacts with absorber i layers that have adjustable bandgaps and low densities of deep lying gap states offers great flexibility in efficiently absorbing and utilising different parts of the solar spectrum.

2.4 Light-Induced Degradation

Although hydrogenated amorphous silicon has many technologically attractive properties, there are several challenges regarding material properties that have yet to be overcome for further advances to be made in the technology. The most problematic of these is the Staebler–Wronski effect , which was discovered in 1977 [7]. It was found that not only did the illumination of glow-discharge deposited a-Si:H films with sunlight lower both the photoconductivity and dark conductivity, but even more striking was that these properties return to their as-deposited values after annealing at temperatures above ~150°C. The reversible changes that occur between the annealed, initial state and 'light-soaked' state have become one of the most investigated phenomena in a-Si:H based material and solar cells [61–64].

However, progress has been relatively slow in obtaining a definitive understanding and systematic control of the SWE and though advances have been made in the understanding of SWE, as of yet there is still no *general consensus* on *the exact nature* of the light-induced defects *or the mechanisms responsible for their creation*. This is in large part because unlike crystalline silicon there is no unique a-Si:H material, and a dependence not only of intrinsic but also of light-induced defects on the deposition conditions and resultant microstructure. However, there is general agreement that hydrogen, which is critical to the passivation of dangling bonds, also plays an important role in the SWE [55, 65–68].

This creation of defects by light severely limits the optoelectronic properties of the a-Si:H based materials since the metastable defects adversely affect not only the electron but also the hole transport, which results in a significant reduction of the performance of solar cells. Extensive studies have been carried out on thin film materials to characterise and understand the mechanisms responsible for the SWE [55, 61, 69, 70], and the results in the vast majority of these studies have been interpreted based on the creation of neutral dangling bonds. Strong evidence has been found in more recent studies carried out on *both* solar cells and corresponding intrinsic thin films that light-induced changes in the charged defects D^-, D^+ states are just as, if not more, important as the D^0 (neutral dangling bond) states [14, 71–74]. Significant progress has been made over the years not only in improving the initial properties of a-Si:H based material but also in reducing the light-induced degradation by optimising the growth conditions to improve the microstructure. The development of protocrystalline a-Si:H materials has resulted in solar cells that not only have higher initial efficiencies, but even more importantly better *end of life* performance [31, 75]. The significant improvements obtained with protocrystalline Si:H are illustrated in Figure 3, which shows the light-induced changes of the fill factor for a 4000 Å thick p–i–n solar cell with a

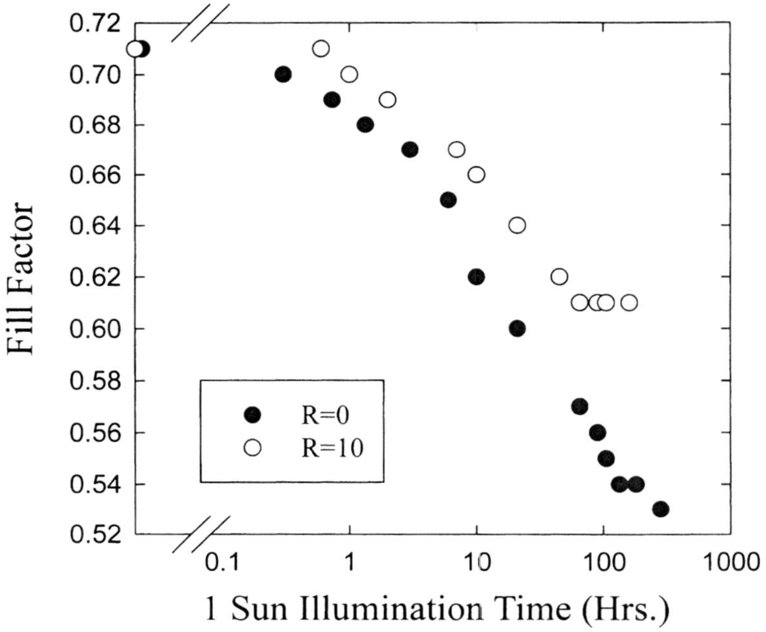

Figure 3 Degradation of the fill factor of a-Si:H p-i-n cells with undiluted (R=0) and protocrystalline (R=10) intrinsic layers under 1 sun illumination at 25°C.

protocrystalline (R = 10) a-Si:H i-layer and for a similar cell deposited under the same conditions with an undiluted (R = 0) intrinsic layer.

However, despite significant advances in improving the properties of a-Si:H based alloys, the fundamental understanding of the SWE *still remains a key issue* that has to be addressed and resolved to assure further improvements in the performance of solar cells based on a-Si:H alloys.

3 Amorphous Silicon Solar Cells

3.1 Physics of Operation

The operation of all solar cells is based on common physical principles. However, since efficient a-Si based solar cells rely on material properties distinctly different from those of crystalline silicon, the basic cell structures are somewhat different. In order to take advantage of the excellent properties of the intrinsic (undoped) a-i:H and a-SiGe:H materials, p–i–n and n–i–p heterojunction cell structures are used rather than the classic n/p junction structures in crystalline silicon. A schematic energy band diagram of an a-Si:H p–i–n structure generated by numerical simulation, using the distribution of defect states illustrated in Figure 2, is shown in Figure 4 under equilibrium conditions in the dark.

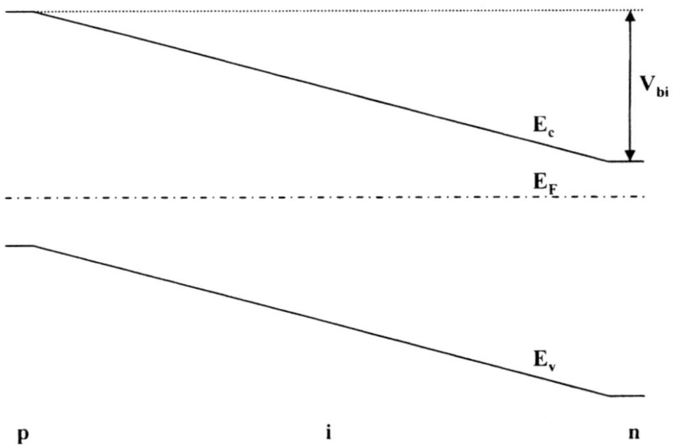

Figure 4 Energy band diagram of an a-Si p-i-n photovoltaic cell in thermodynamic equilibrium in the dark E_c and E_v are the conduction and valence band edges, E_F is the Fermi level and V_{bi} is the built in potential.

The p- and n-layers provide the built in potential of the junction in the device, however due to the short lifetime in the highly defective doped materials [58] the photogenerated carriers in the doped layers are not collected and do not contribute to the cell photocurrents. The fabrication of a p–i–n cell begins with the deposition of a p-type 'window layer' on the transparent conductive oxide (TCO). An a-Si:H intrinsic layer (i-layer) is deposited to form the bulk absorber region of the cell. The final step in forming the single-junction p–i–n cell is the n-layer deposition.

Important considerations for the choice of TCO materials are their optical transmission, conductivities and ability to form a good contact to the p-layers. The ideal TCO should have a low sheet resistance, high optical transparency in the wavelength range 400 to 1000 nm, and result in a small or ideally no potential barrier at the p/TCO interface. The band bending at the interface depends on the front contact material, the p-layer bandgap, doping and densities of states, as well as its thickness. In order to minimise optical absorption the p-type window layers used in high efficiency cells are thin ($\sim$10 nm), which increases the likelihood that they are fully depleted. In order to maximise the cell efficiency it is necessary to achieve a high V_{OC} using a very thin p-layer. If the p-layer is too thick the device performance will be adversely affected by the loss in the photocurrent due to the higher absorption in the thick p-layer.

Most of the important differences in the physics of a-Si based solar cells and crystalline silicon solar cells are a direct result of the most fundamental difference in the materials – the large density of localised gap states in a-Si:H. In a-Si solar cells, light that is absorbed in the i-layer will create electrons and holes, and the collection of these photogenerated carriers is assisted by the internal electric field. Due to the short carrier lifetimes associated with the localised gap states, the photogenerated carriers in a-Si based cells must be collected primarily as a drift current, not as a diffusion current as is the case in crystalline silicon solar cells. These gap states have important ramifications on the cell performance since a large density of photogenerated carriers can become trapped in these states. The native and light-induced defects in a-Si p–i–n devices adversely affect the carrier collection in two ways – they act as recombination centres and also shield the electric field produced by the doped layers – which changes the electric field distribution in the i-layer. Under 1 sun illumination the carrier generation and recombination rates, the electric field distributions, and the carrier profiles in a-Si solar cells are highly non-uniform. The densities of native midgap defect states in a-Si solar cell materials are $\sim 10^{16}$ cm^{-3} in the as-deposited and annealed states. The effects of such a density of defects on the built-in field of an a-Si p–i–n solar cell is illustrated in Figure 5 where the magnitude of the field over the 0.4 μm thick i-layer, determined by the space charge density, is shown for both the equilibrium case of Figure 4 and for the open-circuit condition under 1 sun illumination. It can be seen in Figure 5 that under illumination the large number of excited carriers trapped in localised states exert a strong influence on the electric field distribution with a significant reduction in the field near the centre of the i-layer.

It should be noted that not only the i-layer but also the p/i interface region has a large influence on a-Si solar cell characteristics and the stability of the cells [39, 52, 76, 77]. 'Buffer layers' between the p- and i-layers are generally used to mitigate the detrimental effects of interface defects and thus improve solar cell

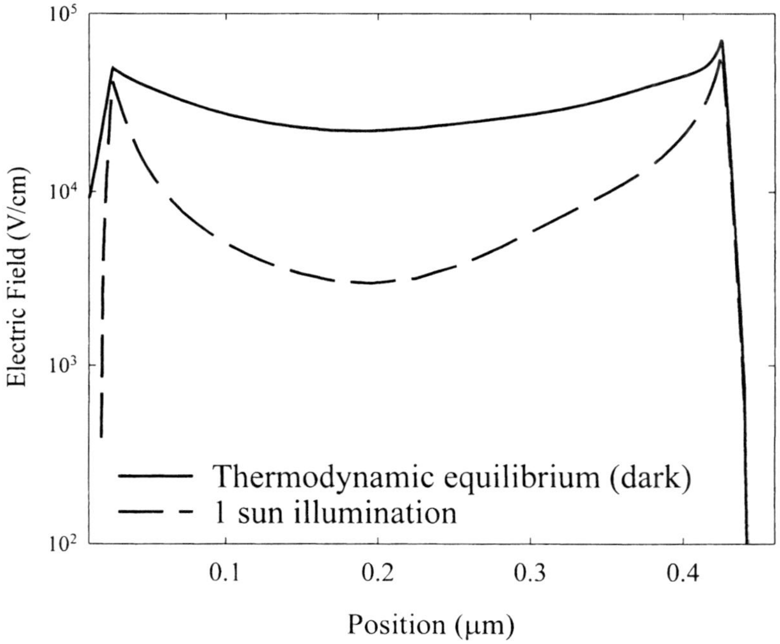

Figure 5 Electric field profile generated by a numerical simulation for a 4000Å thick single-junction a-Si p-i-n cell under thermodynamic equilibrium conditions and under 1 sun illumination at open-circuit conditions.

performance. Highly effective buffer layers have been obtained using protocrystalline Si:H materials fabricated with silane highly diluted with hydrogen [78, 79].

Open-circuit voltages in the amorphous cells just as in crystalline solar cells are determined by the quasi-Fermi level splitting, which depends on the density of photogenerated carriers and the bandgap (E_g); this in turn leads to the well-known dependence of V_{OC} on E_g [80]. Large values of the built-in potential (V_{bi}) are desirable not only in that they can limit the quasi-Fermi splitting but also because they determine the electric field across the i-layer, which is important in the collection of the photogenerated carriers. Over a large range of illumination intensities, the D^0, D^+ and D^- gap states determine the recombination and lifetimes, but at intensities approaching 1 sun illumination the quasi-Fermi level splitting is sufficiently large that the effect of the band tails has to be considered [81]. In addition, carrier recombination in the p/i interface regions can become more important than that in the bulk and can limit the open-circuit voltage under 1 sun illumination, particularly after the introduction of the light induced defects [82].

The short circuit current densities (J_{SC}) are determined by the collection of carriers photogenerated in the i-layer, which depends on its optical absorption and thickness as well as the ability to extract them from the cell. The amount of absorbed sunlight can be readily increased by making the i-layers thicker. However, field-assisted carrier collection is also sensitive to thickness, which results in a negative effect on carrier collection. This adversely affects not only J_{SC}, but generally has an even larger negative effect on the fill factor, since the internal fields are significantly reduced under load (forward bias). The challenge is not only to maximise the optical absorption, such as by increasing the thickness of the i-layer, but also at the same time retain the collection of carriers at a level necessary for high values of the fill factor. A major breakthrough in achieving this result is obtained with optical enhancement based on textured substrates and reflectors [20, 21]. This optical enhancement effect greatly increases the already high optical absorption at longer wavelengths so that significantly higher quantum efficiencies can be obtained at these wavelengths without any increase in the cell thickness. An example of improvements that can be obtained in the spectral response at long wavelengths using different reflectors on p–i–n cells is shown in Figure 6 for two (Cr, Ag) flat metal and two textured (detached, tuned) reflectors.

The fill factor is the cell parameter that is most sensitive to cell thickness as well as the nature and density of native and light-induced defect states in the gap [54, 55, 71, 83]. It is therefore not surprising that there is this strong dependence of the fill factors on the thickness of the i-layer. It should be noted that even though the p/i regions extend over a relatively thin region they can have a large effect on both carrier collection and the electric field distribution, which impacts not only V_{OC} but also the fill factor [40, 78]. The introduction of defects associated with the SWE reduces the free carrier lifetimes and increases the space charge [38, 72, 73]. Since the corresponding redistribution and lowering of the electric fields across the i-layers depend very strongly on their

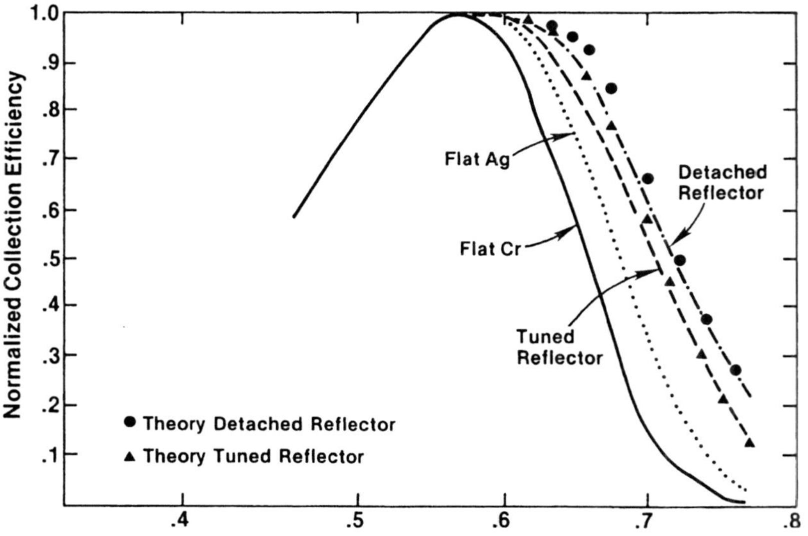

Figure 6 The experimental results for the collection efficiency as a function of wavelength for an a-Si p-i-n solar cell with flat Cr and Ag reflectors as well as a two different textured reflectors. The results are normalized to the peak quantum efficiency at 0.55 μm. Also shown as symbols are the theoretical results for tuned and detached reflectors.

thickness, the large effect on the fill factor is further amplified. This is the underlying reason for making the solar cells as thin as possible.

The development of tandem a-Si:H/a-SiGe:H and triple a-Si:H/a-SiGe:H/a-SiGe:H cell structures allows not only a larger fraction of the incident sunlight to be absorbed, but also allows this increased absorption to be achieved with thin intrinsic absorber layers. Since the optical enhancement is most efficient for the bottom absorber layer [19] and germanium alloying is used to increase the absorption in the bottom absorber layers, one is able to use relatively thin i-layers in the bottom cells. The thickness and optical absorption of each i-layer in a multijunction structure must be adjusted to assure that the same photocurrent is generated in each junction while under 1 sun illumination.

3.2 Device Structures

Amorphous silicon solar cells have been fabricated in the laboratory in a wide variety of different structures [84, 85], but most commercial products utilise p–i–n or n–i–p junction configurations in either single-, double- or triple-junction structures [86]. Most manufacturers offer multijunction structures since they generally exhibit higher stabilised conversion efficiencies.

The device structures used for a-Si based solar cells can also be categorised according to the substrate material. PV Manufacturers such as BP Solar, Energy Photovoltaics, Intersolar, Kaneka, Phototronics, Sanyo, and Sharp use commercial float glass as a substrate while Canon and United Solar Systems use stainless steel foil. In addition, some companies such as Fuji Electric, Iowa Thin Films and Sanyo deposit a-Si solar cells on plastic substrates.

Generally, a p–i–n junction configuration is used with glass substrates so that the light is incident on the glass and passes first through the p-layer side of the cell (this is sometimes referred to a glass superstrate structure). An example of a glass superstrate cell structure is shown in Figure 7. This is a device structure used by BP Solar in the Toano, VA manufacturing plant and has the configuration: glass/textured tin oxide/p–i_1–n/p–i_2–n/zinc oxide/aluminum/EVA/glass where the i_1-layer is an a-Si:H alloy, the i_2-layer is an a-SiGe:H alloy and EVA is ethylene vinyl acetate.

BP Solar buys commercial tin oxide coated soda-lime-silicate glass and has another company seam the edges of the glass, apply a conductive frit and heat-strengthen the glass. The glass companies deposit a layer of silicon dioxide or a similar layer on the glass to obtain uniform nucleation of the textured tin oxide so as to assure a uniform appearance. The textured tin oxide has a surface roughness on the order of a few hundred nm, which scatters the light as it enters the cell leading to increased absorption of the longer-wavelength radiation.

The first junction is formed by depositing a thin p-layer ($\sim$10 nm of a boron-doped a-SiC:H alloy) on the tin oxide, followed by $\sim$160 nm of an a-Si:H i-layer and then $\sim$10 nm of phosphorus-doped microcrystalline Si:H. A tunnel or recombination junction is then formed by depositing $\sim$10 nm of another a-SiC:H p-layer. The second i-layer consists of $\sim$100 nm of an a-SiGe:H alloy where the Ge content is varied so that the band gap of the junction is graded. The second junction is completed by depositing $\sim$20 nm of a phosphorus-doped a-Si:H layer. The back contact is made by first depositing about 100 nm of zinc oxide by low-pressure chemical vapour deposition (CVD) and then sputter depositing about 300 nm of aluminium. The PV modules are then encapsulated using EVA and another sheet of heat-strengthened glass.

An example of an a-Si based multijunction cell structure fabricated on a stainless steel foil is shown in Figure 8. This is a device structure used by United Solar Corporation, which is starting up a new manufacturing facility with an

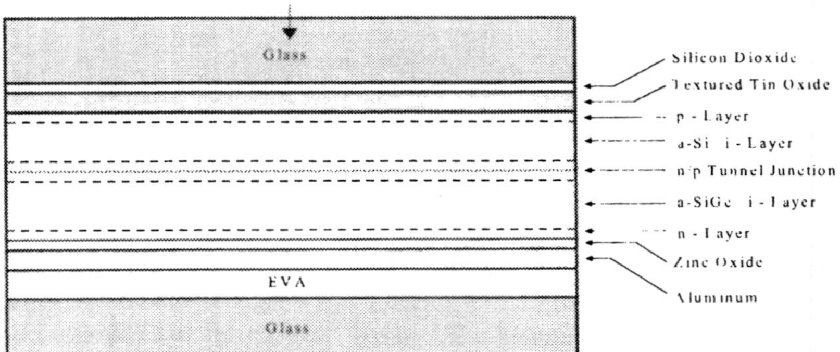

Figure 7 A schematic of an a-Si/a-SiGe tandem device structure fabricated on a glass substrate.

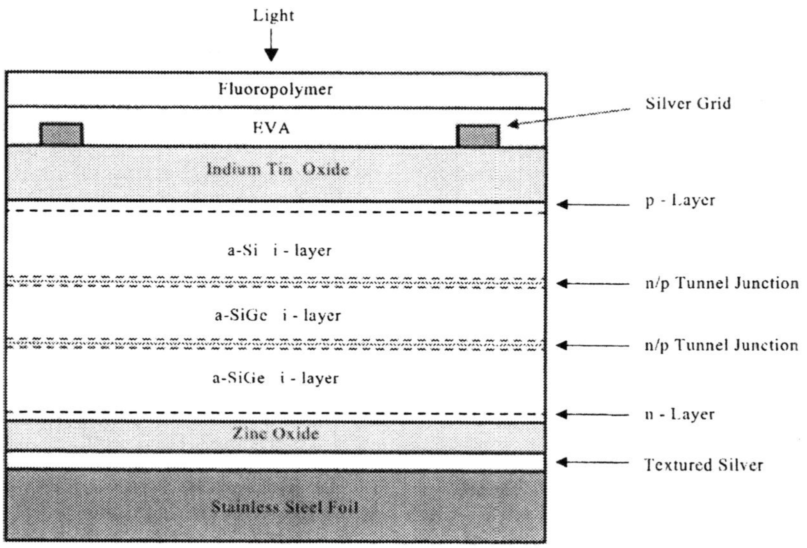

Figure 8 A schematic of a triple-junction device structure fabricated on a stainless steel substrate.

annual capacity of 30 MW_P. This type of device is constructed using an n–i–p configuration where the first a-Si layer deposited on the foil is an n-layer, and the triple-junction device has the configuration: stainless steel foil/textured silver/zinc oxide/n–i_3–p/n–i_2–p/n–i_1–p/ITO/EVA/fluoropolymer where both i_2 and i_3 are a-SiGe:H alloys, i_1 is an a-Si:H alloy, ITO is indium–tin–oxide and the fluoropolymer is typically Tefzel® (a fluoropolymer made by DuPont) [87].

This device structure is fabricated by forming a textured layer of silver or aluminium on the stainless steel foil and coating it with ~100 nm of zinc oxide before depositing the thin silicon semiconductor layers in a continuous process using a roll-to-roll, multi-chamber PECVD system [88]. The first semiconductor junction formed consists of ~20 nm of phosphorus-doped a-Si:H followed by ~130 nm of a graded a-SiGe:H i-layer (i_3) and then ~10 nm of boron-doped microcrystalline Si:H. The next junction is formed by depositing ~10 nm of phosphorus-doped a-S:H (which also forms a tunnel or recombination junction with the underlying p-layer), ~110 nm of another graded a-SiGe:H layer (i_2), which contains less Ge on average than the i_3-layer, and ~10 nm of boron-doped microcrystalline Si:H. The last junction is then formed by depositing ~10 nm of phosphorus-doped a-S:H (which forms another tunnel junction on the underlying p-layer), ~100 nm of a-Si:H (i_1) and ~10 nm of boron-doped protocrystalline Si:H. All the undoped layers (i-layers) are deposited using hydrogen dilution so that the films are close to becoming microcrystalline [88]. The top contact is formed by evaporating a conductive antireflection coating of indium–tin–oxide (ITO), and silver grids are then deposited to help collect the photocurrent. The device structure is completed by laminating the solar cell with EVA and Tefzel®.

Several companies are using other types of device structures that utilise amorphous silicon-based alloys. Phototronics and Energy Photovoltaics manufacture a-Si tandem PV modules on glass where both junctions consist of a-Si:H, and the i-layer thicknesses are adjusted so that they generate approximately the same photocurrent. Kaneka has been manufacturing single-junction a-Si PV modules on glass, and recently started production of a tandem module on glass that utilises a front junction of a-Si:H and a rear junction of microcrystalline silicon (~1–2 microns thick) [89]. Sanyo has developed a single-crystal silicon solar cell that utilises p/i and i/n a-Si:H heterojunction layers that are only 10–20 nm thick, and this type of device exhibits efficiencies as high as 21% in laboratory devices [90].

3.3 Performance and Stability

Present-day commercial a-Si based PV modules typically exhibit stabilised conversion efficiencies in the range of 6–8% while those based on single-crystal or polycrystalline silicon generally exhibit efficiencies in the range of 11–14%. The best stabilised efficiencies of a-Si based PV modules reported by a number of companies are listed in Table 1. In some cases, ranges of efficiencies are shown and were estimated from company product data.

For a small-area (0.25 cm^2) laboratory triple-junction cell, United Solar reported a stabilised conversion efficiency of 13.0% [87], which is the highest efficiency achieved to date in an a-Si based solar cell; the initial efficiency was 15.2%, which is also a record for an a-Si based solar cell.

Triple-junction cells typically exhibit light-induced degradation on the order of about 10–15% while tandem cells exhibit about 12–20% and single-junction cells about 18–30%, depending on a variety of factors such as the thickness of the i-layers and the deposition conditions. The light-induced degradation of a-Si based solar cells is strongly influenced by the thickness of the i-layer [8], and thus triple-junction cells which typically contain i-layers that are on the order of 100 nm thick usually exhibit less light-induced degradation than cell structures

Table 1 Stabilised efficiencies of a-Si-based PV modules

Company	Stabilised efficiency (%)/(aperture area)	Device configuration
BP Solar	8.1%/(0.36 m^2)	a-Si/a-SiGe tandem on glass
BP Solar	7.6%/(0.74 m^2)	a-Si/a-SiGe tandem on glass
Fuji Electric	9.0%/(0.32 m^2)	a-Si/a-SiGe tandem on plastic
Intersolar	~4.5–5.0%/(0.30 m^2)	Single junction on glass
Iowa Thin Films	~4.5–5.5%/(0.45 m^2)	Same gap tandem on plastic
Kaneka	8.1%/(0.41 m^2)	Single junction on glass
Kaneka	~10%/(0.37 m^2)	a-Si/μc-Si tandem on glass
Phototronics	~6.0–6.5%/(0.55 m^2)	Same gap tandem on glass
Sanyo	9.3%/(0.51 m^2)	a-Si/a-SiGe tandem on glass
United Solar	10.1%/(0.09 m^2)	Triple junction on steel foil
United Solar	7.9%/(0.45 m^2)	Triple junction on steel foil

with thicker i-layers (single-junction cells generally contain i-layers that are ~250–350 nm thick).

The effect of i-layer thickness and device structure on the light-induced degradation is shown in Figure 9. As shown in the figure, the tandem cell degraded about 12% after a few hundred hours of light soaking and then stabilised. The degradation of the single-junction cells varied from about 25% for a device with a 0.25 μm thick i-layer to about 50% for a device with a 0.60 μm thick i-layer. It is clear from the data in Figure 9 that most of the degradation occurred in the first week of exposure to simulated sunlight.

Outdoor testing of BP Solar tandem modules has shown that they stabilise in the first several months of outdoor exposure and then exhibit seasonal variations in performance associated with changes in the solar spectrum [91] and with changes in the ambient temperature [92]. Many of the a-Si PV arrays exhibit a sinusoidal variation in performance with a 1-year cycle time, which has been attributed to partial annealing of the light-induced degradation. However, as shown in Figure 10, some arrays such as the one at Montgomery College in Maryland exhibit a variation in performance (open circles) with a cycle time of six months. As also shown in the figure, this is the same cycle time exhibited by the plane of array irradiance (gray vertical lines) [91]. While the data collection at the Montgomery College array did not start until after the array had been exposed to sunlight for several months, it is evident from the figure that the output power of the array has remained essentially constant over the four-year period shown.

While the temperature behaviour of a-Si based solar cells can vary over a wide range depending on the device structure and material quality [93], the temperature coefficient of the output power of commercial a-Si PV modules is

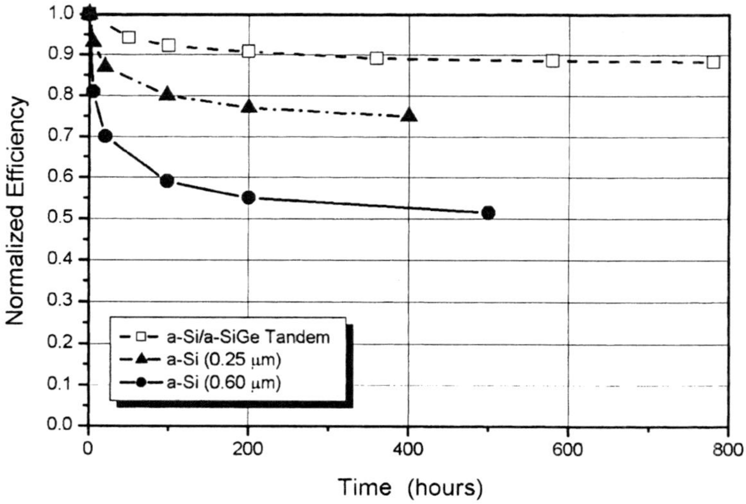

Figure 9 The conversion efficiency of different types of a-Si based solar cells as a function of exposure time to 1 sun illumination. (The initial efficiencies were in the 6 – 8% range.)

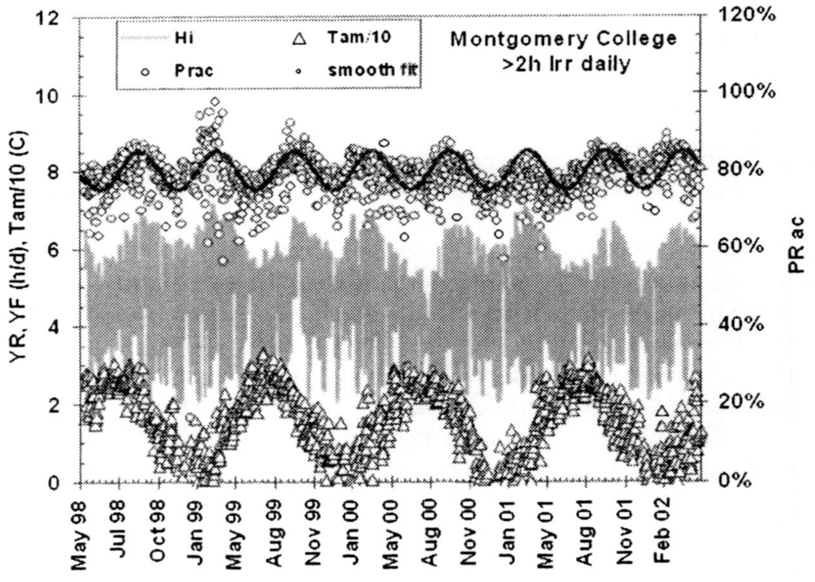

Figure 10 Performance of the BP Solar tandem module array at Montgomery College (Maryland) over a four-year period.

usually in the range of −0.2 to −0.3%/°C, which is about one half the value observed for crystalline silicon modules. However, some a-Si PV arrays have exhibited positive temperature coefficients [93, 94], which could be related to power conditioning effects or spectral effects [91]. Amorphous silicon solar cells that were light soaked and characterised at temperatures of 40 and 80°C in the laboratory exhibited relatively small temperature coefficients [93]. Thus while an a-Si PV module might exhibit a temperature coefficient for the power of about −0.25%/°C for short-term temperature excursions, the temperature dependence of stabilised a-Si PV arrays operating in a steady-state mode appears to be quite small due to annealing effects. However, the actual performance of a-Si PV arrays is also strongly influenced by other factors such as the solar spectrum and the power conditioning.

3.4 Reliability

While there is convincing field data that the latest generation of a-Si PV modules can operate reliably for periods of more than four years (see Figure 10), there are still concerns about the long-term reliability of thin-film modules. There are other factors besides light-induced degradation that can adversely influence the long-term performance of a-Si PV modules and arrays. The performance of all PV arrays can be adversely affected by a number of factors such as malfunctioning power conditioning equipment, failures associated with the wiring or interconnects between modules and damaged or degraded modules. One of the major challenges for thin-film PV technologies is to assure good module reliability over operational periods of 20 years or more.

All PV modules are subjected to a battery of qualification tests, which have been developed to assure that the modules will meet all functional specifications after exposure to a variety of environmental conditions. The tests that simulate these exposures are specified in publications (IEC 1215, IEEE 1262 and UL 1703) issued by the International Electrotechnical Commission (IEC), the Institute of Electrical and Electronics Engineers (IEEE) and the Underwriters Laboratory (UL). The qualification procedure includes tests such as electrical performance, electrical isolation (dry and wet), visual inspection, thermal cycling between −40 and +85°C, light soaking, ultraviolet light exposure, humidity freeze cycling between −40 and +85°C (while maintaining a relative humidity of 85% for temperatures above room temperature), static and dynamic mechanical loading, hail impact tests, surface cut susceptibility, hot-spot endurance and outdoor exposure.

These qualification tests appear to work well for crystalline silicon modules since there have been many crystalline silicon PV arrays that have operated for over two decades or more in a variety of outdoor environments. However, other tests may need to be added to adequately qualify thin-film modules for reliable operation over 20 years of more in all outdoor environments. In general, thin-film PV modules appear to be more susceptible to moisture ingression problems than crystalline silicon modules since the contacting layers and semiconductor layers are very thin, and some of these layers can be delaminated or consumed by electrochemical corrosion [95, 96]. Thus, companies making thin-film PV modules are actively investigating ways to develop improved encapsulants that will not delaminate and that will minimise moisture ingression.

4 Production of Amorphous Silicon Solar Cells

The manufacturing process for a-Si solar cells can vary significantly depending on the type of substrate used to make PV modules. In this section, we will describe the manufacturing process used by BP Solar to make tandem modules on glass, the process used by United Solar to make triple-junction modules on stainless steel foil and the process used by Iowa Thin Films to make single-junction modules on plastic.

4.1 Manufacturing Process

The manufacturing process used by BP Solar to manufacture a-Si/a-SiGe tandem modules on glass is shown schematically in Figure 11. This plant has an annual capacity of 10 MW_P per year and produces PV modules on soda-lime-silicate float glass (each glass plate is 0.8 m^2 in area). BP Solar purchases glass that has been coated with textured tin oxide, conductive silver frit bus bars and has also been edge-seamed and heat strengthened before it is delivered to the plant. The production process starts with washing the glass plates and then transporting

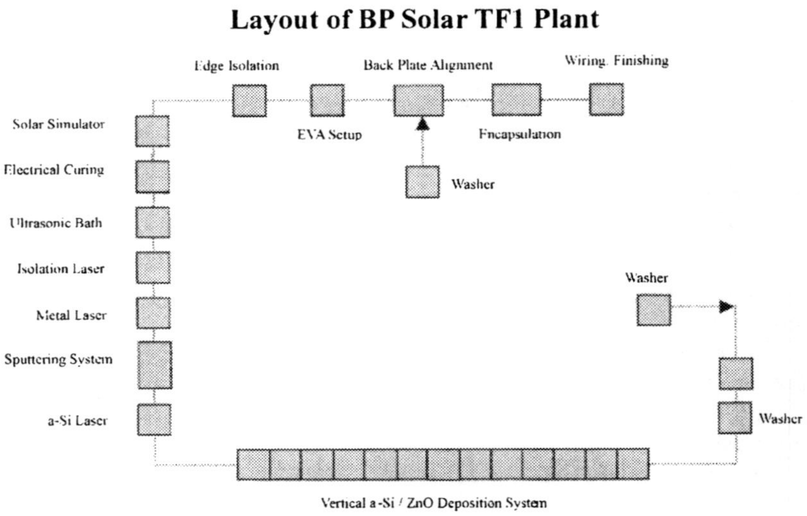

Figure 11 A diagram showing the layout of the BP Solar plant in Toano, VA.

them to a laser scribing station that segments the tin oxide layer and defines the operating characteristics (nominal voltage and current) of the module. The tin oxide is generally scribed into a series of parallel strips ~9 mm wide using a Nd:YAG laser that is frequency doubled to produce a green laser beam that is ~15 microns in diameter.

After the laser scribing process, the plates are washed and loaded in a multi-chamber PECVD system that deposits doped and undoped layers of amorphous silicon and microcrystalline silicon alloys to form an a-Si/a-SiGe tandem structure (see Figure 7). The last deposition chamber in this system uses low-pressure CVD to deposit a thin layer of ZnO. The a-Si and ZnO layers are then scribed using another Nd:YAG laser. After laser scribing, the plates are coated with a thin layer of aluminium using magnetron sputtering. The plates are then laser scribed again in close proximity to the earlier scribes to selectively remove all the layers other than the tin oxide. This laser-scribing step completes the series connection as shown in Figure 12 since the front contact of each strip cell is connected in series to the back contact of the next adjacent cell.

Next a final laser scribe is made at a relatively high power around the perimeter of the plates to ensure electrical isolation. The plates are then cleaned in an ultrasonic bath to remove all debris before passing to a bed-of-nails station that applies a reverse bias to electrically cure cells that are excessively leaky [22]. After electrical curing, the performance of the plates is determined using a solar simulator. Good electrical isolation is assured by using an abrasion wheel to remove all the thin films from the outside edge of the plate. The plates are then fabricated into modules by encapsulating a back plate of float glass to the front plate with ethylene vinyl acetate (EVA), attaching lead wires and mounting the module in a frame. The completed modules then undergo a final power test before they are shipped to customers.

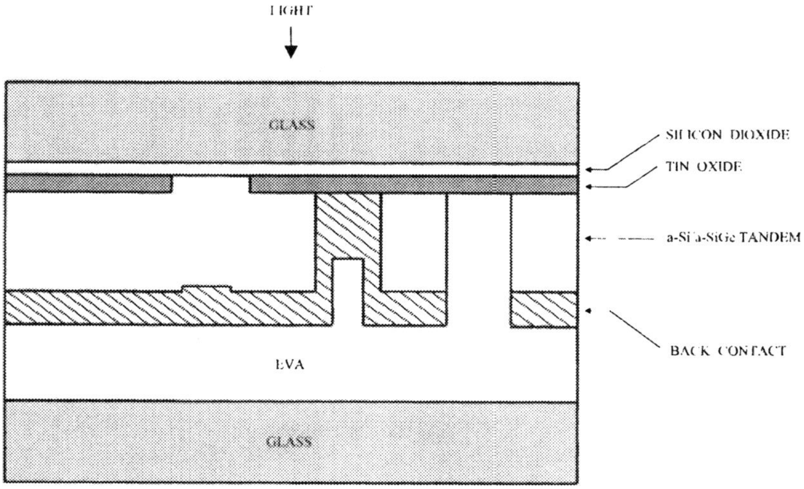

Figure 12 A schematic showing the interconnection region of a tandem module.

The distribution of the initial output power for a run of several thousand modules in the BP Solar TF1 facility is shown in Figure 13. When the light-induced degradation is taken into account, the stabilised power is about 14% less than that shown. The data in Figure 13 show that large-area a-Si PV modules can be produced with relatively high yields (greater than 80%).

Companies such as United Solar that use stainless steel foil as a substrate are using a roll-to-roll process to deposit an a-Si/a-SiGe/a-SiGe triple-junction structure [87]. United Solar has been operating a PV module manufacturing plant in Troy, Michigan since 1997 with an annual capacity of 5 MW_P. The starting substrate for this facility is a roll of flexible stainless steel foil that is about 0.8 km long and 35.6 cm wide on which they sequentially sputter-deposit aluminium and ZnO and then deposit nine a-Si alloy layers (see Figure 8) using a highly automated, multi-chamber PECVD system. They sputter ITO onto the a-Si structure to form a top-contacting layer, and they cut the foil into individual solar cells and apply a current-collecting metal grid onto the ITO. The cells are wired together, laminated with EVA and Tefzel®, fitted with electrical connectors or a junction box and framed to form an environmentally protected module. United Solar has also published data showing that their manufacturing process is capable of producing cells with yields greater than 80% [97].

United Solar is now starting up a new manufacturing plant in Auburn Hills, Michigan, with an annual capacity of 30 MW_P of triple-junction modules per year. This facility contains a multi-chamber PECVD system that is almost 100 m long and will process six rolls of stainless steel foil simultaneously, and each roll of steel foil is about 2.4 km long.

Companies such as Iowa Thin Films use a roll of polyimide plastic (0.05 mm thick) as a starting substrate in the production of another type of flexible a-Si PV module [98]. Each polyimide roll is about 33 cm wide and 0.73 km long. Iowa

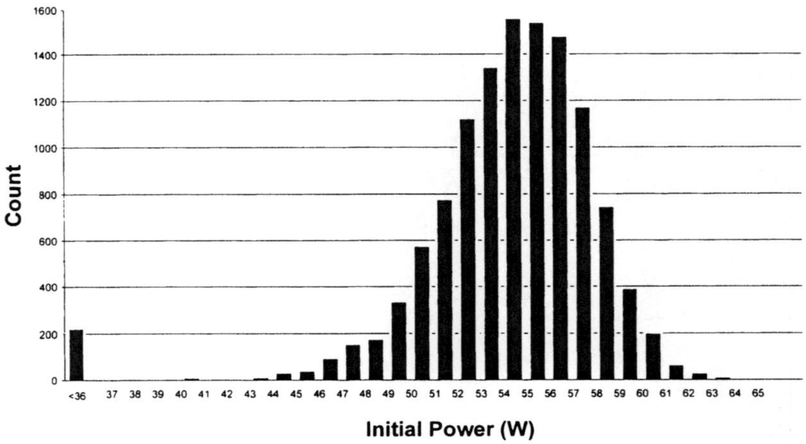

Figure 13 Distribution of the initial power of tandem modules produced at the BP Solar TF1 facility in Toano, VA.

Thin Films sputters aluminium onto the polyimide before depositing a p–i–n a-Si structure by PECVD. They use a laser to scribe through both the a-Si and aluminium layers, and screen-print an insulating ink in the interconnection region before sputter-depositing ZnO. They then screen-print a silver grid onto the ZnO before laser scribing the ZnO. A laser is used to interconnect the cells by fusing a section of the silver grid of one cell to the bottom aluminium contact of the adjacent cell. The coated substrates are laminated to a Tedlar sheet using a roll-based laminator and then cut into individual modules before being framed and wired.

4.2 Manufacturing Costs

A number of studies [99–101] have predicted that manufacturing costs should be on the order of $1/$W_P$ or less when thin-film modules are produced at rates greater than about 10 MW_P per year. Most of these studies assume that the manufacturing plants would be highly automated and that the PV modules would be produced at high yields. The first generation thin-film PV manufacturing plants are costing about $2–$3 per W_P (of annual capacity) to construct, and this capital cost will probably drop to about $1/$W_P$ as the size of the plants increase to 100 MW_P per year and larger. As the plants get larger, the material costs become the major cost component [100].

As shown in Figure 14 for a-Si/a-SiGe tandem modules, more than 60% of the material costs are associated with the glass, frames and packaging, while only 13.3% is associated with the silane, germane and doping gases used to deposit the a-Si alloy layers. Since the utilisation of silane and germane is only on the order of several % in most PECVD reactors, it may be possible to reduce the costs of the semiconductor layers to less than $0.02/$W_P$. While it will be difficult to reduce most of the other material costs (such as glass), the cost of the modules

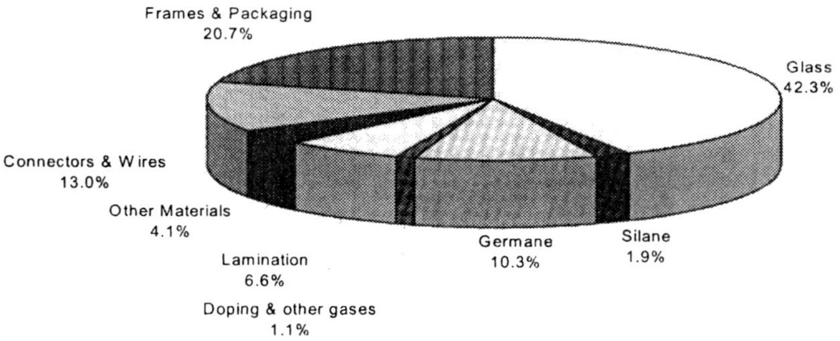

Figure 14 Material cost percentages for a-Si/a-SiGe tandem modules on glass. (The glass cost includes a front plate of tin oxide coated glass and a back plate.)

(in $\$/W_P$) can be reduced by increasing the efficiency of the modules. Increasing the module efficiency also helps to reduce the total cost of a PV system by reducing area-related balance of system costs (such as mounting structures and land).

Another way to significantly reduce the cost of PV systems is to integrate the thin-film PV structure into building materials such as roofs or windows. United Solar has developed a-Si PV modules that function as roofing shingles, standing seam metal roofs and PV laminates that can be installed directly onto metal roofs [102]. BP Solar has developed semi-transparent PV modules (called PowerView™) that are patterned by selective laser ablation so that they allow partial transmission of light (5–10%), and can also be patterned with signs or logos [103]. These PowerView™ modules have been integrated into the canopy over automobile fueling stations and can now be found in over 300 BP service stations worldwide. Building-integrated PV (BIPV) applications reduce the cost of PV systems since the product has dual use. In the case of a PV window, the glass, frame, and installation are all similar to that associated with an architectural window. Thus, the incremental cost of a PV window system over architectural windows is that associated with the thin film PV structure, the wiring and power conditioning equipment.

4.3 Environmental Issues

While PV solar energy is widely viewed as an ideal way to produce electric power from a virtually inexhaustible source without noise or pollution, there are environmental issues that must be addressed. The entire process of mining and refining raw materials, manufacturing PV modules and disposing of obsolete modules must be developed not only for low cost, but must also be environmentally friendly.

If the PV module manufacturing processing requires the use of toxic materials, then systems and procedures must be established to minimise the risk to employees. BP Solar uses toxic doping gases such as diborane and phosphine only in a diluted form (~1–20 vol.% in silane) in the production of a-Si/a-SiGe tandem modules. Trimethylboron (~ 1–5% in silane), which is less toxic than diborane is also used as a p-type dopant source. Since silane is pyrophoric, a leak will cause the silane to ignite, and the dopant gas will be oxidised in the flame. The resulting silicate glass powder is non-toxic.

The silane and germane feedstock gases at the BP Solar TF1 plant in Virginia are stored in an outdoor holding area, and are fed into the facility through stainless steel pipes. All exhaust gases are passed through a burn box and the powder is collected in a bag house for disposal. The powder consists mainly of silicon dioxide fused with small amounts of oxides of germanium, boron and phosphorus.

The process used by BP Solar to manufacture a-Si modules does not use any solvents, acids or other wet chemicals (except for detergents in water for cleaning substrates). As described in Section 4.1, all module patterning and interconnections are performed using lasers, so there are no harmful waste products or effluents produced in the manufacturing process. In addition, since a-Si PV modules do not contain any toxic materials, there are no environmental risks associated with module breakage, fires or long-term disposal in landfills.

5 Future Trends

While the overall market for PV has grown by about 15–20% over the last 20 years (and even more rapidly in the last few years), the market for a-Si solar cells has experienced a different growth pattern. Amorphous silicon solar cells were first introduced commercially by Sanyo in 1980 for use in solar-powered calculators, and shipments increased rapidly to 3.5 MW_P by 1985 (representing about 19% of the total PV market that year). Shipments of a-Si PV modules reached ~40 MW_P in 2001, but this represented only about 11% of the total PV market. This apparent loss of market share is due to the saturation of the low-light consumer application market and the relatively slow acceptance of a-Si PV modules in the terrestrial PV market.

The key drivers behind the terrestrial PV market are module efficiency, selling price and reliability. The conversion efficiencies of crystalline silicon PV modules are generally in the range of 11–14% while the stabilised efficiencies of a-Si PV modules are typically in the range 6–8%. Crystalline silicon PV modules currently sell for about $3.00–$3.30 per W for large-quantity purchases, and a-Si PV modules sell for about 10–20% less on a $/$W_P$ basis. However, the total price of many PV systems (in $/$W_P$) is less with crystalline silicon modules due to the costs associated with the balance of system (the cost of the area-related components are reduced with higher efficiency cells). Crystalline silicon PV modules have exhibited good long-term reliability with many arrays still in operation after more than two decades. Many first generation thin-film modules exhibited reliability problems in outdoor testing, and the reliability of more recent thin-film product has not yet been demonstrated since the product has only been in the field for a few years.

Thus, the future of a-Si PV modules will depend critically on the ability to further increase the stabilised efficiency, lower the manufacturing cost and improve the long-term reliability. The major factor limiting the performance of a-Si PV modules is the light-induced degradation. While a-Si/a-SiGe tandem modules degrade about 12–20% before reaching a steady state, elimination of the light-induced degradation would allow a much larger increase in performance since both junctions could then be made thicker leading to significantly higher photocurrents. At this juncture, it is not clear that the light-induced degradation can be completely eliminated since most investigators believe that the metastable defects are intrinsic to the a-Si alloys. Nonetheless, there is a large R&D effort to develop a better understanding of the light-induced degradation in a-Si alloys and to reduce the degradation. Considering the large potential payoff, this effort is likely to continue for the near future.

One approach that is being aggressively pursued by a number of organisations is to develop new types of solar cells based on microcrystalline silicon (μc-Si) or multijunction structures involving both a-Si and μc-Si [104, 105]. Recently stabilised efficiencies of the order of 10% have been demonstrated for large-area (0.41 m^2) a-Si/μc-Si tandem modules [106].

As mentioned in Section 4.2, the cost of thin-film PV modules and systems can be reduced by integrating thin-film PV into building materials, but this approach requires both large-scale production of suitably sized BIPV products and the integration of the products into the building industry infrastructure. In addition, BIPV products must be designed to be cosmetically appealing since they should not detract from the appearance of the building. The successful implementation of BIPV on a large scale will require a cooperative effort involving PV manufacturers, building materials companies, construction companies, electric utilities and government agencies.

Another factor that affects the cost of a-Si PV modules is the relatively high capital cost of the manufacturing equipment and facilities. This cost can be reduced by increasing the throughput of the production equipment. At the present time, the production bottleneck and major equipment cost are usually associated with the a-Si alloy deposition process where the deposition rate is typically on the order of 0.1–0.2 nm/s. Thus, a number of organisations are engaged in research and development programs to increase the deposition rates for both a-Si and μc-Si alloys.

The reliability of a-Si PV modules has improved over the last 20 years with the development of new processes for electrically isolating and encapsulating the modules. However, more research and development is required to assure that all thin-film PV modules can survive for extended periods ($>$ 20 years) in hostile environments such as high-voltage arrays operating in hot humid climates.

In summary, while a-Si photovoltaics has become a \$100 million dollar business, further research, development and engineering will be required to increase the performance, lower the manufacturing costs and improve the reliability in order to assure that a-Si PV will play a major role in future world energy production.

References

[1] Chittik, R.C., Alexander, J.H. and Sterling, H.E., 1969. The preparation and properties of amorphous silicon. *J. Electrochem. Soc.*, Vol. 116, pp. 77–81.

[2] Carlson, D.E., 1977. Semiconductor device having a body of amorphous silicon. US Patent 4,064,521.

[3] Spear, W.E. and LeComber, P.G., 1975. Substitutional doping of amorphous silicon. *Solid State Communications*, Vol. 17, pp.1193–1196.

[4] Carlson, D.E. and Wronski, C.R., 1976. Amorphous silicon solar cells. *Appl. Phys. Lett.*, Vol. 28, pp. 671–673.

[5] Carlson, D.E., Wronski, C.R., Triano, A. and Daniel, R.E., 1976. Solar cells using Schottky barriers on amorphous silicon. *Proc. 12th IEEE Photovoltaic Specialists Conf.*, Baton Rouge, pp. 893–895.

[6] Brodsky, M.H., Frisch, M.A., Ziegler, J.F. and Lanfard, W.A., 1977. Quantitative analysis of hydrogen in glow discharge amorphous silicon. *Appl. Phys. Lett.*, Vol. 30, pp. 561–563.

[7] Staebler, D.L. and Wronski, C.R., 1977. Reversible conductivity change in discharge produced amorphous silicon. *Appl. Phys. Lett.*, Vol. 31, pp. 292–294.

[8] Hanak, J. and Korsun, V., 1982. Optical stability studies of a-Si solar cells. *Proc. 16th IEEE Photovoltaic Specialists Conf.*, San Diego, pp. 1381–1383.

[9] Kroll, U., Meier, J., Keppner, H., Shah, A., Littlewood, S.D., Kelly, I E. and Giannoules, P., 1995. Origins of atmospheric contamination in amorphous silicon prepared by very high frequency (70 MHz) glow discharge. *J. Vac. Sci. Technol.* Vol. A 13(6), pp. 2742–2746.

[10] Watanabe, T., Azuma, K., Nakatani, M., Suzuki, K., Sonobe, T. and Shimada, T., 1986. Chemical vapor deposition of a-Si:H films utilising a microwave excited Ar plasma stream. *Jpn. J. of Applied Physics*, Vol. 25(12), pp. 1805–1810.

[11] Saito, K., Sano, M., Ogawa, K. and Kajita, I., 1993. High efficiency a-Si:H alloy cell deposited at high deposition rate. *J. Non-Cryst. Solids*, Vols. 164–166, pp. 689–692.

[12] Carlson, D.E., Rajan, K., Arya, R.R., Willing, F. and Yang, L., 1998. Advances in amorphous silicon photovoltaic technology. *J. Materials Research*, Vol. 13(10), pp. 2754–2762.

[13] Tanaka, K. and Matsuda, A., 1987. Glow-discharge amorphous silicon: growth process and structure. *Materials Science Report*, Vol. 2, pp. 139–184.

[14] Wronski, C.R., Pearce, J.M., Koval, R.J., Niu, X., Ferlauto, A.S., Koh, J. and Collins, R.W., 2002. Light induced defect creation kinetics in thin film protocrystalline silicon materials and their solar cells. *Mat. Res. Soc. Proc.*, Vol. 715, p. A13.4.

[15] Zanzucchi, P., Wronski, C.R. and Carlson, D.E., 1977. Optical and photoconductivity properties of discharge produced a-Si. *J. Appl. Phys.*, Vol. 48, pp. 5227–5236.

[16] Tawada, Y., Okamoto, H. and Hamakawa, Y., 1981. a-SiC:H/a-Si:H heterojunction solar cell having more than 7.1% conversion efficiency. *Appl. Phys. Lett.*, Vol. 39(3), pp. 237–239.

[17] Nakamura, G., Sato, K., Ishihara, T., Usui, M., Okaniwa, K. and Yukimoto, Y., 1983. Tandem type amorphous solar cells. *J. Non-Cryst. Solids*, Vol. 59–60, pp. 1111–1114.

[18] Ishihara, T., Terazono, S., Sasaki, H., Kawabata, K., Itagaki, T., Morikawa, H., Deguchi, M., Sato, K., Usui, M, Okaniwa, K., Aiga, M., Otsubo, M. and Fujikawa, K., 1987. High efficiency triple-junction amorphous solar cells. *Proc. 19th IEEE Photovoltaic Specialists Conf.*, New Orleans, pp. 749–755.

[19] Yang, J., Banerjee, A., Lord, K. and Guha, S., 1998. Correlation of component cells with high efficiency amorphous silicon alloy triple-junction solar cells and modules. *Proc. 2nd World Conf. on Photovoltaic Solar Energy Conversion*, Vienna, pp. 387–390.

[20] Yablonovitch, E., and Cody, G.D., 1982. Intensity enhancement in textured optical sheets for solar cells. *IEEE Trans. Electron Dev.*, Vol. 29, pp. 300–305.

[21] Deckman, H., Wronski, C.R., and Yablonovitch, E., 1984. Optical enhancement of solar cells. *Proc. 17th IEEE Photovoltaic Specialists Conf.*, Kissimmee, pp. 955–960.

[22] Nostrand, G. E. and Hanak, J., 1979. Method of removing the effects of electrical shorts and shunts created during the fabrication process of a solar cell. US Patent 4,166,918.

[23] Ayra, R.R. and Carlson, D.E., 2002. Amorphous silicon PV module manufacturing at BP Solar. *Progress in Photovoltaics*, Vol. 10, pp. 69–76.

[24] Hanak, J.J., 1981. Laser processing technique for fabricating series-connected and tandem junction series-connected solar cells into a solar battery. US Patent 4,292,092.

[25] Carlson, D.E., Gleaton, M. and Ganguly, G., 2000. Effects of oil and dopant contaminants on the performance of amorphous silicon solar cells. *Proc. 16th European Photovoltaic Solar Energy Conf.*, Glasgow.

[26] Kinoshita, T., Isomura, M., Hishikawa, Y. and Tsuda, S., 1996. Influence of oxygen and nitrogen in the intrinsic layer of a-Si solar cells. *Jpn. J. Appl. Phys.*, Vol. 35, pp. 3819–3824.

[27] Tsai, C.C., Knights, J.C., Chang, G. and Wacker, B., 1986. Film formation mechanisms in the plasma deposition of hydrogenated amorphous silicon. *J. Appl. Phys.*, Vol. 59, pp. 2998–3001.

[28] Street, R.A., 1991. *Hydrogenated Amorphous Silicon*. Cambridge University Press, New York.

[29] Shimizu, T., Zhang, Q., Nishino, T., Takashima, H. and Kumeda, M., 1996. Influence of hydrogen content and Si-H bond structure on photocreated dangling bonds in hydrogenated amorphous silicon films. *Jpn. J. Appl. Phys.*, Vol. 35, pp. 4409–4412.

[30] Ganguly, G. and Matsuda, A., 1996. Role of hydrogen dilution in improvement of a-SiGe:H alloys. *J. Non-Cryst. Solids*, Vols. 559–562, pp. 98–200.

[31] Lee, Y., Jiao, L., Liu, H., Lu, Z., Collins, R. W. and Wronski, C. R., 1996. Stability of a-Si solar cells and corresponding intrinsic materials fabricated using hydrogen diluted silane. *Proc. 25th IEEE Photovoltaic Specialists Conf.*, Washington, DC, pp. 1165–1168.

[32] Collins, R.W., 1994. Real time spectroscopic ellipsometry studies of the nucleation, growth and optical functions of thin films, Part 1: tetrahedrally bonded materials. In: *Physics of Thin Films*, Academic Press, New York, pp. 49–125.

[33] Koh, J., Ferlauto, A.S., Rovira, P.I, Wronski, C.R. and Collins, R.W., 1999. Evolutionary phase diagrams for plasma-enhanced chemical vapor deposition of silicon thin films from hydrogen diluted silane. *Appl. Phys. Lett.*, Vol. 75, pp. 2286–2288.

[34] Lu, Y., Kim, S., Gunes, M., Lee, Y., Wronski, C.R. and Collins, R.W., 1994. Process-property relationships for a-$Si_{1-x}C_x$:H deposition: excursions in parameters space guided by real time spectroellipsometry. *Mat. Res. Soc. Symp. Proc.*, Vol. 336, pp. 595–600.

[35] Collins, R.W., Koh, J., Ferlauto, A.S., Rovira, P.I., Lee, Y., Koval, R.J. and Wronski, C.R., 2000. Real Time Analysis of Amorphous and Microcrystalline Silicon Film Growth by Multichannel Ellipsometry. *Thin Solid Films*, Vol. 364, pp. 129–137.

[36] Ferlauto, A.S., Ferreira, G.M., Koval, R.J., Pearce, J.M., Wronski, C.R., Collins, R.W., Al-Jassim, M.M. and Jones, K.M., 2002. Thickness Evolution of the Microstructural and Optical Properties of Si:H Films in the Amorphous-to-Microcrystalline Phase Transition Region. *Proc 29th IEEE Photovoltaic Specialists Conf.*, New Orleans, pp. 1076–1081.

[37] Ferlauto, A.S., Rovira, P.I., Koval, R.J., Wronski, C.R. and Collins, R.W., 2000. Effects of H_2-Dilution and Plasma Power in Amorphous Silicon Deposition: Comparison of Microstructural Evolution and Solar Cell Performance. *Proc. 28th IEEE Photovoltaics Specialists Conf.*, Anchorage, pp. 713–716.

[38] Koval, R.J., Koh, J., Lu, Z., Jiao, L., Wronski, C.R. and Collins, R.W., 1999. Performance and Stability of Si:H *p–i–n* Solar Cells with i-Layers Prepared at the Thickness-Dependent Amorphous-to-Microcrystalline Phase Boundary. *Appl. Phys. Lett.*, Vol. 75, pp. 1553–1555.

[39] Koval, R.J., Pearce, J.M., Ferlauto, A.S., Rovira, P.I., Collins, R.W. and Wronski, C.R., 2000. The Role of Phase Transitions in Protocrystalline Si:H on the Performance their of Solar Cells. *Proc. 28th IEEE Photovoltaic Specialists Conf.*, Anchorage, pp. 750–753.

[40] Koval, R.J., Pearce, J.M., Ferlauto, A.S., Collins, R.W. and Wronski, C.R., 2001. Evolution of the Mobility Gap with Thickness in Hydrogen-Diluted Intrinsic Si:H Materials in the Phase Transition Region and Its Effect on *pi–n* Solar Cell Characteristics. *Mat. Res. Soc. Proc.*, Vol. 664, p. A16.4.

[41] Koval, R.J. Chen, C. Ferreira, G.M., Ferlauto, A.S., Pearce, J.M., Rovira, P.I., Wronski, C. R. and Collins, R.W., 2001. Protocrystalline Si:H p-type Layers for Maximization of the Open Circuit Voltage of a-Si:H *n–i–p* Solar Cells. *Mat. Res. Soc. Proc.*, Vol. 715, p. A6.1.
[42] Koval, R.J., Pearce, J.M., Chen, C., Ferreira, G.M., Ferlauto, A.S., Collins, R.W. and Wronski, C.R., 2002. Microstructurally Engineered p-layers for Obtaining High Open-Circuit Voltages in a-Si:H *n–i–p* Solar Cells. *Proc. 29th IEEE Photovoltaic Specialists Conf.*, New Orleans, pp. 1090–1093.
[43] Guha, S., Yang, J., Nath, P. and Hack, M., 1986. Enhancement of open circuit voltage in high efficiency amorphous silicon alloy solar cells. *App. Phys. Lett.*, Vol. 49, pp. 218–219.
[44] Dersch, H., Stuke, J. and Beichler, J., 1981. Light induced dangling bonds in hydrogenated amorphous silicon. *Appl. Phys. Lett.*, Vol. 38, pp. 456–458.
[45] Carlson, D.E. and Wronski, C.R., 1979. Amorphous silicon solar cells, in Topics. In: Brodsky, M.H. Ed., *Applied Physics*, Spring-Verlag, Berlin, pp. 289–329.
[46] Collins, R.W. and Vedam, K., 1995. Optical properties of solids. In: Trigg G.L., Ed., *Encyclopedia of Applied Physics*, VCH Publishers, New York, pp. 285–336.
[47] Luft, W., 1988. Characteristics of hydrogenated amorphous silicon-germanium alloys. *Proc. 20th IEEE Photovoltaic Specialists Conf.*, Vol. 1, Las Vegas, pp. 218–223.
[48] Nevin, W.A., Yamagishi, H., Asaoka, K. and Tawada, Y., 1992. Wide-bandgap hydrogenated amorphous silicon carbide prepared from an aromatic carbon source, *Proc. 22nd IEEE Photovoltaic Specialists Conf.*, Vol. 2, Las Vegas, pp. 1347–1351.
[49] Dawson, R.M., Li, Y., Gunes, M., Nag, S., Collins, R.W., Bennett, M. and Wronski, C.R., 1992. Optical properties of the component materials in multijunction hydrogenated amorphous silicon based solar cells. *Proc. 11th European Photovoltaic Solar Energy Conf.*, Montreux, pp. 680–683.
[50] Roxlo, C., Wronski, C.R., Abeles, B., Cody, G.D., 1983. Comment on optical absorption edge in a-Si:H_x. *Solid State Communications*, Vol. 47, pp. 985–987.
[51] Tiedje, T., 1984. Information about band-tail states from time-of-flight experiments. In: Pankove J.I., Ed., *Semiconductors and Semimetals*, Academic Press, Orlando, Vol. 21C, pp. 207–238.
[52] Koh, J., Lee, Y., Fujiwara, H., Wronski, C.R., and Collins, R.W., 1998. Optimization of hydrogenated amorphous silicon *p–i–n* solar cells with two-step i layers guided by real-time spectroscopic ellipsometry. *Appl. Phys. Lett.*, Vol. 73, pp. 1526–1528.
[53] Branz, H.M. and Silver, M., 1990. Potential fluctuation due to inhomogeneity in hydrogenated amorphous silicon and the resulting charged dangling bond defects. *Phys. Rev. B.*, Vol. 42, pp. 7420–7428.
[54] Jiao, L., Liu, H., Semoushikina, S., Lee, Y. and Wronski, C. R., 1996. Initial, rapid light induced changes in hydrogenated amorphous silicon

materials and solar cell structures: the effect of charged defects. *Appl. Phys. Lett.*, Vol. 69, pp. 3713–3715.

[55] Gunes, M. and Wronski, C.R., 1997. Differences in the densities of charged defect states and kinetics of Staebler–Wronski in undoped (non-intrinsic) hydrogenated amorphous silicon. *J. Appl. Phys.*, Vol. 81, pp. 3526–3536.

[56] Kamei, T., Hata, N., Matsuda, A., Uchiyama, T., Amano, S., Tsukamoto, K., Yoshioa, Y. and Hirao, T., 1996. Deposition and extensive light soaking of highly pure hydrogenated amorphous silicon. *Appl. Phys. Lett.*, Vol. 68, pp. 2380–2382.

[57] Collins, R.W. and Fujiwara, H., 1997. Growth of hydrogenated amorphous and its alloys. *Current Opinion in Solid State & Material Science*, Vol. 2, pp. 417–424.

[58] Wronski, C.R., Abeles, B., Tiedje, T. and Cody, G.D., 1982. Recombination centers in phosphorous doped hydrogenated amorphous silicon. *Solid State Communications*, Vol. 44, pp. 1423–1426.

[59] Tawada, Y., Okamoto, H., and Hamakawa, Y., 1997. a-SiC:H/a-Si:H heterojunction solar cell having more than 7.1% conversion efficiency. *Appl. Phys. Lett.*, Vol. 39, pp. 237–239.

[60] Lee, Y., Ferlauto, A. and Wronski, C.R., 1997. Contributions of bulk, interface and build-in potential to the open circuit voltage of a-Si solar cells. *Proc. 26th IEEE Photovoltaic Specialists Conf.*, Anaheim, pp. 683–686.

[61] Wronski, C.R., 1984. The Staebler–Wronski Effect. *Semiconductors and Semimetals*, Vol. 21C, pp. 347–373.

[62] Fritzsche, H., 1997. Search for explaining the Staebler–Wronski effect. *Mat. Res. Soc. Symp. Proc.*, Vol. 467, pp. 19–31.

[63] Stutzmann, M., 1997.Microscopic aspects of the Staebler–Wronski effect. *Mat. Res. Soc. Symp. Proc.*, Vol. 467, pp. 37–48.

[64] Wronski, C.R., 1997. The light-induced changes in a-Si:H materials and solar cells – Where we are now. *Mat. Res. Soc. Symp. Proc.*, Vol. 467, pp. 7–17.

[65] Yang, L. and Chen, L., 1993. Fast and slow metastable defects in hydrogenated amorphous silicon. *Appl. Phys. Lett.*, Vol. 63, pp. 400–402.

[66] Von Roedern, B., 1993. Shortfall of defect models for amorphous silicon solar cells. *Appl. Phys. Lett.*, Vol. 62, pp.1368–1369.

[67] Sakata, I., Yamanaka, M., Namase, S. and Hayashi, Y., 1992. Deep defect states in hydrogenated amorphous silicon studied by a constant photocurrent method. *J. Appl. Phys.*, Vol. 71, pp. 4344–4353.

[68] Carlson, D.E. and Rajan, K., 1998. Evidence for Proton Motion in the Field-Induced Recovery of Light-Induced Degradation in Amorphous Silicon Solar Cells. *J. Appl. Phys.*, Vol. 83, pp. 1726–1729.

[69] Stutzmann, M., Jackson, W.B. and Tsai, C.C., 1985. Light induced metastable defects in hydrogenated amorphous silicon: a systematic study. *Phys. Rev. B*, Vol. 32(1), pp. 23–47.

[70] Wronski, C.R., Gunes, M. and McMahon, T.J, 1994. Charged defect states in intrinsic hydrogenated amorphous silicon. *J. Appl. Phys.*, Vol. 76(4), pp. 2260–2263.

[71] Lu, Z., Jiao, H., Koval, R., Collins, R.W. and Wronski, C.R., 1999. Characteristics of different thickness a-Si:H/metal Schottky barrier cell structures – results and analysis. *Mat. Res. Soc. Symp. Proc.*, Vol. 557, pp. 785–790.

[72] Koval, R., Niu, X. Jiao, L., Pearce, J., Ganguly, G., Yang, J., Guha, S., Collins, R.W. and Wronski, C.R., 2000. Kinetics of Light Induced Changes in Protocrystalline Thin Film Materials and Solar Cells. *Mat. Res. Soc. Proc.*, Vol. 609, p. A15.5.

[73] Pearce, J., Niu, X., Koval, R., Ganguly, G., Carlson, D., Collins, R.W. and Wronski, C.R., 2001. Contributions of D^{0} and non-D^{0} gap states to the kinetics of light induced degradation of amorphous silicon under 1 sun illumination. *Mat. Res. Soc. Proc.*, Vol. 664, p. A12.3.

[74] Wronski, C.R., Pearce, J.M., Koval, R.M. and Collins, R.W., 2001. Metastability in Hydrogenated Amorphous Based Materials and Solar Cells. *Technical Digest of the 12th International Photovoltaic Science and Engineering Conf.*, Jeju, Korea, pp. 33–36.

[75] Yang, L. and Chen, L.F., 1994. The effect of hydrogen dilution on the stability of a-Si:H based solar cells. *Mat. Res. Soc. Proc.*, Vol. 336, pp. 669–674.

[76] Sakai, H., Yoshida, T., Fujikake, S., Hama, T. and Ichikawa, Y., 1990. Effect of p/i interface layer on dark J–V characteristics and Voc in *p–i–n* a-Si solar cells. *J. Appl. Phys.*, Vol. 67, pp. 3494–3499.

[77] Hack, M. and Shur, M., 1986. Limitations to the open circuit voltage of amorphous silicon solar cells. *Appl. Phys. Lett.*, Vol. 49, pp. 1432–1434.

[78] Lee, Y., Ferlauto, A.S., Lu, Z., Koh, J., Fujiwara, H., Collins, R.W. and Wronski, C.R., 1998. Enhancement of stable open circuit voltage in a-Si:H *P–I–N* solar cell by hydrogen dilution of P/I interface regions. *Proc. 2nd World Conf. on Photovoltaic Solar Energy Conversion*, Vienna, pp. 940–943.

[79] Pearce, J.M., Koval, R.J., Ferlauto, A.S., Collins, R.W., Wronski, C.R., Yang, J. and Guha, S., 2000. Dependence of open circuit voltage in protocrystalline Si:H solar cells on carrier recombination in p/i interface and bulk regions. *Appl. Phys. Lett.*, Vol. 77, pp. 3093–3095.

[80] Crandall, R. and Schiff, E.A., 1985. The correlation of Open Circuit Voltage with Bandgap in Amorphous Silicon-based *P–i–n* Solar Cells. *AIP Conference Proceedings*, Tempe, AZ, pp. 481–486.

[81] Schiff, E.A., 2002. Thermionic emission model for interface effects on the open-circuit voltage of amorphous silicon based solar cells. *Proc. 29th IEEE Photovoltaic Specialists Conf.*, New Orleans, in press.

[82] Pearce, J.M., Koval, R.J., Ferlauto, A.S., Collins, R.W. and Wronski, C.R., 2001. Limitations of Bulk Generation-Recombination on Open Circuit Voltage under 1 Sun Illumination in Amorphous Silicon Solar Cells. *Technical Digest of the 12th International Photovoltaic Science and Engineering Conf.*, Jeju, Korea, pp. 251–252.

[83] Wronski, C.R., Lu, Z., Jiao, L. and Lee, Y., 1997. An approach to self consistent analysis of a-Si:H material and *p–i–n* solar cell properties. *Proc. 26th IEEE Photovoltaic Specialists Conf.*, Anaheim, pp. 587–590.

[84] Carlson, D.E., 1977. Amorphous Silicon Solar Cells. *IEEE Trans. Electron Devices*, ED-24(4), pp. 449–453.

[85] Carlson, D.E., 1984. Solar Energy Conversion. In: Joanopoulos, J.D. and Lucovsky, G., Eds., *Topics in Applied Physics*, Springer-Verlag, Berlin, Heidelberg, Vol. 55, pp. 203–244.

[86] Wronski, C.R. and Carlson, D., 2001. Amorphous Silicon Solar Cells. In: *Photoconversion of Solar Energy*, Vol. 3: Clean Electricity from Photovoltaics. Imperial College Press.

[87] Guha, S., Yang, J., Banerjee, A., Hoffman, K. and Call, J., 1999. Manufacturing issues for large volume production of amorphous silicon alloy photovoltaic modules. *AIP Conf. Proc.*, Vol. 462, pp. 88–93.

[88] Guha, S. and Yang, J., 2002. Amorphous Silicon Technology. *Proc. 29th IEEE Photovoltaic Specialists Conf.*, New Orleans, pp. 1070–1075

[89] Yamamoto, K., Nakajima, A., Yoshimi, M., Sawada, T., Fukuda, S., Hayashi, K., Suezaki, T., Ichikawa, M., Koi, Y., Goto, M., Takata, H. and Tawada, Y., 2002. High efficiency thin film silicon cell and module, *Proc. 29th IEEE Photovoltaic Specialists Conf.*, New Orleans, pp. 1110–1113.

[90] Kawamoto, K., Nakai, T., Baba, T., Taguchi, M., Sakata, H., Tsuge, S., Uchihashi, K., Tanaka, M. and Kiyama, S., 2001. A high-efficiency HIT™ Solar Cell (21.0% – 100 cm^2) with excellent interface properties. *Proc. 12th Photovoltaic Science and Engineering Conf.*, Cheju Island, Korea.

[91] Ransome, S., and Wohlgemuth, J., 2002. Performance of BP Solar Tandem Junction Amorphous Silicon Modules. *Proc. 29th IEEE Photovoltaic Specialists Conf.*, New Orleans, pp. 1420–1423.

[92] King, D., Kratochvil, J. and Boyson, W., 2000. Stabilization and Performance Characteristics of Commercial Amorphous Silicon PV Modules. *Proc. 28th IEEE Photovoltaic Specialists Conf.*, Anchorage, pp. 1446–1449.

[93] Carlson, D.E., Lin, G. and Ganguly, G., 2000. Temperature dependence of amorphous silicon solar cell PV parameters. *Proc. 28th IEEE Photovoltaic Specialists Conf.*, Anchorage, pp. 707–712.

[94] Gottschalg, R., Infield, D.G. and Kearney, M.J., 2002. Experimental Investigation of Spectral Effects on Amorphous Silicon Solar Cells in Outdoor Operation. *Proc. 29th IEEE Photovoltaic Specialists Conf.*, New Orleans, pp. 1138–1141.

[95] Fagnan, D.A., D'Aiello, R.V. and Mongon, J., 1987. Solarex/Philadelphia Electric Amorphous Silicon PV Test Site. *Proc. 19th IEEE Photovoltaic Specialists Conf.*, New Orleans, pp. 1508–1509.

[96] Osterwald, C.R., McMahon, T.J. and del Cueto, J.A., 2002. Electrochemical corrosion of SnO_2:F transparent conducting layers in thin film photovoltaic modules. *Solar Energy Materials and Solar Cells*, in press.

[97] Guha, S., Yang, J., Banerjee, A., Hoffman, K., Sugiyama, S. and Call, J., 1997. Triple-Junction Amorphous Silicon Alloy PV Manufacturing Plant

of 5 MW Annual Capacity. *Proc. 26th IEEE Photovoltaic Specialists Conf.*, Anaheim, pp. 607–610.
[98] Braymen, S., Grimmer, D., Jeffrey, F., Martens, S., Noack, M., Scandrett, B. and Thomas, M., 1999. Monolithic amorphous silicon modules on continuous polymer substrates'. *AIP Conf. Proc.*, Vol. 462, pp. 737–740.
[99] Carlson, D.E., 1989. Low-cost power from thin film photovoltaics. In: Johansson, T.B., Bodlund, B. and Williams, R.H., Eds., *Electricity*, Lund University Press, Lund, Sweden, pp. 595–626.
[100] Woodcock, J.M., Schade, H., Maurus, H., Dimmler, B., Springer, J. and Ricaud, A., 1997. A study of the upscaling of thin film solar cell manufacture towards 500 MW_P per annum. *Proc.14th European Photovoltaic Solar Energy Conf.*, Barcelona, pp. 857–860.
[101] Sabisky, E., Kiss, Z., Ellis, F., Eser, E., Gau, S., Kampas, F., VanDine, J., Weakliem, H. and Varvar, T., 1989. "Eureka" – a 10 MW_P a-Si:H Module Processing Line. *Proc. 9th European Photovoltaic Solar Energy Conf.*, Freiburg, 1989, p. 27.
[102] Nath, P., Vogeli, C., Jones, K., Singh, A., Garcia, I. and Guha, S., 2000. Field-installed peal and stick PV laminates for metal roofs. *Proc. 28th IEEE Photovoltaic Specialists Conf.*, Anchorage, pp. 1543–1544.
[103] Ayra, R.R. and Carlson, D.E., 2002. Amorphous Silicon PV Module Manufacturing at BP Solar. *Progress in Photovoltaics*, Vol. 10, pp. 69–76.
[104] Meier, J., Sptiznagel, J., Fay, S., Bucher, C., Graf, U., Kroll, U., Dubail, S., Shah, A., 2002. Enhanced light-trapping for micromorph tandem solar cells by LP-CVD ZnO. *Proc. 29th IEEE Photovoltaic Specialists Conf.*, New Orleans, pp. 1118–1121.
[105] Yamamoto, K., Yoshimi, M., Suzuki, T., Nakata, T., Sawada, T., Nakajimi, A. and Hayashi, K., 2000. Large-area and high efficiency a-Si/poly-Si stacked solar cell submodule. *Proc. 28th IEEE Photovoltaic Specialists Conf.*, Anchorage, pp. 1428–1432.
[106] Yamamoto, K., Nakajimi, A., Yoshimi, M., Sawada, T., Fukuda, S., Hayashi, K., Suezaki, T., Ichikawa, M., Koi, Y., Goto, M., Takata, H. and Tawada, Y., 2002. High Efficiency Thin Film Silicon Solar Cell and Module. *Proc. 29th IEEE Photovoltaic Specialists Conf.*, New Orleans, pp. 1110–1113.

IIc-2

Microcrystalline Silicon Solar Cells

Kenji Yamamoto, Kaneka Corporation, Otsu, Shiga, Japan

1 Introduction

With raw materials for crystalline solar cells in increasingly short supply, renewed efforts are being made in the study of ways to make practical thin-film solar cells for use in tomorrow's large-scale solar cell installations. In particular, since thin-film polycrystalline silicon solar cells are made of the same materials as frequently used, bulk silicon solar cells, they are being enthusiastically studied at laboratories worldwide due to the abundance of raw materials for such devices, their stability, and the characteristics they are expected to possess as silicon devices. Part of the motivation behind these studies comes from the theoretical finding that if these devices can be constructed so that they trap sufficient light, it should ideally be possible to achieve photoelectric conversion efficiencies in excess of 20%, even with solar cells in which the photoelectric layer is just a few micrometres thick [1].

Historically, the silicon thin-film photo absorbed layers in such devices have been produced by high-temperature processes, chiefly chemical vapour deposition (CVD) [2], liquid-phase epitaxy (LPE) [3], and zone melt recrystallisation (ZMR) [4].

However, it has become apparent that the properties of the underlying substrate are the key to obtaining high-performance silicon thin-film solar cells. That is, a high-quality silicon thin-film photoelectric layer can be obtained by suppressing the diffusion of impurities from below and using epitaxial deposition that exploits the characteristics of the underlying material, and evidently single-crystal silicon has become the optimal material for the substrate. Accordingly, this research has shifted toward recycling single-crystal silicon, and at present research is being actively performed into single-crystal silicon thin-film solar cells, primarily focusing on porous silicon delamination techniques [5].

In a somewhat different direction, attention has recently been focused on thin-film solar cells made with crystalline silicon with a small grain size. These are formed by plasma CVD using inexpensive substrate materials such as glass and by using low temperatures regardless of the type of substrate instead of these high-temperature processes [6–11].

Figure 1 shows the relationship between grain size and V_{oc} (open circuit voltage) as summarised by Werner [12]. Here, V_{oc} can be regarded as a parameter reflecting the cell characteristics and crystalline properties. As Figure 1 shows, superior characteristics are obtained with a grain size of 100 μm, but the characteristics are worse with a grain size of a few tens of μm. Conversely, favourable characteristics have been obtained experimentally both at Neuchâtel University and at Kaneka Corporation's PV Research Division by using thin-film polycrystalline silicon formed at low temperatures by plasma CVD with a submicron grain size (this is generally referred to as microcrystalline silicon due to the small grain size). This is an interesting discovery that has caused recent research to shift toward both extremes.

This chapter discusses these thin-film polycrystalline silicon solar cells with very small grain size formed at low temperatures (referred to as microcrystalline silicon solar cells) that have been the subject of studies conducted by the

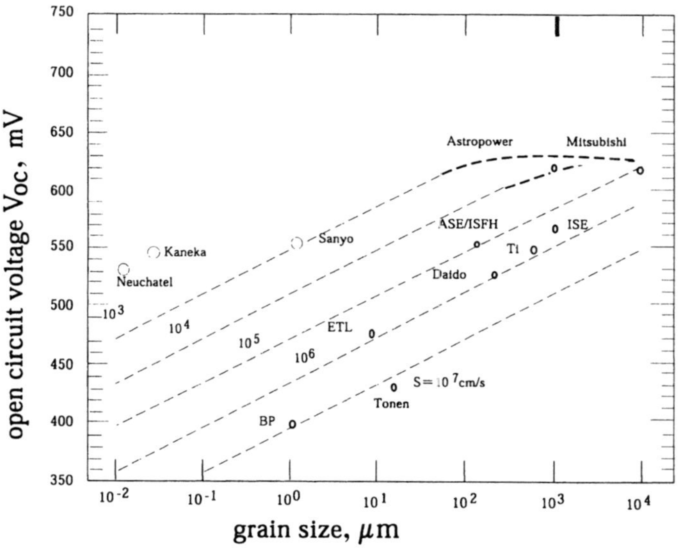

Figure 1 The relationship between grain size and open circuit voltage (V_{oc}) in solar cells. V_{oc} is correlated with the carrier lifetime (diffusion length). In this chapter, microcrystalline silicon cells correspond to a grain size of 0.1 μm or less. S indicates the recombination velocity at the grain boundaries, and Ti, BP, ASE and ISE are abbreviations of the research facilities from which the associated data came.

Kaneka PV Research Division. It will also introduce the results of the research and development of silicon hybrid solar cells with a tandem structure comprising these solar cells and amorphous solar cells, which were put to practical use for the first time in 2001.

2 Thin-Film Polycrystalline (Microcrystalline) Silicon Solar Cells

2.1 Microcrystalline Silicon Thin-Film Solar Cells Formed at Low Temperature

Microcrystalline silicon solar cells formed by plasma CVD at low temperatures are assumed to have a shorter lifetime than single-crystal cells, and it is common to employ a p–i–n structure including an internal electric field in the same way as in amorphous solar cell. A p–i–n type microcrystalline silicon solar cell is formed by a process fairly similar to that of an amorphous solar cell. Strictly speaking, these cells can be divided into p–i–n and n–i–p types according to the film deposition order, although the window layer of the solar cell is the p-type layer in both cases.

The characteristics of a cell having a p–i–n structure were first reported by workers at Neuchâtel University [6, 7]. Unlike an amorphous solar cell, this cell does not deteriorate when exposed to light. The very first reports on the characteristics of an n–i–p cell were made by the Kaneka PV Research

Division using the light trapping structure described below. These cells had an intrinsic conversion efficiency of 10.7% and an apparent efficiency of 10.1% for a film thickness of 2 μm (surface area 1 cm^2, measured by the Japan Quality Assurance Organization (JQA)) (Figure 2) [8–11]. Also, by subjecting the silicon film in the photoelectric layer of this cell to X-ray diffraction (XRD) measurements, it was found to have a preferential (110) orientation. The p–i–n and n–i–p cells have different characteristics due to their different fabrication sequences. A large difference is that the underlying layer of a p–i–n cell is the transparent electrode, whereas the underlying layer of a n–i–p cell is the back electrode.

As a general rule, transparent electrodes are made of oxides, and since there is a risk of these oxides being reduced by the hydrogen atoms that are needed to form microcrystalline cells, there is a smaller process window in the cell formation conditions for the p–i–n type. From the viewpoint of the ease with which integrated structures can be formed, which is a characteristic of thin-film solar cells, an advantage of p–i–n cells is that they can be formed as superstrate modules using integration techniques similar to those used for amorphous silicon as described below. Furthermore, it should be possible to make integrated structures of n–i–p cells by methods equivalent to those used for Cu(InGa)Se_2-based solar cells [13]. At the present time, it is difficult to determine which is better, but it should become possible to arrive at a conclusion in terms of cost, performance and applications through the production of sub-modules in the future.

2.2 Carrier Transport in Microcrystalline Silicon Thin-Film Solar Cells

To improve the efficiency of these solar cells in the future, it is absolutely essential to gain some understanding of the relationship between the crystalline

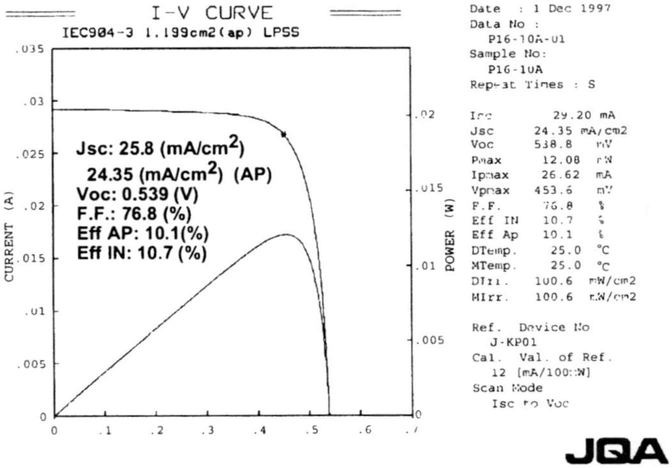

Figure 2 Current–voltage characteristics of an n–i–p type microcrystalline silicon cell (film thickness 2 μm, area 1 cm^2, measured by JQA). Eff$_{AP}$ *is the characteristic for the area including the grid electrode, and* Eff$_{IN}$ *is the characteristic of the effective area not including the grid.*

microstructure including grain boundaries and the carrier transport process. In particular, the carrier lifetime (diffusion length) that determines the solar cell characteristics and the recombination velocity at grain boundaries. However, in a microcrystalline silicon cell, the general non-uniformity that is typical of its grain structure has a direct and complex effect on carrier transport, and thus there have been no conclusive reports relating to these physical properties by direct measurement and analysis. Rather, the efficiency of these solar cells has only recently come to light, so only the first steps have been made in the research of their fundamental properties.

A report by Werner et al. [12] discussed why the very small crystals produced at low temperature (as discussed in the introduction) exhibit relatively favourable cell characteristics. Although the cell characteristics are the product of current, voltage and fill factor, the open circuit voltage is directly related to the cell's lifetime and is used here as a guide to the overall cell characteristics. This is because the current is related to light trapping effects and the cell film thickness as discussed below.

According to Werner et al., the reason why a microcrystalline silicon cell exhibits a characteristic of 10% or more regardless of its small grain size is because of the very low recombination speed at its grain boundaries. This low recombination speed originates from the small barrier height at the grain boundaries, which is guaranteed by (a) the passivation of hydrogen at grain boundary defects, (b) uncharged intrinsic grains (low oxygen density), and (c) the (110) orientation of the crystal grains.

In Figure 1, as can be seen from the relationship between the open circuit voltage, the grain size and the recombination speed already summarised by Werner et al., further improvement of the lifetime of the crystalline silicon layer (i.e., improvement of V_{oc}) in order to make substantial efficiency improvements may require not only a substantially larger grain size but also a lower recombination velocity at the grain boundaries (e.g., 100 cm/s). Or, to put it another way, however much the grain size is increased as shown in Figure 1, the conversion efficiency decreases when the recombination velocity at the grain boundaries increases (corresponding to a film produced at high temperature).

The actual crystalline structure of a microcrystalline silicon solar cell has been investigated by transmission electron microscopy (TEM) and atomic force microscopy (AFM). For example, Meier et al. [14], at Neuchâtel University, have classified the crystalline structures. Of course the microstructures differ depending on the production process and conditions. It has also been reported that diffusion transport mechanisms are dominant within the grains whereas drift transport is more prominent at the grain boundaries [14]. These grain boundaries exhibit amorphous characteristics which give rise to a passivation effect, and it has been pointed out that this may reduce the boundary recombination velocity.

At the Julich Research Laboratory, Rech et al. evaluated the cell characteristics as a function of the silane concentration with respect to the hydrogen dilution [15] and showed that the peak cell performance (where the maximum value of V_{oc} is achieved for the crystal) occurs at the boundary where the resulting film changes from amorphous to crystalline. It is hoped that a more precise analysis of the microstructure and a deeper understanding of its relationship with the boundary recombination velocity will be possible in the future.

To understand the phenomena behind carrier transport, the Kaneka PV Research Division is at the initial stages of a study into the relationship between V_{oc} and film thickness [10, 11]. Figure 3 shows the dependence of V_{oc} on film thickness for a microcrystalline silicon device we produced. The maximum value of V_{oc} we obtained was about 550 mV. As the film thickness decreases, V_{oc} increases slightly. This is thought to be because as the film thickness decreases, the overall number of defects decreases and V_{oc} increases. We thus obtained a result that agrees with the theory stated at the beginning that V_{oc} improves with a thinner film.

Figure 3 also shows the results of simulation by the one-dimensional computer program PCID [16]. For a given carrier lifetime (for the case of diffusion length), V_{oc} can be increased by further increasing the carrier concentration. However, according to Werner's findings, this causes the grain boundaries in microcrystalline layers to become weakly n-type, resulting in a higher recombination velocity. This has the effect of causing V_{oc} to decrease, so it is also essential to control the grain boundaries.

In any case, compared with the V_{oc} of highly efficient single-crystal solar cells (about 700 mV), the V_{oc} of microcrystalline silicon solar cells is at least 20% lower, and the key to improving the cell efficiency is to increase V_{oc} – i.e., to improve their lifetime.

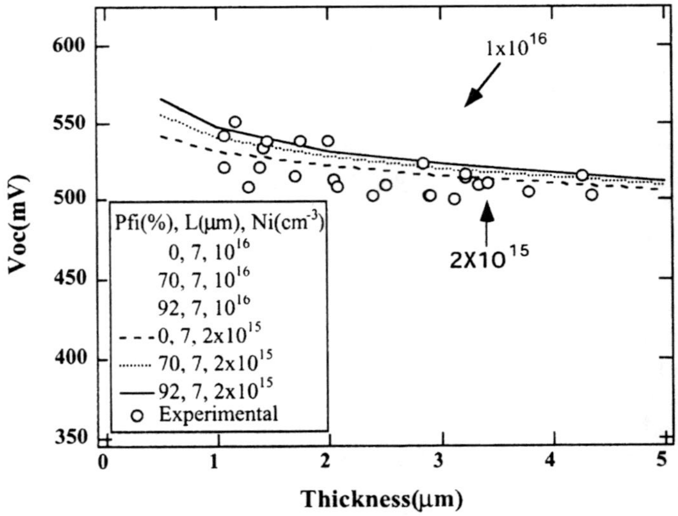

Figure 3 The variation of V_{oc} with film thickness. The open circles represent experimental data, P_{fi}(%) represents the confinement factor (0%, 70% or 92%), L represents the diffusion length, and N_i represents the carrier concentration. A simulation was also performed using PCID. The simulated results agree well with the experimental data at a carrier concentration of $2 \times 10^{15}/cm^3$.

2.3 High-Speed Fabrication of Microcrystalline Silicon Films

To consider the fabrication speed of microcrystalline silicon films, it is essential to understand the film fabrication mechanisms involved. A detailed description of these mechanisms can be found in a report published by Kondo et al. [17].

Based on the large amount of experimental data gained by the Kaneka PV Research Division, the guidelines for forming high-quality microcrystalline silicon by plasma CVD are very simple – as long as it is possible to supply a sufficient quantity of hydrogen atoms to the substrate during crystallisation without causing any damage (most of it is ion damage), then the film fabrication speed can be improved without deteriorating the cell characteristics. Throughput is another important consideration for mass-production, and it is, therefore, essential to develop ways of depositing films at high speed over large areas. At the Kaneka PV Research Division, as stated below, we have devised a technique for fabricating high quality, uniform thin films of microcrystalline silicon at a film deposition rate of 1.1 nm/s on metre-size substrates.

2.4 The Light-Trapping Effect of Microcrystalline Silicon Cells

Light-trapping techniques are a way of increasing the performance of microcrystalline solar cells. This is a core technique for cells made from microcrystalline silicon because – unlike amorphous silicon – it is essentially an indirect absorber with a low absorption coefficient. That is, the thickness of the Si film that forms the active layer in a microcrystalline silicon solar cell is just a few micrometres, so it is not able to absorb enough incident light compared with solar cells using ordinary crystalline substrates. As a result, it is difficult to obtain a high photoelectric current. Light-trapping technology provides a means of extending the optical path of the incident light inside the solar cell by causing multiple reflections, thereby improving the light absorption in the active layer.

Light trapping can be achieved in two ways: (i) by introducing a highly reflective layer at the back surface to reflect the incident light without absorption loss, and (ii) by introducing a textured structure at the back surface of the thin-film Si solar cell. Of course, the carrier lifetime is also important, and it goes without saying that the diffusion length must be at least as long as the film thickness.

At the initial stages, we focused on the light trapping effects of textured structures formed on the silicon surface [8]. We then improved the solar cell characteristics by reducing the film fabrication temperature (although as the film fabrication temperature decreases, the degree of surface texturing also decreases). A solar cell with a textured surface and a back reflective layer for increasing absorption is said to have a STAR (naturally Surface Texture and enhanced Absorption with a back Reflector) structure.

At present, two types of light trapping structure are in use. Figure 4(a) shows a cell that uses a flat highly reflective layer at the back surface, and Figure 4(b) shows a cell that uses a textured type of highly reflective layer. This highly

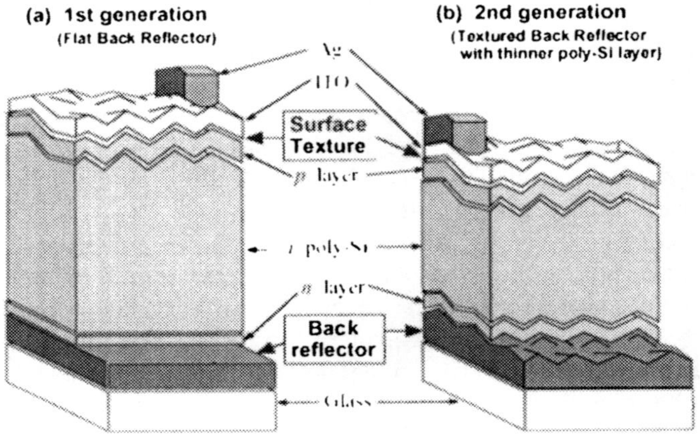

Figure 4 Cross-sections through light-trapping microcrystalline silicon solar cell devices. (a) First generation (flat back reflector); (b) second generation (textured back reflector, thinner polycrystalline silicon layer).

reflective back layer also acts as the solar cell's back electrode. Thin-film polycrystalline silicon forms naturally with a textured structure on its surface, and the size of this texture is strongly dependent on the film thickness. When the film is relatively thick (4 μm or more), the surface texture is suitable for light trapping, but when the film is relatively thin (1.5 μm or less), an adequate surface texture does not form [8–11]. It is thus necessary to use a textured reflective layer at the back surface. Of course, to be precise the texture characteristics depend on the fabrication conditions as well as on the film thickness.

Cells of this sort are fabricated as follows. First, after a back reflector layer has been formed on a glass substrate, a layer of n-type silicon is deposited on it by plasma CVD. Next, the active layer, which is a thin film of intrinsic polycrystalline silicon, is formed by plasma CVD. Finally, a p-type silicon film and an indium tin oxide (ITO) film are deposited in turn, and then a comb grid electrode is formed on top.

Now consider the light trapping effect of a thin film polycrystalline silicon layer with a thickness of 1.5 μm. Figure 5(a) shows the reflection spectra of samples corresponding to two types of 1.5 μm thick cell (one with a flat back reflector and one with a textured back reflector), and Figure 5(b) shows the corresponding collection efficiency spectra.

In Figure 5(a) the minimum that appears in the spectra at around 550 nm arises because the ITO film was deposited under non-reflective conditions. The oscillation seen at wavelengths above 600 nm is caused by interference with the silicon film in the active layer. The reflectivity of the sample with the textured back surface is significantly lower in the infrared region. Figure 5(b) shows the spectral sensitivity spectra corresponding to the flat and textured samples. In the case of the textured sample, it can be seen that the sensitivity is higher at

a)

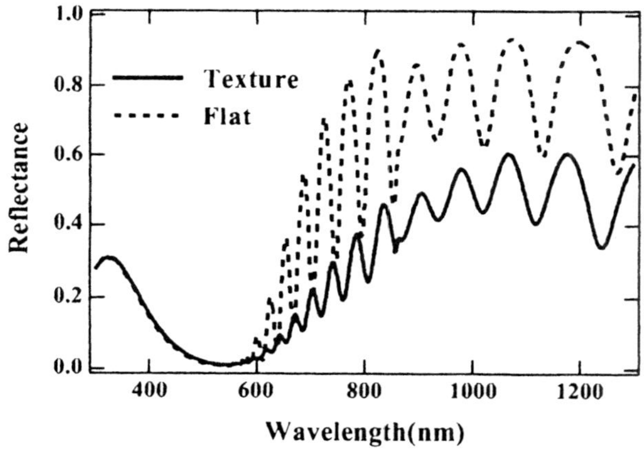

b)

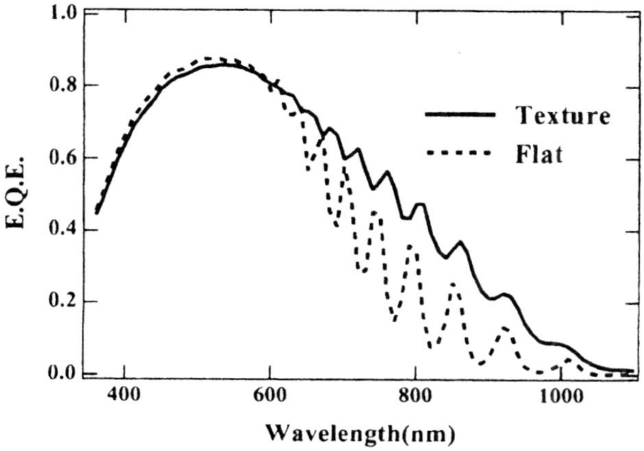

Figure 5 The characteristics of microcrystalline solar cells on glass substrates having flat and textured back reflectors: (a) reflectivity and (b) collection efficiency characteristics (quantum efficiency).

wavelengths above 600 nm, which corresponds to the reduction of the reflectivity spectrum at longer wavelengths.

This result provides experimental verification of the contribution made by the minute textured structures to light trapping. In the future, it will be necessary to aim at increasing the efficiency by obtaining larger currents in the thin film through the formation of better light trapping structures by controlling the profile of the microcrystalline silicon surface and the underlying reflective layer.

3 Application to Stacked (Hybrid) Type Solar Cells

3.1 Silicon Hybrid Solar Cells

Although the microcrystalline silicon cells formed at low temperature as described above have a potential for high efficiency, their efficiency in single-cell structures is currently only about 10%, which is much lower than that of bulk polycrystalline cells. To achieve a substantial improvement of efficiency there needs to be some kind of breakthrough, such as a substantial reduction in the abovementioned grain boundary recombination or the establishment of more advanced light trapping techniques. In an attempt to achieve this, we have investigated the use of two- and three-stacked (hybrid) structures in which multiple cells with different light absorption characteristics are stacked together. This approach allows better characteristics to be obtained with existing materials and processes. The advantages of using a layered structure include the following: (i) it is possible to receive light by partitioning it over a wider spectral region, thereby using the light more effectively; (ii) it is possible to obtain a higher open-circuit voltage; and (iii) it is possible to suppress, to some extent, the rate of reduction in cell performances caused by optical degradation phenomena that are observed when using amorphous silicon based materials.

At the Kaneka PV Research Division, we have recently made further progress with the investigation of stacked (tandem) cells in which the abovementioned amorphous silicon cells are combined with microcrystalline silicon cells. We have been studying structures in which a transparent intermediate layer is provided between the amorphous silicon layer of the top cell ad the microcrystalline silicon layer of the bottom cell (next-generation hybrid solar cell).

Figure 6(a) illustrates the structure of such a device. Compared with a conventional hybrid cell, the presence of a transparent intermediate layer, by its nature, causes some of the light that reaches the microcrystalline silicon layer of the bottom cell by passing through the amorphous silicon layer of the top cell to be reflected back into the top cell (amorphous silicon). As a result, the effective light sensitivity of the top cell is higher than that of a top cell with the same thickness in a conventional hybrid structure, so it is possible to expect an improvement in the short-circuit current J_{sc} of the overall hybrid cell. Figure 6(b) shows how the sensitivity spectrum of the hybrid cell is affected in practice by the addition of a transparent intermediate layer. Even if the amorphous silicon and microcrystalline silicon layers have the same thickness, the cell with a transparent intermediate layer is more sensitive at shorter wavelengths (top cell).

This result seems to suggest the possibility of controlling the respective light sensitivities of the top and bottom cells by introducing a suitable transparent intermediate layer. From the results of Figure 6(b) it can be inferred that by controlling the thickness of the transparent intermediate layer it is possible to reduce the thickness of the top cell needed to obtain the same current compared with a conventional hybrid cell. As a result, it should be possible to

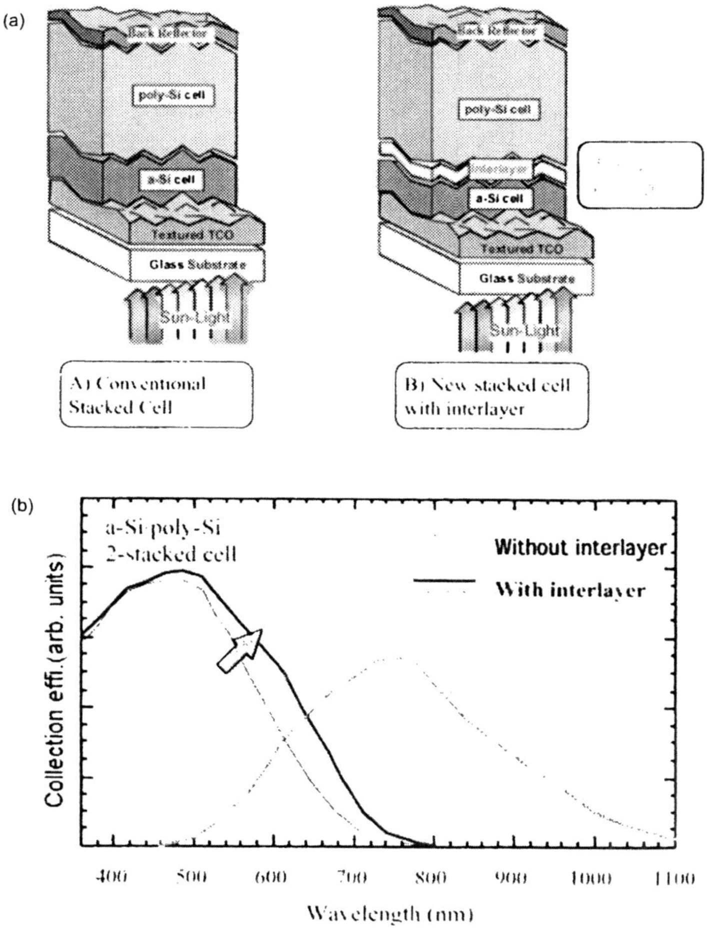

Figure 6 A next-generation tandem model. (a) Device cross-section of an amorphous–microcrystalline silicon layered next-generation tandem cell having a transparent intermediate layer (next-generation silicon hybrid cell). (b) Spectral sensitivity characteristics of the hybrid cell including the intermediate layer (absorption efficiency). An arrow indicates the enhancement of spectral response of top cell by the insertion of intermediate layer.

obtain roughly the same short circuit current J_{sc} even when the amorphous silicon film thickness is reduced. If the thickness of the amorphous silicon film in the top cell can be reduced, then it should also be possible to reduce the photo-degradation of cell.

This structure (amorphous silicon–transparent intermediate layer–microcrystalline silicon) makes it possible to improve not only $J_{sc,}$ but also the cell characteristics compared with a conventional hybrid cell with no transparent intermediate layer without reducing V_{oc} or the fill factor (FF). As Figure 7 shows,

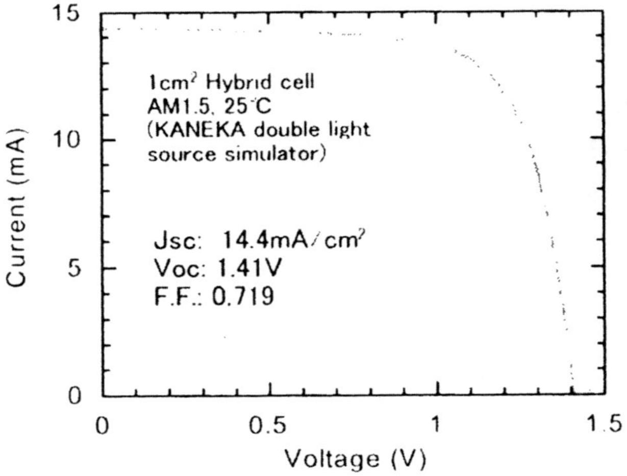

Figure 7 Current-voltage characteristic of an amorphous–microcrystalline silicon stacked next-generation hybrid cell with a transparent intermediate layer (initial efficiency; area 1 cm^2).

an initial conversion efficiency of 14.5% has been achieved with a 1 cm square small area cell of this type [18].

3.2 Large-Area Thin-Film Silicon Hybrid Modules

At the Kaneka PV Research Division, we have begun developing techniques for depositing microcrystalline silicon films on large area substrates, and for the production of silicon hybrid modules based on tandem cells of amorphous silicon and microcrystalline silicon [19, 20]. We have also been developing mass production techniques that were put into commercial operation in 2001. These hybrid modules employ a superstrate structure in which light is incident through a glass substrate on which a transparent electrode is formed, and are integrated by laser scribe processing (Figure 8(a)). By adding a microcrystalline silicon film deposition process, the amorphous silicon module production process can be used to fabricate silicon hybrid modules.

As a result of improving the efficiency of a large area hybrid module measuring 910 × 455 mm, we have achieved an initial efficiency of 12.3% and a peak output power of 47.1 W as shown in Figure 8(b). We have also achieved the goal of a film deposition speed of 1.1 nm/s on metro-size large area substrates. The efficiency of these cells is somewhat lower (11.2%), but still of practical use [21]. Some of these modules are shown in Figure 9.

4 Conclusion

It has been demonstrated that microcrystalline silicon with a very small grain size formed at low temperature by plasma CVD is a promising material for use in

(a)
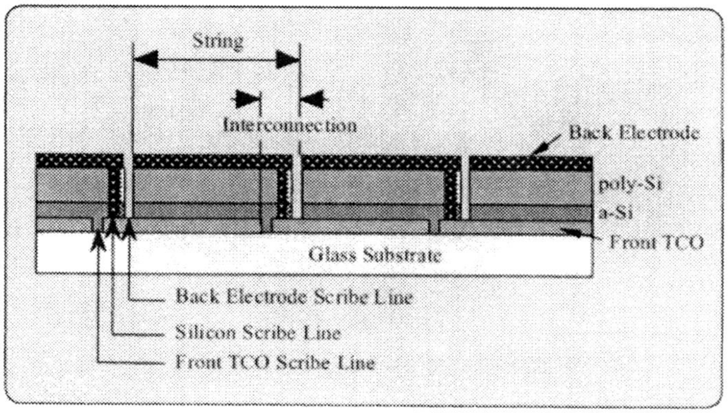

(b)
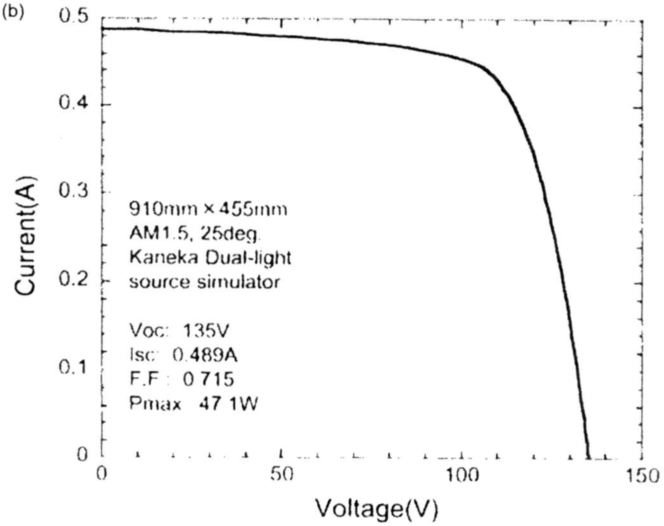

Figure 8. A silicon hybrid cell. (a) Cross-sectional structure of the silicon hybrid module (amorphous–microcrystalline silicon tandem solar cell module; series-connected). (b) Current–voltage characteristics of a large area (910 × 455 mm) silicon hybrid module (initial efficiency).

solar cells. Amorphous solar cells and stacked silicon (hybrid) solar cells have already been put to practical use in power modules. Further study is needed to ascertain and control the relationship between recombination at grain boundaries and the grain boundary microstructure, which is a non-uniform system. This should make it possible to improve the inherent performance of microcrystalline silicon solar cells. There is also a need for the development of crystal deposition methods that allow some degree of control over the texture structure of the silicon surface to allow a suitable level of light trapping to be implemented.

Figure 9 Appearance of the hybrid modules.

References

[1] Spitzer, M., Schewchun, J., Vera, E.S. and Lofersky, J.J., 1980. *Proc. 14th IEEE Photovoltaic Specialists Conf.*, San Diego, p. 375.

[2] Reif, R., 1984. *J. Electrochem. Soc.*, Vol. 131, p. 2430.

[3] Nishida, S., Nakagawa, K., Iwane, M., Iwasaki, Y., Ukiyo, N., Mizutani, M. and Shoji, T., 2001. Si-film growth using liquid phase epitaxy method and its application to thin-film crystalline Si solar cell. *Sol. Energy Mater. Sol. Cells*, Vol. 65, p. 525.

[4] Morikawa, S., Kawama, Y., Matsuno, Y., Hamamoto, S., Imada, K., Ishihara, T., Kojima, K. and Ogama, T., 2001. Development of high-efficiency thin-film Si solar cells using zone-melting recrystallization. *Sol. Energy Mater. Sol. Cells*, Vol. 65, p. 261.

[5] Bergmann, R., Rinke, T., Wagner, T. and Werner, J., 2001. Thin film solar cells on glass based on the transfer of monocrystalline Si films. *Sol. Energy Mater. Sol. Cells*, Vol. 65, p. 355.

[6] Meier, J., Flueckiger, R., Keppner, H. and Shah, A., 1994. Complete microcrystalline *p–i–n* solar cell – Crystalline or amorphous cell behavior? *Appl. Phys. Lett.*, Vol. 65, p. 860.

[7] Meier, J., Torres, P., Platz, R., Dubail, S., Kroll, U., Anna Selvan, J.A., Pellaton Vaucher, N., Hof, Ch., Fischer, D., Keppner, H., Shah, A., Ufert, K.-D., Giannoules, P. and Koehler, J., 1996. *MRS Spring Meeting*, San Francisco, Vol. 131, p. 507.

[8] Yamamoto, K., Nakajima, A., Suzuki, T., Yoshimi, M., Nishio, H. and Izumina, M., 1994. *Jpn. J. Appl. Phys.* Vol. 33, p. L1751.

[9] Yamamoto, K., Yoshimi, M., Suzuki, T., Tawada, Y., Okamoto, Y. and Nakajima, A., 1998. Below 5 μm thin film poly-Si solar cell on glass substrate fabricated at low temperature. *Proc. 2nd World Conf. on Photovoltaic Solar Energy Conversion*, Vienna, p. 1284.

[10] Yamamoto, K., Yoshimi, M., Suzuki, T., Tawada, Y., Okamoto, Y. and Nakajima, A., 1999. Thin-film poly-Si solar cells on glass substrate fabricated at low temperature. *Appl. Phys.* Vol. A69, p. 179.

[11] Yamamoto, K., 1999. Very thin film crystalline silicon solar cells on glass substrate fabricated at low temperature. *IEEE Trans. Electron Devices*, Vol. ED-46, p. 2041.

[12] Werner, J. and Bergmann, R., 1999. *Technical Digest 12th Int. Photovoltaic Science and Engineering Conf.*, Sapporo, p. 2041.

[13] Wieting, R., Delaney, D., Dietrich, M., Fredric, C., Jensen, C. and Willett, D., 1995. Progress in CIS-based photovoltaics through statistical process control, *Proc. 13th European Photovoltaic Solar Energy Conf.*, Nice, p. 1627.

[14] Vallat-Sauvain, E., Kroll, U., Meier, J., Wyrsch, N. and Shah, A., 2000. Microstructure and surface roughness of microcrystalline silicon prepared by very high frequency-glow discharge using hydrogen dilution. *J. Non-Cryst. Solids*, Vol. 125, pp. 266–269.

[15] Repmann, T., Appenzeller, W., Roschek, T., Rech, B. and Wagner, H., 2000. *Proc. 28th IEEE Photovoltaic Specialists Conf.*, Anchorage, p. 912.

[16] Basore, A., 1990. Numerical modeling of textured silicon solar cells using PC-1D, *IEEE Trans. Electron Devices* Vol. ED-37, p. 337.

[17] Kondo, M., Fukawa, M., Guo, L. and Matsuda, A., 2000. *J. Non-Cryst. Solids*, Vol. 84, pp. 226–269.

[18] Fukuda, S., 2002. 49th Spring Meeting of the JSAP and Related Societies, p. 930.

[19] Yamamoto, K., Yoshimi, M., Suzuki, T., Nakata, T., Sawada, T., Nakajima, A. and Hayashi, K., 2000. *Proc. 28th IEEE Photovoltaic Specialists Conf.*, Anchorage, p. 1428.

[20] Yamamoto, K., Yoshimi, M., Sawada, T., Nakajima, A., Hayashi, K., Suezaki, T., Takata, H. and Tawada, Y., 2001. *Tech. Digest 12th Photovoltaic Science and Engineering Conf.*, Jeju, p. 547.

[21] Suezaki, T., 2002. 49th Spring Meeting of the JSAP and Related Societies, p. 931.

IIc-3

CdTe Thin-Film PV Modules

Dieter Bonnet[1], ANTEC Solar GmbH, Arnstadt, Germany

[1] For current address, see List of Contributors on page x

1 Introduction

CdTe is very well suited for use as active material in thin-film solar cells due to four special properties [1]:

- CdTe has an energy gap of 1.45 eV, and therefore is well adapted to the spectrum of solar radiation.
- The energy gap of CdTe is 'direct', leading to very strong light absorption.
- CdTe has a strong tendency to grow as an essentially highly stoichiometric, but p-type semiconductor film and can form an p–n heterojunction with CdS. (CdS has a rather wide energy gap of 2.4 eV and grows n-type material under usual film deposition techniques.)
- Simple deposition techniques have been developed suited for low-cost production.

Current densities of up to 27 mA cm^{-2} and open-circuit voltages of 880 mV, leading to AM 1.5 efficiencies of 18%, can be expected for CdTe cells made under a mature technology.

Figure 1 shows the typical film sequence of this cell. In the preferred arrangement, first a transparent conducting film (typically In_2O_3 or SnO_2 or a combination of both) is deposited onto glass-plate used as transparent substrate. Then an n-CdS film is deposited, followed by the active p-conducting CdTe film. A special treatment improves the p–n junction between CdS and CdTe ('activation'). Finally a low-resistance contact is deposited onto the CdTe, which can be opaque.

Light enters the cell through the glass-substrate. Photons transverse the TCO and CdS layers. These films are not active in the photovoltaic charge generation process although leading to some – unwanted – absorption. The CdTe film is the active absorber of the solar cell. Electron–hole pairs are generated close to the junction. The electrons are driven by the built-in field through the interface into the n-CdS film. The holes remain in the CdTe and join the pool of the holes promoting the p-conduction of this material and finally have to leave the cell via

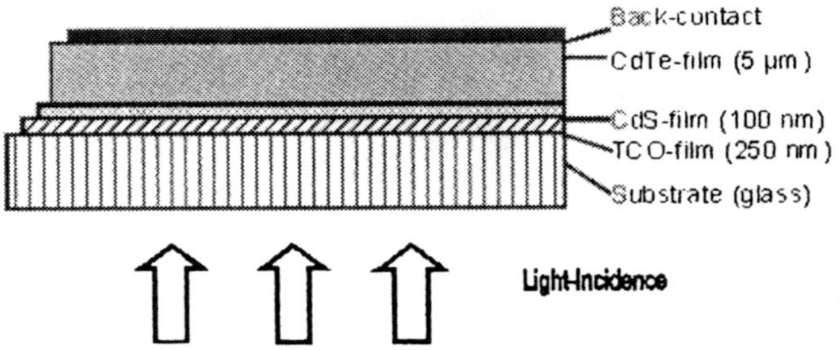

Figure 1 Film sequence of the CdTe thin-film solar cell.

the back contact. Electric power is drawn by metallic contacts attached to the TCO film and the back contact.

Due to the strong light absorption in CdTe of about 10^5 cm^{-1} for light having a wavelength below 800 nm, a film thickness of a few micrometres would be sufficient for complete light absorption. For practical reasons a thickness of about 3–7 μm is often preferred.

Intensive research has shown that this junction can be mastered so that the following basic criteria for solar cells can be fulfilled under conditions of industrial production:

- Effective generation of mobile minority charge carriers in the CdTe film.
- Efficient separation of charge carriers by means of the internal electric field of the p-n junction between n-CdS and p-CdTe.
- Low loss-extraction of the photocurrent by means of ohmic contacts to the TCO and back-contact films.
- Simple fabrication technologies for low-cost, high-volume production.

Solar cells of efficiencies above 16% have been made in research laboratories and industrial efforts have led to the recent start-up of industrial production units at three private companies in the USA and Germany each aiming at large scale production of 100,000 m^2 per annum or more. First large area modules have recently surpassed the 10% efficiency mark.

2 Steps for Making Thin-Film CdTe Solar Cells

2.1 Film Deposition

2.1.1 CdTe

Most techniques to deposit CdTe films rely on one or both of the following properties:

- CdTe, if heated in vacuum up to temperatures above 600°C, sublimes congruently liberating Cd and Te in equal amounts, the residue remaining stoichiometric CdTe.
- In CdTe films condensing on substrates kept above 400°C (or heated up to this temperature after deposition) the stoichiometric compound is the stable solid phase. The constituting elements have a significantly higher vapour pressure than the compound.

These properties make it relatively easy to produce CdTe films suited for thin-film solar cells: No excessive care has to be taken to provide for stoichiometry, as long as the substrate temperature is sufficiently high. CdTe or Cd + Te or decomposable compounds of Cd and Te can be used as starting material. Upon arrival of Cd and Te on the substrate even in a non 1:1 ratio, CdTe condenses (nearly) stoichiometrically as long as the substrate is heated at 400–500°C or higher during or after the actual deposition. The film quality increases with temperature up to 600°C. At higher temperatures the sticking coefficient decreases (re-sublimation). A p-doping effect is achieved due to a small natural nonstoichiometry, in the form of Cd deficiencies, probably vacancies. No additional doping is used. Typical doping levels are around 10^{15} cm^{-3}. These values are somewhat low, but can be tolerated in thin-film cells. If films are deposited at lower temperatures and therefore not necessarily at stoichiometric ratio, they can be heated to create the stoichiometric compound. This allows numerous film deposition technologies to be applied. The only requirement is absence of disturbing impurities, which might jeopardise the native p-doping and charge carrier lifetime. High purity (up to 99.999%) of the elements and the compound can be achieved on an industrial scale as the elements – Cd and Te – can be easily purified by standard metallurgical procedures.

Numerous film deposition processes have been studied in the past and all have led to good cells exhibiting efficiencies above 10%. Only a few processes have properties suited to large-scale production, though. They have been developed by industrial units, as discussed in the following.

Vacuum Deposition – Sublimation/Condensation. Solid CdTe material in form of powder or granulate is sublimed in vacuum and condenses on the substrate maintained at elevated temperatures between 450 and 600°C, using the basic thermodynamic properties of CdTe already mentioned. Commercial processes of different kind have been developed, which can achieve very high deposition rates (>10 $\mu m/min$) and can be applied to continuous-flow in-line processes, using low-cost, rugged vacuum systems. The processes do have high materials yield as the material is forced to condense only on the substrate either by close space between source and substrate (close-spaced sublimation = CSS) or by prevention of deposition on the walls kept at elevated temperatures above 600°C. Two factories using sublimation processes have been built in Germany and the USA. Production and sale of modules has started in both facilities.

Electrodeposition. CdTe films are formed from aqueous solutions of $CdSO_4$ and Te_2O_3 at temperatures of around 90°C. An n-type film of low electronic quality is formed. The basic reaction is as follows:

$$Cd^{2+}_{(aq)} + HTeO_2^+{}_{(aq)} + 3H^+_{(aq)} + 6e^- \rightarrow CdTe_{(s)} + 2H_2O_{(liq)}$$

Grain-size enhancement, doping conversion into p-type and improvement of electronic transport properties are achieved by thermal post-annealing under the influence of Cl-based compounds. The driving electric potential is applied to the transparent conductive film on the substrate and has to be very homogeneous over the whole surface to be coated. This requires low deposition current density, resulting in low deposition rates. This can be compensated for high throughput by coating a large number of substrates in parallel. A production plant has been built in the USA and presently is ramped-up to production quantities.

Chemical Spraying. An aerosol of water droplets containing heat-decomposable compounds of Cd and Te is sprayed onto a heated substrate, forming CdTe from the liberated elements. Processes have been developed, which do not require vacuum and can be applied easily in inline systems by using linear nozzle arrays. A first pilot plant has been built in the USA by an industrial venture, which subsequently has been put for sale and finally was abandoned.

Screen Printing. Slurries containing Cd and Te are screen printed onto the substrate and transformed into CdTe by thermal reaction under the influence of added $CdCl_2$. Due to some porosity of the films, comparatively thick layers are required for good operation of the cells. This technology is presently employed on a commercial scale with production capacity of ~1 MW_p/year. Small modules are manufactured and used in consumer applications. There are some doubts on the suitability of this process for large-scale production of high-efficiency, low-cost modules.

2.1.2 CdS

Like CdTe, CdS has the same strong tendency to form stoichiometric films, but unlike CdTe, CdS films are natively n-doped by a slight non-stoichiometry. CdS films can be deposited by the same processes as CdTe as its basic properties are quite similar to those of CdTe, e.g. its tendency to sublime and condense congruently. The following processes have been studied more intensely in view of production:

- Sublimation/condensation, like close-spaced sublimation and hot-wall sublimation
- Electrodeposition
- Screen printing

Another process is especially suited for CdS: chemical bath deposition (CBD). In this process a metastable solution containing Cd and S leads to spontaneous formation of thin CdS films on surfaces of substrates immersed into the solution at temperatures around 80°C. The chemical reaction basically is as follows:

$$Cd^{2+}(\text{complex})_{(aq)} + (NH_2)CS_{(aq)} + 2OH^-_{(aq)} \rightarrow CdS_{(s)} + H_2O_{(liq)}$$

The CdS films so formed are tightly adherent and very homogeneous even at low thickness.

A potential disadvantage on forming abrupt junctions of high photoelectronic quality between CdTe and CdS is that CdS has a significant lattice mismatch to CdTe. Fortunately post-deposition treatments, described below, allow the amelioration of the junction.

2.1.3 TCO Films

In TCO films a compromise is achieved between high electronic conduction required for low series resistance in cells and high optical transmission for high light input and ensuing high photocurrent. Several materials are presently in use and under development for industrial application:

SnO_2. SnO_2 films can be produced by a spraying process at ambient pressure. $SnCl_4$ is dissolved in water and sprayed onto a heated substrate in air. $SnCl_4$ decomposes under reaction with oxygen forming SnO_2 films and yielding HCl, which is led away. Substrates of this kind are made on a commercial basis and are presently used by several industrial solar cell manufacturers. The films typically have area resistivities of 10 Ω/square and transmission values of around 70–80%. Alternative deposition techniques are cathode sputtering of metallic Sn targets in an oxidising ambient. Although more expensive than spraying, better quality films are achieved.

ITO. In, Sn mixed oxide films have higher performance, when sputtered from an oxide target (either better conductivity or optical transmission than pure SnO_2 films). They are more expensive due to the use of In. As indium may diffuse into the CdS/CdTe film packet and lead to unwanted n-doping of CdTe during high-temperature processing, usually a thin pure SnO_2 film is deposited onto the ITO as diffusion barrier for In.

$CdSnO_4$. This compound can be deposited by co-sputtering of oxides of Cd and Sn. It requires annealing processes at elevated temperatures, which are not suited for use of cheap soda-lime glass. Future process improvements may overcome this setback. The films show better performance than ITO, i.e. higher transmission at equal resistivity or lower resistivity at equal transmission, making them an interesting option for industrial production.

ZnO:Al. This material is routinely used as transparent contact for CIS-based thin-film solar cells. It can easily be fabricated in thin-film form by sputtering of a heterogeneous target containing ZnO and Al. Al acts as donor in ZnO. Unfortunately the film looses its doping during thermal stress ($>500°C$) at deposition of CdTe. There is hope that more stable films can be made eventually, as the material is more cost efficient than ITO.

2.1.4 Substrates

The most common transparent substrate to be used is glass. The cheapest glass – soda-lime glass or windowpane glass – is suitable. It exhibits, if made by the float glass process, a very flat surface well suited to thin-film deposition. It is limited in processing-temperature at 520°C or somewhat higher, if suitably suspended. It is sufficiently cheap ($<10\$/m^2$), and can be bought cut and edge treated in virtually unlimited quantities. It is indeed used by the three production facilities, which have recently become operative.

If higher temperature (which may lead to better quality films) is desired, the second option is borosilicate glass, which can be heated to temperatures above 600°C without softening. The higher cost of this material presently prevents its industrial use. Research groups have made cells of up to 16.2% efficiency on such glass.

2.2 Improvement of Critical Regions of the CdTe Solar Cell

Figure 2 shows an image obtained by scanning electron microscopy of the broken edge of a CdTe solar cell in which the CdTe film has been made by close-spaced sublimation. Figure 3 shows the critical regions, i.e. the CdTe/CdS junction and the back-contact region, which have proven to be the most critical parts of the cell, on which efficiency depends strongly.

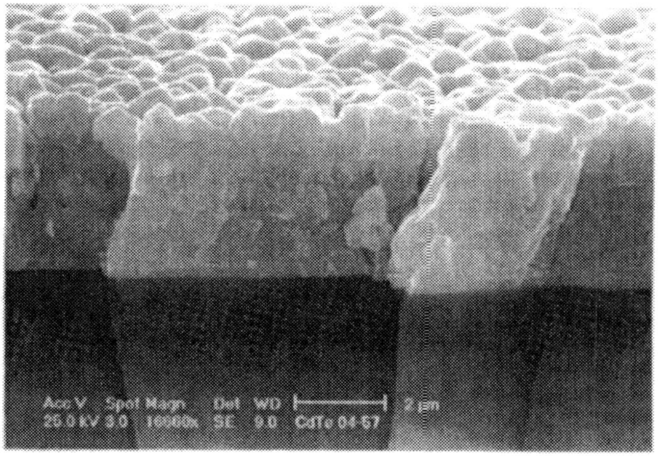

Figure 2 SEM image of the cross-section of a CdTe solar cell.

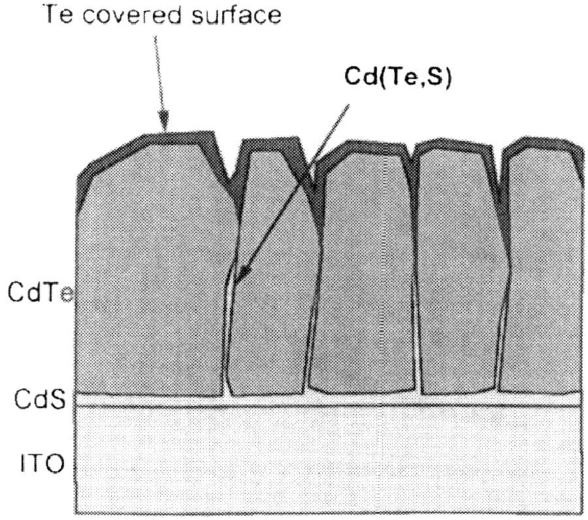

Figure 3 Schematic illustration of the key features of the CdTe solar cell.

2.2.1 The p–n Heterojunction – Improvement by Activation

Although both materials forming the junction are II–VI compounds and have a close chemical relationship, their lattice constants differ by about 5%. This leads to a significant density of interface states which can be expected to result in strong charge carrier recombination. Junctions 'as made' indeed show low charge carrier collection efficiency and thereby low power efficiency of around 2%. Annealing of the system at temperatures of around 400°C leads to some improvement, but only the still not completely understood 'activation' changes the junction so far that efficiencies of up to 16% have been observed. In this activation step, the junction is annealed at temperatures of 400–500°C in the presence of Cl-containing species, generally $CdCl_2$, which is deposited onto the film stack or admitted in vapour form for a time of around a few minutes. Figure 4 shows the strong improvement of the quality of the I–V curves under this treatment [2]. This procedure in the first instance has been developed quite empirically, although based on historic processes for manufacture of CdS photoconductive films [3]. Only recently has light been shed on the basic processes which take place during activation. Three essential effects go hand-in-hand.

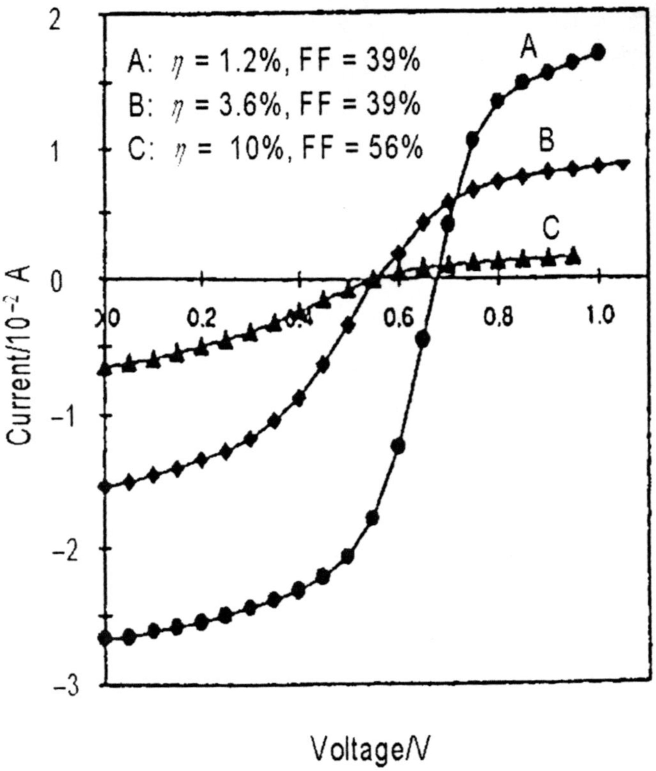

Figure 4 I–V curves of cells showing improvement by thermal treatment processes (from [2]).

Recrystallisation – Grain Growth. Generally for polycrystalline materials in which grain growth may occur there is a limiting grain size in which the driving force is balanced by a retarding force. The net driving force decreases as the grain radius increases. Generally smaller grain sizes are observed for films grown at lower substrate temperatures due to lower mobility of atoms during growth. Upon annealing and activation small grains can start to grow, but only to a certain limit. If on the other hand larger grains are formed already upon film deposition at higher deposition temperatures, such grains will grow less, as they are already closer to the final equilibrium. This has been observed experimentally: Grain size in films deposited at 500°C does not increase upon activation. Figure 5 shows that the grain diameter grows with distance from the junction – at which location nucleation of the films occurs – to an average size of 2 μm [4]. (The film has been deposited by CSS at 500°C.) It does not change even under extended treatment times with $CdCl_2$ species at 400°C. This means that for these samples the grains were at equilibrium directly after material deposition, a satisfying situation. Figure 6 shows a TEM cross-section of a CdS/CdTe film stack deposited at a substrate temperature of 525°C, indicating the typical morphology of such a system, clearly showing the high density and three dimensional distribution of planar defects, mainly stacking faults and twins [5]. The density, however, varies from one grain to the other.

For films deposited at significantly lower temperatures (<400°C) the films show smaller grain sizes directly after deposition, together with strong orientation with the 111-axis perpendicular to the substrate. This orientation is lost upon recrystallisation during $CdCl_2$ activation [6]. Figure7, obtained by AFM (atomic force microscopy), shows directly three steps in the process of recrystallisation: During annealing under $CdCl_2$ new small grains form on the highly oriented films and grow into the final unoriented phase, indicating close

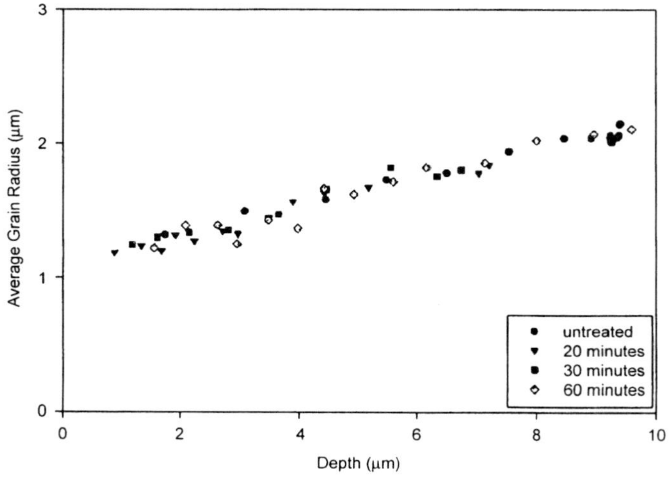

Figure 5 Grain size distribution of a CdTe film as function of the distance from the junction after different thermal treatments [4].

Figure 6 X-ray transmission image of a CdTe film [5]. (Reproduced with permission from IEEE.)

packing. The reference film deposited at 590°C does not grow. The loss of preferential orientation upon annealing of this low temperature film is strikingly illustrated in Figure 8 by X-ray pole diagrams by the same author [7]. The author attributes the grain growth and recrystallisation to lattice strain energy present in the films, which drives the process. In high-temperature-deposited films no strain is present, leading to no significant recrystallisation. This strain present in low-temperature films is clearly illustrated by the author from an effective decrease of lattice constant: After (low-temperature) deposition the lattice constant for a is 6.498 Å, which is reduced on activation to 6.481 Å, the published equilibrium value for crystalline powder.

In another study (Figure 9) grain size distributions were measured [8] for films deposited at relative low temperatures directly after deposition (at 340°C) and after annealing (at 580°C) and alternatively after activation (at 430°C). The final grain size distribution has a maximum at 1 μm.

Immediately at the junction, where the film nucleates, some grain growth has been observed for high-temperature films also [9]. This region is difficult to access. More detailed knowledge may be crucial for an efficiency increase.

Interdiffusion – Intermixing. CdS and CdTe in thermal equilibrium may form mixed compounds CdS_xTe_{1-x} only for quite limited regions close to the single compounds, leaving a miscibility gap between $x = 0.16$ and $x = 0.86$ at 650°C [10]. At lower temperatures the gap widens. In the Te-rich region the material shows a lower band gap than pure CdTe, an interesting feature in some II–VI compounds containing Te.

It is to be expected that CdS and CdTe intermix upon deposition at elevated temperatures to a certain degree. Intermixing at the interface can be expected to reduce the effects of lattice mismatch between CdS and CdTe. This can be analysed by SIMS-depth profiling: SIMS analyses allow the elements to be determined quantitatively at the surface, while it is removed layer by layer, e.g.

Figure 7 Grain reconstruction in a CdTe film made at low temperature during activation (a, b, c) and unchanged film made by high-temperature CSS (d) [6]. (© American Institute of Physics.)

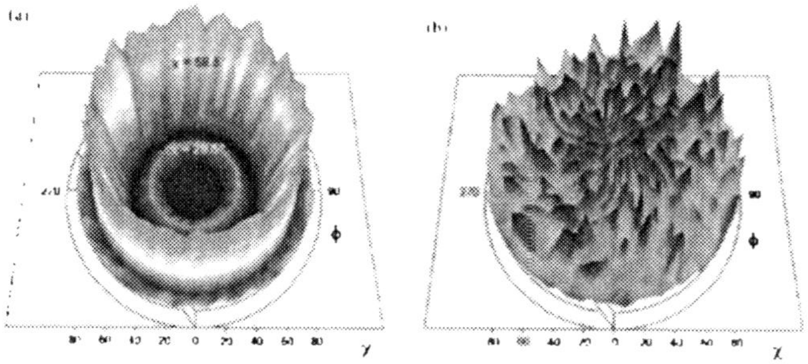

Figure 8 X-ray pole diagrams of 311 planes in a CdTe film: (a) before and (b) after activation at 400°C V. (© American Institute of Physics.)

by sputtering ('depth-profiling'). Such experiments have shown that intermixing of CdS and CdTe is a function of substrate temperature and post-deposition $CdCl_2$ activation. The degree of intermixing indeed has been observed to increase with increase of substrate temperature. Further increase is induced by activation. Excessive interdiffusion leads to deterioration of the device performance. Spectral response curves for devices made from CdTe films deposited at 610°C and activated for different times are shown in Figure 10 [11]. Indeed the intermixing is manifest by a longer wavelength response due to the lower bandgap material

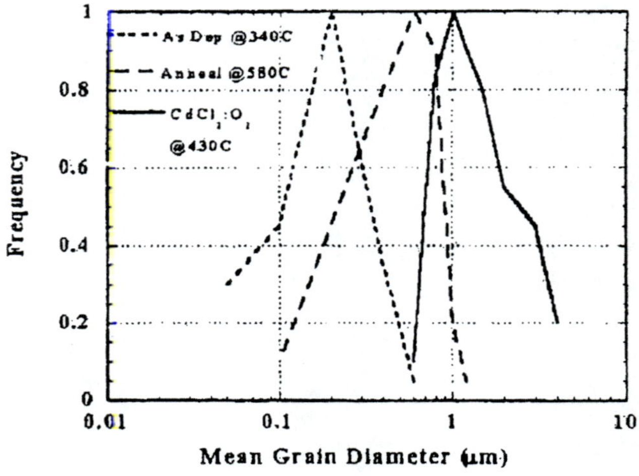

Figure 9 Grain size distribution of a CdTe film directly after deposition, after annealing and after activation [8]. (Reproduced with kind permission of James & James Publ.)

for small values of x, the Te rich mixture. This goes hand in hand with increased total photocurrent. Significant amounts of sulphur (probably CdS_xTe_{1-x}) can be detected both at grain boundaries and within heavily faulted grains (as opposed to grains with low defect density) for films deposited at temperatures between 500 and 600°C [5]. Model calculation yielded three-dimensional distributions of S and Te which have led to isocompositional contour plots of diffused regions [12] (Figure 11). In the case shown for two adjacent grains after activation the different grain sizes result in a different alloy profile. For the narrower grain, no pure CdTe remains at all, while the wider grain exhibits the entire range of alloy composition. The evolution of mixed regions and their progression have been followed by X-ray diffraction studies, impressively showing the emergence of lines corresponding to mixed material. Figure 12 shows initially the line of pure CdTe and after 10, 20 and 40 minutes of activation the emergence and growth of alloy lines. Modelling of these results yielded diffusion coefficients, which have been used to obtain the above three-dimensional profiles of Figure 11. Similar results have been obtained by measuring the lattice constant of films before and after activation [6]. The value for lattice constant a before annealing – as mentioned above – has a value of 6.498 Å (indicating stress in the CdTe film) and changes into two distinct values of 6.481 (relaxed CdTe) and 6.468 Å. The latter value does not occur for films deposited onto ITO film without any CdS. These results can only be explained by the occurrence of a CdS_xTe_{1-x} species with x = 0.2, corresponding to 6.468 Å.

The presence of oxygen traces during CdTe film deposition usually is considered harmless, a fact which eases the requirements on vacuum equipment. Recently [13] it has been observed that oxygen leads to a reduced CdTe/CdS interdiffusion during activation, a fact which may reduce the danger of shunting of cells by CdS-enhanced grain boundaries, as illustrated in Figure 11.

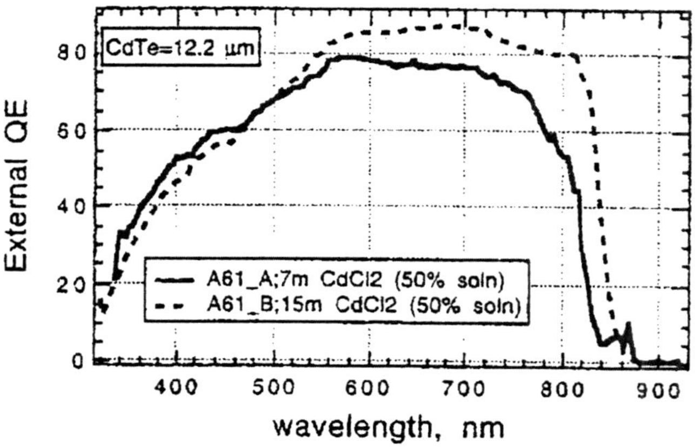

Figure 10 Illustration of intermixing of CdTe and CdS, by long wavelength extension of sensitivity due to lower bandgap mixed compound [11]. (Reproduced with kind permission of the Material Research Society.)

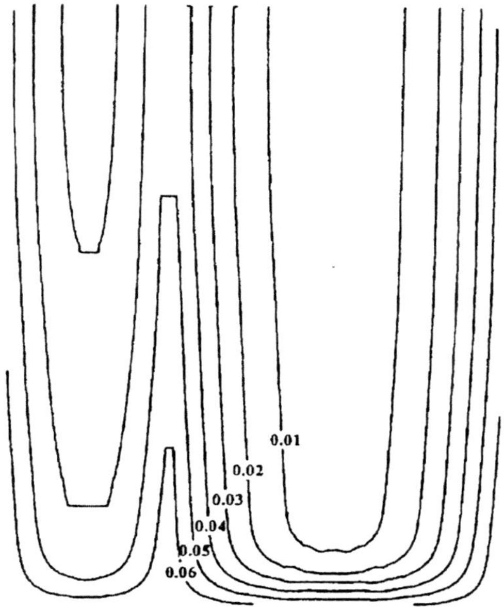

Figure 11 Simulated isocomposition lines of two grains in a CdTe/CdS film. Parameter = x in $CdTe_xS_{1-x}$ [8]. (© American Institute of Physics.)

Increase of Charge-Carrier Lifetime. Electrons generated as minority carriers in the CdTe films have to reach (by diffusion) the field region at the junction and transverse the junction (by drift). Even under low lifetime conditions, electrons generated directly at the junction can transverse the field region: The 'Schubweg' for an electron of 10 cm^2/Vs mobility and a lifetime of 1 ns in an

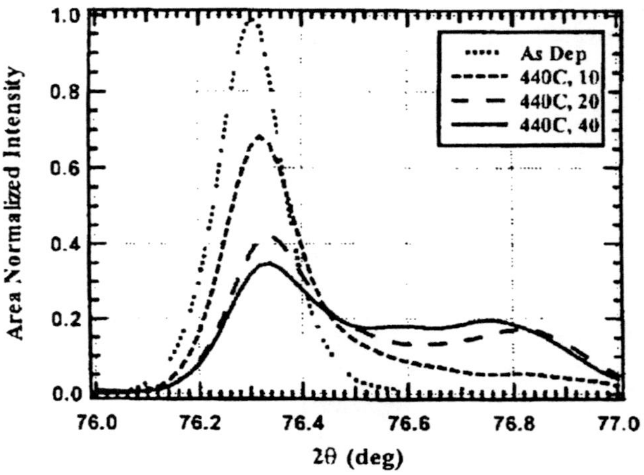

Figure 12 Emergence of mixed CdTe/CdS material after increased thermal stressing identified by X-ray diffraction [8]. (© American Institute of Physics.)

internal field of 10^4 V/cm of 1 μm with at a potential difference of 1 V i: $\mu\tau E = 10^{-4}$ cm. If diffusion in the field-free region is required – for carriers created further away from the junction by red light – lifetime becomes more important. Moutinho et al. [14] have shown that CdTe films of good cells (11%) can show 2 ns lifetime. They have plotted efficiency of cells made similarly as function of lifetime (Figure 13) allowing the conclusion that in CdTe solar cells diffusion of minority carriers also plays an important role for achieving high efficiency.

Conclusions on Activation. All these results indicate that major structural effects in CdTe solar cells occur upon the 'magic' activation process, namely grain growth and interdiffusion, which are stronger for low-temperature deposited films than for films deposited at temperatures of 500°C or more. These ('high-temperature') films directly lead to a stable structure with less recrystallisation and grain grow required.

The main effect of activation on efficiency of the devices, although, is an electronic improvement not so much by morphological effects but by improvement of the crystalline and electronic quality of grains immediately at the junction. Evidently the lifetime of minority carriers in CdTe (electrons) determines the charge carrier collection efficiency of the device.

2.2.2 The Back Contact

It is well-known from semiconductor physics that it is not easy to contact a low-doped p-type semiconductor of relatively high energy gap. There are two general principles for making ohmic contacts to p-type semiconductors:

1 Use of a metal of work function higher than the electron affinity + energy gap of the semiconductor in order to align the top of the valence band

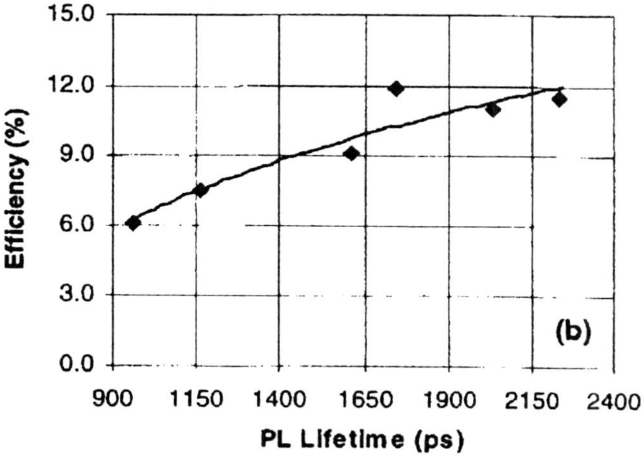

Figure 13 Increase of efficiency of a cell by increase of minority carrier lifetime [14]. (Reproduced with permission from IEEE.)

with the Fermi level of the metal. The electron affinity + energy gap of CdTe is > 6 eV. There is no metal of work function of > 6 eV. This means that all metals lead to a blocking contact, as is illustrated in Figure 14, showing the band diagram of this situation.

2 Generation of a highly doped back-surface layer in the semiconductor. The unavoidable Schottky barrier created by the back-contact metal in the semiconductor will then be thin enough for holes to tunnel through efficiently. Figure 15 shows the band diagram for the second option, a highly p-doped surface region. Consequently it remains to find practical solutions for this.

Efforts for high p-doping in CdTe usually fail due to a strong tendency for self-compensation of acceptors by formation of donors at elevated temperatures as are used here. Furthermore acceptors cannot be introduced by diffusion of atoms or ions from the surface, as diffusion preferentially proceeds along grain boundaries, leading to shunting of the cell by conducting grain boundaries before sufficient doping levels are achieved within the grains. In many cases copper (an acceptor in CdTe) has been added, e.g. in graphite contacts still used for experimental contacting, which upon annealing can diffuse into the CdTe film. Its diffusion coefficient along grain boundaries is 100 times that in bulk CdTe [15]. If Cu reaches the junction, it first reduces the junction width, and then it compensates donors in the CdS layer. Recently, more stable contacts using Cu have been made by depositing Te/Cu double films, which can react to produce Cu_2Te compound films upon annealing, some of the Cu diffusing into the CdTe film, leading to a minute surface doping. [16].

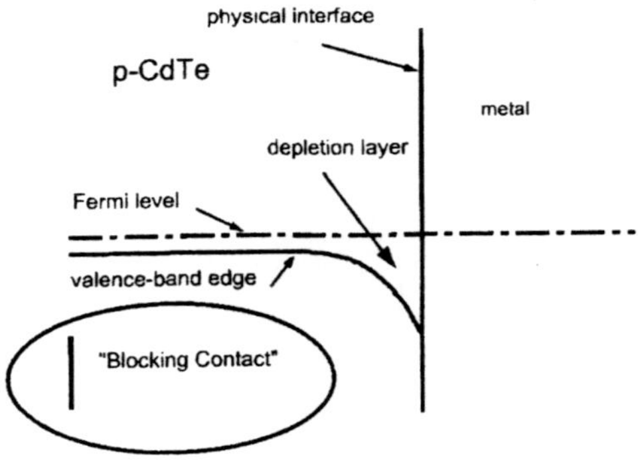

Figure 14 Energy band scheme of the metal–CdTe interface of a CdTe solar cell illustrating blocking contact.

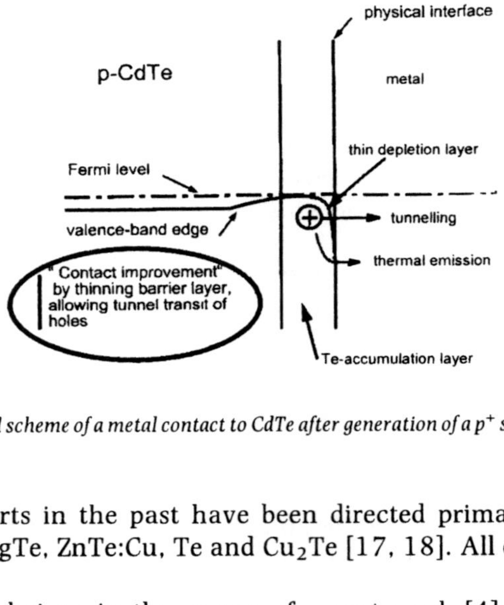

Figure 15 Energy band scheme of a metal contact to CdTe after generation of a p^+ surface by Te enhancement leading to a thin barrier.

Alternative efforts in the past have been directed primarily towards three semiconductors HgTe, ZnTe:Cu, Te and Cu_2Te [17, 18]. All of them have led to unstable contacts.

It has become obvious in the course of recent work [4] that a stable back contact cannot consist simply of a metal film. A new contact system has been developed, consisting of a triple procedure [4]:

- Generation of an *accumulation layer*, e.g. by suitable etching of CdTe a Te-enhanced surface layer can be generated.
- Deposition of a p-type narrow-bandgap, chemically inert semiconductor or semimetal ('*buffer layer*').
- Deposition a metal film for low resistance current collection ('*metal contact*').

The role of the buffer layer essentially consists of protecting the (chemically sensitive) p-type accumulation layer from being corroded by the (reactive) metallic current-collection film. All three components had to be individually optimised and mutually adjusted. Such a triple structure can be highly stable and made by using techniques suited to large-scale production.

The etching process used proceeds into the grain boundaries, leading to 'Te-caps' covering the grains as illustrated in Figure 3. This has been strikingly visualised experimentally [17] by sputter-etching away the first 30 nm of the film and analysing the new surface by scanning Auger analysis. Figure 16 clearly shows grain boundaries having excess Te in the form of lighter coloured regions.

2.3 Stability Issues

Due to the material's strong ionicity (72%) the energy of the bond between Cd and Te is quite high (5.75 eV) [18]. The energy of any photon in the solar spectrum is lower than the binding energy in CdTe or CdS, so that breaking of bonds must not be considered. The strong bonding leads to an extremely high chemical and thermal stability, reducing the risk of degradation of performance or any liberation of Cd to a very low level. No degradation intrinsic to the material can be expected.

The stability risks of back contacts has been virtually eliminated by the triple structure described in Section 2.2.2. Nevertheless careful process development has to be performed to avoid stability risks from other processing steps, like influences of additives in the lamination material. Only dedicated tests of products can yield to assured stability of the product.

2.4 Best Performance of Cells

Using the most advanced techniques, record efficiencies have been achieved by a few groups, indicating the potential of the CdTe thin-film solar cell. In 1984 the magic limit of 10% efficiency was surpassed by a group at Kodak laboratories, using close spaced sublimation [19]. In 1993 an efficiency of 15.8% was achieved [20] by using again CSS for formation of CdTe films on borosilicate

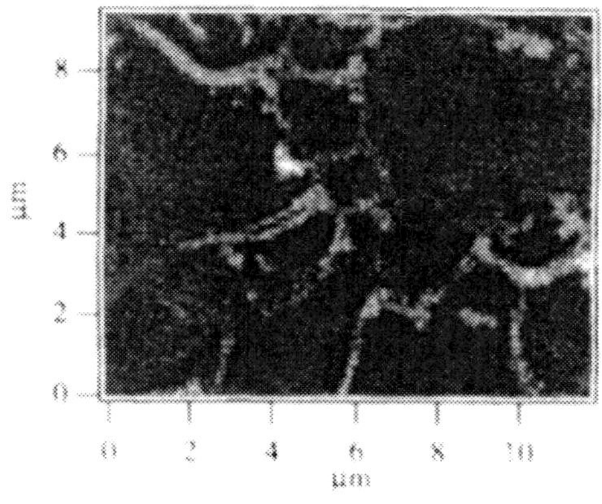

Figure 16 Te-capped CdTe grains by scanning EDX analysis of an ion milled back surface (light areas are Te-enhanced) [17]. (Reproduced with permission from IEEE.)

glass, at temperatures of around 600°C by a group from the University of South Florida. Fine-tuning has been achieved using CdS films made by chemical bath deposition and finally applying an antireflection coating onto the glass surface positioned towards the sun. This value could be surpassed only 7 years later by a group from NREL, which achieved 16.5% efficiency [21]. The important advance in this work has been the use of $CdSnO_4$ deposited onto borosilicate glass.

In both recent cases the higher deposition temperature allowed by use of (expensive) borosilicate glass has been a central issue. For industrial production this type of glass is presently considered too expensive. Therefore in industrial production, low-cost soda-lime glass is used, which is limited in temperature endurance.

3 Making of Integrated Modules

3.1 Interconnection of Cells

Semiconductor solar cells are devices delivering open circuit voltages of less than 1 V. As electric power for commercial applications requires higher voltages, it has proven advantageous to connect a multitude of cells in series in 'modules'. Whereas in the case of silicon solar cells, individual cells have to be series connected by conductors welded onto both sides of wafers, thin-film cells have a strong advantage to allow integrated series connection of numerous cells, which are at the same time defined in area and interconnected. If the different layers of the cells – TCO, p–n film stack and back contact-are individually separated (scribed) into parallel stripes, which overlap asymmetrically the series-connection of one distinct cell with its neighbour can be achieved periodically for all cells so generated, as illustrated in Figure 17. After deposition of the TCO film, a first set of separation scribing lines at a periodic distance of about 1 cm is applied, typically by laser ablation. Subsequently the p-n film sandwich is deposited and separated at the same periodicity, so that this scribing line opens the TCO beneath for the back contact, which is deposited subsequently. If the back contact now is separated by a similar set of lines at a small distance from its contact line to the TCO, the interconnection is achieved. Figure 17 shows the principle for a set of three cells.

In some cases, such as electrodeposition, it is not permissible to separate the TCO film before depositing the semiconductor film. Here a variation in the procedure allows the scribing of the first two lines after the deposition of the semiconductor film [22]. An insulating fill-up of the first scribing line is required in this process, illustrated in Figure 18.

Evidently, using these techniques, the cell width can be adjusted according to technical needs and/or commercial requirements: On the one hand, more interconnection triple scribes lead to higher loss in active area. (Typically a scribing system requires between 0.2 and 0.3 mm.) On the other hand wider cells will lead to increased series resistance (lower fill-factor), as the current density

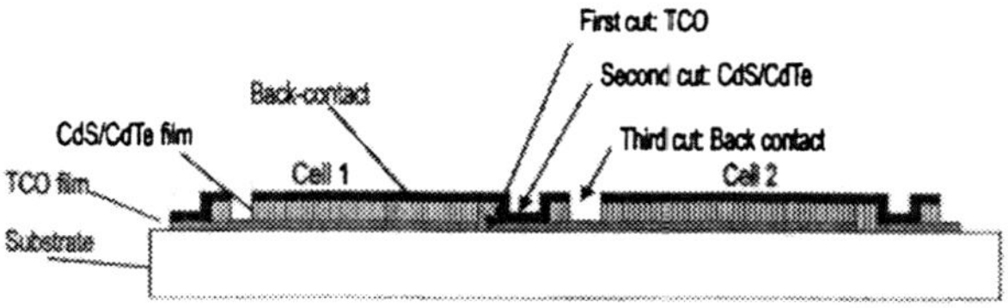

Figure 17 Interconnection principle.

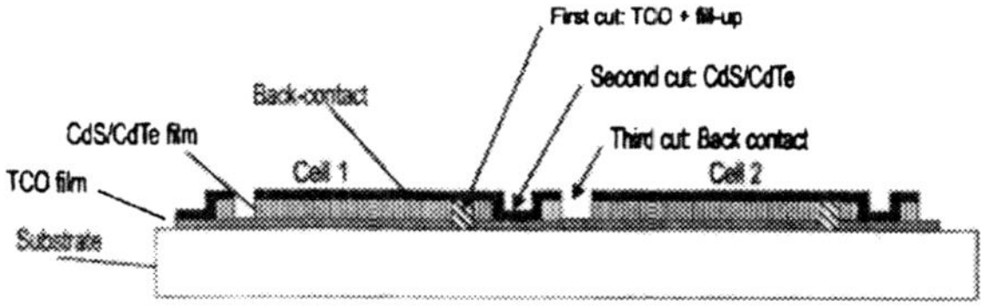

Figure 18 Modified interconnection principle.

being conducted through the TCO film will be higher, leading to higher voltage drop. For CdTe a cell width of 9–10 mm seems to be an optimum value for TCO films of 8–10 Ω per square resistivity. (For CIS cells, due to higher current density and lower voltage of the individual cell, smaller cells – about 6 mm wide – are appropriate; for amorphous silicon conversely wider cells are optimal.)

3.2 Contacting

The photocurrent of a module transverses all series connected cells and is extracted by contacts to the first and last cell, and all individual cell voltages add to the total voltage. Usually metallic conductors are attached to the free contact area of the first and last cell by conducting adhesive and further connected by contact bands towards the point of the module, where it is to transverse the back sealing, typically glass. Usually the contact bands are Sn-plated Cu ribbons. Figure 19 indicates the topology of the contacting conductors. Care has to be taken to avoid shunting at any place.

3.3 Lamination

After contact bus attachment the module needs sealing and protection against external influences. This usually is achieved by laminating a second glass plate to the module-carrying side of the first glass plate. It is required to remove all films at a boundary region of 1–2 cm of the module to provide the required electric insulation of the module. This is usually achieved before contacting by sand blasting or laser ablation.

For historical reasons, in many cases the technology used for silicon modules is also used for CdTe modules, namely sealing by EVA (ethyl-vinyl-acetate),

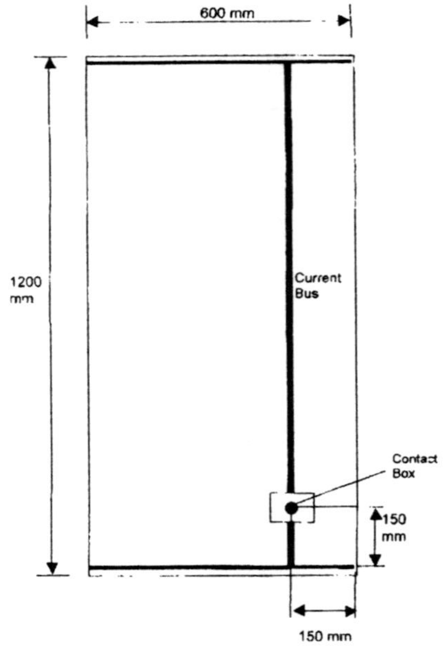

Figure 19 Topology of contact bus system, as seen from the back of the module.

which is used as a monomer, applied as a sheet and polymerised by a thermal annealing step under vacuum. Commercial EVA is readily available. It is 'overqualified' for CdTe, as intense development work has been invested into the commercial product to stabilise it against degradation in the sunlight impinging onto silicon modules. In the case of CdTe this special quality is not required, as sunlight directly enters the junction after having passed the glass and the lamination material is positioned below the second glass and therefore oriented away from the sun.

The cover glass so laminated to the module has a hole, through which the contact bands are guided into a contact box, which is attached to the glass (cf. Figure 19). Two stable current guides with plugs are attached to the box and will be used for connecting the modules so generated to the user-circuit. Evidently all connections and contacts must be extremely well protected against water and water vapour for achieving the expected lifetime of the module.

4 Production of CdTe Thin-Film Modules

4.1 Generalised Production Sequence

The different steps to make cells and modules, described in Sections 2 and 3 can be arranged into a sequence to make ready-to-use modules as a product:

1. Selection of substrate glass, soda-lime glass ('floatglass') as substrate.
2. Deposition of the transparent conductive coating (SnO_2, ITO, etc.).
3. Scribing of the TCO film into parallel bands, defining and separating the cells. For some processes this scribing can be done later, after deposition of the semiconductor films.
4. Deposition of CdS films of lowest possible thickness, typically around 100 nm.
5. Deposition of CdTe by the process of choice for the particular product.
6. Activation of the film stack by influence of $CdCl_2$ at temperatures of around 400°C.
7. Application of the second scribing step, which opens the semiconductor stack for contacting the TCO film. Optionally step 3 can be applied in parallel, the scribing line afterwards has to be filled up for electrical insulation.
8. Application of the back-contact structure, consisting of a set of steps, e.g. etching to achieve a Te accumulation, application of a buffer layer, application of the metallic back contact.
9. Separation cut to separate the back contacts of the neighbouring cells.
10. Attachment of the contact bus structure.
11. Lamination with a second glass (or plastic) using a suitable plastic, such as EVA or a thermoplastic film, the contacts protruding through a hole in the cover glass plate for the next step.
12. Contact box attachment, in which the fragile contact bands from the module are connected to stable cables with suitable plugs for commercial application.
13. Measurement of each module's efficiency using a solar simulator.

This sequence – with some modifications – is the basis of the following industrial efforts into production.

4.3 Industrial Production of Modules

A commercial product can only be manufactured with expectation of cost-covering revenues, if a factory above a certain capacity is built, using the dimensions of scale. It is generally accepted today that a capacity of around 100,000 m^2 per annum is presently appropriate. All units of significantly lower capacity should be called pilot plant as they will require more funds than any return can provide. At around 600,000 m^2 per annum a cost potential of 0.6 / W_p has been estimated by a group of experts [23].

Presently three industrial units are known to actively pursue the target of large-scale production. They basically differ by the deposition technique for CdS and CdTe.

4.3.1 BP Solar Inc. (Fairfield, California, USA)

Work at BP Solar has started in Great Britain in 1984, when British Petroleum took over a galvanic deposition process for CdTe thin-film solar cells from

Monosolar Inc. in the USA. BP Solar continued development of the basic electrodeposition process for CdTe [24]. Upscaling work resulted in a factory, which has been built in the USA. Due to low deposition rates, parallel deposition is used: A large number (40 to 100) plates is immersed into a tank containing a recirculation system for the continuously replenished electrolyte. A constant potential is applied to the plates via the TCO films, already covered by CdS. (The CdS film is deposited by the chemical bath deposition technique, also in parallel onto a larger number of TCO-coated glasses.) The total charge applied to the plates is a measure for film thickness. A charge of about 12000 Cb/m^2 leads to approximately 1.6 μm of CdTe film. Post annealing leads to strain relief, change of conduction polarity from n to p and activation by means of a Cl-compound added to the bath. The grain size grows from 0.1 to 0.2μm to about 0.4 μm during this procedure.

Great care has to be taken to avoid lateral voltage drop over the plate surface during CdTe deposition. Te- or Cd-rich compounds can form upon deviation from this condition. To achieve good quality films, lowest possible surface resistivity of the TCO has to be chosen. The cell-defining and module-generating scribe application follows the principle shown in Figure 18. In order to keep the TCO film intact as long as possible, the TCO scribing lines are applied after CdTe deposition, necessitating a fill-up procedure. Module sizes are around 1 m^2, yielding a power of around 72 W under simulated terrestrial solar light are reported [25]. More than 150,000 m^2 of glass substrates can be processes per annum. The line can produce 0.55 m^2 and 0.94 m^2 modules. The deposition system consists of 8 identical deposition tanks. Each tank is able to simultaneously coat 40 0.55 m^2 or 24 0.94 m^2 substrates. Recently (2002) BP Solar has manufactured the first large area CdTe module of $>$ 10% efficiency[2].

4.3.2 First Solar LLC (Toledo, Ohio, USA)

This company has started work within the scope of a predecessor called Solar Cells Inc. around 1991 in Toledo by a senior shareholder of Glasstech Inc. In 1999 a joint venture with a finance group from Arizona, USA, has been formed, which led to the new name First Solar LLC [26]. The basic process for generating CdS/CdTe junctions relies on the sublimation/condensation properties of CdS and CdTe described above. Substrates coated by SnO_2 films by Libbey Owens Ford using a spray process, enter a vacuum system through a loadlock and move – lying on a roller system, to avoid warping – into a chamber heated to temperatures around 560°C. Vapour sources are positioned above the substrates out of which CdS and CdTe vapours emerge, are directed towards the substrates and condense sequentially on the substrates. The CdS and CdTe material is continuously fed into the evaporators. Module interconnection of cells is performed ex-post, i.e. first and second scribing lines are applied after semiconductor deposition, requiring filling of scribing line 1 (cf. Figure 18).

---

[2] In 2002 the management of BP Solar decided to terminate its CdTe development and production efforts after 18 years of investment and closed down the factory.

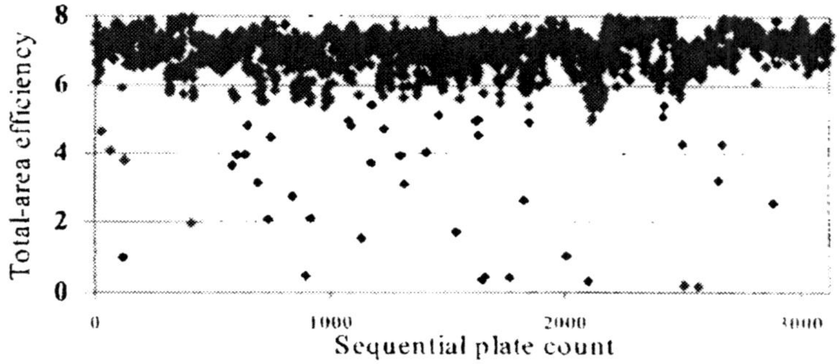

Figure 20 Total-area efficiency for 3128 sequentially deposited modules from First Solar LLC (from [27]).

The standard module size of First Solar is 60 × 120 cm^2 = 0.72 m^2 [27]. Efficiencies of 6–8% can be routinely achieved. Figure 20 shows area efficiency vs. sequential plate count for 3128 sequentially deposited modules.

4.3.3 ANTEC Solar GmbH (Arnstadt, Germany)

The technology used at ANTEC Solar is essentially based on the development started at Battelle Institut around 1970 [28]. After closure of Battelle Institut the know how has been transferred to ANTEC GmbH, a management buy-out from Battelle and developed into a manufacturing technology. ANTEC Solar has been founded in 1996 in order to start production of CdTe thin-film PV modules. A fully automated production plant has been built in Arnstadt (Germany) and production has started in 2001. The basic process used for deposition of CdTe is close spaced sublimation. Glass-substrates carrying scribed TCO films are transported in vacuum above crucibles containing CdS and CdTe at temperatures of 700°C. The semiconductor materials condense at temperatures of 500°C and form the n–p diode structure, which is activated in a $CdCl_2$ atmosphere. Module size is 60 × 120 cm^2, presently efficiencies of modules is about 7%, increase up to 8% is expected to be achieved in 2002. In contrast to BP Solar and First Solar the TCO is also made in the plant in the online system. Definition of interconnected cells is achieved according to Figure 17, i.e. at three different stages of the plant by laser ablation (first scribe) and mechanical ablation (second end third scribe)[3].

Figure 21 shows the total area efficiency of 2000 modules manufactured in one production run. The production plant presently employs 100 persons.

4.4 A 10 MW Production Line

In the following, by means of an example, the production line of ANTEC Solar GmbH is described and illustrated.

---

[3] Because of financial problems, ANTEC Solar had to declare insolvency in 2002. In 2003 the plant has been taken-over by a new owner, who successfully restarted production.

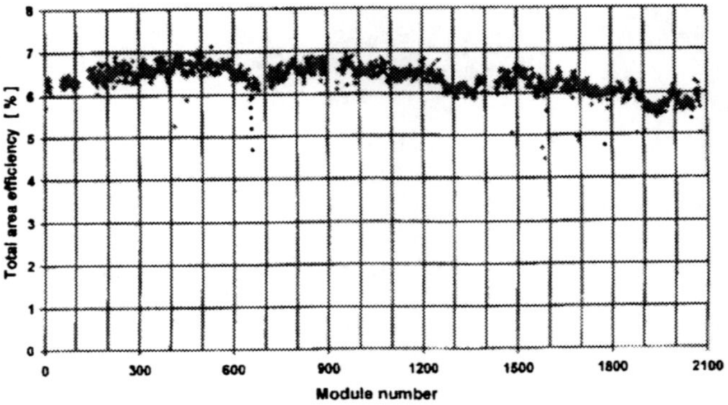

Figure 21 Total area efficiency for 2000 sequentially deposited modules from ANTEC Solar GmbH.

The deposition line was conceived as a two-step production line, consisting of a fully automated deposition line for integrated modules on glass substrates of 60×120 cm^2 and a semi-automatic module line, for hermetical sealing, contacting, measuring and customising of the modules into a marketable product.

The fully automatic in-line deposition procedure consists of nine steps:

1. Cleaning of the substrate (float glass).
2. Deposition of the transparent contact (ITO + SnO_2) at around 250°C.
3. Scribing of the TCO film for cell definition and interconnection).
4. Deposition of CdS and CdTe by Close-Spaced Sublimation (CSS) at around 500°C.
5. Activation (improvement of junction by annealing in Cl-containing atmosphere at around 400°C).
6. Wet-chemical etching for contact preparation (Te accumulation).
7. Scribing by mechanical tools for interconnection of cells.
8. Deposition of two-layer back contact by sputtering.
9. Scribing by mechanical tools for separation and interconnection of cells.

Most of these positions are connected by heating or cooling segments in order to present the plates to the deposition steps at the adequate temperatures They take a large part of the equipment length. The highest temperature (500°C) is reached during deposition of the semiconductor films, compatible with glass stability. The total length of the automated deposition line is 165 m. Glass plates (to become modules) are transported by automated conveyor systems and are not touched by human hands during the processes. They are collected in boxes in sets of 30 and transferred to the module line in an adjacent hall.

In the module line the substrates with sets of interconnected cells (often called 'submodules') are contacted and sealed for convenient use in energy generating systems. This part of the factory consists of the following procedures:

1. Cleaning of the cover glass (float glass).
2. Deposition of the contact buses onto the modules.
3. PV function testing to identify substandard modules, which can be excluded from further processing.
4. Cutting of EVA sheet to size.
5. Joining of module, EVA sheet and cover glass.
6. Lamination (sets of 6 modules).
7. Fill-in of contact hole in cover glass.
8. Attachment of contact box.
9. Quantitative measurement of PV performance (sets of 3 modules).
10. Type-tag attachment.
11. Classification, selection and packaging for dispatch.

The module line requires some manual handling and adjustments, due to the heterogeneity of the processes. Figure 22 shows the geometrical arrangement of the different processes. Each line has been installed in a separate hall. Both halls are connected by an aisle for transfer of modules. Buffer stations allow the removal of partly completed modules in case of an incident in the down-line stations without crashes in the up line stations. The overall primary parameter for the deposition line and also the module line has been the production speed defined at 120,000 modules per annum, which leads to an average linear transportation speed of approximately 1 m/min in the production line.

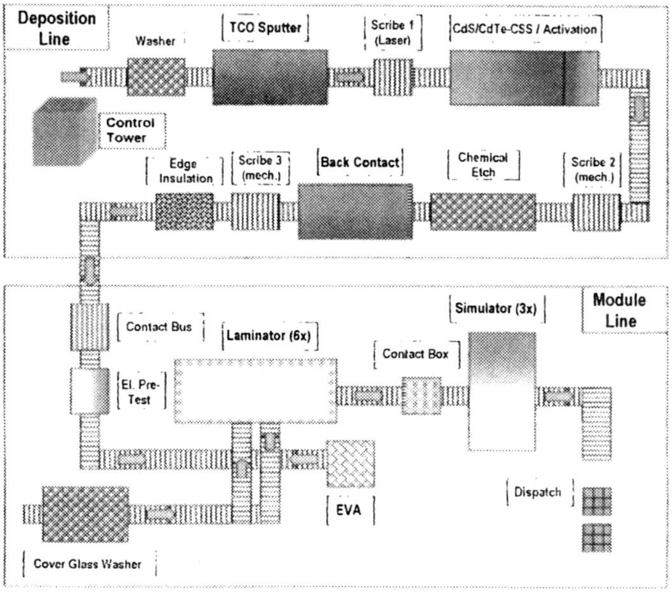

Figure 22 Schematic of the components of the plant.

The procedures described above contain (a) state-of-the-art procedures, (b) new, but simple procedures, and (c) new developed procedures.

Standard procedures are sputter-deposition of ITO, SnO_2, metals. Dedicated equipment has been constructed by experienced equipment manufacturers. In some cases (e.g. ITO deposition) multiple sputtering targets are needed in order to achieve the required film thickness at the given transport-speed of the substrates (around 1 m/min). The final steps of contacting and lamination are also using state-of-the-art techniques. Equipment can be bought custom-made on the market. An unforeseen effort in the set-up and initial operation of the plant had to be dedicated to the transport of the substrate sheets through heating and cooling stages without breakage due to excessive thermal gradients. After some significant modifications breakage under thermal stress could be virtually eliminated.

Simple, but new processes are essentially the core processes for the active semiconductors, namely CSS-deposition of CdS, CdTe, and – if required – back contact buffer-semiconductors. It has been shown, that CdTe can be deposited onto stationary substrates at rates of 10 μm/min and more. For achieving the required production speed, three crucibles containing a week's supply of starting material of CdTe are used. For CdS, due to its low thickness only one crucible is used. As mentioned, the vacuum requirements are rather low, so that no high-vacuum equipment is needed. (For CdTe oxygen-incorporation to some extent indeed is considered advantageous by some researchers.) Laser scribing of the TCO film does not require basic new technologies. Equipment can be built by industry and processing data have to be developed and adjusted. In spite of these arguments, some problems have been incurred by thermal stress conditions of crucibles and substrates. Fast heating invariably leads to breakage of glass substrates.

More involved *new processes* are activation of the junction CdS/CdTe and etching of the CdTe surface as contact preparation procedure. Etching requires the wetting of the substrates carrying the activated CdS/CdTe film stacks by nitric/phosphoric acid plus water solutions ('NP'-etch), rinsing and drying. Although being essentially simple, procedures equipment had to be developed to incorporate these processes under tightly controlled parameters into the production line allowing the target speed to be achieved. Activation requires exposition of the substrates carrying the TCO-CdS-CdTe film-stack to, e.g.$CdCl_2$ at elevated temperatures. Fortunately traces of oxygen do not play a detrimental role. (The activation step in the laboratory can even be executed in air.) Furthermore, $CdCl_2$ at room temperature has extremely low vapour pressure. Some special chemical engineering skill has been needed to conceive an in-line activation stage fitting into the production line. Mechanical scribing has turned out to be more involved, due to the requirement that the production line should have only one service interval per week. Scribing tools have to survive one week under tough tolerances. New solutions had to be found for recognising and identifying the first laser cuts below the CdTe film for precise positioning of the second and third scribing lines. Implementation of automatic image recognition of previously scribed lines in the TCO films for precise positioning of subsequent scribing lines has been a non-trivial task

Figure 23 View of the total deposition line.

After the plant had been built and assembled in 1999 and first functional tests had been passed early 2000, de-bugging and process optimisation has required extensive tests and equipment-modifications by the hardware and software manufacturers. Production has started in 2001. Figure 23 shows the deposition line in its entirety.

4.5 Environmental and Health Aspects

The production of polycrystalline thin-film CdTe solar modules basically employs techniques common in chemical and microelectronic industry. The substances involved are easily manageable by standard processes. Production is possible under existing safety laws without putting into risk health of staff. It is technically and economically possible to design and operate a factory with zero emission. Workers in a production environment have been tested regarding Cd uptake and shown Cd content in blood and urine far below the threshold concern level under periodic medical scrutiny [29]. Smokers have shown a somewhat higher Cd level than non-smokers, but still below any threshold for concern.

A number of studies from third parties [30–33] show negligible risk under use of CdTe solar modules for the environment and humans even under irregular conditions. In case of exposure to fire, the substrate- and cover-glass will melt long before the CdTe decomposes, thereby including the semiconductor into the re-solidifying glass. Incineration experiments conducted by BP Solar in cooperation with a fire research institution using typical household inventory plus CdTe modules have not led to detectable emissions of Cd compounds [30].

During use a CdTe module can be compared to laminated glass similar to that used in cars. Thus modules will not easily break and release their content. At their end of life, modules can be recycled by crushing the whole modules and either returning the debris to the smelters, which can inject the material into their processes without significant additional cost, or dissolving the films by liquid or gaseous etchants.

The system's company debis (a subsidiary of Daimler-Chrysler AG) has established a life cycle inventory for CTS thin-film solar modules of ANTEC Solar GmbH guided by ISO 14040 and 14041. To achieve this aim, the total energy and material flow for the module's life cycle has been accounted with help of the CUMPAN® software system. This allows, for example, for the raw materials the determination of the total energy required for their production and processing.

- The manufacture of one square metre of CTS-module uses 126 kg raw material and primary energy carrier and 70 kWh electric energy.
- Under the climatic conditions of Germany, the generation of 1000 kWh of electricity per annum at a module lifetime of 30 years saves 16,244 kg of carbon dioxide and further undesired materials, such as sulphuric dioxide and nitrogenic oxides.
- The production plant is built not to emit any material. Water is reprocessed and re-used. (Waste heat at this time is emitted to the environment, though.)
- A recycling process is envisaged for spent and reject modules. In view of environmental safety, economical retrieval of valuable raw materials, and securing of hazardous materials, this process can be considered satisfying.
- Compared with alternative thin-film modules (a-Si, CIS), emissions and waste during production amount to similar values.
- The emission of cadmium can be judged as low in comparison to other emission sources (coal-fired power plants, phosphate fertiliser). Even in case of accidents (e.g. fire), by reason of the small quantities of (thermally stable) material per square metre of module area, no environmentally critical emission must be contemplated.
- The total energy used for fabricating a module will be retrieved by the module within 15 months using the actual energy uptake of the ANTEC factory. From then on the module operates in an environmentally benign way.

4.6 Material Resources

More than 99% of the weight of a CdTe modules consist of float glass, EVS and metal connectors, readily available in virtual unlimited quantities. TCO films made of SnO_2 are not considered to by limited by available Sn supplies. In used in ITO films is a more rare resource at an annual production of 120 t [34]. Furthermore it is also used for ITO films in liquid crystal displays and the $CuInSe_2$-based thin-film solar cell. ITO can be substituted by SnO_2 and – possibly – ZnO, Sn and Zn being abundant metals. CdTe and CdS warrant a closer consideration. Presently CdTe (and CdS) are offered in the required purity by 5 industrial enterprises. Cd is presently produced as a by-product of Zn at 20,000 t per year [34]. Te, due to low demand is presently produced at 300 t per year as a by-product of Cu [34]. Growing future demand according to an expert in the field [35] can be satisfied by more efficient extraction from the anodic slurries in Cu

Figure 24 View of a CdTe thin-film PV module.

electrorefining on the one hand and also by exploiting Te-rich ores in South America not yet exploited. S finally is an abundant element.

In summary, no critical material bottleneck is expected for an expanding production. The CdTe thin-film solar cell will be able to take a share in ameliorating future energy shortages and the climate change expected from burning fossil fuels.

5 The Product and Its Application

The photovoltaic modules manufactured in a factory are ready-to-use in a suitable PV installation. CdTe modules by any one manufacturer are mass products made on large scale (100,000 per year or more). Typically the modules are sized about $0.6 \times 1.2\ m^2$. A module 60 cm wide can easily be carried under the arm by one person. Glass-glass laminates, as manufactured, e.g., by First Solar and ANTEC Solar have a weight of 16 kg. A contact box furnishes two cables carrying plugs (male and female for easy series connection), which are long-term stable. Figure 24 shows an ATF module from ANTEC Solar. A very homogeneous appearance of most thin-film modules helps organic and visually pleasing installation.

5.1 Product Qualification

A PV module is a product made for long time deployment in a harsh environment and therefore must be furnished with a warranty for approximately 20 years of useful life. In order to provide such assurance, the international standards agencies (IEC = International Electrotechnic Commission) have designed international norms which are valid worldwide. The norm pertaining to thin-film modules is IEC 61646, 'Thin film terrestrial photovoltaic modules – Design qualification and type approval'. This norm has been published in 1996 and originally has been designed with amorphous silicon modules in mind.

Modules to be classified as fulfilling IEC 61646 have to undergo an extensive set of tests defined in detail in the text of this norm. Briefly these are 4 sets, for which 8 modules have to be provided, typical for the production discussed:

A. Performance tests

- Power output at standard test condition (25°C, 1 sun)
- Power output at nominal operating conditions
- Power output at low sunlight intensity (20%)

B. Endurance tests

- Long-term outdoor exposure
- Light soaking
- Temperature shock tests
- Damp-heat test

C. Mechanical tests

- Mechanical load
- Twist test
- Hail-test

Figure 25 Installation of CdTe modules from First Solar LLC on an office building.

Figure 26 Installation of CdTe modules from ANTEC Solar on the wall of a public administration building.

Figure 27 Installation of CdTe modules on a private residence roof.

D. Electric tests

- Insulation test
- Water immersion test

5.2 Examples of Installation of CdTe Modules

Thin-film modules can show a highly homogeneous surface appearance and very little variation from module to module. This allows the assembly of large,

Figure 28 Close-up of mounting system on metal rafters for roof installation.

highly homogeneous panels on rooftop, facade or on the ground. Figures 25–27 show examples of such installations. Modules in form of glass/glass laminates can be mounted on special structures, which can be invisible and nonetheless watertight. Figure 28 shows how modules can be mounted on metal rafters, which allow water drainage and also seal the interior of the roof by rubber lips.

6 The Future

After more than 20 years of development in various industrial and academic laboratories, the CdTe thin-film solar module has entered the production stage and first experience is gained in this process. This will definitively lead to the next generation of plant at capacities around 1,000,000 m^2 per year. It is expected that the 'learning curve' will be transgressed quite fast which will lead to a mature low-cost product.

References

[1] Zanio, K., 1978. Cadmium Telluride: Materials Preparation, Physics, Defects, Applications. *Semiconductors and Semimetals*, Vol. 13.

[2] Bonnet, D. et al., 1995. The CdTe Thin Film Solar Cell – EUROCAD. Final Report to the Commission of the European Communities, Project No. JOU2-CT92-0243, 1995. Referenced data later also published in: Al Allak, H.M. et al., 1995. The Effect of Processing Conditions on the Electrical and Structural Properties of CdS/CdTe Solar Cells. *Proc. 13th European Photovoltaic Solar Energy Conference*, Nice, pp. 2135–2138.

[3] Bube, R., 1960. *Photoconductivity of Solids*. Wiley, New York, London, p. 94.

[4] Bonnet, D. et al., 2002. CADBACK: The CdTe Thin Film Solar Cell – Improved Back contact. *Final Report to the European Commission*, Contract No. JOR3-CT98-0218.

[5] Dhere, R. et al., 1997. Influence of CdS/CdTe Interface Properties on the Device Properties. *Proc. 26th IEEE Photovoltaic Specialists Conference*, Anaheim, pp. 435–437.

[6] Moutinho R.H. et al., 1998. Effects of $CdCl_2$-treatment on the recrystallization and electro-optical properties of CdTe films. *J. Vac. Sci. Technol.*, Vol. A16, pp. 1251–1257.

[7] Moutinho H.R. et al., 2000. Alternative procedure for the fabrication of close-spaced sublimated CdTe solar cells. *J. Vac. Sci. Technol.*, Vol. A18, pp.1599–1603.

[8] McCandless, B. and Birkmire, R., 2000. Diffusion in CdS/CdTe Thin-Film Couples. *Proc. 16th European Photovoltaic Solar Energy Conf.*, Glasgow, pp. 349–352.

[9] Durose, K., private communication.

[10] Nunoue, Sh., Hemmi, T. and Kato, E., 1990. Mass Spectrometric Study of the Phase Boundaries of the CdS/CdTe System. *J. Electrochem. Soc.*, Vol. 137, pp.1248–1251.

[11] Dhere, R.G. et al., 1966. Intermixing at the CdS/CdTe interface and its Effect on Device Performance. *Mat. Res. Soc. Symp. Proc.*, Vol. 426, pp. 361–366.

[12] McCandless, B., Engelman, M.G and Birkmire, R.W., 2001. Interdiffusion of CdS/CdTe Thin Films: Modelling X-Ray Diffraction Line Profiles. *J. Appl. Phys.*, Vol. 89, pp. 988–994.

[13] Yan, Y., Albin, D.S. and Al-Jassim, M.M., 2001. The effect of oxygen on junction properties in CdS/CdTe solar cells. *Proc. NCPV Program Meeting*, pp. 51–52.

[14] Moutinho, H.R. et al., 2000. Study of CdTe/CdS solar cells using CSS CdTe deposited at low temperature. *Proc. 28th IEEE Photovoltaic Specialists Conf.*, Anchorage, pp. 646–649.

[15] Hegedus, S.S., McCandless, B.E. and Birkmire, R.W., 2000. Analysis of stress-induced degradation in CdS/CdTe solar cells. *Proc. 28th IEEE Photovoltaic Specialists Conf.*, Anchorage, pp. 535–538.

[16] Hegedus, S.S., McCandless, B.E. and Birkmire, R.W., 2001. Initial and Stressed Performance of CdTe Solar Cells: Effect of Contact Processing. *Proc. NCPV Program Review Meeting*, pp. 119–120.

[17] Levi, D.H. et al., 1997. Back contact effects on junction photoluminescence in CdTe/CdS solar cells. *Proc. 26th IEEE Photovoltaic Specialists Conf.*, Anaheim, pp. 351–354.

[18] Hartmann, H., Mach, R. and Selle, B., 1981. Wide gap II–VI compounds as electronic materials. In: Kaldis, E., Ed., *Current Topics in Materials Science*, Amsterdam, pp. 1–414.

[19] Tyan, Y.-S. and Perez-Albuerne, E.A., 1982. Efficient Thin Film CdS/CdTe Solar Cells. *Proc. 16th IEEE Photovoltaic Specialists Conf.*, San Diego, pp. 794–800.

[20] Britt, J. and Ferekides, C., 1993. Thin film CdS/CdTe solar cell with 15.8% efficiency. *Appl. Phys. Lett.*, Vol. 62, pp. 2851–2852.

[21] Wu, X. et al., 2001. 16.5% efficient CdS/CdTe polycrystalline thin film solar cell. *Proc. 17th European Photovoltaic Solar Energy Conf.*, Munich, pp. 995–1000.

[22] Rose, D. et al., 2000. R&D of CdTe-absorber photovoltaic cells, modules and manufacturing equipment: Plan and progress to 100 MW/yr. *Proc. 28th IEEE Photovoltaic Specialists Conf.*, Anchorage, pp. 428–431.

[23] Woodcock, J.M. et al., 1997. A study on the upscaling of thin film solar cell manufacture towards 500 MWp per annum. *Proc. 14th European Photovoltaic Solar Energy Conf.*, Barcelona, pp. 857–860.

[24] Cunningham, D.W. et al., 2000. Advances in Large Area Apollo Module Development. *NCPV Program Review Meeting*, pp. 261–262.

[25] Cunningham, D.W. et al., 2000. Large Area Appollo Thin Film Module Development. *Proc. 16th European Photovoltaic Solar Energy Conf.*, Glasgow, pp. 281–285.

[26] McMaster, A. et al., 2000. PVMat Advances in CdTe Product Manufacturing. *NCPV Program Review Meeting*, pp. 101–102.

[27] Rose, D., Powell, R., 2001. Research and Progress in High-Throughput Manufacture of Efficient, Thin-Film Photovoltaics. *NCPV Program Review Meeting*, pp. 209–210.

[28] Bonnet, D. and Rabenhorst, H., 1972. New Results on the Development of a Thin Film p-CdTe–n-CdS Heterojunction Solar Cell. *Proc. 9th IEEE Photovoltaic Specialists Conf.*, Silver Springs, pp.129–131.

[29] Bohland, J.R. and Smigielski, K., 2000. First Solar's Module Manufacturing Experience; Environmental, Health and Safety Results. *Proc. 28th IEEE Photovoltaic Specialists Conf.*, Anchorage, pp. 575–578.

[30] Alsema, E.A. and van Engelenburg, B.C.W., 1992. Environmental Risks of CdTe and CIS Solar Cell Modules. *Proc. 11th European Photovoltaic Solar Energy Conf.*, Montreux, pp. 995–998.

[31] Patterson, M.H., Turner, A.K., Sadeghi, M. and Marshall, R.J., 1994. Health, Safety and Environmental Aspects of the Production and Use of CdTe Thin Film Photovoltaic Modules. *Proc. 12th European Photovoltaic Solar Energy Conf.*, Amsterdam, pp. 951–953.

[32] Moskowitz, P.D., Steinberger, H. and Thumm, W., 1994. Health and Environmental Hazards of CdTe Photovoltaic Module Production, Use and Decommissioning. *Proc. First World Conf. on Photovoltaic Energy Conversion*, Hawaii, pp. 115–118.

[33] Steinberger, H., 1998. Health and Environmental Risks from the Operation of CdTe- and CIS Thin Film Modules. *Proc. 2nd World Conf. on Photovoltaic Solar Energy Conversion*, Vienna, pp. 2276–2278.

[34] US Bureau of Mines, 1992. Mineral Commodity Summary.

[35] Daub, G., PPM Pure Metals GmbH., personal communication.

IIc-4

$Cu(In,Ga)Se_2$ Thin-Film Solar Cells

U. Rau and H. W. Schock, Institut für Physikalische Elektronik (IPE), Universität Stuttgart, Germany

1 Introduction

With a power conversion efficiency of 18.8% on a 0.5 cm^2 laboratory cell [1] and 16.6% for mini-modules with an area of around 20 cm^2 [2] Cu(In,Ga)Se_2 is today by far the most efficient thin-film solar cell technology. The start of production at several places provides a new challenge for research on this material. However, these recent achievements are based on a long history of research and technological development.

$CuInSe_2$ was synthesised for the first time by Hahn in 1953 [3]. In 1974, this material was proposed as a photovoltaic material [4] with a power conversion efficiency of 12% for a single-crystal solar cell. In the years 1983–84, Boeing Corp. reported efficiencies in excess of 10% from thin polycrystalline films obtained from a three-source co-evaporation process [5]. In 1987 Arco Solar achieved a long-standing record efficiency for a thin-film cell of 14.1% [6]. It took a further ten years, before Arco Solar, at that time Siemens Solar Industries (now Shell Solar), entered the stage of production. In 1998, the first commercial Cu(In,Ga)Se_2 solar modules were available [7]. In parallel, a process which avoids the use of H_2Se is being developed by Shell Solar in Germany [8] (see also Chapter V-1). Other companies in the USA, Global Solar and ISET, plan to commercialise modules prepared on other than glass substrates. In Europe, the long-term development efforts of the EUROCIS consortium on the co-evaporation process resulted in the activity of Würth Solar with pilot production envisaged in 2003 [9, 10]. In Japan, two lines for film preparation are planned by Showa Shell (selenisation by H_2Se) [11] and Matshushita (co-evaporation) [12].

In this chapter, we give a short overview on the present knowledge of Cu(In,Ga)Se_2-based heterojunction thin film solar cells. We focus on four points: (i) The description of the basic material properties such as crystal properties, phase diagram, and defect physics. (ii) Description of the cell technology starting from the growth of the polycrystalline Cu(In,Ga)Se_2 absorber up to device finishing by heterojunction formation and window layer deposition. This section

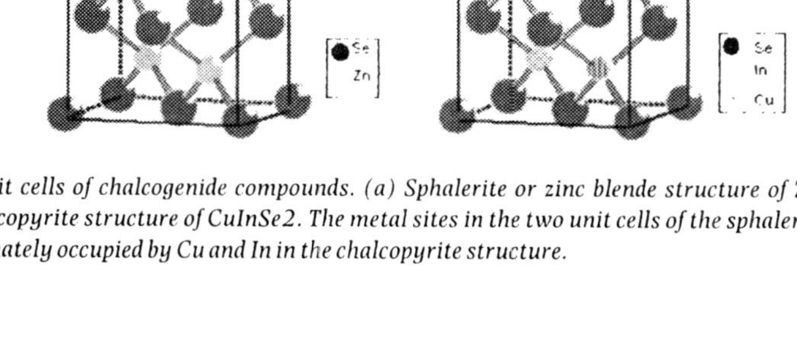

Figure 1 Unit cells of chalcogenide compounds. (a) Sphalerite or zinc blende structure of ZnSe (two unit cells); (b) chalcopyrite structure of CuInSe2. The metal sites in the two unit cells of the sphalerite structure of ZnSe are alternately occupied by Cu and In in the chalcopyrite structure.

also discusses basic technologies for module production. (iii) The electronic properties of the finished heterostructure. (iv) Finally, Section 5 discusses the photovoltaic potential of wide-gap chalcopyrites, namely $CuGaSe_2$ and $CuInS_2$, as well as that of the pentenary alloy system $Cu(In,Ga)(S,Se)_2$ and the possibility of building graded-gap structures with these alloys.

This chapter can only briefly cover those scientific issues that are relevant for photovoltaic applications. More detailed information can be found in two recent review articles by the present authors [13, 14] as well as in references [15–19].

2 Material Properties

2.1 Chalcopyrite Lattice

$CuInSe_2$ and $CuGaSe_2$, the materials that form the alloy $Cu(In,Ga)Se_2$, belong to the semiconducting I–III–VI_2 materials family that crystallise in the tetragonal chalcopyrite structure. The chalcopyrite structure of, for example, $CuInSe_2$ is obtained from the cubic zinc blende structure of II–VI materials like ZnSe by occupying the Zn sites alternately with Cu and In atoms. Figure 1 compares the two unit cells of the cubic zinc blende structure with the chalcopyrite unit cell. Each I (Cu) or III (In) atom has four bonds to the VI atom (Se). In turn each Se atom has two bonds to Cu and two to In. Because the strengths of the I–VI and III–VI bonds are in general different, the ratio of the lattice constants c/a is not exactly two. Instead, the quantity $2 - c/a$ (which is -0.01 in $CuInSe_2$, $+0.04$ in $CuGaSe_2$) is a measure of the tetragonal distortion in chalcopyrite materials.

2.2 Band Gap Energies

The system of copper chalcopyrites $Cu(In,Ga,Al)(Se,S)_2$ includes a wide range of band-gap energies E_g from 1.04 eV in $CuInSe_2$ up to 2.4 eV in $CuGaS_2$, and even 2.7 eV in $CuAlS_2$, thus, covering most of the visible spectrum. All these compounds have a direct band gap making them suitable for thin film photovoltaic absorber materials. Figure 2 summarises lattice constants a and band-gap energies E_g of this system. Any desired alloys between these compounds can be produced as no miscibility gap occurs in the entire system. We will discuss the status and prospects of this system in Section 5 in more detail.

2.3 The Phase Diagram

Compared with all other materials used for thin-film photovoltaics, $Cu(In,Ga)Se_2$ has by far the most complicated phase diagram. Figure 3 shows the phase diagram of $CuInSe_2$ given by Haalboom et al. [20]. This investigation had a special focus on temperatures and compositions relevant for the preparation of thin-films. The phase diagram in Figure 3 shows the four different phases which

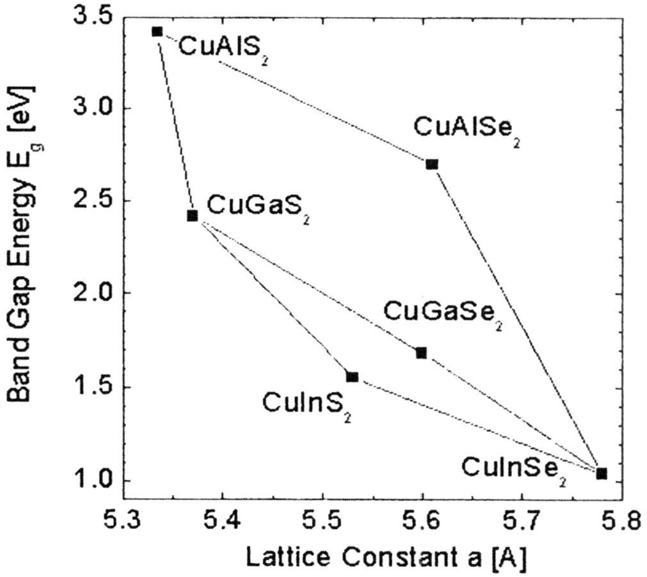

Figure 2 Band-gap energies Eg *vs. the lattice constant a of the Cu(In,Ga,Al)(S,Se)$_2$ alloy system.*

have been found to be relevant in this range: the α-phase ($CuInSe_2$), the β-phase ($CuIn_3Se_5$), the δ-phase (the high-temperature sphalerite phase) and Cu_ySe. An interesting point is that all neighbouring phases to the α-phase have a similar structure. The β-phase is actually a defect chalcopyrite phase built by ordered arrays of defect pairs (Cu vacancies V_{Cu} and In–Cu antisites In_{Cu}). Similarly, Cu_ySe can be viewed as constructed from the chalcopyrite by using Cu–In antisites Cu_{In} and Cu interstitials Cu_i. The transition to the sphalerite phase arises from disordering the cation (Cu, In) sub-lattice, and leads back to the zinc blende structure (cf. Figure 1(a)).

The existence range of the α-phase in pure $CuInSe_2$ on the quasi-binary tie line Cu_2Se-In_2Se_3 extends from a Cu content of 24 to 24.5 at.%. Thus, the existence range of single-phase $CuInSe_2$ is relatively small and does not even include the stoichiometric composition of 25 at.% Cu. The Cu content of absorbers for thin-film solar cells varies typically between 22 and 24 at.% Cu. At the growth temperature this compositional range lies within the single-phase region of the α-phase. However, at room temperature it lies in the two-phase $\alpha + \beta$ region of the equilibrium phase diagram [20]. Hence one would expect a tendency for phase separation in photovoltaic-grade $CuInSe_2$ after deposition. Fortunately, it turns out that partial replacement of In with Ga, as well as the use of Na-containing substrates, considerably widens the single-phase region in terms of (In + Ga)/(In + Ga + Cu) ratios [21]. Thus, the phase diagram hints at the substantial improvements actually achieved in recent years by the use of Na-containing substrates, as well as by the use of Cu(In,Ga)Se_2 alloys.

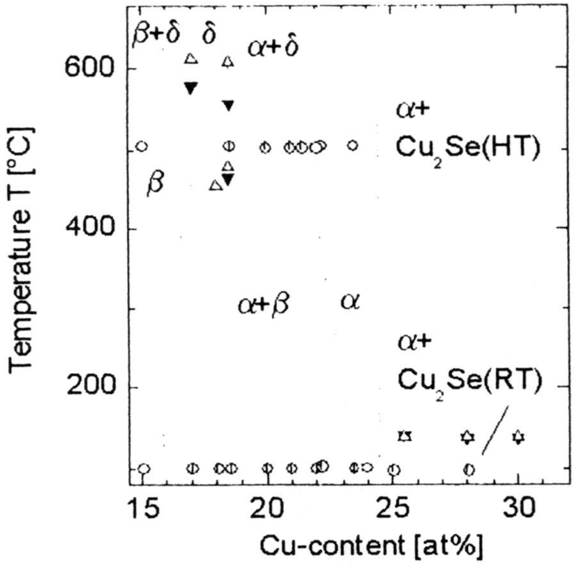

Figure 3 Quasi-binary phase diagram of $CuInSe_2$ along the tie-line that connects the binary compounds In_2Se_3 and Cu_2Se established by Differential Thermal Analysis (DTA) and microscopic phase analysis (After Haalboom et al. [20]). Note that at 25 at. % Cu no single phase exists.

2.4 Defect Physics of $Cu(In,Ga)Se_2$

The defect structure of the ternary compounds $CuInSe_2$, $CuGaSe_2$, $CuInS_2$, and their alloys, is of special importance because of the large number of possible intrinsic defects and the role of deep recombination centres for the performance of the solar cells. The features that are somewhat special to the Cu-chalcopyrite compounds are the ability to dope these compounds with native defects, their tolerance to large off-stoichiometries, and the electrically neutral nature of structural defects in these materials. It is obvious that the explanation of these effects significantly contributes to the explanation of the photovoltaic performance of these compounds. Doping of $CuInSe_2$ is controlled by intrinsic defects. Samples with p-type conductivity are grown if the material is Cu-poor and annealed under high Se vapour pressure, whereas Cu-rich material with Se deficiency tends to be n-type [22, 23]. Thus, the Se vacancy V_{Se} is considered to be the dominant donor in n-type material (and also the compensating donor in p-type material), and the Cu vacancy V_{Cu} the dominant acceptor in Cu-poor p-type material.

By calculating the metal-related defects in $CuInSe_2$ and $CuGaSe_2$, Zhang et al. [24] found that the defect formation energies for some intrinsic defects are so low that they can be heavily influenced by the chemical potential of the components (i.e., by the composition of the material) as well as by the electrochemical potential of the electrons. For V_{Cu} in Cu-poor and stoichiometric material, a negative formation energy is even calculated. This would imply the spontaneous formation of large numbers of these defects under equilibrium conditions. Low (but positive) formation energies are also found for the antisite Cu_{In} in Cu-rich material (this defect is a shallow acceptor which could be responsible for the p-type conductivity of Cu-rich, non-Se-deficient $CuInSe_2$). The dependence of the defect formation energies on the electron Fermi level could explain the strong tendency of $CuInSe_2$ to self-compensation and the difficulties of achieving extrinsic doping. The results of Zhang et al. [24] provide a good theoretical model of defect formation energies and defect transition energies, which exhibits good agreement with experimentally obtained data. Table 1 summarises the ionisation energies and the defect formation energies of the 12 intrinsic defects in $CuInSe_2$. The energies (bold values in Table 1) for V_{Cu}, V_{In}, Cu_i, Cu_{In}, In_{Cu}, are obtained from a first principle calculation [24] whereas the formation energies in italics (Table 1) and for the other defects are calculated from the macroscopic cavity model [25]. The ionisation energies used in Table 1 are either taken from Zhang et al. [24] or from the data compiled in reference [26]. Note that the data given in references [25, 26] for the cation defects differ significantly from those computed in reference [24].

Further important results in reference [24] are the formation energies of *defect complexes* such as ($2V_{Cu}$,In_{Cu}), (Cu_{In},In_{Cu}) and ($2Cu_i$,Cu_{In}), where Cu_i is an interstitial Cu atom. These formation energies are even lower than those of the corresponding isolated defects. Interestingly, the ($2V_{Cu}$,In_{Cu}) complex does not exhibit an electronic transition within the forbidden gap, in contrast to the isolated In_{Cu}-anti-site, which is a deep recombination centre. As the ($2V_{Cu}$,In_{Cu}) complex is most likely to occur in In-rich material, it can accommodate a large amount of excess In (or likewise deficient Cu) and, at same time, maintain the

Table 1 Electronic transition energies and formation energies ΔU of the 12 intrinsic defects in $CuInSe_2$. Source: the ionisation energies in italics are derived from reference [26], and the formation energies in italics are from reference [25]. All the numbers in bold type are from reference [24]

Transition	Defect transition energies[a] and formation energies[b] (eV)											
	V_{Cu}	V_{In}	V_{se}	Cu_i	In_i	Se_i	In_{Cu}	Cu_{In}	Se_{Cu}	Cu_{Se}	Se_{In}	In_{Se}
(−/0)	**0.03**	**0.17**						**0.29**				
	0.03	*0.04*	*0.04*[c]			*0.07*		*0.05*		*0.13*	*0.08*	
(−/2−)		**0.41**						**0.58**				
(2−/3−)		**0.67**										
(0/+)				**0.2**			**0.25**					
			0.11[d]	*0.08*	*0.07*		*0.04*		*0.06*			*0.09*
(+/2+)							**0.44**					
ΔU/ eV	**0.60**	**3.04**		**2.88**			**3.34**	**1.54**				
	2.9	*2.8*	*2.6*	*4.4*	*9.1*	*22.4*	*1.4*	*1.5*	*7.5*	*7.5*	*5.5*	*5.0*

[a] Difference between the valence/conduction band energy for acceptor/donor states.
[b] Formation energy ΔU of the neutral defect in the stoichiometric material.
[c] Covalent.
[d] Ionic.

electrical performance of the material. Furthermore, ordered arrays of this complex can be thought as the building blocks of a series of Cu-poor Cu–In–Se compounds such as $CuIn_3Se_5$ and $CuIn_5Se_8$ [24].

Let us now concentrate on the defects experimentally detected in photovoltaic grade (and thus In-rich) polycrystalline films. In-rich Cu(In,Ga)Se_2 is in general highly compensated, with a net acceptor concentration of the order of 10^{16} cm^{-3}. The shallow acceptor level V_{Cu} (which lies about 30 meV above the valence band) is assumed to be the main dopant in this material. As compensating donors, the Se-vacancy V_{Se} as well as the double donor In_{Cu} are considered. The most prominent defect is an acceptor level at about 270–300 meV above the valence band, which is reported by several groups from deep-level transient spectroscopy [27] and admittance spectroscopy [28, 29]. This defect is also present in single crystals [30]. The importance of this transition results from the fact that its concentration is related to the open-circuit voltage of the device [31–33,]. Upon investigating defect energies in the entire Cu(In,Ga)(Se,S)$_2$ alloy system, Turcu et al. [34] found that the energy distance between this defect and the valence band maximum remains constant when alloying $CuInSe_2$ with Ga, whereas the energy distance increases under S/Se alloying. Assuming that the defect energy is independent from the energy position of the band edges, like the defect energies of transition metal impurities in III–V and II–IV semiconductor alloys [35, 36], the authors of reference [34] extrapolate the valence band offsets $\Delta E_v = -0.23$ eV for the combination $CuInSe_2/CuInS_2$ and $\Delta E_v = 0.04$ eV for $CuInSe_2/CuGaSe_2$. Recently, transient photocapacitance studies by Heath et al. [37] unveiled an additional defect state in Cu(In,Ga)Se_2 at about 0.8 eV from the valence band. Again, the defect energy is independent of the Ga content in the alloy. Figure 4 summarises the energy positions of bulk defects in the $Cu(In_{1-x}Ga_x)Se_2$ and the $CuIn(Se_{1-y}S_y)_2$ alloy system with the defect energy of the bulk acceptor used as a reference energy to align the valence and conduction band energy [38]. Additionally, Figure 4 shows the activation energy of an interface donor. This energy corresponds to the energy difference ΔE_{Fn} between the Fermi energy and the conduction band minimum at the buffer absorber interface [39]. Notably this energy difference remains small upon alloying $CuInSe_2$ with Ga, whereas ΔE_{Fn} increases when alloying with S.

3 Cell and Module Technology

3.1 Structure of the Heterojunction Solar Cell

The complete layer sequence of a ZnO/CdS/Cu(In,Ga)Se_2 heterojunction solar cell is shown in Figure 5. The device consists of a typically 1 μm thick Mo layer deposited on a soda-lime glass substrate and serving as the back contact for the solar cell. The Cu(In,Ga)Se_2 is deposited on top of the Mo back electrode as the photovoltaic absorber material. This absorber layer has a thickness of 1–2 μm. The heterojunction is then completed by chemical bath deposition (CBD) of CdS

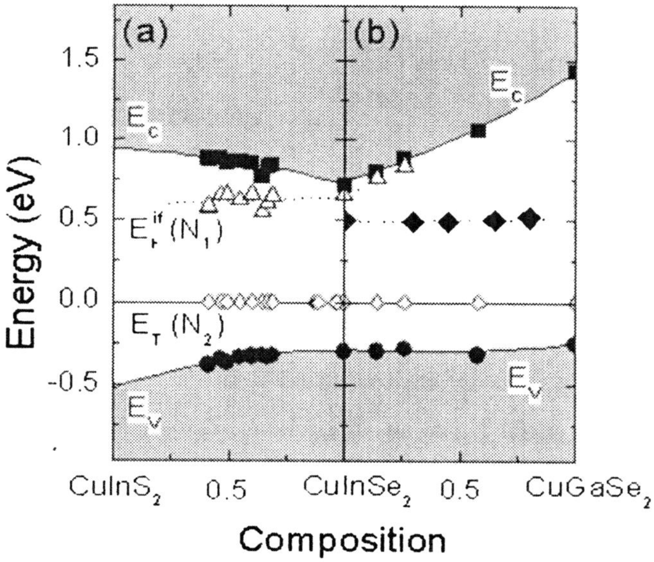

Figure 4 Band gap evolution diagram of the CuIn(Se,S)$_2$ (a) and the Cu(In,Ga)Se$_2$ (b) alloy system with the trap energy E_T(N_2, open diamonds) taken as an internal reference to align the conduction band and the valence band energies E_c and E_v. The energy position of an additional defect state in Cu(In,Ga)Se$_2$ (full diamonds) as well as that of an interface donor (open triangles) in Cu(In,Ga)(Se,S)$_2$ is also indicated.

(typically 50 nm thick) and by the sputter deposition of a nominally undoped (intrinsic) *i*-ZnO layer (usually of thickness 50–70 nm) and then a heavily doped ZnO layer. As ZnO has a band-gap energy of 3.2 eV it is transparent for the main part of the solar spectrum and therefore is denoted as the window layer of the solar cell.

3.2 Key Elements for High-Efficiency Cu(In,Ga)Se$_2$ Solar Cells

We first mention four important technological innovations which, during the decade 1990–2000, have led to a considerable improvement of the efficiencies and finally to the record efficiency of 18.8% [1]. These steps are the key elements of the present Cu(In,Ga)Se$_2$ technology:

- The film quality has been substantially improved by the crystallisation mechanism induced by the presence of Cu$_y$Se ($y < 2$). This process is further supported by a substrate temperature close to the softening point of the glass substrate [40].
- The glass substrate has been changed from Na-free glass to Na-containing soda-lime glass [40, 41]. The incorporation of Na, either from the glass substrate or from Na-containing precursors, has led to an enormous improvement of the efficiency and reliability of the solar cells, as well as to a larger process tolerance.

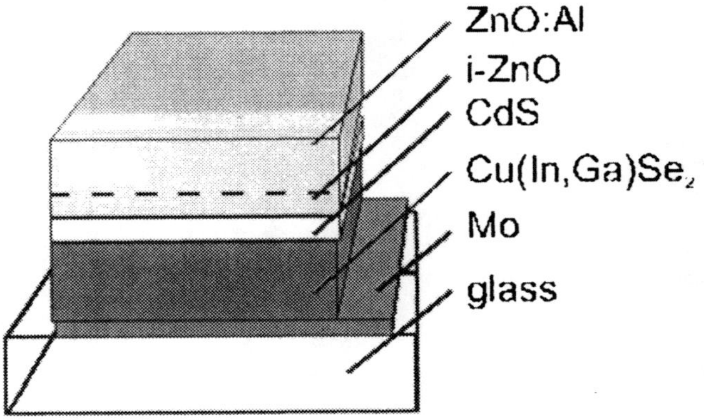

Figure 5 Schematic layer sequence of a standard ZnO/CdS/Cu(In,Ga)Se_2 thin-film solar cell.

- Early absorbers consisted of pure $CuInSe_2$. The partial replacement of In with Ga [42] is a further noticeable improvement, which has increased the band gap of the absorber from 1.04 eV to 1.1–1.2 eV for the high-efficiency devices. The benefit of 20–30% Ga incorporation stems not only from the better band gap match to the solar spectrum but also from the improved electronic quality of Cu(In,Ga)Se_2 with respect to pure $CuInSe_2$ [21, 32].
- The counter electrode for the $CuInSe_2$ absorber of the earlier cells was a 2 μm thick CdS layer deposited by Physical Vapour Deposition (PVD). This has been replaced by a combination of a 50 nm thin CdS buffer layer laid down by chemical bath deposition [43, 44] and a highly conductive ZnO window layer.

3.3 Absorber Preparation Techniques

3.3.1 Basics

The preparation of Cu(In,Ga)Se_2-based solar cells starts with the deposition of the absorber material on a Mo-coated glass substrate (preferably soda-lime glass). The properties of the Mo film and the choice of the glass substrate are of primary importance for the final device quality, because of the importance of Na, which diffuses from the glass through the Mo film into the growing absorber material. In the past, some processes used blocking layers such as SiN_x, SiO_2 or Cr between the glass substrate and the Mo film to prevent the out-diffusion of Na. Instead, Na-containing precursors like NaF [45], Na_2Se [46] or Na_2S [47] are then deposited prior to absorber growth to provide a controlled, more homogeneous, incorporation of Na into the film. The control of Na incorporation in the film from precursor layers allows the use of other substrates like metal or polymer foils. The most obvious effects of Na incorporation are better film morphology and higher conductivity of the films [48]. Furthermore, the incorporation of Na induces beneficial changes in the defect distribution of the absorber films [49, 50].

The explanations for the beneficial impact of Na are manifold, and it is most likely that the incorporation of Na in fact results in a *variety* of consequences. During film growth, the incorporation of Na leads to the formation of $NaSe_x$ compounds. This slows down the growth of $CuInSe_2$ and could at same time facilitate the incorporation of Se into the film [51]. Also the widening of the existence range of the α-($CuInSe_2$) phase in the phase diagram, discussed above, as well as the reported larger tolerance to the Cu/(In + Ga) ratio of Na-containing thin films, could be explained in this picture. Furthermore, the higher conductivity of Na-containing films could result from the diminished number of compensating V_{se} donors.

During absorber deposition, a $MoSe_2$ film forms at the Mo surface [52, 53]. $MoSe_2$ is a layered semiconductor with p-type conduction, a band gap of 1.3 eV and weak van der Waals bonding along the *c*-axis. The *c*-axis is found to be in parallel with, and the van der Waals planes thus perpendicular to, the interface [53]. Because of the larger band gap of the $MoSe_2$ compared with that of standard $Cu(In,Ga)Se_2$ films, the $MoSe_2$ layer provides a low-recombinative back surface for the photogenerated minority carriers (electrons) in the $Cu(In,Ga)Se_2$ absorber and at the same time provides a low-resistance contact for the majority carries (holes).

Photovoltaic-grade $Cu(In,Ga)Se_2$ films have a slightly In-rich overall composition. The allowed stoichiometry deviations are astonishingly large, yielding a wide process window with respect to composition. Devices with efficiencies above 14% are obtained from absorbers with (In + Ga)/(In + Ga + Cu) ratios between 52 and 64% if the sample contains Na [48]. Cu-rich $Cu(In,Ga)Se_2$ shows the segregation of a secondary Cu_ySe phase preferentially at the surface of the absorber film. The metallic nature of this phase does not allow the formation of efficient heterojunctions. Even after *removal* of the secondary phase from the surface by etching the absorber in KCN, the utility of this material for photovoltaic applications is limited. However, the importance of the Cu-rich composition is given by its role during film growth. Cu-rich films have grain sizes in excess of 1 μm whereas In-rich films have much smaller grains. A model for the film growth under Cu-rich compositions comprises the role of Cu_ySe as a flux agent during the growth process of co-evaporated films [54]. For $Cu(In,Ga)Se_2$ prepared by selenisation, the role of Cu_ySe is similar [55].Therefore, growth processes for high quality material have to go through a copper-rich stage but have to end up with an indium-rich overall composition.

3.3.2 Co-evaporation Processes

The absorber material yielding the highest efficiencies is $Cu(In,Ga)Se_2$ with a Ga/(Ga + In) ratio of 20–30%, prepared by co-evaporation from elemental sources. Figure 6 sketches a co-evaporation set-up as used for the preparation of laboratory-scale solar cells and mini-modules. The process requires a maximum substrate temperature of about 550°C for a certain time during film growth, preferably towards the end of growth. One advantage of the evaporation route is that material deposition and film formation are performed during the same processing step. A feedback loop based on a quadrupole mass spectrometer or an

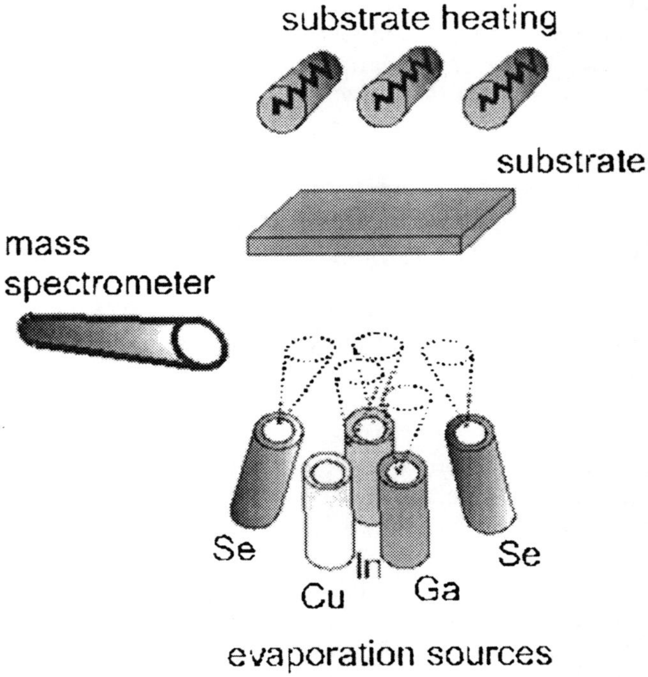

Figure 6 Arrangement for the deposition of Cu(In,Ga)Se_2 films on the laboratory scale by co-evaporation on a heated substrate. The rates of the sources are controlled by mass spectrometry.

atomic absorption spectrometer controls the rates of each source. The composition of the deposited material with regard to the metals corresponds to their evaporation rates, whereas Se is always evaporated in excess. This precise control over the deposition rates allows for a wide range of variations and optimisations with different sub-steps or stages for film deposition and growth. These sequences are defined by the evaporation rates of the different sources and the substrate temperature during the course of deposition.

Advanced preparation sequences always include a Cu-rich stage during the growth process and end up with an In-rich overall composition in order to combine the large grains of the Cu-rich stage with the otherwise more favourable electronic properties of the In-rich composition. The first example of this kind of procedure is the so called Boeing or *bilayer process* [5], which starts with the deposition of Cu-rich Cu(In,Ga)Se_2 and ends with an excess In rate. The most successful co-evaporation process is the so-called *three-stage process* [56]. This process starts with the deposition of In,Ga, and Se at relatively low temperatures, then uses a Cu-rich growth stage by evaporating Cu in excess at elevated temperature, and at the end again deposits only In, Ga, and Se to ensure the overall In-rich composition of the film. The three-stage process currently leads to the Cu(In,Ga)Se_2 solar cells with the highest efficiencies.

3.3.3 Selenisation Processes

The second class of absorber preparation routes is based on the separation of deposition and compound formation into two different processing steps. High efficiencies are obtained from absorber prepared by selenisation of metal precursors in H_2Se [57–59] and by rapid thermal processing of stacked elemental layers in a Se atmosphere [60]. These sequential processes have the advantage that approved large-area deposition techniques like sputtering can be used for the deposition of the materials. The $Cu(In,Ga)Se_2$ film formation then requires a second step, the selenisation.

The very first large-area modules were prepared by the selenisation of metal precursors in the presence of H_2Se more than ten years ago [6]. Today, a modification of this process yields the first commercially available $Cu(In,Ga)Se_2$ solar cells manufactured by Shell Solar Industries. This process is schematically illustrated in Figure 7. First, a stacked layer of Cu, In and Ga is sputter-deposited on the Mo-coated glass substrate. Then selenisation takes place under H_2Se. To improve the device performance, a second thermal process under H_2S is added, resulting in an absorber that is $Cu(In,Ga)(S,Se)_2$ rather than $Cu(In,Ga)Se_2$.

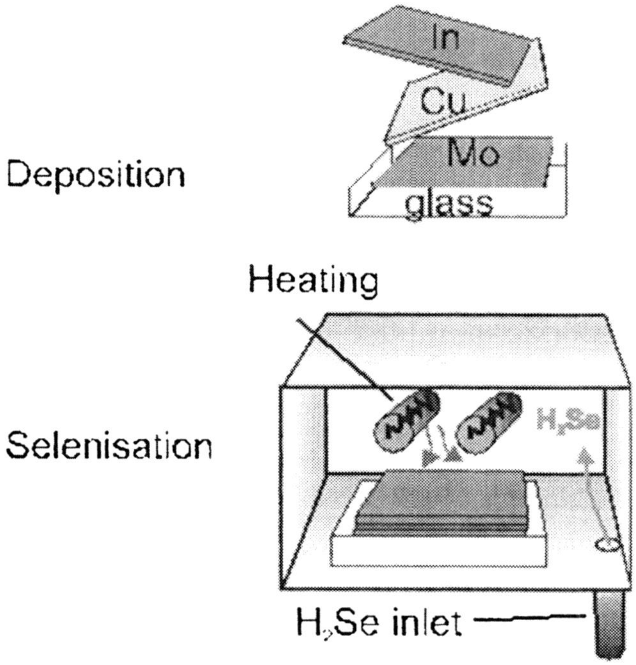

Figure 7 Illustration of the sequential process. First a stack of metal (Cu,In,Ga) layers deposited by sputtering onto a Mo-coated glass. In the second step, this stack is selenised in H_2Se atmosphere and converted into $CuInSe_2$.

A variation of the method that avoids the use of the toxic H_2Se during selenisation is the rapid thermal processing of stacked elemental layers [60]. Here the precursor includes a layer of evaporated elemental Se. The stack is then selenised by a rapid thermal process (RTP) in either an inert or a Se atmosphere. The highest efficiencies are obtained if the RTP is performed in an S-containing atmosphere (either pure S or H_2S).

On the laboratory scale, the efficiencies of cells made by these preparation routes are smaller by about 3% (absolute) as compared with the record values. However, on the module level, co-evaporated and sequentially prepared absorbers have about the same efficiency. Sequential processes need two or even three stages for absorber completion. These additional processing steps may counterbalance the advantage of easier element deposition by sputtering.

3.3.4 Other Absorber Deposition Processes

Besides selenisation and co-evaporation, other deposition methods have been studied, either to obtain films with very high quality or to reduce the cost of film deposition on large areas. Methods that are used to grow epitaxial III–V compound films, such as molecular beam epitaxy (MBE) [61] or metal organic chemical vapour deposition (MOCVD) [62] have revealed interesting features for fundamental studies like phase segregation and defect formation, but could not be used to form the absorber material for high-efficiency solar cells.

Attempts to develop so-called low-cost processes include electrodeposition [63–66], screen printing and particle deposition [67, 68]. Electrodeposition can be done either in one or two steps. The crucial step is the final film formation in a high-temperature annealing process. The recrystallisation process competes with the decomposition of the material. Therefore, process optimisation is quite difficult. Cells with high efficiencies were obtained by electrodeposition of a Cu-rich $CuInSe_2$ film and subsequent conditioning by a vacuum evaporation step of In(Se) [69]. Particle deposition by printing of suitable inks and subsequent annealing lead to absorber layers with good quality enabling the fabrication of solar cells with efficiencies over 13% [70].

3.3.5 Post-deposition Air Anneal

Air annealing has been an important process step, crucial for the efficiency especially of the early solar cells based on $CuInSe_2$. Though often not mentioned explicitly, an oxygenation step is still used for most of the present-day high-efficiency devices. The beneficial effect of oxygen was explained within the defect chemical model of Cahen and Noufi [71]. In this model, the surface defects at grain boundaries are positively charged Se vacancies V_{Se}. During air annealing, these sites are passivated by O atoms. Because of the decreased charge at the grain boundary, the band bending as well as the recombination probability for photogenerated electrons is reduced. The surface donors and their neutralisation by oxygen are important for the free $Cu(In,Ga)Se_2$ surface as well as for the formation of the $CdS/Cu(In,Ga)Se_2$ interface [72, 73].

3.4 Heterojunction Formation

3.4.1 The Free $Cu(In,Ga)Se_2$ Surface

The surface properties of $Cu(In,Ga)Se_2$ thin films are especially important as this surface becomes the active interface of the completed solar cell. However, the band diagram of the $ZnO/CdS/Cu(In,Ga)Se_2$ heterojunction, especially the detailed structure close to the $CdS/Cu(In,Ga)Se_2$ interface, is still under debate (for recent reviews see references [74, 75]).

The free surfaces of as-grown $Cu(In,Ga)Se_2$ films exhibit two prominent features:

(i) The valence band-edge energy E_V lies below the Fermi level E_F by about 1.1 eV for $CuInSe_2$ films [76]. This energy is larger than the band-gap energy E_g^{bulk} of the bulk of the absorber material. This finding was taken as an indication for a *widening of band gap* at the surface of the film. A recent direct measurement of the surface band gap of polycrystalline $CuInSe_2$ by Morkel et al. [77] proved that the band gap energy E_g^{surf} at the surface of the film is about 1.4 eV, i.e., more than 0.3 eV larger than $E_g^{bulk} \approx 1.04$ eV. In $Cu(In_{1-x},Ga_x)Se_2$ alloys the distance $E_F - E_V$ was found to be 0.8 eV (almost independently of the Ga content if $x > 0$) [78].

(ii) The surface composition of Cu-poor $CuInSe_2$, as well as that of $Cu(In,Ga)Se_2$ films, corresponds to a surface composition of (Ga + In)/(Ga + In + Cu) of about 0.75 for a range of bulk compositions of $0.5 <$ (Ga + In)/(Ga + In + Cu) < 0.75 [76].

Both observations (i) and (ii) have led to the assumption that a phase segregation of $Cu(In,Ga)_3Se_5$, the so-called Ordered Defect Compound (ODC), occurs at the surface of the films. From the fact that bulk $Cu(In,Ga)_3Se_5$ exhibits n-type conductivity [79] it was argued that Cu-poor $Cu(In,Ga)Se_2$ thin films automatically generate a rectifying, buried junction. However, the existence of a separate phase on top of standard $Cu(In,Ga)Se_2$ thin films has not yet been confirmed by structural methods such as X-ray diffraction, high resolution transmission electron microscopy or electron diffraction. Furthermore, if the surface phase exhibited the weak n-type conductivity of bulk $Cu(In,Ga)_3Se_5$, simple charge neutrality estimations [21] show that this would not be sufficient to achieve type inversion.

Based on these arguments, another picture of the surface of $Cu(In,Ga)Se_2$ thin films and of junction formation has emerged [21, 80]. Within the classical Bardeen model [81] of Fermi level pinning by electronic states at semiconductor surfaces, a density of surface states of about 10^{12} cm^{-2} eV^{-1} is sufficient to pin the Fermi level at the neutrality level of the free semiconductor surfaces. Positively charged surface donors are expected in the metal terminated (112) surface of $CuInSe_2$ owing to the dangling bond to the missing Se [71]. Thus, these surface states, rather than a distinct n-type surface phase determines the type inversion of the surface.

Surface states play also an important role in the completed heterostructure, where they become *interface states* at the absorber/buffer interface. *The defect layer model* [21, 80] takes into account a modification of the band structure due to the Cu deficiency of the surface *as well as* the presence of positively charged surface states due to the missing surface Se. However, the defect layer model considers the surface layer not as n-type bulk material (as does the ODC model) but as a p^+-layer. Furthermore, the defect layer is viewed *not as the origin but rather as the consequence* of the natural surface type inversion [21, 80]. Surface states are responsible for the surface band bending that leads to the liberation of Cu from its lattice sites and to Cu migration towards the neutral part of the film. The remaining copper vacancies V_{Cu}^{-} close to the surface result in a high density of acceptor states, i.e., the p^+-defect layer at the film surface. Recent photoelectron spectroscopy experiments of Klein and Jaegermann [82] suggest that the band bending occurring during junction formation leads to a loss of Cu atoms from the surface of $CuInSe_2$ and $CuGaSe_2$, whereas a similar effect was not observed in $CuInS_2$.

3.4.2 Buffer Layer Deposition

Surface passivation and junction formation is most easily achieved by the CBD deposition of a thin CdS film from a chemical solution containing Cd ions and thiourea [83]. The benefit of the CdS layer is manifold:

- CBD deposition of CdS provides complete coverage of the rough polycrystalline absorber surface at a film thickness of only 10 nm [84].
- The layer provides protection against damage and chemical reactions resulting from the subsequent ZnO deposition process.
- The chemical bath removes the natural oxide from the film surface [83] and thus re-establishes positively charged surface states and, as a consequence, the natural type inversion at the CdS/Cu(In,Ga)Se_2 interface.
- The Cd ions, reacting first with the absorber surface, remove elemental Se, possibly by the formation of CdSe.
- The Cd ions also diffuse to a certain extent into the Cu-poor surface layer of the absorber material [85, 86], where they possibly form Cd_{Cu} donors, thus providing additional positive charges enhancing the type inversion of the buffer/absorber interface.
- Open-circuit voltage limitations imposed by interface recombination can be overcome by a low surface recombination velocity in addition to the type inversion of the absorber surface. Thus one might conclude that interface states (except those shallow surface donors responsible for the type inversion) are also passivated by the chemical bath.

Due to the favourable properties of CdS as a heterojunction partner and the chemistry of the CBD process it is difficult to find a replacement. Avoiding CdS and the chemical bath step would be advantageous from the production point of view. On the one hand, a toxic material such as CdS requires additional safety regulation; on the other hand, the chemical bath deposition does not comply with the vacuum deposition steps of an in-line module fabrication. Therefore, research and development in this area relates to two issues: (i) the search for alternative materials for a chemical deposition, and (ii) the development of ways to deposit the front electrode without an intermediate step in a chemical bath.

Promising materials to replace CdS are In(OH,S) [78], Zn(OH,S) [88, 89], ZnS [90, 91] and ZnSe [92–94]. However, all these materials require additional precautions to be taken for the preparation of the absorber surface or front electrode deposition.

3.4.3 Window Layer Deposition

The most commonly used material for the preparation of the front electrode is ZnO doped with B or Al. The first large-area modules produced by ARCO Solar (later Siemens Solar Industries, now Shell Solar Industries) had a ZnO:B window layer deposited by chemical vapour deposition (CVD). Later production facilities at Boeing and EUROCIS use sputtering processes. Present pilot production lines also favour sputtering. As mentioned above, an undoped *i*-ZnO layer with a thickness of about 50–100 nm is needed at the heterojunction in order to achieve optimum performance.

3.5 Module Production and Commercialisation

3.5.1 Monolithic Interconnections

One inherent advantage of thin-film technology for photovoltaics is the possibility of using monolithic integration for series connection of individual cells within a module. In contrast, bulk Si solar cells must be provided with a front metal grid, and each of these front contacts has to be connected to the back contact of the next cell for series connection. The interconnect scheme, shown in Figure 8, has to warrant that the front ZnO layer of one cell is connected to the back Mo contact of the next one. In order to obtain this connection, three different patterning steps are necessary. The first one separates the Mo back contact by a series of periodical scribes and thus defines the width of the cells, which is of the order of 0.5–1 cm. For Mo patterning, a laser is normally used. The second patterning step is performed after absorber and buffer deposition, and the final one after window deposition (cf. Figure 8). Scribing of the semiconductor layer is done by mechanical scribing or laser scribing. The total width of the interconnect depends not only on the scribing tools, but also on the reproducibility of the scribing lines along the entire module. The length of the cells and, accordingly, that of the scribes can be more than 1 m. The typical interconnect width is of the order of 300 μm. Thus, about 3–5% of the cell area must be sacrificed to the interconnects.

3.5.2 Module Fabrication

The technologies for absorber, buffer and window deposition used for module production are the same as those discussed above for the production of small laboratory cells. However, the challenge of producing modules is to transform

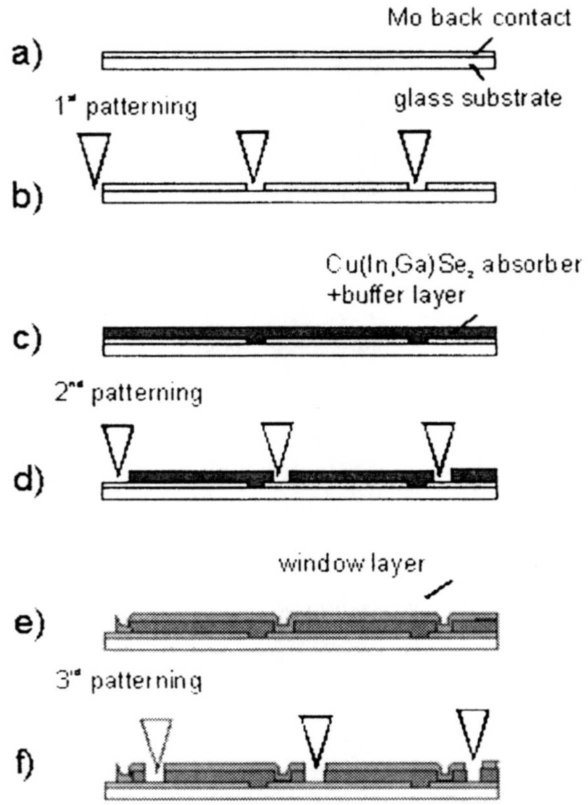

Figure 8 Deposition and patterning sequence to obtain an integrated interconnect scheme for $Cu(In,Ga)Se_2$ thin-film modules.

the laboratory-scale technologies to much larger areas. The *selenisation* process uses as much off-the-shelf equipment and processing as possible (e.g., sputtering of the metal precursors) for fabricating $Cu(In,Ga)Se_2$ films. For *co-evaporation* on large areas, the Centre for Solar Energy and Hydrogen Research in Stuttgart (ZSW) has designed its own equipment, schematically shown in Figure 9, for an in-line co-evaporation process. Line-shaped evaporation sources allow continuous deposition of large-area, high-quality $Cu(In,Ga)Se_2$ films. The relatively high substrate temperatures that are necessary for high-quality material impose problems in handling very large area glass sheets. Future process optimisation therefore implies reduction of the substrate temperature.

A bottleneck for the production is the deposition of the buffer layer in a chemical bath. On the one hand, it is not straightforward to integrate this process in a line consisting mainly of dry physical vapour deposition processes, on the other, it would be favourable for environmental reasons to replace the currently used CdS layer by a Cd-free alternative.

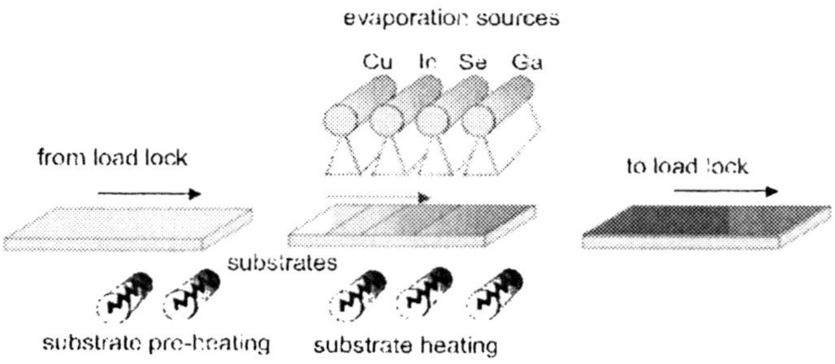

Figure 9 Sketch of an in-line deposition system for co-evaporation of $Cu(In,Ga)Se_2$ absorber films from line-sources.

The ZnO transparent front electrode is put down either by CVD or sputtering. Each method has its specific advantages with respect to process tolerance, throughput, cost and film properties. The widths of the cells within the module – and therefore the relative losses from the patterning – mainly depend on the sheet resistance of the ZnO.

Module encapsulation is an important issue because module stability depends on proper protection against humidity. Low-iron cover glasses provide good protection against ambient influences. Hermetic sealing of the edges is mandatory to obtain stable modules (see below).

3.5.3 Up-Scaling Achievements

$Cu(In,Ga)Se_2$ has the best potential to reach more than 15% module efficiency in the near future. Mini-modules ranging in area from 20 to 90 cm^2 that use the process sequence anticipated for larger area commercial modules have already reached efficiencies around 14–15%. Recently, Siemens Solar Industries fabricated a 1 ft × 4 ft power module (~44 watts) with an independently verified efficiency of 12.1%. Using a totally different approach to the deposition of the absorber layer, the ZSW fabricated a 30 cm × 30 cm module with a verified efficiency of 12.7%. Based on the same co-evaporation process, Würth Solar GmbH, Stuttgart, reported an efficiency of 12.5% for a module of aperture area 5932 cm^2 [95]. More results of the different processes are summarised in Table 2. Note that, because of the promising results from the laboratory scale and the first approaches of up-scaling, several companies other than those mentioned in Table 2 now plan commercial production.

A further challenge is to develop CIGS cells on flexible substrates and hence to extend their area of applications. There are ongoing efforts to produce cells on various kinds of substrates like stainless steel, polymide and at the same time to retain the performance achieved with devices on soda-lime glass [96]. For space applications it is important to reduce the weight by depositing the cells on lightweight foil substrates. Highest small area efficiencies on polymide films formed by spin coating on a glass substrate reach 12.8% [97]. Roll to roll coating on metal foils [98] and polymer films [99, 100] has already reached the stage of pilot production.

Table 2 Comparison of efficiencies η and areas A of laboratory cells, mini-modules, and commercial-size modules achieved with $Cu(In,Ga)Se_2$ thin films based on the co-evaporation and the selenisation process. NREL denotes the National Renewable Energy Laboratories (USA), ZSW is the Center for Solar Energy and Hydrogen Research (Germany), EPV is Energy Photovoltaics (USA), ASC is the Angstrom Solar Centre (Sweden)

Process	Laboratory cell		Mini-module		Module		Laboratory/company
	η (%)	A (cm^2)	η (%)	A (cm^2)	η (%)	A (cm^2)	
Co-evap.	18.8	0.45					NREL [1]
	16.1	0.5	13.9	90	12.7	800	ZSW
					12.3	5932	Würth Solar [95]
			9.6	135			EPV
	11.5	0.5	5.6	240			Global Solar[a]
			16.6	20			ASC [2]
Selenis.	>16	0.5	14.7	18	12.1	3600	Shell Solar
			14.2	50	11.6	864	Showa, Japan

[a] Flexible cells.

3.5.4 Stability

The long-term stability is a critical issue of any solar cell technology because the module lifetime contributes as much to the ratio between produced energy and invested cost as does the initial efficiency. $Cu(In,Ga)Se_2$ modules fabricated by Shell Solar Industries more than 10 years ago show until today very good stability during outdoor operation [101, 102]. However, intense accelerated lifetime testing is made for the now commercially available $Cu(In,Ga)Se_2$ modules. Careful sealing and encapsulation appears mandatory, especially because of the sensitivity of $Cu(In,Ga)Se_2$ to humidity. For non-encapsulated modules, corrosion of the molybdenum contact and the degradation of zinc oxide were found to be the dominating degradation mechanisms [103] during the so-called damp heat test (1000 hours in hot (85°C) and humid (85% humidity) atmosphere). Investigations of non-encapsulated cells [104–106] unveiled further a humidity induced degradation of the $Cu(In,Ga)Se_2$ absorber material. Despite of the sensibility of $Cu(In,Ga)Se_2$ with respect to humidity, well encapsulated modules pass the damp heat test [107].

Recent work of Guillemoles et al. [108, 109] investigates the chemical and electronic stability of $Cu(In,Ga)Se_2$ based solar cells and possible fundamental instabilities of the material system, namely, interface reactions, defect metastability, and constituent element (Cu) mobility. Guillemoles et al. conclude that all reasonably anticipated detrimental interface reactions at the Mo/$Cu(In,Ga)Se_2$, the $Cu(In,Ga)Se_2$ /CdS, or the CdS/ZnO interface are either thermodynamically or kinetically limited. Furthermore, Cu mobility does not contradict long-term stability [108, 109].

3.5.5 Radiation Hardness and Space Applications

One important prospective application for $Cu(In,Ga)Se_2$ cells is in space, where the main power source is photovoltaics. Satellites in low-earth orbits for communication systems require solar cells with high end-of-life efficiencies, despite the high flux of high-energy electrons and protons in that ambient. The radiation hardness of Cu(In,Ga)Se$_2$ has been recognized as early as 1984/85 [110, 111], but only recently have systematic investigations on the radiation response of Cu(In,Ga)Se$_2$ solar cells been undergone using high enough electron and proton fluences to allow quantitative conclusions regarding the defect generation rates [112]. The radiation resistance of Cu(In,Ga)Se$_2$ against high-energy (0.3–3 MeV) electrons turns out to be far better than that of any other photovoltaic material [112]. The radiation hardness of Cu(In,Ga)Se$_2$ against high-energy (0.4–10 MeV) protons is also high, though the difference from other materials is not as high as in case of electron irradiation. Walters et al. [113] have analysed Cu(In,Ga)Se$_2$ in the frame of the so-called damage dose model and found that the damage coefficients for Cu(In,Ga)Se$_2$ are comparable to those of InP and considerably smaller than those of Si and GaAs.

The high mobility of Cu in the Cu(In,Ga)Se$_2$ lattice was proposed to be one important ingredient for a defect healing mechanism that could explain the high radiation resistance of Cu(In,Ga)Se$_2$ [108, 114]. Recent thermal annealing experiments of electron-irradiated [115] and proton-irradiated [116] Cu(In,Ga)Se$_2$ solar cells unveiled a thermally activated healing process with an activation energy of around 1 eV leading to a complete recovery of the device performance. Illumination of the solar cell enhances this annealing process further [116, 117].

The challenge for developing CIGS space cells is to reduce the weight by depositing the cells on foil substrates, and at the same time to retain the performance achieved with devices on soda-lime glass. Recently, Tuttle et al. [118] reported a Cu(In,Ga)Se$_2$ solar cell on lightweight metal foil with a power conversion efficiency of 15.2% (under AM0 illumination) and a specific power of 1235 W/kg. Other approaches to flexible and lightweight Cu(In,Ga)Se$_2$ solar cells embrace Cu(In,Ga)Se$_2$ deposited on plastic foil such as polyimide [119] or the use of a lift-off technique to remove the absorber from the glass substrate after device fabrication [120].

4 Device Physics

4.1 The Band Diagram

The band diagram of the ZnO/CdS/Cu(In,Ga)Se$_2$ heterostructure in Figure 10 shows the conduction and valence band energies E_c and E_v of the Cu(In,Ga)Se$_2$ absorber, the CdS buffer layer and the ZnO window. Note that the latter consists of a highly Al-doped (ZnO:Al) and an undoped (i-ZnO) layer. For the moment, we neglect the polycrystalline nature of the semiconductor materials, which in principle requires a two- or three-dimensional band diagram. Even in the one-dimensional model, important details of the band diagram are still not perfectly clear. The diagram in Figure 10 concentrates on the heterojunction and does not show the contact between the Mo and Cu(In,Ga)Se$_2$ at the back side of the absorber.

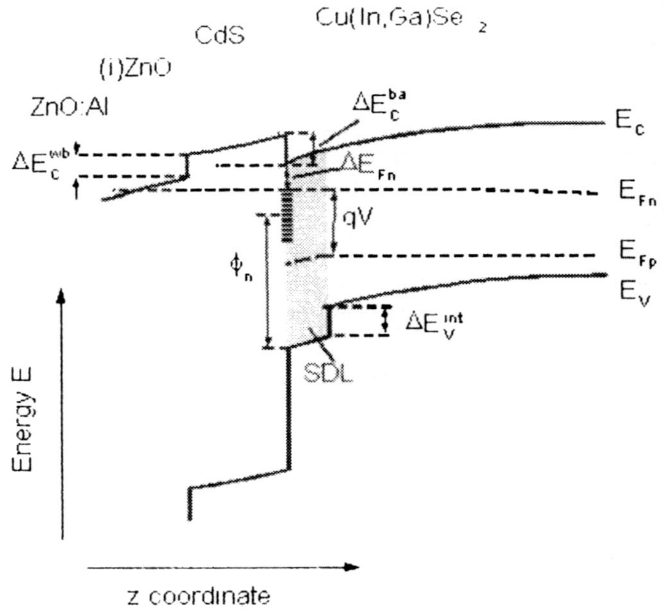

Figure 10 Band diagram of the ZnO/CdS/Cu(In,Ga)Se_2 heterojunction under bias voltage showing the conduction and valence band-edge energies ΔE_c and E_v. The quantities $\Delta E_c^{wb/ba}$ denote the conduction band offsets at the window/buffer and buffer/absorber interfaces, respectively. An internal valence band offset ΔE_V^{int} exists between the bulk Cu(In,Ga)Se_2 and a surface defect layer (SDL) on top of the Cu(In,Ga)Se_2 absorber film. The quantity ΔE_{Fn} denotes the energy distance between the electron Fermi level E_{Fn} and the conduction band at the CdS buffer/Cu(In,Ga)Se_2 absorber interface, and ϕ_n denotes the neutrality level of interface states at this heterointerface.

An important feature in Figure 10 is the 10–30 nm thick surface defect layer (SDL) on top of the Cu(In,Ga)Se_2 absorber, already discussed in Section 3.4. The physical nature of this SDL is still under debate. However, the fact that the band gap at the surface of Cu(In,Ga)Se_2 thin films (as long as they are prepared with a slightly Cu-poor final composition) exceeds the band gap energy of the bulk material [76, 77], has important implications for the contribution of interface recombination to the overall performance of the solar cell. A simplified approach to the mathematical description of the ZnO/CdS/Cu(In,Ga)Se_2 heterojunction including the consequence of the surface band gap widening is given in reference [121].

The most important quantities to be considered in the band diagram are the band discontinuities between the different heterojunction partners. Band discontinuities in terms of valence band offsets ΔE_v^{ab} between semiconductor *a* and *b* are usually determined by photoelectron spectroscopy (for a discussion with respect to Cu-chalcopyrite surfaces and interfaces, see [74]). The valence band offset between a (011)-oriented Cu(In,Ga)Se_2 single crystal and CdS deposited by PVD at room temperature is determined as [122, 123] $\Delta E_v^{ab} = -0.8$ (± 0.2) eV (and, therefore, $\Delta E_c^{ab} = E_g^{CdS} - E_g^{CIS} + \Delta E_v^{ab} \approx 0.55$ eV, with the band gaps $E_g^{CdS} \approx 2.4$ eV and $E_g^{CIS} \approx 1.05$ eV of CdS and $CuInSe_2$, respectively). Several authors have investigated the band discontinuity between polycrystalline $Cu(In,Ga)Se_2$ films and CdS, and found values between −0.6 and −1.3 eV with a clear centre of mass around −0.9 eV, corresponding to a conduction band offset of 0.45 eV [74]. Wei and Zunger [124] calculated a theoretical value of $\Delta E_v^{ab} = -1.03$ eV, which would lead to $\Delta E_C^{ab} \approx 0.3$ eV. Recently, Morkel et al. [77] found a valence band offset between the surface of polycrystalline, Cu-poor prepared $CuInSe_2$ and chemical bath deposited CdS of $\Delta E_v^{ab} = -0.8$ eV. By combining their photoelectron spectroscopy results with measured surface band gap energies of CdS and $CuInSe_2$ from inverse photoemission spectroscopy, the authors of reference [77] conclude that the conduction band offset ΔE_c^{ab} is actually zero. This is because the deposited CdS has $E_g = 2.2$ eV due to S/Se intermixing and the $CuInSe_2$ film exhibits a surface band gap of 1.4 eV, thus $\Delta E_c^{ab} = E_g[\mathrm{Cd(Se,S)}] - E_g^{surf}[\mathrm{CuInSe_2}] + \Delta E_V^{ab} \approx 0$.

The band alignment of polycrystalline $CuInSe_2$ and $Cu(In,Ga)Se_2$ alloys was examined by Schmid et al. [76, 78] who found that the valence band offset is almost independent of the Ga content. In turn, the increase of the absorber band gap leads to a change of ΔE_C^{ab} from positive to negative values. The conduction band offset between the CdS buffer and the ZnO window layer was determined by Ruckh et al. to be 0.4 eV [125].

4.2 Short-Circuit Current

The short circuit current density J_{sc} that can be obtained from the standard 100 mW cm^{-2} solar spectrum (AM1.5) is determined, on the one hand, by *optical losses*, that is, by the fact that photons from a part of the spectrum are either not absorbed in the solar cell or are absorbed *without* generation of electron-hole pairs. On the other hand, not all photogenerated electron-hole pairs contribute to J_{sc} because they recombine before they are collected. We denote these latter limitations as *recombination losses*. Figure 11 illustrates how much from an incoming photon flux from the terrestrial solar spectrum contributes to the final J_{sc} of a highly efficient $Cu(In,Ga)Se_2$ solar cell [126] and where the remainder gets lost. The incoming light, i.e., that part of the solar spectrum with photon energy $h\nu$ larger than the band gap energy $E_g = 1.155$ eV of the specific absorber would correspond to a (maximum possible) J_{sc} of 41.7 mA cm^{-2}. By reflection at the surface and at the three interfaces between the MgF_2 anti-reflective coating, the ZnO window, the CdS buffer, and the $Cu(In,Ga)Se_2$ absorber layer we loose already 0.6 mA cm^{-2}. A further reflection loss of 0.1 mA cm^{-2} is due to those low-energy photons that are reflected at the metallic back and leave the solar cell after having traversed the absorber twice. Due to the high absorption coefficient of $Cu(In,Ga)Se_2$ for $h\nu > E_g$, this portion is very small. More important are parasitic absorption losses by free carrier absorption in the highly doped part of the ZnO window layer (0.9 mA cm^{-2}) and at the absorber/Mo interface (0.5 mA cm^{-2}). Thus, the sum of all *optical losses* amounts to 2.1 mA cm^{-2}. Note that in solar cells and modules there are additional optical losses due to the grid shadowing or interconnect areas, respectively. These losses are not discussed here.

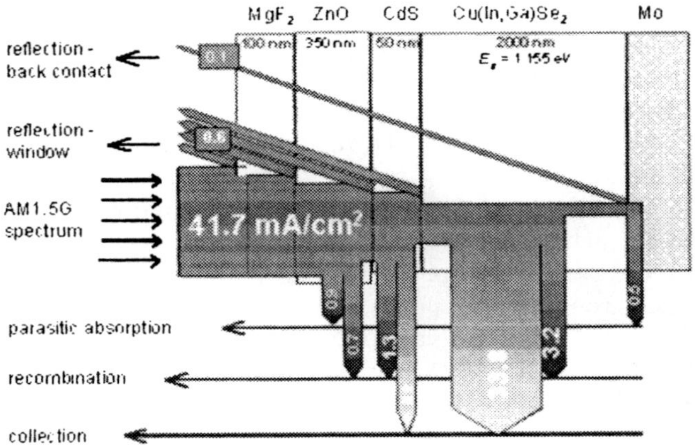

Figure 11 Optical and electronic losses of the short circuit current density J_{sc} of a high-efficiency ZnO/CdS/Cu(In,Ga)Se$_2$ heterojunction solar cell. The incident current density of 41.7 mA/cm^2 corresponds to the range of the AM 1.5 solar spectrum that has a photon energy larger than the band gap energy E_g = 1.155 eV of the Cu(In,Ga)Se$_2$ absorber. Optical losses consist of reflection losses at the ambient/window, at the window/buffer, the buffer/absorber, and at the absorber/back contact interface as well as of parasitic absorption in the ZnO window layer (free carrier absorption) and at the Mo back contact. Electronic losses are recombination losses in the window, buffer, and in the absorber layer. The finally measured J_{sc} of 34.6 mA/cm^2 of the cell stems almost exclusively from the Cu(In,Ga)Se$_2$ absorber and only to a small extend from the CdS buffer layer.

Next, we have to consider that electron hole pairs photogenerated in the ZnO window layer are not separated. Therefore, this loss affecting photon energies $h\nu > E_g$(ZnO) = 3.2 eV contributes to the *recombination losses.* As shown in Figure 11, this loss of high-energy photons costs about 0.7 mA cm^{-2}. Another portion of the solar light is absorbed in the CdS buffer layer (E_g (CdS) $\approx$ 2.4 eV). However, a part of the photons in the energy range 3.2 eV $> h\nu >$ 2.4 eV contributes to J_{sc} because the thin CdS layer does not absorb all those photons and a part of the electron-hole pairs created in the buffer layer still contributes to the photocurrent [127]. In the present example 1.3 mA cm^{-2}, get lost by recombination and 0.8 mA cm^{-2} are collected. However, the major part of J_{sc} (33.8 mA cm^{-2}) stems from electron hole pairs photogenerated in the absorber. Finally, the collection losses in the absorber amount to 3.2 mA cm^{-2}.

The above analysis shows that, accepting the restrictions that are given by the window and the buffer layer, this type of solar cell makes extremely good use of the solar spectrum. There is however still some scope for improving J_{sc} by optimising carrier collection in the absorber and/or by replacing the CdS buffer layer by an alternative material with a higher E_g.

4.3 Open-Circuit Voltage

At open circuit no current flows across the device, and all photogenerated charge carriers have to recombine within the solar cell. The possible recombination paths for the photogenerated charge carriers in the Cu(In,Ga)Se$_2$ absorber are indicated in the band diagram of Figure 12. Here we have considered recombination at the back surface of the absorber (A′) and in the neutral bulk (A), recombination in the space-charge region (B), and recombination at the buffer/absorber interface (C). Note that due to the presence of high electrical fields in the junction region, the latter two mechanism may be enhanced by tunnelling.

The *basic* equations for the recombination processes (A–C) can be found, for example, in [128]. Notably, all recombination current densities J_R can be written in the form of a diode law

$$J_R = J_O\left\{\exp\left(\frac{qV}{n_{id}k_BT}\right) - 1\right\} \tag{1}$$

where V is the applied voltage, n_{id} the diode ideality factor, and k_BT/q the thermal voltage. The saturation current density J_O in general is a thermally activated quantity and may be written in the form

$$J_O = J_{OO}\exp\left(\frac{-E_a}{n_{id}k_BT}\right) \tag{2}$$

where E_a is the activation energy and the prefactor J_{OO} is only weakly temperature dependent.

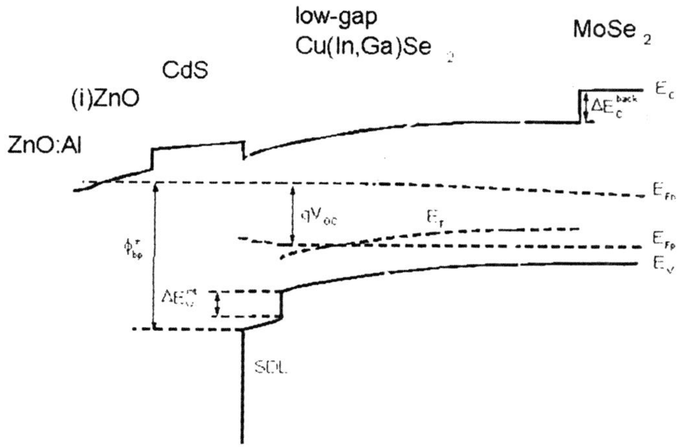

Figure 12 Recombination paths in a ZnO/CdS/(low-gap) Cu(In,Ga)Se2 junction at open circuit. The paths A represent recombination in the neutral volume, A′ recombination at the back contact, B recombination in the space-charge region, and C recombination at the interface between the Cu(In,Ga)Se2 absorber and the CdS buffer layer. Back contact recombination is reduced by the conduction band offset ΔE_c^{back} between the Cu(In,Ga)Se$_2$ absorber and the MoSe2 layer that forms during absorber preparation on top of the metallic Mo back contact. Interface recombination (C) is reduced by the internal valence band offset ΔE_V^{int} between the bulk of the Cu(In,Ga)Se2 absorber and the Cu-poor surface layer. The quantity ϕ^*_{bp} denotes the energy barrier at the CdS/absorber interface and E_T indicates the the energy of a recombination centre in the bulk of the Cu(In,Ga)Se$_2$.

The activation energy E_a for recombination at the back surface, in the neutral zone, and in the space charge recombination is the absorber band gap energy E_g, whereas in case of interface recombination E_a equals the barrier ϕ_{bp} that hinders the holes from the absorber to come to the buffer/absorber interface (cf. Figure 12). In the simplest cases, the diode ideality factor n_{id} is unity for back surface and neutral zone recombination as well as for recombination at the buffer/absorber interface, whereas for space charge recombination $n_{id} = 2$. Equalising the short circuit current density J_{sc} and the recombination current density J_R in Equation (1) for the open circuit voltage situation (i.e., $V = V_{oc}$ in Equation (1)) we obtain with the help of Equation (2)

$$V_{oc} = \frac{E_a}{q} - \frac{n_{id} k_B T}{q} \ln\left(\frac{J_{00}}{J_{sc}}\right) \tag{3}$$

Note that Equation (3) yields the open circuit voltage in a situation where there is single mechanism clearly dominating the recombination in the specific device. Note that Equation (3) is often used for the analysis of the dominant recombination path. After measuring V_{oc} at various temperatures T, the extrapolation of the experimental $V_{oc}(T)$ curve to $T = 0$ yields the activation energy of the dominant recombination process, e.g. $V_{oc}(0) = E_g/q$ in case of bulk recombination or $V_{oc}(0) = \phi_{bp}/q$ in case of interface recombination.

We emphasise that, in practice, measured ideality factors considerably deviate from that textbook scheme and require more refined theories (see e.g. [14] and references therein). However, for a basic understanding of the recombination losses in thin-film solar cells, a restriction to those textbook examples is sufficient.

The band diagram in Figure 12 contains two important features that appear important for the minimisation of recombination losses in $Cu(In,Ga)Se_2$ solar cells. The first one is the presence of a considerably widened band gap at the surface of the $Cu(In,Ga)Se_2$ absorber film. This surface band gap widening implies a valence band offset ΔE_v^{int} at the internal interface between the absorber and the surface defect layer (discussed in Section 4.1). This internal offset directly increases the recombination barrier ϕ_{bp} to an effective value

$$\phi_{bp}^* = \phi_{bp} + \Delta E_v^{int} \tag{4}$$

Substituting Equation (4) in Equation (3) we obtain

$$V_{oc} = \frac{\phi_{bp}}{q} + \frac{\Delta E_v^{int}}{q} - \frac{n_{id} k_B T}{q} \ln\left(\frac{J_{00}}{J_{sc}}\right) \tag{5}$$

It is seen that the internal band offset directly adds to the open circuit voltage that is achievable in situations where only interface recombination is present. However, the open circuit voltage of most devices that have this surface band gap widening (those having an absorber with a final Cu-poor composition) is then limited by bulk recombination.

The second important feature in Figure 12 is the conduction band offset ΔE_c^{back} between the Cu(In,Ga)Se_2 absorber and the thin $MoSe_2$ film. The back surface recombination velocity at the metallic Mo back contact is reduced from the value S_b (without the $MoSe_2$) to a value $S_b^* = S_b \exp(-\Delta E_c/k_B T)$.

4.4 Fill Factor

The fill factor *FF* of a solar cell can be expressed in a simple way as long as the solar cell is well described by a diode law. Green [129] gives the following phenomenological expression for the fill factor

$$FF_0 = \frac{v_{oc} - \ln(v_{oc} + 0.72)}{v_{oc} + 1} \tag{6}$$

It is seen that, through the dimensionless quantity $v_{oc} = qV_{oc}/n_{id}k_BT$, the fill factor depends on the temperature T as well as on the ideality factor n_{id} of the diode (see also Chapter IIa-1). The fill factor FF_0 results solely from the diode law form of Equation (1). In addition, effects from series resistance R_s and shunt resistance R_p add to the fill factor losses. A good approximation is then given by

$$FF = FF_0(1 - r_S)\left[1 - \frac{(v_{oc} + 0.7)}{v_{oc}}\frac{FF_0(1 - r_S)}{r_P}\right] \tag{7}$$

with the normalised series and parallel resistances given by $r_s = R_sJ_{sc}/V_{oc}$ and $r_p = R_PJ_{sc}/V_{oc}$, respectively. The description of Cu(In,Ga)Se_2 solar cells in terms of Equations (6) and (7) works reasonably well, e.g., the world record cell [1] has a fill factor of 78.6% and the value calculated from Equations (6) and (7) is 78.0% (V_{oc} = 678 mV, J_{sc} = 35.2 mA cm^{-2}, R_s = 0.2 Ω cm^2, R_p = 10^4 Ω cm^2, n_{id} = 1.5).

Factors that can further affect the fill factor are (i) the voltage bias dependence of current collection [130], leading to a dependence of $J_{sc}(V)$ on voltage V in Equation (1), (ii) recombination properties that are spatially inhomogeneous [131], or (iii) unfavourable band offset conditions at the heterointerface [132].

4.5 Electronic Metastabilities

The long time increase (measured in hours and days) of the open circuit voltage V_{OC} of Cu(In,Ga)Se_2 based solar cells during illumination is commonly observed phenomenon [133, 134]. In some cases it is not only V_{oc} but also the fill factor that improves during such a light soaking procedure. Consequently, light soaking treatments are systematically used to re-establish the cell efficiency after thermal treatments [107, 135].

For the present day ZnO/CdS/Cu(In,Ga)Se_2 heterojunctions it appears established that there exist at least three types of metastablities with completely different fingerprints [136, 137].

Type I: A continuous increase of the open circuit voltage during illumination and the simultaneous increase of the junction capacitance. Both phenomena are satisfactorily explained as a consequence of persistent photoconductivity in the $Cu(In,Ga)Se_2$ absorber material [138, 139], i.e., the persistent capture of photogenerated electrons in traps that exhibit large lattice relaxations like the well-known DX centre in (Al,Ga)As [140]. This type of metastability affects exclusively the open circuit voltage and can vary from few mV up to 50 mV, especially if the sample has been stored in the dark at elevated temperatures.

Type II: A decrease of the fill factor after the cell has been exposed to reverse voltage bias. In extreme cases, this type of metastability leads to a hysteresis in the IV characteristics, e.g., the fill factor of an illuminated IV curve becomes dependent on whether the characteristic has been measured from negative voltages towards positive ones or vice versa. The application of reverse bias also leads to a metastable increase of the junction capacitance and to significant changes of space charge profiles as determined from capacitance vs. voltage measurements [136, 137]. The type II metastability is especially important for devices with non-standard buffer/window combinations (e.g., Cd-free buffer layers) [141, 142].

Type III: An increase of the fill factor upon illumination with light that is absorbed in the buffer layer or in the extreme surface region of the $Cu(In,Ga)Se_2$ absorber, i.e., the blue part of the solar spectrum. This type of instability counterbalances, to a certain extend, the consequences of reverse bias, i.e., it restores the value of the fill factor after it has been degraded by application of reverse bias.

Our overall understanding of metastabilities, especially of type II and III, is still incomplete. Fortunately, all metastabilities observed in $Cu(In,Ga)Se_2$ so far, tend to improve the photovoltaic properties as soon as the device is brought under operating conditions.

5 Wide-Gap Chalcopyrites

5.1 Basics

The alloy system $Cu(In,Ga,Al)(S,Se)_2$ provides the possibility of building alloys with a wide range of band-gap energies E_g between 1.04 eV for $CuInSe_2$ up to 3.45 eV for $CuAlS_2$ (cf. Figure 2). The highest efficiency within the chalcopyrite system is achieved with the relatively low-band-gap energy E_g of 1.12 eV, and attempts to maintain this high efficiency level at $E_g > 1.3$ eV have so far failed (for a recent review, see [143]). Practical approaches to wide-gap Cu-chalcopyrites comprise (i) alloying of $CuInSe_2$ with Ga up to pure $CuGaSe_2$ with $E_g = 1.68$ eV, (ii) $Cu(In,Al)Se_2$ alloys with solar cells realised up to an Al/(Al + In) ratio of 0.6 and $E_g \approx 1.8$ eV [144], (iii) $CuIn(Se,S)_2$ [145] and $Cu(In,Ga)(Se,S)_2$ [146] alloys with a S/(Se + S) ratio up to 0.5, and (iv) $CuInS_2$ [147] and $Cu(In,Ga)S_2$ [148] alloys. Note that the approaches (i)–(iii) are realised with a final film composition that is slightly Cu-poor, whereas approach (iv) uses films that are Cu-rich. In the latter case, CuS segregates preferably at the film surface. This secondary phase has to be removed prior to heterojunction formation.

The advantage of higher voltages of the individual cells by increasing the band gap of the absorber material is important for thin-film modules. An ideal range for the band gap energy would be between 1.4 and 1.6 eV because the increased open circuit voltage and the reduced short circuit current density would reduce the number of necessary scribes used for monolithic integration of the cells into a module. Also, the thickness of front and back electrodes can be reduced because of the reduced current density.

Table 3 compares the solar cell output parameters of the best chalcopyrite-based solar cells. This compilation clearly shows the superiority of $Cu(In,Ga)Se_2$ with a relatively low Ga content which leads to the actual world champion device. The fact that the best $CuInSe_2$ device has an efficiency of 3 % below that of the best $Cu(In,Ga)Se_2$ device is due not only to the less favourable band-gap energy but also to the lack of the beneficial effect of small amounts of Ga on the growth and on the electronic quality of the thin film, as discussed above.

The difficulty of obtaining wide-gap devices with high efficiencies is also illustrated by plotting the absorber band gap energies of a series of chalcopyrite alloys vs. the attained open-circuit voltages. Figure 13 shows that below $E_g = 1.3$ eV, the V_{OC} data follow a straight line, indicating the proportional gain of $V_{oc} = E_g/q - 0.5$V, whereas at$E_g > 1.3$ eV the gain is much more moderate. At the high band-gap end of the scale the differences of the band-gap energies and the open-circuit voltages of $CuInS_2$ and $CuGaSe_2$ amount to 840 mV and 820 mV, respectively, whereas $E_g - qV_{oc}$ is only 434 eV in the record $Cu(In,Ga)Se_2$ device.

One reason for these large differences $E_g - qV_{oc}$ in wide-gap devices is the less favourable band offset constellation at the absorber/CdS-buffer interface. Figure 14 shows the band diagram of a wide-gap Cu chalcopyrite-based heterojunction with (Figure 14(a)) and without (Figure 14(b)) the surface defect layer. As the increase of band gap in going from $CuInSe_2$ to $CuGaSe_2$ takes place almost exclusively by an increase of the conduction band energy E_c, the positive or zero band offset ΔE_c^{ab} between the absorber and the buffer of a low-gap device (cf.

Table 3 Absorber band-gap energy E_g, efficiency η, open-circuit voltage V_{oc}, short-circuit current density J_{sc}, fill factor *FF*, and area *A* of the best $Cu(In,Ga)Se_2$, $CuInSe_2$, $CuGaSe_2$, $Cu(In,Al)Se_2$, $CuInS_2$, $Cu(In,Ga)S_2$, and $Cu(In,Ga)(S,Se)_2$ solar cells

Material	E_g (eV)	η (%)	V_{oc} (mV)	J_{sc} (mA cm^{-2})	*FF* (%)	*A* (cm^2)	Ref.
$Cu(In,Ga)Se_2$	1.12	18.8	678	35.2	78.6	0.45	[1][a]
$CuInSe_2$	1.04	15.4	515	41.2	72.6	0.38	[40][b]
$CuGaSe_2$	1.68	8.3	861	14.2	67.9	0.47	[151][a]
$Cu(In,Al)Se_2$	1.16	16.6	621	36.0	75.5	?	[157][a]
$Cu(In,Ga)(S,Se)_2$	1.36	13.9	775	24.3	74.0	0.5	[146][b]
$CuInS_2$	1.5	11.4	729	21.8	72.0	0.5	[159][a]
$Cu(In,Ga)S_2$	1.53	12.3	774	21.6	73.7	0.5	[148][a]

[a] Confirmed total area values.

[b] Effective area values (not confirmed).

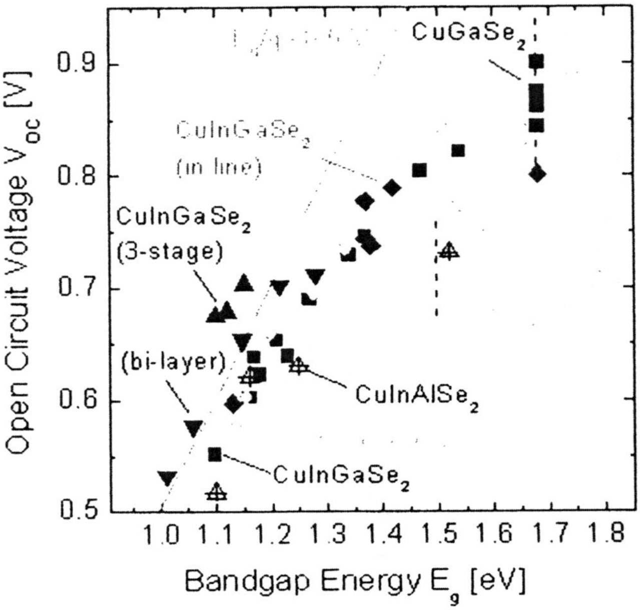

Figure 13 Open-circuit voltages of different Cu-chalcopyrite based solar cells with various band-gap energies of the absorber layers. Full symbols correspond to Cu(In,Ga)Se_2 alloys prepared by a simple single layer process (squares), a bi-layer process (triangles down), and the three-stage process (triangles up). Cu(In,Ga)Se_2 cells derived from an in-line process as sketched in Fig. 9 are denoted by diamonds. Open triangles relate to Cu(In,Ga)S_2, open circles to Cu(In,Ga)(S,Se)$_2$, and the crossed triangles to Cu(In,Al)Se_2 cells.

Figure 12) turns into a negative one in Figure 14. This effect should be weaker for $CuInS_2$ and $CuAlSe_2$ as the increase of E_g in these cases is due to an upwards shift of E_c *and* a downwards shift of E_v. However, *any* increase of E_c implies that the barrier ϕ_{bp} that hinders the holes from the absorber from recombination at the heterointerface does not increase proportionally with the increase of the band-gap energy. Thus in wide-gap absorbers, the importance of interface recombination (determined by the barrier ϕ_{bp}) grows considerably relative to that of bulk recombination (determined by E_g of the absorber) [149]. Using the same arguments with respect to the $MoSe_2$/absorber, the back-surface field provided by this type of heterojunction back contact, as shown in Figure 12, vanishes when the conduction-band energy of the absorber is increased.

Up to now, all arguments for explaining the relatively low performance of wide-gap chalcopyrite alloys deal with the changes in the band diagram. However, recent work focuses on the differences in the electronic quality of the absorber materials. Here, it was found that the concentration of recombination active defects increases when increasing the Ga content in Cu(In,Ga)Se_2 above a Ga/(Ga + In) ratio of about 0.3 [32] and the S content in Cu(In,Ga)(Se,S)$_2$ alloys over a S/(Se+S) ratio of 0.3 [34]. Moreover, a recent systematic investigation of

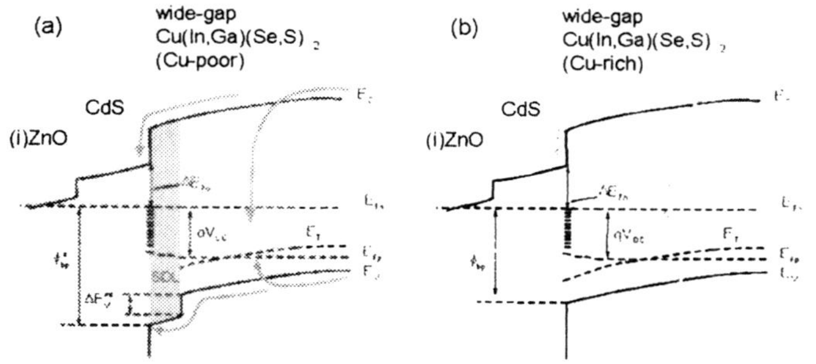

Figure 14: Energy band diagram of a ZnO/CdS/(wide-gap) Cu(In,Ga)(Se,S)2 heterojunction. The band diagram (a) that includes the surface defect layer (SDL) of a Cu-poor prepared film shows that the interface recombination barrier $\phi^*_{bp} = \phi_{bp} + \Delta E^{int}_V$ *is larger than the barrier* ϕ_{bp} *in the device that was prepared Cu-rich (b). The difference is the internal valence band offset* ΔE^{int}_V *between the SDL and the bulk of the absorber. The larger value of* ϕ^*_{bp} *reduces interface recombination.*

the dominant recombination mechanism in large series of Cu(In,Ga)$(Se,S)_2$ based solar cells with different compositions [150] suggests that bulk recombination prevails in all devices that were prepared with a Cu-poor final film composition. Only devices that had a Cu-rich composition (before removing Cu-rich secondary phases) showed signatures of interface recombination. The absence of interface recombination in Cu-poor devices could be a result of the presence of the Cu-poor surface defect layer in these devices. Comparison of Figures 14(a) and (b) illustrates that interface recombination is much more likely to dominate those devices that lack this feature.

5.2 $CuGaSe_2$

$CuGaSe_2$ has a band gap of 1.68 eV and therefore would represent an ideal partner for $CuInSe_2$ in an all-chalcopyrite tandem structure. However, a reasonable efficiency for the top cell of any tandem structure is about 15%, far higher than has been reached by the present polycrystalline $CuGaSe_2$ technology. The record efficiency of $CuGaSe_2$ thin-film solar cells is only 8.3% (9.3% active area) [151] despite of the fact that the electronic properties of $CuGaSe_2$ are not so far from those of $CuInSe_2$. However, in detail, all the differences quantitatively point in a less favourable direction. In general, the net doping density N_A in $CuGaSe_2$ appears too high [152]. Together with the charge of deeper defects, the high doping density leads to an electrical field in the space-charge region, which enhances recombination by tunnelling [153]. The high defect density in $CuGaSe_2$ thin films absorbers additionally leads to a low diffusion length and, in consequence, to a dependence of the collected short circuit current density J_{sc} on the bias voltage V. Because of the decreasing width of the space charge region, $J_{sc}(V)$ decreases with increasing V affecting significantly the fill factor of the solar cell [130]. Note that, on top of the limitation by the unfavourable bulk properties, interface recombination also plays a certain role in $CuGaSe_2$ [154] because of the unfavourable band diagram shown in Figure 14. Therefore, substantial improvement of the performance requires simultaneous optimisation of bulk *and* interface properties. Notably, $CuGaSe_2$ is the only Cu-chalcopyrite material where the record efficiency of cells based on bulk crystals (with η = 9.4%, total area) [155] exceeds that of thin-film devices.

5.3 Cu(In,Al)Se_2

As can be seen in Figure 2, the bandgap change within the Cu(In,Al)Se_2 alloy system is significant even if a small amount of Al is added to $CuInSe_2$ [144]. This fact allows to grow graded structures with only small changes in the lattice constant. Al–Se compounds tend to react with water vapour to form oxides and H_2Se. Furthermore, there is a tendency to phase segregations [144]. However, cells with very good performance have been achieved by a co-evaporation process of absorbers in a band gap range between 1.09 and 1.57 eV, corresponding to Al/(In + Al) ratio x between 0.09 and 0.59 [144, 156]. The highest confirmed efficiency of a Cu(In,Al)Se_2 solar cell is 16.9% [157] (see also Table 3). This device has about the same band gap energy as the record Cu(In,Ga)Se_2 device [1].

5.4 $CuInS_2$ and Cu(In,Ga)S_2

The major difference between $CuInS_2$ and Cu(In,Ga)Se_2 is that the former cannot be prepared with an overall Cu-poor composition. Cu-poor $CuInS_2$ displays an extremely low conductivity, making it almost unusable as a photovoltaic absorber material [145]. Even at small deviations from stoichiometry on the In-rich side, segregation of the spinel phase is observed [158]. Instead, the material of choice is Cu-rich $CuInS_2$. As in the case of $CuInSe_2$, a Cu-rich preparation route implies the removal of the unavoidable secondary Cu–S binary phase by etching the absorber in KCN solution [147]. Such an etch may involve some damage of the absorber surface as well as the introduction of shunt paths between the front and the back electrode. However, as shown in Table 3, the best $CuInS_2$ device [159] has an efficiency above 11%. This record efficiency for $CuInS_2$ is achieved by a sulphurisation process rather than by co-evaporation.

As we have discussed above, interface recombination dominates the open circuit voltage V_{oc} of Cu(In,Ga)(Se,S)$_2$ devices that are prepared with a Cu-rich absorber composition (prior to the KCN etch), like the $CuInS_2$ and Cu(In,Ga)S_2 devices discussed here. It was found recently, that alloying $CuInSe_2$ with moderate amounts of Ga enhances the open circuit voltage V_{oc} [160]. This increase of V_{oc} can counterbalance the loss of short circuit current density J_{sc} resulting from the increased band gap. Apparently, addition of Ga to $CuInS_2$ reduces interface recombination by increasing the interfacial barrier ϕ_{bp}(cf. Figure 14(b)) [150]. However, for Cu(In,Ga)S_2 devices prepared by the sulphurisation route, the benefit of Ga alloying is limited, because, during preparation, most of the Ga added to the precursor ends up confined to the rear part of the absorber layer and, therefore, remains ineffective at the absorber surface [160]. In contrast, when preparing $Cu(In,Ga)S_2$ with co-evaporation, Ga is homogeneously distributed through the depth of the absorber and the Ga content at the film surface is well controlled. Recent work of Kaigawa et al. [148] represents a major progress in wide-gap Cu-chalcopyrites with the preparation of $Cu(In,Ga)S_2$ solar cells with a confirmed efficiency of 12.3% at a band gap energy $E_g = 1.53$ eV. In the same publication [148] an efficiency of 10.1% is reported for a device with $E_g = 1.65$ eV. The open circuit voltage V_{oc} of this device is 831 mV, i.e., only slightly lower than V_{oc} of the best $CuGaSe_2$ cell having an efficiency of 8.3%.

5.5 $Cu(In,Ga)(Se,S)_2$

One possible way of overcoming the disadvantages of the ternary wide-gap materials $CuInS_2$ and $CuGaSe_2$ is to use the full pentenary alloy system $Cu(In_{1-x}Ga_x)(Se_{1-y}S_y)_2$ [146]. Among the materials listed in Table 3, the pentenary system is the only one with an open-circuit voltage larger than 750 mV and an efficiency above 13%, outperforming $CuInS_2$ in both these respects. The advantage of $Cu(In,Ga)(Se,S)_2$ could be due to the mutual compensation of the drawbacks of $CuGaSe_2$ (too high charge density) and that of (Cu-poor) $CuInS_2$ (too low conductivity, if prepared with a Cu-poor film composition).

5.6 Graded-Gap Devices

An interesting property of the $Cu(In,Ga,Al)(S,Se)_2$ alloy system is the possibility of designing graded-gap structures in order to optimise the electronic properties of the final device [161–164]. Such band-gap gradings are achieved during co-evaporation by the control of the elemental sources, but selenisation/sulphurisation processes also lead to beneficial compositional gradings. The art of designing optimum band-gap gradings is to push back charge carriers from critical regions, i.e., regions with high recombination probability within the device. Such critical regions are (i) the interface between the back contact and the absorber layer and (ii) the heterojunction interface between the absorber and the buffer material. Figure 15 shows a band diagram of a graded structure that fulfils the requirements for minimising recombination losses.

(i) To keep the back contact region clear from electrons, one can use a Ga/In grading. The increase of the Ga/(Ga + In) ratio causes a movement of the conduction band minimum upward with respect to its position in a non-graded $CuInSe_2$ device. Thus, back surface grading leads to a gradual increase of the conduction-band energy, as illustrated in Figure 15, and therefore drives photogenerated electrons away from the metallic back contact into the direction of the buffer/absorber junction.
(ii) The minimisation of junction recombination, both at the point of equal capture rates of holes and electrons in the space charge region (recombination path B in Figure 15) as well as at the absorber/buffer interface (path C), requires an increased band gap towards the absorber

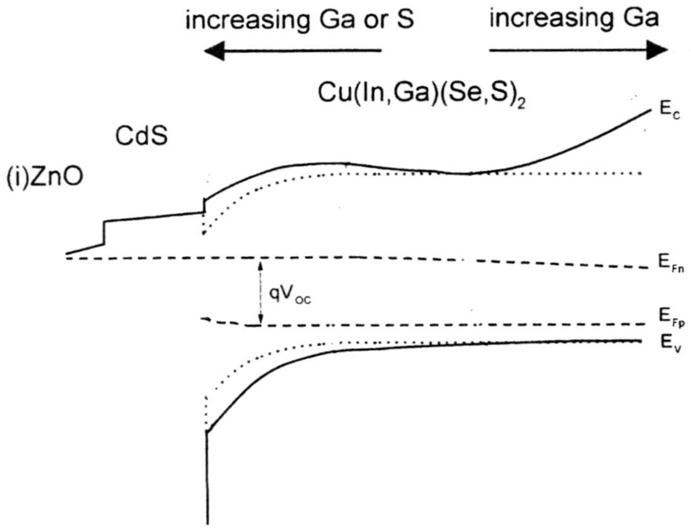

Figure 15 Band diagram of a ZnO/CdS/Cu(In,Ga)(Se,S)$_2$ heterojunction with a graded-gap absorber. The minimum band gap energy is in the quasi neutral part of the absorber. An increasing Ga/In ratio towards the back surface and an increasing Ga/In or S/Se-ratio towards the front minimise recombination in critical regions at the back contact (recombination path A'), in the space charge region (path B), and at the heterointerface (path C). The dotted lines correspond to the conduction and valence band edge energies of a non-graded device.

surface. If one had the choice, one would clearly favour a decrease of the valence-band energy, as shown in Figure 15, over an increase of the conduction band energy. This favours a grading with the help of S/Se alloying, as at least a part of the increasing band-gap energy is accommodated by a decrease of the energy of the valence-band edge that should minimise interface recombination.

6 Conclusions

The objective of this chapter is the description of the recent achievements on Cu(In,Ga)Se$_2$-based solar cells as well as an account of our present understanding of the physical properties of the materials involved and the electronic behaviour of the devices. Cu(In,Ga)Se$_2$ is in a leading position among polycrystalline thin-film solar cell materials because of the benign, forgiving nature of the bulk materials and interfaces. Nevertheless, we want to guide the attention of the reader also in the direction of the work that has still to be done.

Three of the four cornerstones 1–4 for the recent achievements mentioned in Section 3.2 concern the *growth of the films*: the optimised deposition conditions, and the incorporation of Na and Ga. However, no detailed model is available to describe the growth of Cu(In,Ga)Se$_2$, and especially the impact of Na which, in our opinion, is the most important of the different ingredients available to tune the electronic properties of the absorber. A clearer understanding of Cu(In,Ga)Se$_2$ growth would allow us to find optimised conditions in the wide parameter space available, and thus to reduce the number of recombination centres and compensating donors and optimise the number of shallow acceptors.

The deposition of the buffer layer, or more generally speaking, the formation of the heterojunction, is another critical issue. The surface chemistry taking place during heterojunction formation, and also during post-deposition treatments, is decisive for the final device performance. Both processes greatly affect not only the surface defects (i.e., recombination *and* charge), and therefore the charge distribution in the device, but also the defects in the bulk of the absorber. Concentrated effort and major progress in these tasks would not only allow us to push the best efficiencies further towards 20%, but would also provide a sound knowledge base for the various attempts at the commercialisation of Cu(In,Ga)Se$_2$ solar cells.

Acknowledgements

The authors thank all our colleagues at the IPE for various discussions and fruitful collaboration. We are especially grateful to J. H. Werner for his continuous support of the Cu(In,Ga)Se$_2$ research at the *ipe*. We are grateful to M. Turcu for a critical reading of the manuscript. Support of our work during many years by the German Ministry of Research, Technology, and Education (BMBF), by the Ministry of Economics, and by the European Commission is especially acknowledged.

References

[1] Contreras, M., Egaas, B., Ramanathan, K., Hiltner, J., Swartzlander, A., Hasoon, F. and Noufi, R., 1999. Progress toward 20% efficiency in Cu(In,Ga)Se$_2$ polycrystalline thin-film solar cells. *Prog. Photovolt. Res. Appl.* Vol. 7, p. 311.

[2] Kessler, J., Bodegard, M., Hedström, J. and Stolt, L., 2000. New world record Cu(In,Ga)Se$_2$ based mini-module: 16.6%. *Proc.16th European Photovoltaic Solar Energy Conf.*, Glasgow, p. 2057.

[3] Hahn, H., Frank, G., Klingler, W., Meyer, A. and Störger, G., 1953. über einige ternäre Chalkogenide mit Chalkopyritstruktur. *Z. Anorg. u. Allg. Chemie*, Vol. 271, p.153.

[4] Wagner, S., Shay, J.L., Migliorato, P. and Kasper, H.M., 1974. CuInSe$_2$/CdS heterojunction photovoltaic detectors. *Appl. Phys. Lett.* Vol. 25, p. 434.

[5] Mickelsen, R.A. and Chen, W.S., 1980. High photocurrent polycrystalline thin-film CdS/CuInSe$_2$ solar cell. *Appl. Phys. Lett.* Vol. 36, p. 371.

[6] Mitchell, K.C., Ermer, J. and Pier, D., 1988. Single and tandem junction $CuInSe_2$ cell and module technology. *Proc. 20th IEEE Photovoltaic Specialists Conf.*, Las Vegas, p. 1384.

[7] Wieting, R.D., 2002. CIS Manufacturing at the Megawatt Scale. *Proc. 29th IEEE Photovoltaic Specialists Conf.*, New Orleans, p. 480.

[8] Probst, V., Stetter, W., Riedl, W., Vogt, H., Wendl, M., Calwer, H., Zweigart, S., Ufert, K.D., Freienstein, B., Cerva, H. and Karg, F.H., 2001. Rapid CIS-process for high efficiency PV-modules: development towards large area processing. *Thin Solid Films*, Vol. 387, p. 262.

[9] Dimmler, B., Powalla, M. and Schock, H.W., 2002. CIS-based thin-film photovoltaic modules: potential and prospects. *Prog. Photovolt. Res. Appl.* Vol. 10, p.149.

[10] Powalla, M., Lotter, E., Waechter, R., Spiering, S., Oertel, M. and Dimmler, B., 2002. Pilot Line Production of CIGS Modules: First Experience in Processing and Further Developments. *Proc. 29th IEEE Photovoltaic Specialists Conf.*, New Orleans, p. 571.

[11] Kushiya, K., 2001. Improvement of electrical yield in the fabrication of CIGS-based thin-film modules. *Thin Solid Films*, Vol. 387, p. 257.

[12] Negami, T., Satoh, T., Hashimoto, Y., Nishiwaki, S., Shimakawa, S. and Hayashi, S., 2001. Large-area CIGS absorbers prepared by physical vapor deposition. *Solar Energy Mat. Solar Cells*, Vol. 67, p. 1.

[13] Rau, U. and Schock, H.W., 1999. Electronic properties of $Cu(In,Ga)Se_2$ heterojunction solar cells- recent achievements, current understanding, and future challenges. *Appl. Phys. A*, Vol. 69, p. 191.

[14] Rau, U. and Schock, H.W., 2001. $Cu(In,Ga)Se_2$ Solar Cells. In: Archer, M.D. and Hill, R., Eds., *Clean Electricity from Photovoltaics*, Imperial College Press, London, p. 270.

[15] Shay, J.L. and Wernick, J.H., 1975. *Ternary Chalcopyrite Semiconductors: Growth, Electronic Properties, and Applications*, Pergamon Press, Oxford.

[16] Kazmerski, L.L. and Wagner, S., 1985. Cu-ternary chalcopyrite solar cells. In: Coutts, T.J. and Meakin, J.D., Eds., *Current Topics in Photovoltaics*, Academic Press, Orlando, p. 41.

[17] Coutts, T.J., Kazmerski, L.L. and Wagner, S., 1986. *Ternary Chalcopyrite Semiconductors: Growth, Electronic Properties, and Applications*, Elsevier, Amsterdam.

[18] Rockett, A. and Birkmire, R.W., 1991. $CuInSe_2$ for photovoltaic applications. *J. Appl. Phys.* Vol. 70, p. 81.

[19] Stanbery, B.J., 2002. Copper indium selenides and related materials for photovoltaic devices. *Crit. Rev. Solid State*, Vol. 27, p. 73.

[20] Haalboom, T., Gödecke, T., Ernst, F., Rühle, M., Herberholz, R., Schock, H.W., Beilharz, C. and Benz, K.W., 1997. Phase relations and microstructure in bulk materials and thin films of the ternary system Cu-In-Se. *Inst. Phys. Conf. Ser.*, Vol. 152E, p. 249.

[21] Herberholz, R., Rau, U., Schock, H.W., Haalboom, T., Gödecke, T., Ernst, F., Beilharz, C., Benz, K.W., and Cahen, D., 1999. Phase segregation, Cu

migration and junction formation in $Cu(In,Ga)Se_2$. *Eur. Phys. J. Appl. Phys.*, Vol. 6, p. 131.

[22] Migliorato, P., Shay, J.L., Kasper, H.M. and Wagner, S., 1975. Analysis of the electrical and luminescent properties of $CuInSe_2$. *J. Appl. Phys.* Vol. 46, p. 1777.

[23] Noufi, R., Axton, R., Herrington, C. and Deb, S.K., 1984. Electronic properties versus composition of thin films of $CuInSe_2$. *Appl. Phys. Lett.*, Vol. 45, p. 668.

[24] Zhang, S.B., Wei, S.H., Zunger, A. and Katayama-Yoshida, H., 1998. Defect physics of the $CuInSe_2$ chalcopyrite semiconductor. *Phys. Rev B*, Vol. 57, p. 9642.

[25] Neumann, H., 1983. Vacancy formation in $A^IB^{III}C_2{}^{VI}$ chalcopyrite semiconductors. *Cryst. Res. Technol.* Vol. 18, p. 901.

[26] Abou-Elfotouh, F.A., Moutinho, H., Bakry, A., Coutts, T.J. and Kazmerski, L.L., 1991. Characterization of the defect levels in copper indium diselenide. *Solar Cells*, Vol. 30, p. 151.

[27] Igalson, M. and Schock, H.W., 1996. The metastable changes of the trap spectra of $CuInSe_2$-based photovoltaic devices. *J. Appl. Phys.*, Vol. 80, p. 5765.

[28] Schmitt, M., Rau, U. and Parisi, J., 1995. Investigation of deep trap levels in $CuInSe_2$ solar cells by temperature dependent admittance measurements. *Proc. 13th European Photovoltaic Solar Energy Conf.*, Nice, p. 1969.

[29] Walter, T., Herberholz, R., Müller, C. and Schock, H.W., 1996. Determination of defect distributions from admittance measurements and application to $Cu(In,Ga)Se_2$ based heterojunctions. *J. Appl. Phys.*, Vol. 80, p. 4411.

[30] Igalson, M., Bacewicz, R. and Schock, H.W., 1995. 'Dangling bonds' in CuInSe2 and related compounds. *Proc. 13th. European Photovoltaic Solar Energy Conf.*, Nice, p. 2076.

[31] Herberholz, R., Nadenau, V., Rühle, U., Köble, C., Schock, H.W. and Dimmler, B., 1997. Prospects of wide-gap chalcopyrites for thin film photovoltaic modules. *Solar Energy Mater. Solar Cells*, Vol. 49, p. 227.

[32] Hanna, G., Jasenek, A., Rau, U. and Schock, H.W., 2000. Open circuit voltage limitations in $CuIn_{1-x}Ga_xSe_2$ thin- film solar cells – dependence on alloy composition. *Phys. Stat. Sol. A*, Vol. 179, p. R7.

[33] Rau, U., Schmidt, M., Jasenek, A., Hanna, G. and Schock, H.W., 2001. Electrical characterization of $Cu(In,Ga)Se_2$ thin-film solar cells and the role of defects for the device performance. *Solar Energy Mater. Solar Cells*, Vol. 67, p.137.

[34] Turcu, M., Kötschau, I.M. and Rau, U., 2002. Composition dependence of defect energies and band alignments in the $Cu(In_{1-x}Ga_x)(Se_{1-y}S_y)_2$ alloy system. *J. Appl. Phys.*, Vol. 91, p.1391.

[35] Caldas, M., Fazzio, A. and Zunger, A., 1984. A universal trend in the binding energies of deep impurities in semiconductors. *Appl. Phys. Lett.*, Vol. 45, p. 67.

[36] Langer, J.M. and Heinrich, H., 1985. Deep-level impurities: a possible guide to prediction of band-edge discontinuities in semiconductor heterojunctions. *Phys. Rev. Lett.*, Vol. 55, p.1414.

[37] Heath, J.T., Cohen, J.D., Shafarman, W.N., Liao, D.X. and Rockett, A.A., 2002. Effect of Ga content on defect states in $CuIn_{1-x}Ga_xSe_2$ photovoltaic devices. *Appl. Phys. Lett.*, Vol. 80, p. 4540.

[38] Turcu, M. and Rau, U., 2003. Compositional trends of defect energies, band alignments, and recombination mechanisms in the $Cu(In,Ga)(Se,S)_2$ alloy system. *Thin Solid Films*, Vol. 431–432, p. 158.

[39] Herberholz, R., Igalson, M. and Schock, H.W., 1998. Distinction between bulk and interface states in $CuInSe_2$/CdS/ZnO by space charge spectroscopy. *J. Appl. Phys.*, Vol. 83, p. 318.

[40] Stolt, L., Hedström, J., Kessler, J., Ruckh, M., Velthaus, K.O. and Schock, H.W., 1993. ZnO/CdS/$CuInSe_2$ thin-film solar cells with improved performance. *Appl. Phys. Lett.*, Vol. 62, p. 597.

[41] Hedström, J., Ohlsen, H., Bodegard, M., Kylner, A., Stolt, L., Hariskos, D., Ruckh, M. and Schock, H.W., 1993. ZnO/CdS/Cu(In,Ga)Se/sub 2/ thin film solar cells with improved performance. *Proc. 23rd IEEE Photovoltaic Specialists Conf.*, Lousville, p. 364.

[42] Devaney, W.E., Chen, W.S., Steward, J.M. and Mickelson, R.A., 1990. Structure and properties of high efficiency ZnO/CdZnS/$CuInGaSe_2$ solar cells. *IEEE Trans. Electron Devices*, Vol. ED-37, p. 428.

[43] Potter, R.R., Eberspacher, C. and Fabick, L.B., 1985. Device analysis of $CuInSe_2$ /(Cd,Zn)S solar cells. *Proc. 18th IEEE Photovoltaic Specialists Conf.*, Las Vegas, p. 1659.

[44] Birkmire, R.W., McCandless, B.E., Shafarman, W.N. and Varrin, R.D., 1989. Approaches for high efficiency $CuInSe_2$ solar cells. *Proc. 9th. European Photovoltaic Solar Energy Conf.*, Freiburg, p.134.

[45] Contreras, M.A., Egaas, B., Dippo, P., Webb, J., Granata, J., Ramanathan, K., Asher, S., Swartzlander, A. and Noufi, R., 1997. On the role of Na and modifications to Cu(In,Ga)Se absorber materials using thin-MF (M = Na, K, Cs) precursor layers. *Proc. 26th IEEE Photovoltaic Specialists Conf.*, Anaheim, p. 359.

[46] Holz, J., Karg, F. and Phillipsborn, H.V., 1994. The effect of substrate impurities on the electronic conductivity in CIGS thin films. *Proc. 12th. European Photovoltaic Solar Energy Conf.*, Amsterdam, p. 1592.

[47] Nakada, T., Mise, T., Kume, T. and Kunioka, A., 1998. Superstrate type $Cu(In,Ga)Se_2$ thin film solar cells with ZnO buffer layers - a novel approach to 10% efficiency. *Proc. 2nd. World Conf. on Photovoltaic Solar Energy Conversion*, Vienna, p. 413.

[48] Ruckh, M., Schmid, D., Kaiser, M., Schäffler, R., Walter, T. and Schock, H.W., 1994. Influence of substrates on the electrical properties of $Cu(In,Ga)Se_2$ thin films. *Proc. First World Conf. on Photovoltaic Solar Energy Conversion*, Hawaii, p. 156.

[49] Keyes, B.M., Hasoon, F., Dippo, P., Balcioglu, A. and Abouelfotouh, F., 1997. Influence of Na on the elctro-optical properties of Cu(In,Ga)Se_2. *Proc. 26th. IEEE Photovoltaic Specialists Conf.*, Anaheim, p. 479.
[50] Rau, U., Schmitt, M., Engelhardt, F., Seifert, O., Parisi, J., Riedl, W., Rimmasch, J. and Karg, F., 1998. Impact of Na and S incorporation on the electronic transport mechanisms of Cu(In,Ga)Se_2 solar cells. *Solid State Commun.*, Vol. 107, p. 59.
[51] Braunger, D., Hariskos, D., Bilger, G., Rau, U. and Schock, H.W., 2000. Influence of sodium on the growth of polycrystalline Cu(In,Ga)Se_2 thin films. *Thin Solid Films*, Vol. 361, p. 161.
[52] Takei, R., Tanino, H., Chichibu, S. and Nakanishi, H., 1996. Depth profiles of spatially resolved Raman spectra of a $CuInSe_2$- based thin-film solar cell. *J. Appl. Phys.*, Vol. 79, p. 2793.
[53] Wada, T., Kohara, N., Negami, T. and Nishitani, M., 1996. Chemical and structural characterization of Cu(In,Ga)Se_2/Mo interface in Cu(In,Ga)Se_2 solar cells. *Jpn. J. Appl. Phys.*, Vol. 35, p. 1253.
[54] Klenk, R., Walter, T., Schock, H.W. and Cahen, D., 1993. A model for the successful growth of polycrystalline films of $CuInSe_2$ by multisource physical vacuum evaporation. *Adv. Mat.*, Vol. 5, p. 114.
[55] Probst, V., Rimmasch, J., Stetter, W., Harms, H., Riedl, W., Holz, J. and Karg, F., 1995. Improved CIS thin film solar cells through novel impurity control techniques. *Proc. 13th European Photovoltaic Solar Energy Conf.*, Nice, p. 2123.
[56] Gabor, A.M., Tuttle, J.R., Albin, D.S., Contreras, M.A., Noufi, R. and Hermann, A.M., 1994. High-efficiency $CuIn_xGa_{1-x}Se_2$ solar cells from $(In_x,Ga_{1-x})_2Se_3$ precursors. *Appl. Phys. Lett.*, Vol. 65, p. 198.
[57] Binsma, J.J.M. and Van der Linden, H.A., 1982. Preparation of thin $CuInS_2$ films via a two-stage process. *Thin Solid Films*, Vol. 97, p. 237.
[58] Chu, T.L., Chu, S.C., Lin, S.C. and Yue, J., 1984. Large grain copper indium diselenide films. *J. Electrochem. Soc.*, Vol. 131, p. 2182.
[59] Kapur, V.K., Basol, B.M. and Tseng, E.S., 1987. Low-cost methods for the production of semiconductor films for $CuInSe_2$/CdS solar cells. *Solar Cells*, Vol. 21, p. 65.
[60] Probst, V., Karg, F., Rimmasch, J., Riedl, W., Stetter, W., Harms, H. and Eibl, O., 1996. Advanced stacked elemental layer progress for Cu(InGa)Se_2 thin film photovoltaic devices. *Mat. Res. Soc. Symp. Proc.*, Vol. 426, p. 165.
[61] Niki, S., Fons, P.J., Yamada, A., Suzuki, R., Ohdaira, T., Ishibashi, S. and Oyanagai, H., 1994. High quality $CuInSe_2$ epitaxial films – molecular beam epitaxial growth and intrinsic properties. *Inst. Phys Conf. Ser.*, Vol. 152E, p. 221.
[62] Gallon, P.N., Orsal, G., Artaud, M.C. and Duchemin, S., 1998. Studies of CuInSe2 and $CuGaSe_2$ thin films grown by MOCVD from three organometallic sources. *Proc. 2nd World Conf. on Photovoltaic Solar Energy Conversion*, Vienna, p. 515.

[63] Guillemoles, J.-F., Cowache, P., Lusson, A., Fezzaa, K., Boisivon, F., Vedel, J. and Lincot, D., 1996. High quality $CuInSe_2$ epitaxial films – molecular beam epitaxial growth and intrinsic properties. *J. Appl. Phys.*, Vol. 79, p. 7293.

[64] Abken, A., Heinemann, F., Kampmann, A., Leinkühler, G., Rechid, J., Sittinger, V., Wietler, T. and Reineke-Koch, R., 1998. Large area electrodeposition of Cu(In,Ga)Se_2 precursors for the fabrication of thin film solar cells. *Proc. 2nd World Conf. on Photovoltaic Solar Energy Conversion*, Vienna, p.1133.

[65] Lincot, D., Guillemoles, J.-F., Cowache, P., Marlot, A., Lepiller, C., Canava, B., Yousfi, F.B. and Vedel, J., 1998. Solution deposition technologies for thin film solar cells: status and perspectives. *Proc. 2nd World Conf. on Photovoltaic Solar Energy Conversion*, Vienna, p. 440.

[66] Guimard, D, Grand, P.P., Boderau, N., Cowache, P., Guillemoles, J.-F., Lincot, D., Taunier, S., Farah, M. B., and Mogensen, P. 2002. Copper indium diselenide solar cells prepared by electrodeposition, *Proc. 29th IEEE Photovoltaic Specialists Conf.*, New Orleans, p. 692.

[67] Eberspacher, C., Pauls, K.L. and Fredric, C.V., 1998. Improved processes for forming $CuInSe_2$ films. *Proc. 2nd World Conf. on Photovoltaic Solar Energy Conversion*, Vienna, p. 303.

[68] Eberspacher, C., Pauls, K. and Serra, J., 2002. Non-vacuum processing of CIGS solar cells. *Proc. 29th IEEE Photovoltaic Specialists Conf.*, New Orleans, p. 684.

[69] Ramanathan, K., Bhattacharya, R.N., Granata, J., Webb, J., Niles, D., Contreras, M.A., Wiesner, H., Haason, F.S. and Noufi, R., 1998. Advances in the CIS research at NREL, *Proc. 26th IEEE Photovoltaic Specialists Conf.*, Anaheim, p. 319.

[70] Kapur, V.K., Bansal, A., Le, P., and Asensio, O. I. 2002. Non-vacuum printing process for CIGS Solar cells on rigid and flexible substrates. *Proc. 29th IEEE Photovoltaic Specialists Conf.*, New Orleans, p. 688.

[71] Cahen, D. and Noufi, R., 1989. Defect chemical explanation for the effect of air anneal on CdS/$CuInSe_2$ solar cell performance. *Appl. Phys. Lett.*, Vol. 54, p. 558.

[72] Rau, U., Braunger, D., Herberholz, R., Schock, H.W., Guillemoles, J.-F., Kronik, L. and Cahen, D., 1999. Oxygenation and air-annealing effects on the electronic properties of Cu(In,Ga)Se_2 films and devices. *J. Appl. Phys.*, Vol. 86, p. 497.

[73] Kronik, L., Rau, U., Guillemoles, J.-F., Braunger, D., Schock, H.W. and Cahen, D., 2000. Interface redox engineering of Cu(In,Ga)Se_2-based solar cells: oxygen, sodium, and chemical bath effects. *Thin Solid Films*, Vols. 361–362, p. 353.

[74] Scheer, R., 1997. Surface and interface properties of Cu-chalcopyrite semiconductors and devices. *Research Trends in Vacuum Sci. Technol.*, Vol. 2, p. 77.

[75] Rau, U., 2000. Role of defects and defect metastabilities for the performance and stability of Cu(In,Ga)Se_2 based solar cells. *Jpn. J. Appl. Phys.*, Vol. 39 (Suppl. 39-1), p. 389.

[76] Schmid, D., Ruckh, M., Grunwald, F. and Schock, H.W., 1993. Chalcopyrite/defect chalcopyrite heterojunctions on the basis of $CuInSe_2$. *J. Appl. Phys.*, Vol. 73, p. 2902.
[77] Morkel, M., Weinhardt, L., Lohmüller, B., Heske, C., Umbach, E., Riedl, W., Zweigart, S. and Karg, F., 2001. Flat conduction-band alignment at the CdS/$CuInSe_2$ thin-film solar-cell heterojunction. *Appl. Phys. Lett.*, Vol. 79, p. 4482.
[78] Schmid, D., Ruckh, M., and Schock, H.W., 1996. A comprehensive characterization of the interfaces in Mo/CIS/CdS/ZnO solar cell structures. *Solar Energy Mater. Solar Cells*, Vol. 41, p. 281.
[79] Contreras, M.A., Wiesner, H., Niles, D., Ramanathan, K., Matson, R., Tuttle, J., Keane, J. and Noufi, R., 1996. Defect chalcopyrite $Cu(In_{1-x}Ga_x)_3Se_5$ materials and high Ga-content Cu(In,Ga)Se_2-based solar cells. *Proc. 25th IEEE Photovoltaic Specialists*, Washington, DC, p. 809.
[80] Niemeegers, A., Burgelman, M., Herberholz, R., Rau, U., Hariskos, D. and Schock, H.W., 1998. Model for electronic transport in Cu(In,Ga)Se_2 solar cells. *Prog. Photovolt. Res. Appl.*, Vol. 6, p. 407.
[81] Bardeen, J., 1947. Surface states and rectification at a metal semiconductor contact. *Phys. Rev.*, Vol. 71, p. 717.
[82] Klein, A. and Jaegermann, W., 1999. Fermi-level-dependent defect formation in Cu-chalcopyrite semiconductors. *Appl. Phys. Lett.*, Vol. 74, p. 2283.
[83] Kessler, J., Velthaus, K.O., Ruckh, M., Laichinger, R., Schock, H.W., Lincot, D., Ortega, R. and Vedel, J., 1992. Chemical bath deposition of CdS on $CuInSe_2$, etching effects and growth kinetics. *Proc. 6th. Int. Photovoltaic Solar Energy Conf.*, New Delhi, India, p. 1005.
[84] Friedlmeier, T.M., Braunger, D., Hariskos, D., Kaiser, M., Wanka, H.N. and Schock, H.W., 1996. Nucleation and growth of the CdS buffer layer on Cu(In,Ga)Se_2 thin films. *Proc. 25th IEEE Photovoltaic Specialists Conf.*, Washington DC, p. 845.
[85] Ramanathan, K., Wiesner, H., Asher, S., Niles, D., Bhattacharya, R.N., Keane, J., Contreras, M.A. and Noufi, R., 1998. High efficiency Cu(In,Ga)Se_2 thin film solar cells without intermediate buffer layers. *Proc. 2nd World Conf. on Photovoltaic Solar Energy Conversion*, Vienna, p. 477.
[86] Wada, T., Hayashi, S., Hashimoto, Y., Nishiwaki, S., Sato, T. and Nishitina, M., 1998. High efficiency Cu(In,Ga)Se_2 (CIGS) solar cells with improved CIGS surface. *Proc. 2nd World Conf. on Photovoltaic Solar Energy Conversion*, Vienna, p. 403.
[87] Hariskos, D., Ruckh, M., Rühle, U., Walter, T., Schock, H.W., Hedström, J. and Stolt, L., 1996. A novel cadmium free buffer layer for Cu(In,Ga)Se_2 based solar cells. *Solar Energy Mat. Solar Cells*, Vols. 41/42, p. 345.
[88] Kushiya, K., Nii, T., Sugiyama, I., Sato, Y., Inamori, Y., Takeshita, H., 1996. Application of Zn-compound buffer layer for polycrystalline $CuInSe_2$-based thin-film solar cells. *Jpn. J. Appl. Phys.*, Vol. 35, p. 4383.
[89] Kushiya, K., Tachiyuki, M., Kase, T., Nagoya, Y., Miura, T., Okumura, D., Satoh, M., Sugiyama, and I., Yamase, O., 1997. Improved FF of CIGS

thin-film mini-modules with $Zn(O,S,OH)_x$ buffer by post-depostion light soaking. *Proc. 26th IEEE Photovoltaic Specialists Conf.*, Anaheim, p. 327.

[90] Nakada, T., Furumi, K. and Kunioka, A., 1999. High-efficiency cadmium-free Cu(In,Ga)Se-2 thin-film solar cells with chemically deposited ZnS buffer layers. *IEEE Trans. Electron. Devices*, Vol. ED-46, p. 2093.

[91] Nakada, T. and Mizutani, M., 2002. 18% efficiency Cd-free $Cu(In, Ga)Se_2$ thin-film solar cells fabricated using chemical bath deposition (CBD)-ZnS buffer layers. *Jpn. J. Appl. Phys.*, Vol. 41, p. L165.

[92] Ohtake, Y., Kushiya, K., Ichikawa, M., Yamada, A. and Konagai, M., 1995. Polycrystalline $Cu(InGa)Se_2$ thin-film solar cells with ZnSe buffer layers. *Jpn. J. Appl. Phys.*, Vol. 34, p. 5949.

[93] Ohtake, Y., Ichikawa, M., Yamada, A. and Konagai, M., 1995. Cadmium free buffer layers for polycrystalline $Cu(In,Ga)Se_2$ thin film solar cells. *Proc.13th European Photovoltaic Solar Energy Conf.*, Nice, p. 2088.

[94] Konagai, M., Ohtake, Y. and Okamoto, T., 1996. Development of $Cu(InGa)Se_2$ thin film solar cells with Cd-free buffer layers, *Mat. Res. Soc. Symp. Proc.*, Vol. 426, p.153.

[95] Powalla, M., Lotter, E., Waechter, R., Spiering, S., Oertel, M., Dimmler, B., 2002. Pilot line production of CIGS modules: first experience in processing and further developments. *Proc. 29th IEEE Photovoltaic Specialists Conf.*, New Orleans, p. 571.

[96] Hartmann, M., Schmidt, M., Jasenek, A., Schock, H.W., Kessler, F., Herz, K. and Powalla, M., 2000. Flexible and light weight substrates for $Cu(In,Ga)Se_2$ solar cells and modules. *Proc. 28th IEEE Photovoltaic Specialists Conf.*, Anchorage, p. 638.

[97] Tiwari, A.N., Krejci, M., Haug, F.-J. and Zogg, H., 1999. 12.8% Efficiency $Cu(In,Ga)Se_2$ solar cell on a flexible polymer sheet. *Prog. Photovolt.*, Vol. 7, p. 393.

[98] Wiedemann, S., Beck, M.E., Butcher, R., Repins, I., Gomez, N., Joshi, B., Wendt, R.G. and Britt, J.S., 2002. *Proc. 29th IEEE Photovoltaic Specialists Conf.*, New Orleans, p. 575.

[99] Fabick, L.B., Jehle, A., Scott, S., Crume, B., Jensen, G., and Armstrong, J., 2002. A new thin-film space PV module technology. *Proc. 29th IEEE Photovoltaic Specialists Conf.*, New Orleans, p. 971.

[100] Hanket, G.M., Singh, U.P., Eser, E., Shafarman, W.N. and Birkmire, R.W., 2002. Pilot-scale manufacture of $Cu(InGa)Se_2$ films on a flexible polymer substrate. *Proc. 29th IEEE Photovoltaic Specialists Conf.*, New Orleans, p. 567.

[101] Gay, R.R., 1997. Status and prospects for CIS-based photovoltaics. *Solar Energy Mater. Solar Cells*, Vol. 47, p. 19.

[102] Karg, F., Kohake, D., Nierhoff, T., Kühne, B., Grosser, S. and Lux-Steiner, M.C., 2002. Performance of grid-coupled PV arrays based on CIS solar modules. *Proc. 17th European Photovoltaic Solar Energy Conf.*, Munich, p. 391.

[103] Powalla, M. and Dimmler, B., 2001. Process development of high performance CIGS modules for mass production. *Thin Solid Films*, Vol. 387, p. 251.

[104] Schmidt, M., Braunger, D., Schäffler, R., Schock, H.W. and Rau, U., 2001. Influence of damp heat on the electrical properties of $Cu(In,Ga)Se_2$ solar cells. *Thin Solid Films*, Vols. 361–362, p. 283.

[105] Igalson, M., Wimbor, M. and Wennerberg, J., 2002. The change of the electronic properties of CIGS devices induced by the 'damp heat' treatment. *Thin Solid Films*, Vols. 403–404, p. 320.

[106] Deibel, C., Dyakonov, V., Parisi, J., Palm, J., Zweigart, S. and Karg, F., 2002. Electrical characterisation of damp-heat trated $Cu(In,Ga)(S,Se)_2$ solar cells. *Proc. 17th European Photovoltaic Solar Energy Conf.*, Munich, p. 1229.

[107] Karg, F., Calwer, H., Rimmasch, J., Probst, V., Riedl, W., Stetter, W., Vogt, H. and Lampert, M., 1998. Development of stable thin film solar modules based on $CuInSe_2$, *Inst. Phys. Conf. Ser.*, Vol. 152E, p. 909.

[108] Guillemoles, J.-F., Kronik, L., Cahen, D., Rau, U., Jasenek, A. and Schock, H.W., 2000. Stability issues of $Cu(In,Ga)Se_2$-based solar cells. *J. Phys. Chem. B*, Vol. 104, p. 4849.

[109] Guillemoles, J.F., 2002. The puzzle of $Cu(In,Ga)Se_2$ (CIGS) solar cells stability. *Thin Solid Films*, Vols. 403–404, p. 405.

[110] Gay, C.F., Potter, R.R., Tanner, D.P. and Anspaugh, B.E., 1984. Radiation effects on thin film solar cells, *Proc. 17th IEEE Photovoltaic Specialists Conf.*, Kissimmee, p. 151.

[111] Mickelsen, R.A., Chen, W.S., Stanbery, B.J., Dursch, H., Stewart, J.M., Hsiao, Y.R. and Devaney, W., 1985. Development of $CuInSe_2$ cells for space applications, *Proc. 18th IEEE Photovoltaic Specialists Conf.*, Las Vegas, p. 1069.

[112] Jasenek, A. and Rau, U., 2001. Defect generation in $Cu(In,Ga)Se_2$ heterojunction solar cells by high-energy electron and proton irradiation. *J. Appl. Phys.*, Vol. 90, p. 650.

[113] Walters, R.J., Summers, G.P., Messenger, S.R., Jasenek, A., Schock, H.W., Rau, U., Nocerino, J., Tringe, J. and Reinhardt, K., 2002. Displacement damage dose analysis of proton irradadiated CIGS solar cells on flexible substrates. *Proc. 17th European Photovoltaic Solar Energy Conf.*, Munich, p. 2191.

[114] Guillemoles, J.-F., Rau, U., Kronik, L., Schock, H.W. and Cahen, D., 1999. $Cu(In,Ga)Se_2$ solar cells: device stability based on chemical flexibility. *Adv. Mat.*, Vol. 8, p. 111.

[115] Jasenek, A., Schock, H.W., Werner, J.H. and Rau, U., 2001. Defect annealing in $Cu(In,Ga)Se_2$ heterojunction solar cells after high-energy electron irradiation. *Appl. Phys. Lett.*, Vol. 79, p. 2922.

[116] Kawakita, S., Imaizumi, M., Yamaguchi, M., Kushia, K., Ohshima, T., Itoh, H. and Matsuda, S., 2002. Annealing enhancement effect by light illumination on proton irradiated $Cu(In,Ga)Se_2$ thin-film solar cells. *Jpn. J. Appl. Phys.*, Vol. 41(2), p. L797.

[117] Jasenek, A., Rau, U., Weinert, K., Schock, H.W. and Werner, J.H., 2002. Illumination-enhanced annealing of electron irradiated $Cu(In,Ga)Se_2$ solar cells. *Proc. 29th IEEE Photovoltaic Specialists Conf.*, New Orleans, p. 872.

[118] Tuttle, J.R., Szalaj, A. and Keane, J., 2000. A 15.2% AMO/1433 W/kg thin-film Cu(In,Ga)Se_2 solar cell for space applications. *Proc. 28th IEEE Photovoltaic Specialists Conf.*, Anchorage, p. 1042.

[119] Hartmann, M., Schmidt, M., Jasenek, A., Schock, H.W., Kessler, F., Herz, K. and Powalla, M., 2000. Flexible and light weight substrates for Cu(In,Ga)Se_2 solar cells and modules. *Proc. 28th IEEE Photovoltaic Specialists Conf.*, Anchorage, p. 838.

[120] Tiwari, A.N., Krejci, M., Haug, F.J. and Zogg, H., 1999. 12.8% Efficiency Cu(In,Ga)Se_2 solar cell on a flexible polymer sheet. *Prog. Photov.*, Vol. 7, p. 393.

[121] Turcu, M. and Rau, U., 2003. Fermi level pinning at CdS/ Cu(In,Ga)$(Se,S)_2$ interfaces: effect of chalcopyrite alloy composition, *J. Phys. Chem. Solids*, Vol. 64, p. xxx.

[122] Nelson, A.J., Schwerdtfeger, C.R., Wei, S.-H., Zunger, A., Rioux, D., Patel, R., and Höchst, H., 1993. Theoretical and experimental studies of the ZnSe/Cu$InSe_2$ heterojunction band offset, *Appl. Phys. Lett.*, Vol. 62, p. 2557.

[123] Löher, T., Jaegermann, W. and Pettenkofer, C., 1995. Formation and electronic properties of the CdS/Cu$InSe_2$ (011) heterointerface studied by synchrotron-induced photoemission. *J. Appl. Phys.*, Vol. 77, p. 731.

[124] Wei, S.-H. and Zunger, A., 1993. Band offsets at the CdS/Cu$InSe_2$ heterojunction. *Appl. Phys. Lett.*, Vol. 63, p. 2549.

[125] Ruckh, M., Schmid, D., and Schock, H.W., 1994. Photoemission studies of the ZnO/CdS interface. *J. Appl. Phys.*, Vol. 76, p. 5945.

[126] Orgassa, K., Nguyen, Q., Kötschau, I.M., Rau, U., Schock, H.W. and Werner, J.H., 2002. Optimized reflection of CdS/ZnO window layers in Cu(In,Ga)Se2 thin film solar cells. *Proc. 17th European Photov. Solar Energy Conf.*, Munich, p. 1039.

[127] Engelhardt, F., Bornemann, L., Köntges, M., Meyer, Th., Parisi, J., Pschorr-Schoberer, E., Hahn, B., Gebhardt, W., Riedl, W. and Rau, U., 1999. Cu(In,Ga)Se_2 solar cells with a ZnSe buffer layer: interface characterization by quantum efficiency measurements. *Prog. Photovolt. Res. Appl.*, Vol. 7, p. 423.

[128] Bube, R.H., 1992. *Photolectronic Properties of Semiconductors*, Cambridge University Press, Cambridge, UK.

[129] Green, M.A., 1986. *Solar Cells*, University of New South Wales, Sydney, Australia, p. 96.

[130] Shafarman, W.N., Klenk, R. and McCandless, B.E., 1996. Device and material characterization of Cu(InGa)Se_2 solar cells with increasing band gap. *J. Appl. Phys.*, Vol. 79, p. 7324.

[131] Rau, U. and Schmidt, M., 2001. Electronic properties of ZnO/CdS/ Cu(In,Ga)Se_2 solar cells – aspects of heterojunction formation. *Thin Solid Films*, Vol. 387, p. 141.

[132] Niemegeers, A., Burgelman, M. and De Vos, A., 1995. On the CdS/ Cu$InSe_2$ conduction band discontinuity. *Appl. Phys. Lett.*, Vol. 67, p. 843.

[133] Ruberto, M.N. and Rothwarf, A., 1987. Time-dependent open-circuit voltage in $CuInSe_2$/CdS solar cells: Theory and experiment. *J. Appl. Phys.*, Vol. 61, p. 4662.
[134] Sasala, R.A. and Sites, J.R., 1993. Time dependent voltage in $CuInSe_2$ and CdTe solar cells. *Proc. 23rd IEEE Photovoltaic Specialists Conf.*, Louisville, p. 543.
[135] Kushia, K., Tachiyuki, M., Kase, T., Sugiyama, I., Nagoya, Y., Okumura, D., Sato, M., Yamase, O., and Takeshita, H. 1997. Fabrication of graded band-gap Cu(InGa)Se_2 thin-film mini-modules with a $Zn(O,S,OH)_x$ buffer layer, *Solar Energy Mat. Solar Cells*, Vol. 49, p. 277.
[136] Rau, U., Jasenek, A., Herberholz, R., Schock, H.W., Guillemoles, J.-F., Lincot, D. and Kronik, L., 1998. The inherent stability of Cu(In,Ga)Se_2-based solar cells. *Proc. 2nd World Conf. on Photovoltaic Energy Conversion*, Vienna, p. 428.
[137] Zabierowski, P., Rau, U., and Igalson, M., 2001. Classification of metastabilities in the electrical characteristics of ZnO/CdS/Cu(In,Ga)Se_2 solar cells, *Thin Solid Films*, Vol. 387, p. 147.
[138] Rau, U., Schmitt, M., Parisi, J., Riedl, W. and Karg, F., 1998. Persistent photoconductivity in Cu(In,Ga)Se_2 heterojunctions and thin films prepared by sequential deposition. *Appl. Phys. Lett.*, Vol. 73, p. 223.
[139] Meyer, Th., Schmidt, M., Engelhardt, F., Parisi, J. and Rau, U., 1999. A model for the open circuit voltage relaxation in Cu(In,Ga)Se_2 heterojunction solar cells. *Eur. Phys. J. App. Phys.*, Vol. 8, p. 43.
[140] Lang, D.V. and Logan, R.A., 1977. Large-lattice-relaxation model for persistent photoconductivity in compound semiconductors, *Phys. Rev. Lett.*, Vol. 39, p. 635.
[141] Delahoy, A.E., Ruppert, A. and Contreras, M., 2000. Charging and discharging of defect states in CIGS/ZnO junctions. *Thin Solid Films*, Vols. 161–162, p. 140.
[142] Rau, U., Weinert, K., Nguyen, Q., Mamor, M., Hanna, G., Jasenek, A. and Schock, H.W., 2001. *Mat. Res. Soc. Symp. Proc.*, Vol. 668, p. H9.1.1.
[143] Siebentritt, S., 2002. Wide gap chalcopyrites: material properties and solar cells. *Thin Solid Films*, Vols. 403–404, p. 1.
[144] Paulson, P.D., Haimbodi, M.W., Marsillac, S., Birkmire, R.W. and Shafarman, W.N., 2002. $CuIn_{1-x}Al_xSe_2$ thin films and solar cells. *J. Appl. Phys.*, Vol. 91, p. 10153.
[145] Walter, T., Content, A., Velthaus, K.O. and Schock, H.W., 1992. Solar-cells based on $CuIn(Se,S)_2$. *Sol. Energy Mater. Sol. Cells*, Vol. 26, p. 357.
[146] Friedlmeier, T.M. and Schock, H.W., 1998. Improved voltages and efficiencies in Cu(In,Ga)(S,Se)2 solar cells. *Proc. 2nd. World Conf. on Photovoltaic Solar Energy Conversion*, Vienna, p. 1117.
[147] Scheer, R., Walter, T., Schock, H.W., Fearhailey, M.L. and Lewerenz, H.J., 1993. $CuInS_2$ based thin film solar cell with 10.2% efficiency. *Appl. Phys. Lett.*, Vol. 63, p. 3294.

[148] Kaigawa, R., Neisser, A., Klenk, R., Lux-Steiner, M.-Ch., 2002. Improved performance of thin film solar cells based on $Cu(In,Ga)S_2$. *Thin Solid Films*, Vol. 415, p. 266.

[149] Klenk, R., 2001. Characterisation and modelling of chalcopyrite solar cells. *Thin Solid Films*, Vol. 387, p. 135.

[150] Turcu, M., Pakma, O. and Rau, U., 2002. Interdependence of absorber composition and recombination mechanism in $Cu(In,Ga)(Se,S)_2$ heterojunction solar cells. *Appl. Phys. Lett.*, Vol. 80, p. 2598.

[151] Nadenau, V., Hariskos, D. and Schock, H.W., 1977. $CuGaSe_2$ based thin-film solar cells with improved performance, *Proc. 14th European Photovoltaic Solar Energy Conf.*, Barcelona, p. 1250.

[152] Jasenek, A., Rau, U., Nadenau, V. and Schock, H.W., 2000. Electronic properties of $CuGaSe_2$-based heterojunction solar cells. Part II. Defect spectroscopy. *J. Appl. Phys.*, Vol. 87, p. 594.

[153] Nadenau, V., Rau, U., Jasenek, A. and Schock, H.W., 2000. Electronic properties of $CuGaSe_2$-based heterojunction solar cells. Part I. Transport analysis. *J. Appl. Phys.*, Vol. 87, p. 584.

[154] Reiß, J., Malmström, J., Werner, A., Hengel, I., Klenk, R. and Lux-Steiner, M.Ch., 2001. Current Transport in $CuInS_2$ solar cells depending on absorber preparation, *Mat. Res. Soc. Symp. Proc.*, Vol. 668, p. H9.4.1.

[155] Schön, J.H., Klenk, M., Schenker, O. and Bucher, E., 2000. Photovoltaic properties of $CuGaSe_2$ homodiodes. *Appl. Phys. Lett.*, Vol. 77, p. 3657.

[156] Shafarman, W.N., Marsillac, S., Paulson, P.D., Haimbodi, M.W. and Birkmire, R.W., 2002. Material and device characterization of thin film $Cu(InAl)Se_2$ solar cells. *Proc. 29th IEEE Photovoltaic Specialists Conf.*, New Orleans, p. 519.

[157] Marsillac, S., Paulson, P.S., Haimbodi, M.W., Birkmire, R.W. and Shafarman, W.N., 2002. High-efficiency solar cells based on $Cu(InAl)Se_2$ thin films. *Appl. Phys. Lett.*, Vol. 81, p. 1350.

[158] Walter, T. and Schock, H.W., 1993. Structural and electrical investigations of the anion-exchange in polycrystalline CuIn(S,Se)2 thin-films. *Jpn. J. Appl. Phys.*, Vol. 32(3), p. 116.

[159] Siemer, K., Klaer, J., Luck, I., Bruns, J., Klenk, R., Bräunig, D., 2001. Efficient $CuInS_2$ solar cells from a rapid thermal process (RTP). *Solar Energy Mat. Solar Cells*, Vol. 67, p. 159.

[160] Hengel, I., Neisser, A., Klenk, R. and Lux-Steiner, M.C., 2000. Current transport in $CuInS_2$:Ga/Cds/ZnO – solar cells. *Thin Solid Films*, Vols. 361–362, p. 458.

[161] Gray, J.L. and Lee, Y.J., 1994. Numerical modeling of graded band gap CIGS solar cells. *Proc. First World Conf. on Photovoltaic Solar Energy Conversion*, Hawaii, p. 123.

[162] Dhingra, A. and Rothwarf, A., 1996. Computer simulation and modeling of graded bandgap CuInSe2/CdS based solar cells. *IEEE Trans. Electron Devices*, Vol. ED-43, p. 613.

[163] Gabor, A.M., Tuttle, J.R., Bode, M.H., Franz, A., Tennant, A.L., Contreras, M.A., Noufi, R., Jensen, D.G. and Hermann, A.M., 1996. Band-gap engineering in Cu(In,Ga)Se_2 thin films grown from $(In,Ga)_2Se_3$ precursors. *Solar Energy Mater. Solar Cells*, Vol. 41, p. 247.
[164] Dullweber, T., Rau, U., Contreras, M., Noufi, R. and Schock, H.W., 2000. Photogeneration and carrier recombination in graded gap Cu(In, Ga)Se_2 solar cells. *IEEE Trans. Electron. Dev.*, Vol. ED-47, p. 2249.

Part IId

Space and Concentrator Cells

IId-1

GaAs and High-Efficiency Space Cells

V. M. Andreev, Ioffe Physico-Technical Institute,
St. Petersburg, Russia

1 Historical Review of III–V Solar Cells

Since the first solar-powered satellites Vanguard-1 and Sputnik-3 were launched in the spring of 1958, solar cells had become the main source of energy on the spacecrafts. The first space arrays were based on single crystal silicon solar cells with an efficiency of about 10%. During the 1960 and 1970s, the improvements in the Si cell design and technology, such as fabrication of 'violet' cells with an increased short-wavelength photosensitivity, back surface field formation, application of photolithography to make optimal front grid fingers, reduction of optical losses owing to front surface texturing and improvement of the anti-reflection coating, allowed an increase of efficiency of up to 14% (1 sun, AM0). In the last two decades, the Si space cell efficiencies were increased by up to 18%. These advanced Si cells are used for space missions that do not strictly require III–V cells with both higher efficiency and better radiation stability [1, 2].

At the beginning of the 1960s, it was found that GaAs-based solar cells with the Zn-diffused p–n junction ensured the better temperature stability and higher radiation resistance. One of the first scaled applications of the temperature-stable GaAs solar cells took place on the Russian spacecrafts Venera-2 and Venera-3 launched in November 1965 to the 'hot' planet Venus. The area of each GaAs solar array constructed by the Russian Enterprise KVANT for these spacecrafts was 2 m^2. Then the Russian moon-cars were launched in 1970 (Lunokhod-1) and in 1972 (Lunokhod-2) with GaAs 4 m^2 solar arrays in each. The operating temperature of these arrays on the illuminated surface of the Moon was about 130°C. Therefore, silicon-based solar cells could not operate effectively in these conditions. GaAs solar arrays have shown efficiency of 11% and have provided the energy supply during the lifetime of these moon-cars.

The first AlGaAs/GaAs solar cells with passivating wide bandgap window were created in 1970 [3,4]. In the following decades, by means of the liquid-phase-epitaxy (LPE) of AlGaAs/GaAs heterostructures [4–12], their AM0 efficiency was increased up to 18–19% [10–14] owing to the intensive investigations in the fields of physics and technology of space solar cells [15–18]. These investigations were stimulated and supported by ambitious space programmes in the former USSR and in the USA [1, 2, 18]. Owing to the high efficiency and improved radiation hardness of the AlGaAs/GaAs solar cells, the LPE technology was utilised in the high scale production of AlGaAs/GaAs space arrays for the spacecrafts launched in the 1970s and 1980s. For example, an AlGaAs/GaAs solar array with a total area of 70 m^2 was installed in the Russian space station MIR launched in 1986. During 15 years in orbit, the array degradation appeared to be lower than 30%, despite being under severe conditions such as appreciable shadowing, effects of numerous docking and the ambient environment of the station. At that time, it was the most large-scale demonstration of the AlGaAs/GaAs solar cell advantages for space applications. The further improvement of the LPE technology [19, 20] led to increased efficiencies of 24.6% (AM0, 100 suns) on the base of the heterostructures with an ultra thin window AlGaAs layer and a back surface field layer.

Since the late 1970s, AlGaAs/GaAs heterostructures were also produced by the metal organic chemical vapour deposition (MOCVD) technique using metal organic compounds of Group III elements and hydrides of Group V elements [21, 22]. The advantage of MOCVD is a possibility to fabricate multilayer structures in high-yield reactors with layers of a specified composition and a precise thickness varying from 1 to 10 nm to several microns. AlGaAs/GaAs heterostructures with an ultra thin (0.03 μm) top window layer and with a back surface wide-bandgap barrier were fabricated by MOCVD for space cells. AlGaAs/GaAs 4 cm^2 1 sun space solar cells with efficiencies of 21% [23] and 21.7% [24] were fabricated on the base of these structures.

Enhanced light absorption was provided in the cells with an internal Bragg reflector [25–27]. This dielectric mirror increases the effective absorption length of sunlight within the long-wavelength part of photoresponse spectrum and allows the base layer to be made thinner. In this case, the cell efficiency is more tolerant to a reduction of the carrier diffusion length, and, as a result, these cells are more radiation resistant [27].

Owing to the fact that MOCVD is capable of producing single crystal layers on silicon and germanium substrates, it has a potential for fabrication of low-cost, high efficiency III–V solar cells on these substrates. Growing GaAs on Si of a sufficient quality is not possible as a result of the mismatch of 4% between Si and GaAs. However, there is progress in improving the GaAs/Si structure quality by using special structures and growth techniques: strained superlattice, thermo cyclic growth and cyclic structure annealing.

Ge is a quite good lattice match to the GaAs material. Therefore, epitaxial growth of GaAs with a high quality was realised by MOCVD and this is now the basic technique for growth of multilayer AlGaAs/GaAs/Ge single-junction and GaInP/GaAs/Ge multi-junction solar cells. This method provides a good crystal quality of epitaxial structures on Ge substrates, high productivity and reproducibility.

Among other single junction cells, the InP-based cells are rather promising for space applications owing to that fact that InP has a higher radiation resistance [2, 28] than GaAs. However, there are some obstacles for the scaled application of InP-based cells in space arrays. Firstly, there is no lattice-matched wide bandgap window for InP to make stable passivation of the front surface. Secondly, it is difficult to grow this material to a high quality on the Ge and Si substrates due to lattice-mismatches that are as high as 8% between InP and Si and 4% between InP and Ge. Thirdly, InP is a quite expensive material. In spite of these obstacles there are some possible applications of InP cells in the arrays for satellites expected to be launched towards the intermediate orbits, high Earth orbits or orbit-transfer, which are characterised by very high radiation fluencies.

Multi-junction (cascade) cells ensured the further increase of III–V solar cell efficiencies. Despite a large number of theoretical studies of cascade solar cells [1, 29–32], their efficiencies remained low enough for a long time, since the ohmic and optical losses in available designs were unacceptably large. Monolithic and mechanically stacked tandem cells with increased efficiencies were developed in the beginning of the 1990s. In mechanically stacked tandems with GaAs top cells and GaSb (or InGaAs) bottom cells [33–35, 38, 41–43], efficiencies of about 30% were achieved under the concentrated AM0 sunlight. Monolithic cascade cells have been developed and fabricated by MOCVD on the structures GaInAs/InP [40], Si/AlGaAs [44, 64], AlGaAs/GaAs [45, 42], GaAs/Ge [47–51], GaInP/GaAs [36, 52, 53] GaInP/GaAs/Ge [54–60], GaInP/GaInAs [61] and GaInP/GaInAs/Ge [57, 62, 63] heterostructures. Table 1 presents the best reported efficiencies of the cells based on the different structures under one sun and concentrated AM0 sunlight illumination.

2 Single-Junction III–V Space Solar Cells

2.1 Solar Cells Based on AlGaAs/GaAs Structures

Among different investigated heterostructures based on III–V heterojunctions appropriate for fabrication of single junction solar cells, AlGaAs/GaAs heterostructures have found the first application due to the well-matched lattice parameters of GaAs and AlAs, and because GaAs has an optimal bandgap for effective sunlight conversion. In the first solar cells based on AlGaAs/GaAs heterojunctions [3, 4], the basic narrow bandgap material was GaAs. A wide bandgap window was made of AlGaAs close in the composition to AlAs, which is almost completely transparent to sunlight, making solar cells very sensitive in the short wavelength range of the sun spectrum. Such a cell is illuminated through the window, and the light with photon energy exceeding the bandgap value of GaAs is absorbed in it, while the generated minority carriers are separated by the p–n junction field located in GaAs. Because of the close lattice parameters of the contacting materials, the interface in AlGaAs/GaAs

Table 1 Reported efficiencies for III–V space solar cells under AM0 conditions at T = 25–28°C

Cell material	Cell type	Sunlight concentration	Area, cm^2	Eff., %	Ref.
AlGaAs/GaAs	single junction	No concentration	4	21.7	[24]
GaAs/Si	single junction		4	18.3	[64]
InP	single junction		4	19.9	[2]
AlGaAs/GaAs	monolithic 2-junction		0.5	23.0	[2]
GaInP/GaAs	monolithic 2-junction		4	27.2	[56]
GaInP/GaAs/Ge	monolithic 3-junction		4	29.3	[56]
GaInP/GaInAs/Ge	monolithic 3-junction		4	29.7	[57]
GaInP/GaAs/Ge	monolithic 3-junction		26.6	29.0	[56]
AlGaAs/GaAs	single junction	100×	0.07	24.6	[41]
GaAs/GaSb	mechanical stack 2-junction	100×	0.05	30.5	[2]
GaAs/Ge	monolithic 2-junction	9×	0.136	23.4	[50]
GaInP/GaInAs/Ge	monolithic, 3-junction	7.6×	4.1	31.1	[58]
GaInP/GaAs/GaSb	monolithic/mechanical stack 3-junction	15×	circuit	34.0	[65]

heterojunctions is characterised by a low density of surface states, providing a highly effective accumulation of carriers.

The composition n-GaAs/p-GaAs/p-AlGaAs (Figure 1(a)) was the first mostly used heterostructure. The structures were grown by LPE [3–14] or by MOCVD [21, 22]. For example, a 0.5 μm thick p-GaAs layer is either grown epitaxially (MOCVD) or formed by zinc or beryllium diffusion during the growth (LPE) of the $Al_xGa_{1-x}As$ solid solution doped with one of these impurities. The diffusion produces a quasi-electric field that arises as a result of the acceptor concentration gradient (Figure 1(a)) which enhances the effective diffusion length of electrons generated by light in the p-GaAs layer. Using the LPE technique, the highest conversion efficiencies of about 19% (1 sun; AM0) have been obtained in the structures with the wide-bandgap layer of smaller thickness [10–14]. A way of enhancing the short-wavelength photosensitivity is to use in the front layer a solid solution of graded composition with the bandgap (Figure 1(b)) increasing towards the illuminated surface [10]. The strong built-in electric field significantly enhances the value of the effective electron diffusion length and suppresses the surface recombination of the electron–hole pairs generated near the surface by short-wavelength light.

Introduction of a potential barrier at the back surface of the cell photoactive region assists the collection of minority carriers generated in the base. This barrier is made either by growing a buffer layer (Figure 1(d)) of n^+-GaAs, doped to a level exceeding that in the active layer, or by growing an n-AlGaAs layer (Figure 1(c)). The one-sun AM0 efficiency of 21.7% and the concentrator AM0 efficiency of 24.5% at 170 suns have been measured [24] in the cells based on these structures grown by MOCVD with the n-$Al_xGa_{1-x}As$ layer as the back surface barrier. Similar results have been obtained during development of a low-temperature LPE modification for the growth of AlGaAs/GaAs structures [19]

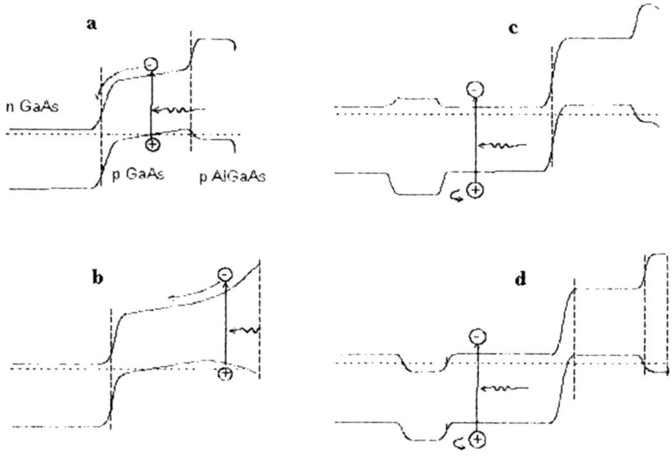

Figure 1 Band diagrams of AlGaAs/GaAs heterostructures developed for space solar cells: (a) p-AlGaAs/p-GaAs/n-GaAs; (b) structure with a graded p-AlGaAs front layer; (c, d) structures with a back-surface field, made of n-AlGaAs (c) and n⁺GaAs (d).

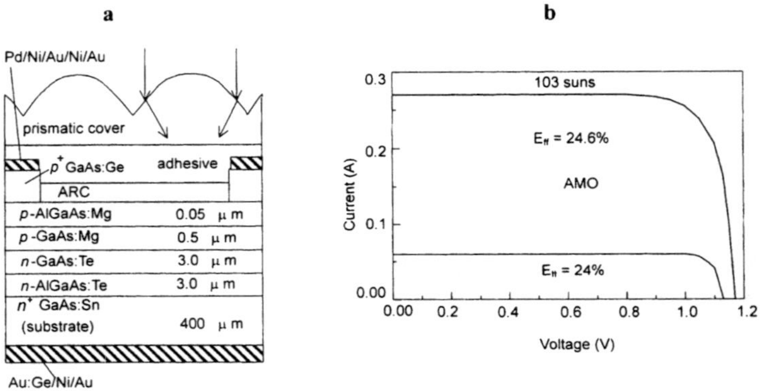

Figure 2 Schematic diagram (a) and illuminated I–V curves (b) of the LPE grown single junction AlGaAs/GaAs concentrator solar cell with a prismatic cover [19]. The cell area is 0.07 cm².

that resulted in fabrication of high efficiency solar cells with a structure shown in Figure 2(a). Silicone prismatic covers optically eliminate the grid line obscuration losses in concentrator cells. Owing to the high crystal quality of the LPE material and optimised optical parameters, a high quantum yield is obtainable in a wide spectral range. The AM0 efficiency of these cells with a prismatic cover was 24.6% (Figure 2(b)) under 103 suns at 25°C.

2.2 Solar Cells With Internal Bragg Reflector

The Bragg reflector (BR) made of semiconductor layers is widely used in lasers and other optical devices. By using a multiple layer composed of two materials with different refractive indices, nearly 100% reflectance can be achieved over a restricted wavelength range. The thickness of each of the two materials is chosen for quarter-wavelength reflection for the given wavelength. These multilayer dielectric stacks selectively reflect a part of the unabsorbed photons providing a second pass through the photoactive region, thereby increasing the photocurrent.

Epitaxial (MOCVD) Bragg reflectors in solar cells [25–27] were based (Figure 3(a)) on the pairs of $Al_xGa_{1-x}As$ and GaAs layers. By increasing the number (N) of pairs, the BR reflectance increases, asymptotically tending to unity and reaches 96% at $N = 12$ [26, 27]. A reflector of this type allows us to increase the effective absorption within the long-wavelength part of the photosensitivity spectrum and to make the base layer thinner. The cell efficiency in this case becomes more tolerant to decreasing the diffusion lengths caused by the high-energy particle irradiation. The solar cell structure with a Bragg reflector (Figure 3(a)) was grown [26, 27] by MOCVD using equipment with a low-pressure horizontal reactor. BR was optimised for reflectance in the 750–900 nm spectral region and consisted of 12 pairs of AlAs and GaAs layers with a thickness of 72 nm for AlAs layers and 59 nm for GaAs. The photocurrent density of 32.7 mA/cm² (AM0, 1 sun, 25°C) and the efficiency of 23.4% (AM0, 18 suns, 25°C) were

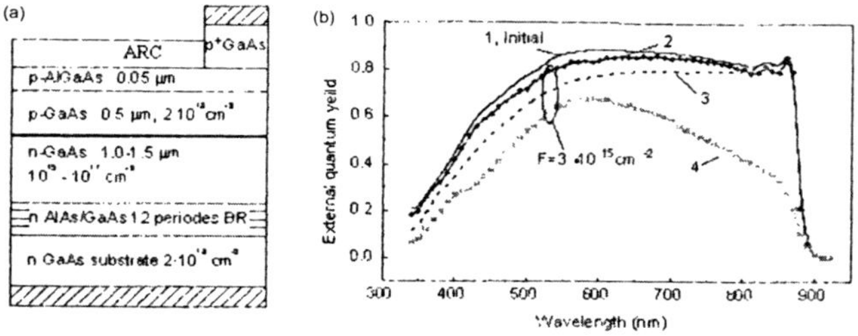

Figure 3 Schematic cross-section of a AlGaAs/GaAs solar cell with internal Bragg reflector (BR) (a) and spectral responses (b), which are shown for the cells with BR (1–3) and without BR (4) before (1) and after (2, 3, 4) irradiation by 3.75 MeV electrons with fluence of 3 × 10^{15} cm^{-2}.

registered for this cell. These values are fairly good, taking into account the smaller thickness of the n-GaAs base layer. The long-wavelength response of the cell with a 1.5 μm n-GaAs layer is nearly the same as for the cell with a 3 μm n-GaAs layer without BR. Reduction of the base thickness improves the cell radiation resistance. Figure 3(b) shows spectral responses of the cells with BR (curves 1–3) and without BR (4). Base thickness in the cells with BR was reduced to 1.1 μm (curves 1, 2) and to 1.3 μm (3), and the base doping level was reduced to 10^{15} cm^{-3} (curves 1, 2) and to 7 × 10^{15} cm^{-3} (curve 3). The base thickness of 3.5 μm and the base doping level of 10^{17} cm^{-3} were in the cell without BR (curve 4). It is seen from Figure 3(b) that reduction of the base thickness and the base doping level in the cells with a Bragg reflector allowed an increase in the radiation resistance. The remaining power factor was 0.84–0.86 in these cells after 1 MeV electron irradiation with a fluence of 10^{15} cm^{-2}.

2.3 GaAs-Based Cells on Ge Substrates

Intensive investigations of single and dual junction GaAs-based heterostructures MOCVD grown on Ge substrates were carried out [47–50]. Ge is less expensive and more mechanically strong than GaAs. Therefore, Ge substrates can be thinned down to 100–150 μm, and cells can be made larger reducing the weight and cost of space arrays. Owing to a good lattice matching between Ge and GaAs, the structures based on GaAs can be grown with a good crystal quality. An increase of the output voltage was observed in GaAs/Ge cells with a photoactive Ge/GaAs interface. A disadvantage of this photoactive Ge results from mismatching of the photocurrents generated in Ge and in GaAs active regions. Usually, the photocurrent from a Ge-sub cell was lower. Therefore, the I–V curve of these cells was 'kinked' with reduced FF and efficiency [48, 49, 51]. Reproducible growth conditions were developed to form GaAs cells on Ge substrates with an inactive interface [49, 51]. MOCVD equipment produced by EMCORE (USA) and AIXTRON (Germany) provides the GaAs-based solar cell structures on Ge-substrates with high productivity (up to 0.2 m^2 per run in AIX-3000 reactor) and with excellent uniformity and reproducibility of the cell performance. AlGaAs/GaAs structure production on the 4-inch Ge wafers allows fabricating the large area (up to 36 cm^2) cells with an average efficiency of 19% (1 sun, AM0, 25°C).

3 Multi-junction Space Solar Cells

3.1 Mechanically Stacked Cells

Optimum bandgaps for multi-junction solar cells were calculated in a number of works [1, 2, 29–40]. Figure 4 illustrates the bandgaps and lattice constants for III–V compounds and their solid solutions. The hatched areas represent the theoretical optimum E_g values for the current matched bottom and top cells: E_{g1} = 1.65–1.8 eV for the top cells and E_{g2} = 1.0–1.15 eV for the bottom cells. The theoretical 1 sun AM0 efficiency for tandem solar cells with these bandgap values is about 32.5% (Figure 5, curve 1).

Mechanical stacks ensure more possibilities for the material choice than monolithic tandems owing to the use of lattice- and current-mismatched combinations of semiconductors. An obstacle for high-scale applications of stacks in space is a more complicated design. It is one reason why stacks are only used in concentrator array applications. In spite of the higher equilibrium temperature in the cells operating under concentrated sunlight, theoretical efficiencies of multi-junction, mechanically stacked cells under 10–100 suns at temperatures of 70–80°C are higher by 2–4% than 1-sun solar cells at 25°C. The use of sunlight concentration for space arrays offers additional advantages: better cell shielding from space radiation, and a potentially low cost of concentrator arrays owing to the low semiconductor material consumption.

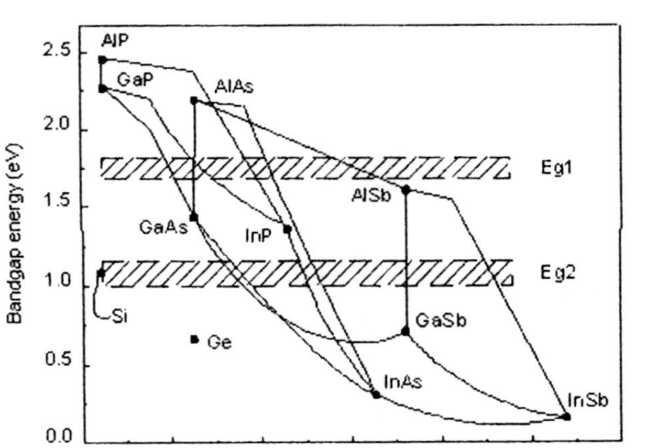

Figure 4 Energy bandgap versus lattice constant for Ge, Si, III–V compounds and ternary solid solutions. Hatched boxes correspond to the highest efficiency current-matched two-junction solar cells.

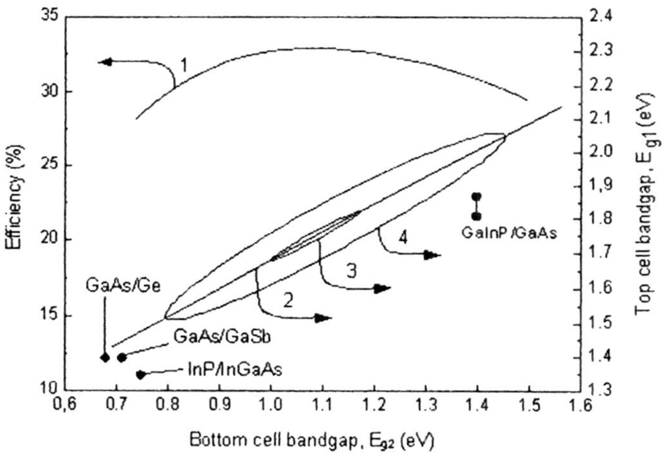

Figure 5 Curve 1: one sun AMO efficiency (curve 1) for two-junction, two-terminal solar cells as a function of E_{g2} under conditions of matched photocurrents in the top and bottom cells. Line 2: relationship between E_{g1} and E_{g2} values, for which the condition of matched photocurrents is fulfilled. Curve 3: iso-efficiency contour ensuring the highest AMO efficiencies of about 32.5%. Curve 4: iso-efficiency contour, for which efficiency decreases down to 30%. The bold points show the bandgaps E_{g1} and E_{g2} of GaAs/Ge, GaAs/GaSb, InP/InGaAs and GaInP/GaAs tandems.

The first high efficiency, mechanically stacked tandems consisted of AlGaAs/GaAs infrared transparent top cells and GaSb infrared sensitive bottom cells were fabricated [35]. GaAs-based cells were made transparent to the infrared part ($\lambda > 0.9$ μm) of sunlight in these stacked cells as n-GaAs substrate doping level was reduced to 10^{17} cm^{-3}. GaSb cells for stacks were fabricated mainly using Zn-diffusion. Maximum photocurrent densities in GaSb cells for the AM0-spectrum were about 30 mA/cm^2 behind a GaAs filter, and efficiencies of about 6% (AM0, 100x) were achieved behind a transparent GaAs cell. The best efficiencies of 29–30% (AM0, 100x) were obtained in two-junction stacks based on AlGaAs/GaAs top and GaSb (or InGaAs) bottom cells [35, 41], which have promise for use in space concentrator arrays [66, 67].

The further efficiency increase up to 34% (AM0, 25°C, 15 suns) was obtained in the triple junction, mechanically stacked voltage-matched circuits based on the monolithic GaInP/GaAs two-junction top cell and the GaSb bottom cell [65]. The efficiencies of 27.5% in the GaInP/GaAs top cell and of 6.5% in the bottom cell were obtained at 15 suns AM0 concentration. To obtain two-terminal circuits, seven GaSb cells were connected in series ensuring output voltage (V_{mp}) of 7×0.375 V = 2.63 V slightly exceeding V_{mp} of 2.4 V for the InGaP/GaAs two-junction cells which were connected in parallel.

Efficiency of 30% is expected in the mini-modules based on these stacks in the ultralight stretched lens array [66], characterised by overall efficiency of 26% and array power density of 350 W/m^2. The further efficiency increase is expected in four-junction stacks based on the monolithic two-junction GaInP/GaAs top cells and the monolithic two-junction AlGaAsSb/GaSb bottom cells (or with another type of the cascade bottom cells).

3.2 Monolithic Multi-junction Solar Cells

In monolithic AlGaAs/GaAs tandems consisting of an $Al_{0.37}Ga_{0.63}As$ ($E_g = 1.93$ eV) upper cell and a GaAs lower cell were grown by MOCVD [45]. The component cells were electrically connected by a metal contact fabricated during the post-growth processing. The efficiency of 25.2% measured under AM0 one-sun illumination was achieved in AlGaAs/GaAs/InGaAsP three-junction cells consisted of a monolithic AlGaAs/GaAs tandem mechanically stacked with InGaAsP ($E_g = 0.95$ eV) single-junction cell.

Monolithic two-terminal GaAs/Ge tandem space concentrator cells with efficiency of 23.4% (9 suns AM0, 25°C) were developed [50]. MOCVD growth of n-GaAs formed a bottom cell in Ge owing to simultaneous diffusion of As and Ga into Ge. Series resistance of the n^+GaAs–p^+Ge tunnel junction formed on the interface limited the effective operation of these tandem cells to 10 suns only.

As is seen in Figures 4 and 5, silicon is a material with an optimum bandgap for the fabrication of bottom cells for two-junction cells with a theoretical efficiency exceeding 30%. Wide bandgap cells in these tandems, however, can be made only from such materials as AlGaAs, GaInP, and GaPAs, which are not lattice-matched to silicon. A considerable advance was realised in the fabrication of GaAs-based epitaxial layers on Si substrates [44, 64]. The results obtained hold a promise for high efficiency monolithic cascade cells on Si substrates, costing less than those on Ge substrates.

Cascade cells based on GaInP/GaAs heterostructures were at first proposed and fabricated at NREL [52, 53]. Then this technology was successfully applied for high-scale production of space arrays based on dual and triple junction GaInP/GaAs/Ge in Spectrolab [54–58], Tecstar [59] and Emcore [60]. Figure 6 shows two of the developed triple junction cell structures, consisting of a (Al)GaInP top cell connected in series by tunnel junction to a GaAs (Figure 6(a)) or InGaAs (Figure 6(b)) middle cell, connected in turn by tunnel junction with a bottom Ge cell. A 1 sun AM0 efficiency as high as 29.3% was achieved in Spectrolab [56] in a three-junction GaInP/GaAs/Ge cell. Large area (26.6 cm^2) three-junction cells have reached 29% AM0 efficiency. The high efficiencies of these cells are compatible with the high radiation hardness. The power remaining factor $P/P_0 = 0.83$ at 10^{15} e^-/cm^2 was measured in typical GaInP/GaAs/Ge cells, ensured mainly by the high radiation resistance of the top GaInP cell.

As mentioned above, the sub cells with $E_g = 1$–1.15 eV ensure a higher theoretical efficiency in cascade cells. Suitable Ge substrates can only be employed for the growth of lattice-mismatched GaInAs epilayers of an optimal composition. GaInAs layers on GaAs or Ge substrates of a satisfactory quality were grown and used for a subsequent epitaxial growth of GaInP/GaInAs cascade structures [61–63]. Efficiency of 27.3% (1 sun AM0, 28°C) was measured [63] in the triple junction $Ga_{0.43}In_{0.57}P/Ga_{0.92}In_{0.08}As$/Ge cells with

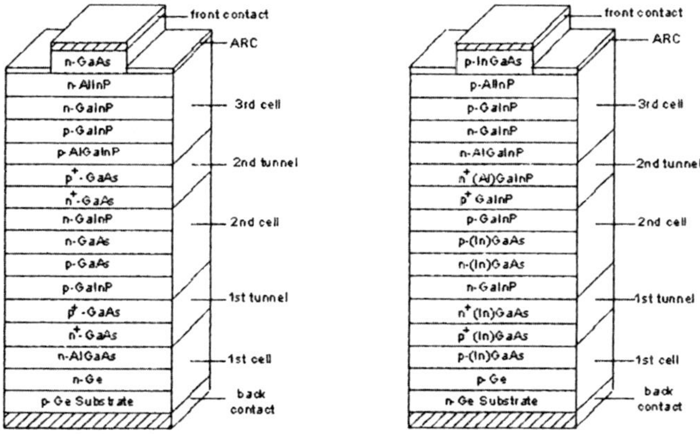

Figure 6 Cross-section of the developed triple junction solar cells: (a) (Al)GaInP/GaAs/Ge cascade cell (n-on-p) with GaAs-based first and second tunnel junctions and second cell; (b) (Al)GaInP/(In)GaAs/Ge cascade cell (p-on-n) with InGaAs-based first tunnel junction and second cell, and (Al)GaInP-based second tunnel junction.

the 0.5% lattice-mismatch to Ge substrate, similar to high-efficiency conventional lattice-matched GaInP/GaAs/Ge cells. The record efficiency of 29.7% [57] at 1 sun AM0 was achieved in a triple junction cell based on the GaInP/GaInAs/Ge structure with improved bandgap control: the bandgap of the GaInP top cell and tunnel junction layers, the bandgap of the GaInAs structure and simultaneous reduction of dislocation density in the structure. Owing largely to this very high efficiency at the start of life, the prototype cells have demonstrated end-of-life AM0 efficiency over 24.4% after irradiation with 1-MeV electrons at fluence of $1 \times 10^{15}\ e^-/cm^2$.

The next step for the efficiency increase was proposed in [36]: the development of monolithic four-junction (Al)GaInP/GaAs/GaInNAs/Ge cells, which contain a 1 eV GaInNAs sub cell lattice-matched to GaAs between the Ge and GaAs sub cells. However, GaInNAs layer has not been obtained with parameters acceptable for incorporation in such a four-junction cells until now, in spite of the intensive investigations of this material. The further efficiency increase was predicted for monolithic five-junction cells based on the (Al)GaInP/GaInP/GaAs/GaInAs/Ge structures with the lattice-mismatched GaInAs layers in a second (from Ge substrate) cell [62].

Recently, new III–V low-dimensional structures based on supperlattice and multi-quantum wells [68, 69], as well as metallic and quantum dot intermediate bands [70, 71] were proposed for solar cells. The main idea in these works is to use low-dimensional heterostructures in order to extend the sunlight absorption to longer wavelengths, but to conserve the high output voltage corresponding to the wide bandgap bulk semiconductor. It was predicted that these structures could obtain higher theoretical efficiencies than the multi-junction solar cells reviewed above. However, new materials, new technologies, and, maybe, new approaches should be developed to realise these predictions.

Acknowledgements

The author expresses his thanks to colleagues from the Photovoltaics Laboratory of the Ioffe Physico-Technical Institute for the help and valuable discussions, Zh. I. Alferov for the continuous interest and support and to all researchers for the permissions to use the copyright material.

References

[1] Flood, D. and Brandhorst, H., 1987. Space solar cells. In: *Current Topics in Photovoltaics*, Coutts, T.J., Meakin, J.D., Eds, Vol. 2, Academic Press, New York, London, pp.143–202.

[2] Bailey, S.G. and Flood, D.J., 1998. Space Photovoltaics. *Prog. Photovolt: Res. Appl.*, Vol. 6(1), pp. 1–14.

[3] Andreev,V.M., Grilikhes,V.A. and Rumyantsev,V.D., 1997. *Photovoltaic Conversion of Concentrated Sunlight*. John Wiley & Sons Ltd.

[4] Alferov, Zh.I., Andreev, V.M., Kagan, M.B./ Protasov, I.I. and Trofim, V.G., 1970. Solar cells based on heterojunction p-AlGaAs–n-GaAs. *Sov. Phys. Semicond.*, Vol. 4(12).

[5] Hovel, H.J. and Woodall, J.M., 1972. High-efficiency AlGaAs-GaAs solar cells. *Appl. Phys. Lett.*, Vol. 21, pp. 379–381.

[6] Andreev, V.M., Golovner, T.M., Kagan, M.B., Koroleva, N.S., Lubochevskaya, T.A., Nuller, T.A. and Tret'yakov, D.N., 1973. Investigation of high efficiency AlGaAs–GaAs solar cells. *Sov. Phys. Semicond.*, Vol. 7(12).

[7] Andreev,V.M., Kagan,M.B., Luboshevskaia,T.L., Nuller,T.A. and Tret'yakov, D.N., 1974. Comparison of different heterophotoconverters for achievement of highest efficiency. *Sov. Phys. Semicond.*, Vol. 8(7).

[8] Alferov, Zh.I., Andreev, V.M., Daletskii, G.S., Kagan, M.B., Lidorenko, N.S. and Tuchkevich, V.M., 1977. Investigation of high efficiency AlAs–GaAs heteroconverters. *Proc. World Electrotechn. Congress*, Moscow, Section 5A, report 04.

[9] Hovel, H.J., 1975. Solar cells. In: *Semiconductors and Semimetals*. Vol. 11, Willardson, R.K. and Beer, A.C., Eds, Academic Press, New York, London.

[10] Woodall, J.M. and Hovel, H.J., 1977. An isothermal etchback–regrowth method for high efficiency $Ga_{1-x}Al_xAs$–GaAs solar cells. *Appl. Phys. Lett.*, Vol. 30, pp. 492–493.

[11] Andreev, V.M., Larionov, V.R., Rumyantsev, V.D., Fedorova, O.M. and Shamukhamedov, Sh.Sh., 1983. P-AlGaAs–pGaAs–nGaAs solar cells with efficiencies of 19% at AM0 and 24% at AM1.5. *Sov. Tech. Phys. Lett.*, Vol. 9(10), pp. 537–538.

[12] Hovel,H.J., 1978. Novel materials and devices for sunlight concentrating systems. *IBM Journal of Research and Development*, Vol. 22, pp. 112–121.

[13] Fanetti, E., Flores, C., Guarini, G., Paletta, F. and Passoni, D., 1981. High efficiency 1.43 and 1.69 eV band gap $Ga_{1-x}Al_xAs$–GaAs solar cells for multicolor applications. *Solar Cells*, Vol. 3, pp. 187–194.
[14] Knechtly, R.C., Loo, R.Y. and Kamath, G.S., 1984. High-efficiency GaAs solar cells. *IEEE Trans. Electron Dev.*, ED-31 (5), pp. 577–588.
[15] Rauschenbach, H.S., 1980. *Solar Cell Array Design Handbook. The Principles and Technology of Photovoltaic Energy Conversion*, Litton Educational Publishing, Inc., New York.
[16] Luque, A., 1989. *Solar Cells and Optics for Photovoltaic Concentration.* Adam Hilger, Bristol, Philadelphia.
[17] Partain, L.D. Ed., 1995. *Solar Cells and Their Application*, John Wiley & Sons.
[18] Iles, P.A., 2000. Future of Photovoltaics for space applications. *Prog. Photovolt: Res. Appl.*, Vol. 8, pp. 39–51.
[19] Andreev, V.M., Kazantsev, A.B., Khvostikov, V.P., Paleeva, E.V., Rumyantsev, V.D. and Shvarts, M.Z., 1994. High-efficiency (24.6%, AM0) LPE Grown AlGaAs/GaAs Concentrator Solar Cells and Modules. *Proc. First World Conf. on Photovoltaic Energy Conversion*, Hawaii, pp. 2096–2099.
[20] Andreev, V.M. and Rumyantsev, V.D, 1996. A^3B^5 based solar cells and concentrating optical elements for space PV modules. *Sol. Energy Mater. Sol. Cells*, Vol. 44, pp. 319–332.
[21] Dupuis, R.D., Dapkus, P.D., Vingling, R.D. and Moundy, L.A., 1977. High-efficiency GaAlAs/GaAs heterostructure solar cells grown by metalorganic chemical vapor deposition. *Appl. Phys. Lett.*, Vol. 31, pp. 201–203.
[22] Nelson, N.J., Jonson, K.K., Moon, R.L., Vander Plas, H.A. and James, L.W., 1978. Organometallic- sourced VPE AlGaAs/GaAs concentrator solar cells having conversion efficiencies of 19%. *Appl. Phys. Lett.*, Vol. 33, pp. 26–27.
[23] Werthen, J.G., Virshup, G.F., Ford, C.W., Lewis, C.R. and Hamaker, H.C., 1986. 21% (one sun, air mass zero) 4 cm^2 GaAs space solar cells. *Appl. Phys. Lett.*, Vol. 48, pp. 74–75.
[24] Tobin, S.P., Vernon, S.M., Woitczuk, S.J., Baigar, C., Sanfacon, M.M. and Dixon, T.M., 1990. Advanced in high-efficiency GaAs solar cells. *Proc. 21st IEEE Photovoltaic Specialists Conf.*, pp.158–162.
[25] Tobin, S.P., Vernon, S.M., Sanfacon, M.M. and Mastrovito, A., 1991. Enhanced light absorption in GaAs solar cells with internal Bragg reflector. *Proc. 22nd IEEE Photovoltaic Specialists Conf.*, pp.147–152.
[26] Andreev, V.M., Komin, V.V., Kochnev, I.V., Lantratov, V.M. and Shvarts, M.Z., 1994. High-efficiency AlGaAs–GaAs Solar Cells with Internal Bragg Reflector. *Proc. First World Conf. on Photovoltaic Energy Conversion*, Hawaii, pp.1894–1897.
[27] Shvarts, M.Z., Chosta, O.I., Kochnev, I.V., Lantratov, V.M. and Andreev, V.M., 2001. Radiation resistant AlGaAs/GaAs concentrator solar cells with internal Bragg reflector. *Sol. Energy Mater. Sol. Cells*, Vol. 68, pp. 105–122.

[28] Yamaguchi, M., Space solar cell R&D activities in Japan. 1997. *Proc. 15th Space Photovoltaic Research and Technology Conf.*, pp. 1–10.
[29] Fan, C.C., Tsaur, B.-Y. and Palm, B.J., 1982. Optimal design of high-efficiency tandem cells. *Proc. 16th IEEE Photovoltaic Specialists Conf.*, San Diego, pp. 692–698.
[30] Green, M.A., 1982. *Solar Cells*. Prentice-Hall Inc., New Jersey.
[31] Lamorte, M.F. and Abbott, D.H., 1980. Computer modeling of a two-junction, monolithic cascade solar cell. *IEEE Trans. Electron. Dev.*, ED-27, pp. 231–249.
[32] Spitzer, M.B. and Fan, C.C., 1990. Multijunction cells for space applications. *Solar Cells*, Vol. 29, pp. 183–203.
[33] Fraas, L.M., 1995. High-efficiency III–V multijunction solar cells. In: *Solar Cells and Their Applications*. Partain, L.D., Ed., John Wiley & Sons, pp.143–162.
[34] Jain, R.K. and Flood, D.J., 1993. Monolithic and mechanical multijunction space solar cells. *Journal of Solar Energy Engineering*, Vol. 115, pp. 106–111.
[35] Fraas, L.M., Avery, J.E., Martin, J., Sundaram, V.S., Giard, G., Dinh, V.T., Davenport, T.M., Yerkes, J.W. and O'Neil, M.J., 1990. Over 35-percent efficient GaAs/GaSb tandem solar cells. *IEEE Trans. Electron. Dev.*, Vol. ED-37, pp. 443–449.
[36] Kurtz, S.R., Myers, D. and Olson, J.M., 1997. Projected performance of three- and four-junction devices using GaAs and GaInP. *Proc. 26th IEEE Photovoltaic Specialists Conf.*, Anaheim, pp. 875–878.
[37] Yamaguchi, M., 2001. Multi-junction solar cells: present and future. *Technical Digest 12th Int. Photovoltaic Solar Energy Conf.* pp. 291–294.
[38] Bett, A.W., Dimroth, F., Stollwerk, G. and Sulima, 1999. O.V., III–V compounds for solar cell applications. *Appl. Phys.*, A69, pp. 119–129.
[39] Yamaguchi, M. and Luque, A., 1999. High efficiency and high concentration in photovoltaics. *IEEE Trans. Electron Devices*, Vol. ED-46(10), pp. 41–46.
[40] Wanlass, M.W., Ward, J.S., Emery, K.A., Gessert, T.A., Osterwald, C.R. and Coutts, T.J., 1991. High performance concentrator tandem solar cells based on IR-sensitive bottom cells. *Solar Cells*, Vol. 30, pp. 363–371.
[41] Andreev, V.M., Karlina, L.B., Kazantsev, A.B., Khvostikov, V.P., Rumyantsev, V.D., Sorokina, S.V. and Shvarts, M.Z., 1994. Concentrator Tandem Solar Cells Based on AlGaAs/GaAs–InP/InGaAs (or GaSb) Structures. *Proc. First World Conf. on Photovoltaic Energy Conversion*, Hawaii, pp. 1721–1724.
[42] Andreev, V.M., Khvostikov, V.P., Paleeva, E.V., Rumyantsev, V.D., Sorokina, S.V., Shvarts, M.Z. and Vasil'ev, V.I., 1996. Tandem solar cells based on AlGaAs/GaAs and GaSb structures. *Proc. 23d Int. Symposium on Compound Semiconductors*, pp. 425–428.
[43] Andreev, V.M., 1999. R&D of III–V compound solar cells in Russia. *Technical Digest 11th Int. Photovoltaic Solar Energy Conf.* pp. 589–592.

[44] Umeno, M., Kato, T., Yang, M., Azuma, Y., Soga, T. and Jimbo, T. 1994. High efficiency AlGaAs/Si tandem solar cell over 20%. *Proc. First World Conf. on Photovoltaic Energy Conversion*, Hawaii, pp.1679–1684.

[45] Chung, B.-C., Virshup, G.F., Ristow, M. Ladle, Wanlass, M.W., 1991. 25.2%-efficiency (1-sun, air mass 0) AlGaAs/GaAs/InGaAsP three-junction, two-terminal solar cells, *Proc. 22nd IEEE Photovoltaic Specialists Conf.*, Las Vegas, pp. 54–57.

[46] Andreev,V.M., Khvostikov, V.P., Rumyantsev, V.D., Paleeva, E.V., Shvarts, M.Z., Monolithic two-junction AlGaAs/GaAs solar cells. 1997. *Proc. 26th IEEE Photovoltaic Specialists Conf.*, Anaheim, pp. 927–930.

[47] Timmons, M.L., Hutchley, J.A., Wagner, D.K. and Tracy, J.M., 1988. Monolithic AlGaAs/Ge cascade cell. *Proc. 21st IEEE Photovoltaic Specialists Conf.*, Kissimmee, pp. 602–606.

[48] Tobin, S.P., Vernon, S.M., Bajgar, C., Haven, V.E., Geoffroy, L.M., Sanfacon, M.M., Lillington, D.R., Hart, R.E., Emery, K.A. and Matson, R.L., 1988. High efficiency GaAs/Ge monolithic tandem solar cells. *Proc. 20th IEEE Photovoltaic Specialists Conf.*, Las Vegas, pp. 405–410.

[49] Iles, P.A., Yeh, Y.-C.M., Ho, F.N., Chu, C.L. and Cheng, C., 1990. High-efficiency (> 20% AM0) GaAs solar cells grown on inactive Ge substrates. *IEEE Electron Device Letters*, Vol. 11(4), pp. 140–142.

[50] Wojtczuk, S., Tobin, S., Sanfacon, M., Haven, V., Geoffroy, L. and Vernon, S., 1991. Monolithic two-terminal GaAs/Ge tandem space concentrator cells. *Proc. 22nd IEEE Photovoltaic Specialists Conf.*, Las Vegas, pp. 73–79.

[51] Iles, P.A. and Yeh, Y.-C.M., 1995. Silicon, gallium arsenide and indium phosphide cells: single junction, one sun space. In: *Solar Cells and Their Applications*, Partain, L.D., Ed., John Wiley & Sons, pp. 99–121.

[52] Olson, J.M., Kurtz, S.R., Kibbler, A.E. and Faine, P., 1990. Recent advances in high efficiency $GaInP_2$/GaAs tandem solar cells. *Proc. 21st IEEE Photovoltaic Specialists Conf.*, Kissimmee, pp. 24–29.

[53] Bertness, K.A., Kurtz, S.R., Friedman, D.J., Kibbler, A.E., Kramer, C. and Olson, J.M., 1994. High-efficiency GaInP/GaAs tandem solar cells for space and terrestrial applications. *Proc. First World Conf. on Photovoltaic Energy Conversion*, Hawaii, pp.1671–1678.

[54] Chiang, P.K., Krut, D.D., Cavicchi, B.T., Bertness, K.A., Kurtz, S.R. and Olson, J.M., 1994. Large area GaInP/GaAs/Ge multijunction solar cells for space application. *Proc. First World Conf. on Photovoltaic Energy Conversion*, Hawaii, pp. 2120–2123.

[55] Chiang, P.K., Ermer, J.H., Niskikawa, W.T., Krut, D.D., Joslin, D.E., Eldredge, J.W. and Cavicchi, B.T., 1996. Experimental results of $GaInP_2$/GaAs/Ge triple junction cell development for space power systems. *Proc. 25th IEEE Photovoltaic Specialists Conf.*, Washington, D.C., pp. 183–186.

[56] King, R.R., Karam, N.H., Ermer, J.H., Haddad, M., Colter, P., Isshiki, T., Yoon, H., Cotal, H.L., Joslin, D.E., Krut, D.D., Sudharsanan, R., Edmondson, K., Cavicchi, B.T. and Lillington, D.R., 2000. Next-generation, high-efficiency III–V multijunction solar cells. *Proc. 28th IEEE Photovoltaic Specialists Conf.*, Anchorage, pp. 998–1005.

[57] King, R.R., Fetzer, C.M., Colter, P.C., Edmondson, K.M., Ermer, J.H., Cotal, H.L., Yoon, H., Stavrides, A.P., Kinsey, G., Krut, D.D. and Karam, N.H., 2002. High-efficiency space and terrestrial multijunction solar cells trough bandgap control in cell structures. *Proc. 29th IEEE Photovoltaic Specialists Conf.*, New Orleans, pp. 776–781.

[58] Stavrides, A., King, R.R., Colter, P., Kinsey, G., McDanal, A.J., O'Neill, M.J. and Karam, N.H., 2002. Fabrication of high efficiency, III–V multijunction solar cells for space concentrators. *Proc. 29th IEEE Photovoltaic Specialists Conf.*, New Orleans, pp. 920–922.

[59] Chiang, P.K., Chu, C.L., Yeh, Y.C.M., Iles, P., Chen, G., Wei, J., Tsung, P., Olbinski, J., Krogen, J., Halbe, S. and Khemthong, S. 2000. Achieving 26% triple junction cascade solar cell production. 2000. *Proc. 28th IEEE Photovoltaic Specialists Conf.*, Anchorage, pp.1002–1005.

[60] Hou, H.Q., Sharps, P.R., Fatemi, N.S., Li, N., Stan, M.A., Martin, P.A., Hammons, B.E. and Spadafora, F., 2000. Very high efficiency InGaP/GaAs dual-junction solar cell manufacturing at Emcore Photovoltaics. *Proc. 28th IEEE Photovoltaic Specialists Conf.*, Anchorage, pp.1173–1176.

[61] Bett, A.W., Dimroth, F., Lange, G., Meusel, M., Beckert, R., Hein, M., Riesen, S.V. and Schubert, U., 2000. 30% monolithic tandem concentrator solar cells for concentrations exceeding 1000 suns. *Proc. 28th IEEE Photovoltaic Specialists Conf.*, Anchorage, pp. 961–964.

[62] Dimroth, F., Schubert, U., Bett, A.W., Hilgarth, J., Nell, M., Strobl, G., Bogus, K. and Signorini, C., 2001. Next generation GaInP/GaInAs/Ge multijunction space solar cells. *Proc. 17th European Photovoltaic Solar Energy Conf.*, Munich, pp. 2150–2154.

[63] King, R.R., Haddad, M., Isshiki, T., Colter, P., Ermer, J., Yoon, H., Joslin, D.E. and Karam, N.H., 2000. Metamorphic GaInP/GaInAs/Ge solar cells. *Proc. 28th IEEE Photovoltaic Specialists Conf.*, Anchorage, pp. 982–985.

[64] Yamaguchi, M., Ohmachi, Y., O'Hara, T., Kadota, Y., Imaizumi, M. and Matsuda, S., 2000. GaAs-on-Si solar cells for space use. *Proc. 28th IEEE Photovoltaic Specialists Conf.*, Anchorage, pp. 1012–1015.

[65] Fraas, L.M., Daniels, W.E., Huang, H.X., Minkin, L.E., Avery, J.E., O'Neill, M.J., Mc.Danal, A.J. and Piszczor, M.F., 2001. 34% efficient InGaP/GaAs/GaSb cell-interconnected-circuit for line-focus concentrator arrays. *Proc. 17th European Photovoltaic Solar Energy Conf.*, Munich, pp. 2300–2303.

[66] O'Neill, M.J., McDanal, A.J., George, P.J., Piszczor, M.F., Edwards, D.L., Hoppe, D.T., Eskenazi, M.I., Botke, M.M., Jaster, P.A. and Brandhorst, H.W., 2002. Development of the ultra-light stretched lens array. *Proc. 29th IEEE Photovoltaic Specialists Conf.*, New Orleans, pp. 916–919.

[67] Andreev, V.M., Larionov, V.R., Lantratov, V.M., Grilikhes, V.A., Khvostikov, V.P., Rumyantsev, V.D., Sorokina, S.V. and Shvarts, M.Z., 2000. Space concentrator module based on short focus linear Fresnel lenses and GaAs/GaSb tandem stacks. *Proc. 28th IEEE Photovoltaic Specialists Conf.*, Anchorage, pp.1157–1160.

[68] Green, M.A., 2000. Prospects for photovoltaic efficiency enhancement using low-dimensional structures. *Nanotechnology*, Vol. 11, pp. 401–405.

[69] Ekins-Daukes, N.J., Barnes, J.M., Barnham, K.W.J., Connolly, J.P., Mazzer, M., Clark, J.C., Grey, R., Hill, G., Pate, M.A. and Roberts, J.S., 2001. Strained and strain-balanced quantum well devices for high-efficiency tandem solar cells, *Sol. Energy Mater. Sol. Cells* , Vol. 68, pp. 71–87.

[70] Luque, A. and Marti, A., 2001. A metallic intermediate band high efficiency solar cell. *Prog. Photovolt: Res. Appl.*, Vol. 9, pp. 73–86.

[71] Marti, A. Cuadra, L. and Luque, A., 2001. Partial filling of a quantum dot intermediate band for solar cells. *IEEE Trans. Electron Devices*, Vol. ED-48(10), pp. 2394–2399.

IId-2

High-Efficiency Concentrator Silicon Solar Cells

Pierre J. Verlinden, Origin Energy, Adelaide, Australia

1 Introduction

Concentrating sunlight for photovoltaic conversion has always been a very attractive solution. Since one can easily acknowledge that the cost of photovoltaic energy conversion is driven by the fabrication cost of the solar cells, and particularly the cost of the semiconductor material, it becomes obvious that much less expensive concentrating lenses or mirrors can replace the expensive solar cells area. The concentration ratio can be increased several hundred fold, to the point where the cost of fabricating the solar cells becomes insignificant in the overall PV system cost. However, this benefit does not come without cost. Concentrating PV systems need solar trackers with a tracking precision that increases with the concentration ratio, and a more expensive module design with a well-engineered and low-cost cooling system for the solar cells.

Concentration not only increases the energy productivity of the solar cell material and device, but also increases its efficiency, since both current and voltage increase with the light intensity. However, in order to compare efficiencies of concentrator PV systems with flat plate PV systems, we have to remember that concentrator systems only use direct sunlight, about 85% of the incident power density on a clear sunny day. Therefore, a 20% efficient concentrator system would produce about the same amount of energy as a 17% efficient flat plate system, if they are both mounted on the same tracking system.

The series resistance of the cell limits the concentration ratio to which the solar cells can be used and the efficiency advantage of concentration systems over flat plates. To collect a current that is, for example at 500× concentration, almost 20 A/cm^2 from a solar cell that has an open-circuit voltage of 0.800 V requires a series resistance that is less than 0.001 Ω cm^2. To achieve such low series resistance, the concentrator solar cell requires a well-engineered double-level, solderable metallisation scheme [4, 5] (see Figure 1).

The carrier recombination in commercial one-sun solar cells for flat-plate application is usually dominated by bulk Shockley-Read-Hall (SRH) or surface recombination. In high-efficiency (> 18%) solar cells or when medium-concentration (< 100×) is applied, the carrier recombination is usually

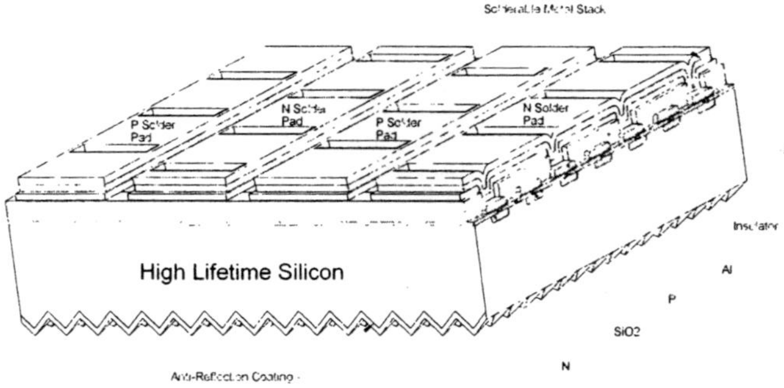

Figure 1 Structure of IBC silicon solar cell.

dominated by junction recombination. Auger recombination usually dominates in high-concentration silicon solar cells. Auger recombination occurs when an electron from the conduction band recombines with a hole from the valence band giving its energy to another electron. This is the opposite mechanism of impact ionisation. The Auger recombination rate increases as the cube of the carrier density, and is generally the dominant recombination mechanism when the carrier concentration exceeds 10^{17} cm^{-3}. In order to reduce the Auger recombination rate, concentrator solar cells must be as thin as possible (typically 120 μm or less) and, therefore, require a good light trapping.

Because the solar cell cost represents only around 10% of the total concentrator PV system cost, a high efficiency solar cell provides a great leverage for reducing the cost of solar electricity; the higher the concentration ratio the greater the leverage. The most efficient silicon solar cell, both for laboratory cells and for production scale, is currently the interdigitated back contact (IBC) and, in particular, the point-contact (PC) solar cell [5–9]. The structure of such a cell is shown in Figure 1. It has attained a conversion efficiency of 28.3% [2, 3] in the laboratory, and 26.8% at 100× (AM1.5D, 10 W/cm^2, 25°C) at the production scale [1]. It is, at the present time, the only high-intensity silicon solar cell commercially available. Figures 2 and 3 show examples of concentrator silicon solar cells manufactured by SunPower Corporation of Sunnyvale, California. Two examples of a reflective concentrator dish, built by Solar Systems Pty Ltd of Hawthorn, Australia and using high-efficiency concentrator silicon solar cells, are shown in Figures 4 and 5. The 20 m^2 concentrator system represented in Figure 5 was the first silicon-based concentrator system to reach an overall system efficiency of 20% under normal operating conditions [6]. Figure 6 shows a picture of the receiver of such concentrator system in operation.

2 The Family of Backside Contact Silicon Solar Cells

2.1 IBC Solar Cells

Schwartz et al. [7, 8] introduced the IBC design in 1975. The main reason that the IBC solar cell design is particularly suitable for high concentration is that both metal contacts are made on the backside of the cell (Figure 1). Therefore, there is no shadowing effect on the front side. The trade-off between the shadowing effect and the series resistance is eliminated, which is particularly interesting for concentration applications. Furthermore, the design of the front side (optical side) and the backside (electrical side) can easily be optimised separately. In a conventional solar cell, the doping of the front side emitter is a trade-off between series resistance and efficiency: a lighter doped front side emitter would improve the quantum efficiency and reduce the emitter saturation current density, but it would increase the series resistance of the cell. On the other hand, in IBC solar cells, the front side could be optimised for maximum

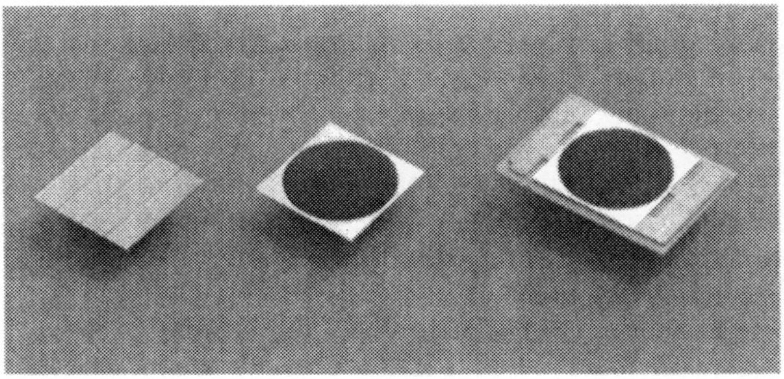

Figure 2 A point-contact silicon solar cell for Fresnel concentrator application. (Courtesy of SunPower Corporation.)

Figure 3 A point-contact silicon solar cell for dense array concentrator application. (Courtesy of SunPower Corporation.)

quantum efficiency, reduced recombination and reduced sub-linearity, without compromising the series resistance of the cell.

The main requirements for obtaining a high efficiency with IBC cells are:

- Long recombination lifetime in the bulk. After the solar cell fabrication the diffusion length must be at least 5 times longer than the solar cell thickness. For this reason, Float-Zone (FZ), high-resistivity, n-type substrates with carrier lifetime greater than 1 ms are generally preferred. The typical solar cell thickness is 100–150 μm.
- Low front surface recombination velocity. The front surface must be passivated with a thin, thermally grown, silicon dioxide layer, grown in a very clean environment.

2.2 Front-Surface-Field, Tandem-Junction and Point-Contact Solar Cells

Over the years, several variants in the IBC design have been introduced. The front-surface-field (FSF) solar cell has a high-low (n+/n or p+/p) junction on the

Figure 4 Picture of a 25 kW dish concentrating PV system with point-contact silicon solar cells. (Courtesy of Solar Systems Pty Ltd.)

Figure 5 A dish concentrating PV system with point-contact silicon solar cells. This PV system, installed in White Cliff, Australia, demonstrated 20% overall efficiency under normal operating conditions [6]. (Courtesy of Solar Systems Pty Ltd.)

front side of the IBC solar cell (Figure 7(a)). The function of the front surface field is to reduce the effective front-surface recombination velocity for the carriers in the bulk of the device. It behaves the same way as a back surface field (BSF) in a conventional n+/p/p+ solar cell. However, since the high-low junction is now applied on the front side of the cell and since the solar cell is operating under high injection, the effective recombination velocity increases as the injection level increases and the effect of the FSF decreases. In high injection, the front-surface recombination current density, $J_{fs,rec}$, can be expressed as a function of the

Figure 6 A dense array receiver with point-contact silicon solar cells. (Courtesy of Solar Systems Ltd Pty.)

saturation current density of the high-low junction, J_{0+}, or as a function of an effective surface recombination velocity, S_{eff}:

$$J_{fs,rec} = \frac{n^2}{n_i^2} J_{0+} = qnS_{eff} \tag{1}$$

where n is the carrier concentration (note that, in high-injection, the electron and hole concentrations are equal, $n = p$) at the front surface of the cell and n_i is the intrinsic carrier concentration.

Tandem-junction (TJ) solar cells are identical to FSF solar cell, except that a floating p-n junction replaces the high-low junction at the front side of the cell (Figure 7b). The front surface recombination current density, $J_{fs,rec}$, is determined by Equation (1) in the same way as in the FSF solar cell, and there is practically no difference between a TJ and a FSF solar cell (the only difference resides in the space charge recombination current at a very low intensity). The designer choice between a TJ and a FSF solar cell is fully determined by the ability to make a front side emitter with the lowest emitter saturation current density, J_{0+}, as possible. For this reason, n-type (phosphorous doped) emitters are usually preferred.

As we can see from Equation (1), the front surface recombination current increases as n^2 and the effective surface recombination velocity increases as the injection level increases:

$$S_{eff} = \frac{n}{qn_i^2} J_{0+} \tag{2}$$

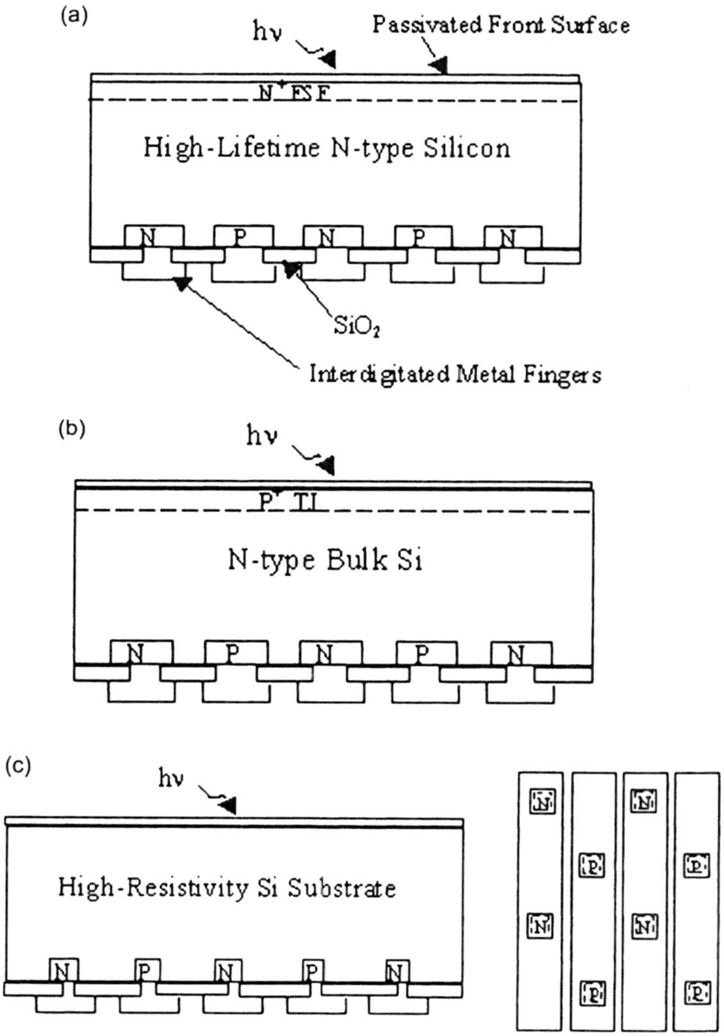

Figure 7 (a) Structure of a front-surface-field silicon solar cell. (b) Structure of a tandem junction silicon solar cell. (c) Structure of a point-contact silicon solar cell.

These characteristics result in a significant sub-linearity of the responsivity of the cell. The responsivity of a solar cell is, for a given spectrum, the total photo-generated current per unit of incident power. The responsivity has the dimension of amp/watt. The typical value of responsivity for high-efficiency concentrator solar cells is 0.395 A/W. If it varies with the concentration ratio, the solar cell is called sub-linear or super-linear. Therefore, a doped front side surface (TJ or FSF) is usually reserved to low-concentration applications ($< 10\times$).

Point-contact solar cells are IBC cells with a reduced emitter area and reduced metal contact area on the backside of the cell (Figure 7(c)) in order to decrease as much as possible the emitter recombination. For one-sun or low-concentration applications ($< 5\times$), the backside emitter coverage fraction could be kept at almost 100% since the dominant recombination mechanisms are SRH and surface recombinations, and a simple IBC design is preferred. However, the metal contact area is kept very low, typically less than 1%, depending on the metal-Si contact resistance. For concentration ratios between $5\times$ and $200\times$, the backside emitter coverage fraction could be reduced to typically around 10%, whereas the metal contact area is typically 5% of the backside of the cell.

3 Modelling of Backside Contact Solar Cells

The simplest way to model IBC solar cells is to consider the integral approach as proposed by R.M. Swanson [9, 10] and, for a precise simulation of an IBC or PC solar cell, we refer the reader to the model developed in these papers which includes 3D effects close to the backside contacts. However, we will give below a simple formulation.

The current, I, at the terminals of a solar cell can be written in the form:

$$I = I_{ph} - I_{b,rec} - I_{s,rec} - I_{em,rec} \quad (3)$$

where I_{ph} is the photogenerated current (the maximum photogenerated current within the silicon material with a defined thickness, including reflection losses and light trapping), $I_{b,rec}$ is the sum of all the bulk recombination currents (including SRH and Auger recombination), $I_{s,rec}$ is the sum of all the surface recombination currents (including front side, backside and edge surface recombination) and $I_{em,rec}$ is the sum of all the emitter recombination currents (including backside emitters, the front TJ or FSF emitter if it exists, and also including recombination at the contacts). The recombination currents increase as the carrier concentration increases and, therefore, increase as the terminal voltage increases. Eventually, at the open-circuit condition, the sum of all the recombination currents will be equal to the photocurrent, I_{ph}, and the terminal current will be equal to zero. Therefore, at V_{oc}

$$I_{ph} = I_{b,rec} + I_{s,rec} + I_{em,rec} \quad (4)$$

Calculating the different components of the recombination current is the most interesting and the most powerful approach to simulate, optimise or analyse the performance of solar cells. Table 1 gives a summary of the different relevant recombination mechanisms in silicon, their controlling parameters, typical values in high-efficiency concentrator silicon solar cells and the corresponding recombination currents under high-level injection.

If an IBC solar cell is to be modelled, in a simple manner, as a backside illuminated conventional solar cell, one has to realise that:

- in most of the device, the electron and holes are flowing in the same direction, from the front side of the cell toward the backside where the collecting n-type and p-type junctions are;

Table 1 Recombination mechanism and the corresponding recombination current in highly injected silicon solar cells.

Carrier recombination mechanism	Controlling parameter	Unit	Typical values	Recombination current density (A/cm^2)
Bulk (SRH, trap assisted)	τ_B	s	1–10 ms	$\frac{qnW}{\tau_B}$
Surface (trap assisted)	S	cm/s	1–4 cm/s	qnS
Emitter	J_o	A/cm^2	50–200 fA/cm^2	$\frac{n^2}{n_i^2} J_o$
Auger	C_A	cm^6/s	1.66×10^{-30}	$qn^3 C_A W$

The ambipolar Auger coefficient value is from Sinton [11, 12]. In this chapter, τ_B represents the bulk lifetime and, since we are considering high-injection conditions only, we have assumed that $\tau_n = \tau_p = \tau_B$, where τ_n and τ_p are the electron and hole lifetimes.

- since the cell is under high injection, and to maintain electrical neutrality in the device, the electron and hole concentrations are equal throughout the device;
- for the same reasons as above, electrons and holes are flowing with the same ambipolar diffusion constant, $D_a = 2D_nD_p/(D_n + D_p)$, where D_n and D_p are the electron and hole diffusion constants;
- in most of the device, from the front-side to very close to the emitter area where 3D current flow starts to appear, the current flow is almost unidirectional, perpendicular to the surface:

$$J_n = -J_p = qD_a \frac{dn}{dx} = qD_a \frac{dp}{dx} \tag{5}$$

where J_n and J_p are the electron and hole current densities, and, therefore, the total current along the wafer J_T is negligible:

$$J_T \approx 0 \tag{6}$$

Since the backside carrier concentration is determined by the terminal voltage of the solar cell

$$n_{back} = n_i \exp(\frac{qV}{2k_BT}) \tag{7}$$

we can use Equation (3) and the expression of recombination currents in Table 1 to calculate the cell current at any bias voltage. We can take, as a simple example, a high-efficiency IBC or PC solar cell under high injection, with long carrier lifetime ($\tau_n = \tau_p \gg W^2/D_a$, where W is the thickness of the cell) and low surface recombination velocity ($S_0 \ll D_a/W$, where S_o is the surface recombination velocity at the front surface).

In short-circuit condition, we can consider that, in the first approximation:

- $n_{back} = 0$,

- the emitter and surface recombination currents at the back of the cell can be neglected, as well as Auger recombination,
- the electron and hole concentrations linearly decrease from front to back.

Equation (3) then becomes:

$$J_{sc} = J_{ph} - \frac{qn_0W}{2\tau_B} - qn_0S_0 = qD_a\frac{n_0}{W} \tag{8}$$

where J_{sc} is the short circuit current density, n_0 is the front-surface electron concentration and currents were replaced by current densities in view of the effectively planar geometry of the cell as a result of the perpendicular direction of the current. A few iterations are necessary to determine the short-circuit current and the front-surface carrier concentration.

It becomes immediately apparent that, if the thickness of the cell, W, increases, the front-surface carrier concentration, n_0, increases, which, in turn, results in an increased bulk and front-surface recombination current.

In open-circuit condition, we can consider that the electron and hole concentrations are constant throughout the device. The open-circuit voltage is given by:

$$V_{oc} = 2\frac{k_BT}{q}\ln(\frac{n}{n_i}) \tag{9}$$

after solving the following equation:

$$J_{ph} = \frac{qnW}{\tau_B} + qn^3C_aW + qnS_0 + qnS_{back}(1 - A_n - A_p) + \frac{n^2}{n_i^2}(A_nJ_{0n} + A_pJ_{0p}) \tag{10}$$

where J_{on} and J_{op} are the saturation current densities at the n and p junctions, A_n and A_p are the n-type and p-type emitter coverage fractions respectively, and S_{back} is the backside surface recombination velocity.

The maximum power point and efficiency of the solar cell can be determined the same way, considering that the backside carrier concentration is given by Equation (7) and that the carrier concentration is still linearly distributed from front to back. Equation (3) becomes:

$$\begin{aligned} J_{mp} = J_{ph} - \frac{qn_{avg}W}{\tau_B} - q\int_0^W n(x)^3C_a dx - qn_0S_0 \\ - qn_{back}S_{back}(1 - A_n - A_p) - \frac{n_{back}^2}{n_i^2}(A_nJ_{0p} + A_pJ_{0p}) = qD_a\frac{n_0 - n_{back}}{W} \end{aligned} \tag{11}$$

where $n_{avg} = (n_0 + n_{back})/2$ is the average carrier concentration and

$$n(x) = n_{back} + (n_0 - n_{back})(1 - \frac{x}{W}) \tag{12}$$

Several iterations are necessary to reach the solution.

This is a very simple way to analyse the IBC or PC solar cells. It allows a quick determination of the carrier concentration in the device and an analysis of the different recombination mechanisms. Note that Equation (1) must be used for the front-side recombination current expression in FSF or TJ solar cells. For a more precise modelling, including 3D effects at the backside of the solar cell, the reader is referred to the Swanson model [9, 10].

4 Perimeter and Edge Recombination

So far, we have not considered the edge recombination. It has been recently demonstrated that, in most high-efficiency silicon solar cells, the dominant recombination mechanism is a recombination current at the unpassivated surface at the edge of the silicon die [13]. Two cases need to be considered here:

- aperture illuminated solar cells (e.g. cells for Fresnel lens modules, Figure 2),
- totally illuminated solar cells (e.g. cells for dense array receivers, dish and thermo-photovoltaic (TPV) applications, Figure 3).

In order to reduced the edge recombination, the aperture illuminated IBC solar cells must have its active area as far as possible from the edge of the silicon die, in theory at a distance of least three times the ambipolar diffusion length at the considered carrier concentration. In practice, economical considerations prevent manufacturers to increase this distance too much and a typical distance between the edge of the silicon die and the nearest emitter at the backside of the cell is 500 μm. For totally illuminated solar cells, there is an optimal distance that can be calculated as explained in [13]. The width of the border region is optimal when the illuminated border region generates just enough current to supply the recombination current at the edge of the silicon die. A wider or narrower border region than the optimal width would result in a lower efficiency. If d is the width of the border region, P the cell perimeter, and n the average carrier density at the middle of the cell, at the maximum power point (V_m, I_m) and at the considered concentration ration, the current generated by the border region is:

$$I_{border} = dPJ_{sc}$$

and the edge recombination current is:

$$I_{edge} = qPWD_a n/d$$

The border width is optimal when:

$$I_{border} = I_{edge}$$

and

$$d = \sqrt{\frac{qWD_a n}{J_{sc}}} \tag{13}$$

For a fixed geometry and cell thickness, the carrier concentration and short-circuit current are more or less proportional to the concentration ratio. Therefore, the optimal border width is generally independent of the concentration ratio. However, since the average carrier concentration increases with the thickness of the cell, the optimal border width is roughly proportional to the thickness of the cell. For a typical IBC solar cell for dense array application at 400× and with a thickness of 100 μm ($n \approx 10^{17}\text{cm}^{-3}$), the optimal border width is about 135 μm.

Equation (13) is valid if the edge is unpassivated such as in silicon solar cells diced with a dicing saw. New techniques to passivate the edges of solar cells are needed but, so far, there has been no satisfactory development in low-temperature passivation of silicon surfaces.

5 Manufacturing Process for Backside Contact Solar Cells

The typical process flow to manufacture IBC solar cells is presented in Figure 8 although actual processes may significantly differ in practice. A typical process requires 6 photolithography steps. Additional photolithography steps are required for:

- inverted pyramid texturisation instead of random texture,
- local thinning instead of uniform wafer thinning.

6 Stability of Backside Contact Cells

In the same way as many other high-efficiency silicon solar cells, the IBC cells are subject to efficiency degradation. If float-zone (FZ) wafers are used, the degradation is limited to the silicon-silicon dioxide interface [1, 14–16]. The degradation of the interface could be due to the loss of hydrogen atoms passivating the dangling bonds at the silicon-silicon dioxide interface, or the creation of new interface states. The degradation of the interface is manifested by:

- an increase of the surface recombination velocity, which, in turn, results in a reduction of the short-circuit current over the entire range of concentration ratio or,
- by an increase of the front emitter current Jo which results in an increase of the sub-linearity (the cell responsivity decreases as the concentration ratio increases) if the front surface is lightly doped or inverted due to charges in the dielectric layer.

Both mechanisms will result in a significant decrease in efficiency. The degradation has been observed so far due to the following individual or combined conditions:

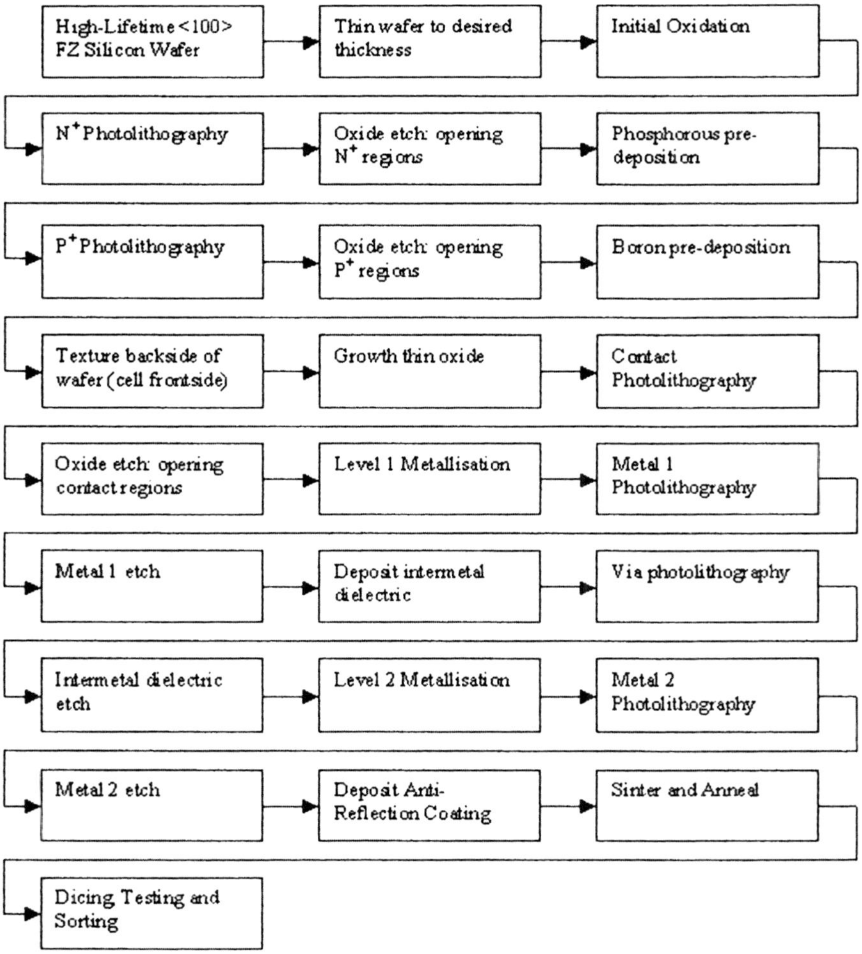

Figure 8 Process flow of manufacturing IBC solar cells.

- UV light, which creates hot electrons that can overcome the potential barrier at the silicon-silicon dioxide interface and create new interface states,
- elevated temperature (> 100°C), which allows hydrogen atoms to escape from the interface,
- mechanical stress.

It is also assumed that the following conditions enhance the degradation:

- highly concentrated sunlight which creates a large concentration of carriers at the surface of the solar cell;

- the presence of moisture;
- the presence of atomic hydrogen which could break an existing Si-H bond to form a hydrogen molecule H_2.

There is still much research to be done to understand the degradation mechanisms and to find a solution to this issue.

The easiest way to prevent the strong influence of the degradation of the surface recombination velocity to solar cell efficiency is to isolate the surface from the rest of the cell with FSF or a floating tandem junction. Unfortunately, although it makes the cell more stable, the front side emitter has a significant impact on the sub-linearity of the solar cell [1, 15], which makes this solution unsuitable for concentration application. In any case, the front emitter must be lightly doped in such way that the front emitter J_o is kept as low as possible. Indeed, we can see from Equation (1) that the greater the front emitter J_o, the more significant the sub-linearity of the solar cell will be.

There are other ways to ensure stability of the IBC solar cells:

- It is very important that, during the solar cell fabrication, the front passivating oxide layer is grown with the smallest interface state density. 1,1,1-Trichloroethane (TCA), dichloroethane (DCA), trichloroethylene (TCE) or HCl oxide growth, as well as an aluminium anneal, should be avoided as they have been recognised as giving very unstable Si–SiO_2 interfaces [16, 17].
- UV light with wavelength below 400 nm should be filtered from the solar spectrum, either at the solar cell level or at the concentrator module level. For example, the anti-reflection coating (ARC) can include a UV filter such as TiO_2 or a very thin polysilicon layer [14, 16].
- The front surface of the solar cell should be coated with a thin layer of silicon nitride (as part of the ARC), or another material, hopefully compatible with the ARC, forming a great hydrogen and moisture diffusion barrier (for example, a very thin amorphous silicon or polysilicon layer) [16].

7 Toward 30% Efficiency Silicon Cells

In 1985, Gray and Schwartz presented a paper entitled 'Why don't we have a 30% efficient silicon solar cell?' [18]. At that time, the highest reported efficiency for silicon solar cell was 22% under concentrated sunlight. In that paper, the authors concluded that 30% efficiency would be attainable with a silicon solar cell if:

- good light trapping was developed which does not degrade the open-circuit voltage of the solar cell,
- the contact and grid series resistance could be reduced enough to allow for high concentration ratio,
- the current crowding effects are reduced by a judicious design,

- a novel heterojunction or heteroface contact with small contact resistance and low emitter saturation current was developed.

A few years later, the first three steps to attain the 30% efficiency target had been addressed and resolved: the PC solar cell with a thin substrate and long carrier lifetime, with a textured and passivated front surface, proved to be the best design for high efficiency concentrator solar cell [19, 20]. In 1989, the highest reported efficiency for silicon PC solar cell was 28.3% [21]. The same year, Swanson responded to Gray's paper in a publication entitled "Why we will have a 30% efficient silicon solar cell" [22] where he demonstrates that all the pieces are now in place to fabricate a 30% efficient silicon solar cell. The last milestone to reach the 30% breakthrough, namely the development of polysilicon emitters with a low contact resistance, and low emitter saturation currents had just been demonstrated at Stanford [23, 24]. Swanson announced: Within one year, cells will be reported with efficiency in excess of 30% [22].

We believe that 30% efficient silicon solar cells are possible in a manufacturing environment. In order to achieve this goal, polysilicon or heterojunction emitter technology needs to be implemented, along with several other improvements to the existing technology (thinner cells, improved light trapping, and reduced dimension geometries).

8 How to Improve the Efficiency of Backside Contact Solar Cells

Campbell and Green [25] discussed the efficiency limit of silicon solar cells under concentrated sunlight. They showed that the limit of efficiency for a silicon cell is between 30 and 35%. These very high efficiencies have never been demonstrated so far with a single-junction silicon solar cell. Commercially available PC concentrator cells have efficiencies around 26.5%. In order to reach such high efficiencies, the recombination mechanisms such as trap-assisted SRH recombination, in the bulk or at the surface, and emitter recombination must be negligible compared to Auger recombination. The Auger recombination is intrinsic to the material and, for example, cannot be improved by using a purer starting material or a cleaner fabrication process. Therefore, the only way to

Table 2 Strategy for 30% efficient backside contact solar cells.

Step	Parameter	Target value	Acceptable value	Unit
1	Emitter J_o	3×10^{-14}	5×10^{-14}	A/cm^2
2	Emitter contact resistance	10^{-6}	10^{-4}	Ω cm^2
3	Cell thickness with manufacturable yield	30	80	μm
4	Light trapping: number of passes for long wavelengths	50	35	passes
5	Unit cell geometry	25	40	μm
6	Total series resistance	5×10^{-4}	10^{-3}	Ω cm^2

reduce the Auger recombination rate inside the cell is to reduce its thickness. However, using a very thin silicon solar cell requires the use of a very effective scheme for light trapping. For example, Campbell and Green [25] suggest that the optimum cell would be less than 1 μm thick and could reach 36–37% under concentration.

By comparison, the recombination in the PC solar cell at 250× (25 W/cm^2) is almost equally dominated by emitter (40%) and Auger (40%) recombination. The surface and bulk SRH recombination represents less than 20% of the overall recombination rate. In the PC design, the emitter coverage fraction has already been reduced to the minimum acceptable, and the only way to further reduce the emitter recombination rate is to reduce the emitter saturation current density Jo. Therefore, in order to reach 30% or greater efficiency, the following strategy needs to be used:

1. Reduce the emitter recombination by reducing the emitter saturation current density from 2×10^{-13} to 5×10^{-14} A/cm^2. Polysilicon heterojunction emitters with less than 5×10^{-14} A/cm^2 as saturation current density have already been demonstrated [23, 24].
2. Demonstrate low contact resistance between the bulk and the emitters. A desired value for the contact resistance between bulk and emitter is less than $10^{-5}\,\Omega\,\text{cm}^2$.
3. Reduce the cell thickness in order to reduce the Auger recombination rate. For a practical concentrator for which the angle of incidence of the light to the cell is between 0 and 30 degrees, the optimal thickness is about 30 μm [25]. This is very difficult to achieve in a manufacturing environment. Manufacturing techniques, such as local thinning, need to be implemented.
4. Improve light trapping by designing new textured surfaces on both sides of the cell.
5. Reduce the unit cell geometry down to a dimension equal or smaller than the thickness of the cell.
6. Reduce the external series resistance (metal and interconnect).

8.1 Reduce Emitter Saturation Current Density

The saturation current density of an emitter J_o represents the sum of all the recombination mechanisms inside the emitter. It includes the SRH, surface, contact and Auger recombination mechanisms, as well as heavy doping effects such as band gap narrowing. The best diffused emitter J_o that can be achieved is around 2×10^{-14} A/cm^2. However, such an emitter is very transparent which means that, if a metal contacts it, its saturation current density dramatically increases about 100 fold. For contacting PC solar cell, the emitter must be opaque and the best saturation current density of such emitter is around 2×10^{-13} A/cm^2.

An ideal emitter for silicon should be transparent to majority carriers and a mirror for minority carriers. Therefore, it should have a wide band gap such that an additional potential barrier will appear in the minority band when it is doped (see Figure 9). Although there are many semiconductors with a larger band gap than silicon, so far none of the large band gap materials have proven to be ideal for contacting silicon. Indeed, the potential barrier at the emitter heterojunction blocks the minority carriers, they still can recombine at the interface due to the presence of interface traps.

There are only two materials that were reported to have low J_o: polysilicon emitters and semi-insulating polysilicon emitters (SIPOS), both used with an interfacial thin oxide layer [23, 24, 26, 27]. However, some very promising results have recently been reported on non-concentrator solar cells with amorphous silicon emitters [28] that do not require an interfacial thin oxide to block the minority carriers.

8.2 Demonstrate Low Contact Resistance

The growth of a thin (10 to 20 Å) interfacial layer of silicon dioxide is critical to obtain low emitter saturation current density with polysilicon. However, in order to achieve a low contact resistance, it is necessary to anneal the polysilicon emitter to break up the oxide layer on about 1% of the contact area [23]. The trade-off between J_o and a high contact resistance is highly dependent on the oxide thickness and the anneal (or breakup) step.

8.3 Reduce the Cell Thickness

Once the emitter recombination mechanism has been reduced, it is possible to reduce the Auger recombination rate by using a thinner substrate for the solar cell. Solar cells made on 100 μm thick wafers are can be manufactured in large volumes but thinner cells present a real challenge for manufacturing. Since both sides of the cells must be processed for texturing, passivation and anti-reflection coating, the thinning of the wafers cannot just be done by lapping the wafers at the end of the process. Solar cells of 30–80 μm thickness must be fabricated on locally thinned substrates.

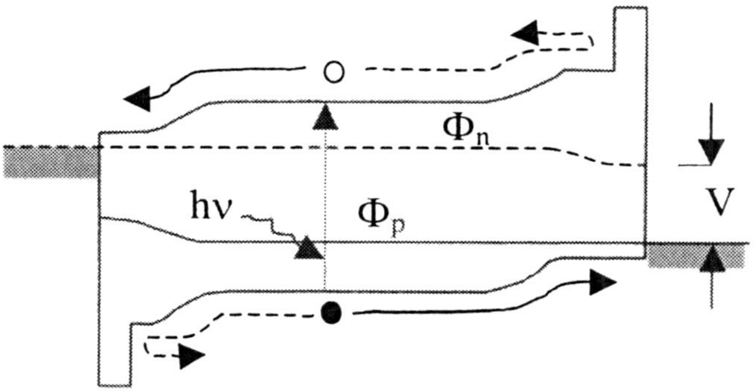

Figure 9 Band diagram of a p–i–n solar cell with heterojunction emitters.

8.4 Improve Light Trapping

When very thin silicon solar cells are designed, the light trapping property of the cells becomes a significant factor for the efficiency. Present cells have only one-side texturisation. Unfortunately, this is not enough for cells thinner than 100 μm. In this case, a double side texturisation or perpendicular grooving must be designed in order to insure good light trapping. In current PC solar cells, the analysis of the internal quantum efficiency near the edge of the band gap shows that the effective number of passes for long wavelength light is greater than 30 [1]. The light trapping can be improved with double side texture, perpendicular slats, or parquet grooves in order to attain up to 50 passes of the light.

8.5 Shrink Geometries

The present PC solar cells have a significant current crowding loss and large internal series resistance accounting for about 0.002 Ω cm^2 which represents about 3% power loss at 25 W/cm^2 or 250×. The unit cell dimension should be reduced from the current 140 μm to 40 μm, and if possible to 25 μm. The metallisation and inter-metal dielectric technology usually prevents the shrinking of the unit cell. In order to be able to shrink the unit-cell dimension to 40 or even 25 μm, a state-of-the-art plasma etching process for the metal, as well as a plasma-enhanced chemical vapour deposition (PECVD) SiO_2-based dielectric as inter-metal dielectric layer must be adopted.

8.6 Reduce Series Resistance

Shrinking the geometries of the unit cell in order to reduce the internal series resistance may have a negative effect on the external series resistance. This problem is due to the fact that, if the gap between metal lines is kept constant, the metal coverage fraction decreases with decreasing the dimensions of the unit cell. A new metallisation design, similar to what has been proposed in previous papers [4, 5], must be implemented in order to reduce the series resistance. As in step 8.5, the new metal design requires a plasma etching process for metal and a SiO_2-based dielectric layer. The target is to reduce the total series resistance to less than 0.001 Ω cm^2, or about a 2.5% power loss at 500× concentration.

Table 3 Comparison of performance of a commercial PC solar cell (HECO335 from Sun-Power Corporation) cell with the projected new high-performance cell at 100× and 275× (AM1.5D, 10 and 27.5 W/cm^2, T_c = 25°C). The values at 275× concentration are in brackets.

Parameter	Commercially available PC solar cell	New high-performance cell	Unit
Responsivity	0.403 (0.379)	0.416 (0.416)	A/W
Open-circuit voltage	807 (825)	826 (845)	mV
Fill factor	82.6 (79)	87 (85.9)	%
Efficiency	26.8 (24.7)	29.9 (30.2)	%

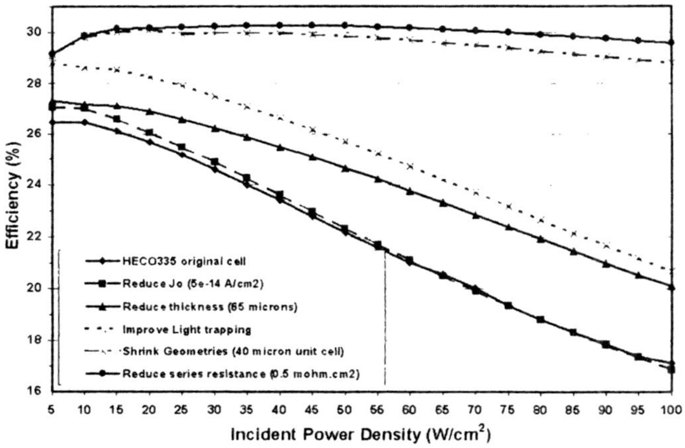

Figure 10 Efficiency (AM1.5D, 25°C) of the point-contact silicon solar cell vs. incident power density after each step of this strategy.

8.7 Target Performance

Table 3 shows the target performance of a 30% efficient silicon solar cell compared to the commercially available point-contact solar cells. Figure 10 shows the improvement in efficiency based on the above strategy. Of course, the full benefit of the polysilicon emitter with low J_o is observed when all the other improvements (thinner cells, improved light trapping and shrunk geometries) are in place.

9 Conclusions

Interdigitated back contact and point-contact silicon solar cells have demonstrated to be the most efficient and the most suitable silicon solar cells for high-concentration applications. Commercially available PC solar cells have demonstrated efficiencies up to 26.8% (at 10 W/cm^2, AM1.5D, 25°C) in large volume production. This chapter has described the structure of IBC, FSF, TJ and PC cells, as well as the process to fabricate them. A simple model for the simulation, optimisation and analysis of the different recombination mechanisms in concentrator IBC cells has been presented. Finally, we have presented a plan for the development of 30% efficiency concentrator cells.

Acknowledgements

The author would like to thank A. Terao and S. Daroczi of SunPower Corporation, and J. Lasich of Solar Systems Pty Ltd for supplying pictures of the point-contact solar cells and the concentrator PV systems.

References

[1] Verlinden, P.J., Swanson, R.M., Crane, R.A., Wickham K. and Perkins, J. 1995. A 26.8% Efficient Concentrator, Point-Contact Solar Cell, *Proc. 13th European Photovoltaic Solar Energy Conf.*, Nice, pp. 1582–1585.

[2] Sinton R.A. and Swanson, R.M. 1992. Development Efforts on Silicon Solar Cells, Final Report, Electric Power Research Institute, Palo Alto, CA, February, pp. 2–44

[3] Swanson, R.M., Sinton, R.A., Midkiff, N. and Kane, D.E. 1988. Simplified Designs for High-Efficiency Concentrator Solar Cells, *Sandia Report, SAND88-0522*, Sandia National Laboratories, Albuquerque, NM, July.

[4] Verlinden, P.J., Swanson, R.M., Sinton, R.A. and Kane, D.E. 1988. Multilevel Metallization for Large Area Point-Contact Solar Cells, *Proc. 20th IEEE Photovoltaic Specialists Conf.*, Las Vegas, pp. 532–537.

[5] Verlinden, P.J., Sinton, R.A. and Swanson, R.M. 1988. High-Efficiency Large-Area Back Contact Concentrator Solar Cells with a Multilevel Interconnection, *Int. Journal of Solar Energy*, Vol. 6, pp. 347–365.

[6] Verlinden, P.J., Terao, A., Smith, D.D., McIntosh, K., Swanson, R.M., Ganakas, G. and Lasich, J. 2001. Will We Have a 20%-efficient (PTC) Photovoltaic System?, *Proc. 17th European Photovoltaic Solar Energy Conf.*, Munich, pp. 385–390.

[7] Schwartz, R.J. and Lammert, M.D. 1975. *IEEE International Electron Devices Meeting*, Washington DC, pp.350–351.

[8] Lammert, M.D. and Schwartz, R.J. 1977. *IEEE Trans. Electron Devices*, Vol. ED-24(4), pp.337–342.

[9] Swanson, R.M. 1985 Point Contact Silicon Solar Cells, Theory and Modeling, *Proc. 18th IEEE Photovoltaic Specialist Conf.*, Las Vegas, pp. 604–610.

[10] Swanson, R.M. 1988. Point Contact Solar Cells, Modeling and Experiment, *Solar Cells*, Vol. 7(1), pp.85–118.

[11] Sinton, R.A. 1987. Device Physics and Characterization of Silicon Point-Contact Solar Cells, Ph.D. Thesis, Stanford University, Stanford, CA.

[12] Sinton, R.A. and Swanson, R.M. 1987. Recombination in Highly Injected Silicon, *IEEE Trans. Electron Devices*, ED-34(6), p. 1380.

[13] Sinton, R.A., Verlinden, P.J., Swanson, R.M., Crane, R.A., Wickham K. and Perkins, J. 1995. Improvements in Silicon Backside-Contact Solar Cells for High-Value One-Sun Applications, *Proc. 13th European Photovoltaic Solar Energy Conf.*, Nice, pp.1586–1589.

[14] Gruenbaum, P.E., Gan, J.Y., King, R.R. and Swanson, R.M. 1990. Stable Passivations for High-Efficiency Silicon Solar Cells, *Proc. 21st IEEE Photovoltaic Specialists Conf.*, Kissimmee, pp. 317–322.

[15] Verlinden, P.J., Swanson, R.M., Sinton, R.A., Crane, R.A., Tilford, C., Perkins, J. and Garrison, G. 1993. High-Efficiency Point-Contact Silicon Solar Cells for Fresnel Lens Concentrator Modules, *Proc. 23rd IEEE Photovoltaic Specialists Conf.*, Louisville, pp. 58–64.

[16] Gruenbaum, P.E. 1990. Photoinjected Hot-Electron Damage at the Silicon/Silicon Dioxide Interface in Point-Contact Solar Cells, Stanford University, Stanford.
[17] Cudzinovic, M., Pass, T., Terao, A., Verlinden, P.J. and Swanson, R.M. 2000. Degradation of Surface Quality due to Anti-Reflection Coating Deposition on Silicon Solar Cells, *Proc. 28th IEEE Photovoltaic Specialists Conf.*, Anchorage, pp. 295–298.
[18] Gray, J.L. and Schwartz, R.J. 1985. Why don't we have a 30% efficient Silicon Solar Cell?, *Proc. 18th IEEE Photovoltaic Specialists Conf.*, Las Vegas, pp. 568–572.
[19] Verlinden, P., Van de Wiele, F., Stehelin, G. and David, J.P. 1986. An Interdigitated Back Contact Solar Cell with High Efficiency under Concentrated Sunlight, *Proc. 7th European Photovoltaic Solar Energy Conf.*, Seville, pp. 885–889.
[20] Sinton, R.A. et al., 1986. 27.5 Percent Silicon Concentrator Solar Cells, *IEEE Electron Device Letters*, Vol. EDL-7(10), pp. 567–569.
[21] Sinton, R.A. and Swanson, R.M. 1987. An Optimization study of Si Point-Contact Concentrator Solar Cells, *Proc. 19th IEEE Photovoltaic Specialists Conf.*, New Orleans, pp. 1201–1208.
[22] Swanson, R.M. 1989. Why we will have a 30% efficient Silicon Solar Cell, *Proc. 4th International Photovoltaic Science and Engineering Conf.*, Sydney, pp. 573–580.
[23] Gan, J.Y. 1990. Polysilicon Emitters for Silicon Concentrator Solar Cells, Ph.D. Thesis, Stanford University, Stanford.
[24] Gan, J.Y. and Swanson, R.M. 1990. Polysilicon Emitters for Silicon Concentrator Solar Cells, *Proc. 21st IEEE Photovoltaic Specialists Conf.*, Kissimmee, pp. 245–250.
[25] Campbell, P. and Green, M. 1986. The Limiting Efficiency of Silicon Solar Cells under Concentrated Sunlight, *IEEE Trans. Electron Devices*, Vol. ED-33(2), pp. 234–239.
[26] Christel, L.A. 1987. Polysilicon-Contacted P+ Emitter for Silicon Solar Cell Applications, *Sandia National Laboratories Report SAND87-7021*.
[27] Kwark, Y.H. and Swanson, R.M. 1985. SIPOS Heterojunction Contacts to Silicon, *Sandia National Laboratories Report SAND85-7022*.
[28] Sakata, H., Kawamoto, K., Taguchi, M., Baba, T., Tsuge, S., Uchihashi, K., Nakamura, N. and Kiyama, S. 2000. 20.1% Highest Efficiency Large Area (101 cm^2) HIT Cell, *Proc. 28th IEEE Photovoltaic Specialists Conf.*, Anchorage, pp. 7–12.

Part IIe

Organic and Dye Sensitised Cells

IIe-1

Photoelectrochemical Solar Cells

A. J. McEvoy, Institute for Molecular and Biological Chemistry, Faculty of Basic Sciences, Ecole Polytechnique Fédérale de Lausanne, Switzerland

1 Origins of Photoelectrochemistry

It has already been made evident in previous chapters that the photovoltaic effect results from a non-ohmic contact between two phases with different conduction mechanisms, typically a metal and a semiconductor to provide a Schottky barrier, or two semiconductors of opposite carrier polarity for a p–n junction device. In each case the excitation of the charge carrier pair, a conduction band electron and the positive hole in the valence band, is consequent on the absorption of a photon by the semiconductor, the energy of the photon being greater than the bandgap. The contact then serves to separate the photoexcited charge carriers, the electrons and holes, so that a potential difference can be maintained across it under illumination and a current can flow in an external circuit. In all of these cases, the materials are solids and the conduction is by electronic processes, the mobility of holes or electrons. These solid-state semiconductor-based devices dominate photovoltaic science and technology. However, it should be remembered that the first observation of the photovoltaic effect by Becquerel almost 200 years ago [1] was of the asymmetry of behaviour of two identical electrodes, one illuminated, both immersed in an a liquid electrolyte. We can now recognise the material he used, a silver halide, as a semiconductor, and the Becquerel device would at present be classified as a photoelectrochemical cell. The contact of the semiconductor with the electrolyte, in which the conduction mechanism is the mobility of ions rather than of electrons or holes, therefore forms a photoactive junction functionally equivalent to those later discovered for solid-state photovoltaic devices. There has been the suggestion, therefore, given the ease of formation of semiconductor–electrolyte junctions, that they would offer a low-cost practical alternative in the field of solar energy conversion.

In the conventional solid-state cell the Fermi levels of the two materials forming the junction equilibrate in darkness, this being associated with a displacement of majority carriers within the semiconductor and the establishment of a depletion layer near the interface. Where the contacting phase is a redox electrolyte in a photoelectrochemical system, there is an analogy with the Fermi level of a Schottky contact metal. The redox electrolyte is so named because ions of the same species but different oxidation states are present, the more reduced (red-) being convertible to an oxidised state (-ox) by the acquisition of a positive charge, at an anodic electrode for example. The process is reversible at a cathode. The potential at which no current passes at a reversible electrode, able to donate or acquire charge in contact with a given redox system, is the redox potential, at which electrode and both ionic components are at equilibrium. In the photoelectrochemical system it is the redox potential of the ions in the electrolyte which is functionally equivalent to the Fermi level of the photovoltaic junction materials. The range of potential associated with the oxidised component, more positively charged and therefore depleted in electrons, corresponds to the conduction band of a semiconductor, while the filled electron states of the valence band find their analogue in potential distribution of the reduced component. As with the permitted electron energy levels in condensed

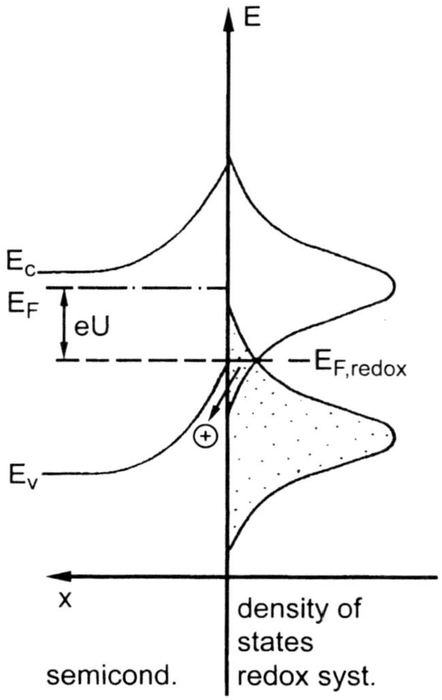

Figure 1 In darkness the Fermi level of the semiconductor equilibrates with the redox level of the electrolyte; under illumination the electron quasi-Fermi level is raised, giving rise to a photovoltage.

matter, these ionic species also form bands and with the ionic levels dispersed due to effects such as solvation, their interaction with the electrolyte environment gives rise to a reorientation energy which may be of the order of 1 eV.

Under illumination, as in the case of a solid state junction, photons of energy greater than the bandgap of the semiconductor are absorbed, their energy then serving to excite an electron to the conduction band, leaving a mobile positive hole in the valence band. The gradient of the bands then orients the separation of the charge carrier pairs, the hole being directed towards the electrolyte interface in the case of an n-type semiconductor, where it is available to oxidise an ion in the reduced state of the redox system. The electrode, here functioning as a photoanode since an oxidation reaction takes place on its surface, then delivers the accumulated electrons through an ohmic contact to an external circuit, whence they are returned to the electrochemical system though a counterelectrode, normally metallic and therefore photoinert. On the counterelectrode, the cathode surface in the present example, a reduction reaction takes place, thereby regenerating the electrolyte whose overall redox composition remains unaltered. The photogeneration of the charge carriers of course splits the Fermi level into quasi-Fermi levels representative of the occupation of states by holes and electrons respectively, the extent of the splitting representing the generation of a photovoltage between the electrolyte redox level (and the counterelectrode) and the ohmic collector contact to the photoanode. The parallel between the photovoltaic device and photoelectrochemical energy conversion by a regenerative cell is therefore complete, the system remaining invariant while a part of the radiant energy of the absorbed incident light appears as electrical energy in the external circuit. Figure 1 shows schematically the relationship on the energy scale of the factors relevant to the photoelectrochemical process, based on this model as developed by Gerischer [2]. A 'mirror image' process can also be presented for a regenerative cell with a p-type semiconductor photocathode operated with a metal anode as oxidising counterelectrode.

2 Photoelectrolysis

Since in a photoelectrochemical cell a photoactive electrode is effectively integrated into an electrolytic cell, it is not a requirement that a reversible redox process constitute the electrochemical activity. Otherwise stated, the cathodic reaction is not necessarily the reverse reduction of the oxidation taking place at the anode. The electrochemistry can therefore also give rise to a photoelectrolytic or a photosynthetic effect. For example, the redox system of hydrogen oxidation to water in an acid environment is commonly accepted as the reference level in the electrochemical scale, with respect to which the electrolytic evolution of oxygen from water takes place at +1.23 V. If the valence band edge is more positive than this level, the photogenerated holes (h^+) may transfer directly to an aqueous electrolyte for the reaction:

$$2H_2O + 4h^+ \Longleftrightarrow O_2 + 4H^+ \qquad (1)$$

With a sufficiently wide band-gap, the selected semiconductor can also deliver electrons via the external circuit to the metal cathode, whose Fermi level can then lie negative of the hydrogen evolution potential, and which can then sustain the reaction:

$$2H^+ + 2e^- \Longleftrightarrow H_2 \qquad (2)$$

An efficient photoelectrolytic system of this type as been the 'philosopher's stone' of electrochemists since the first report of photoassisted water splitting into its component elements by Fujishima and Honda some 30 years ago [3] using a titania electrode. The necessity of such a wide bandgap semiconductor to obtain photoelectrolysis points up the two major problems in photoelectrochemical materials selection, namely efficiency and stability. The band gap of titania is over 3 eV, so the material generates charge carrier pairs only by the absorption of ultraviolet light, being insensitive in the visible spectrum. The efficiency of solar energy conversion is therefore low, since as is already known, the optimum spectral matching bandgap is 1.4 eV. As Memming [4] pointed out also, it is relatively easy to obtain the hydrogen evolution reaction at a chosen electrocatalytic counterelectrode; the oxygen evolution process however, as is evident from Equation (1) is a multistep process for which the photoanode surfaces seldom offer a catalytic reaction route with correspondingly fast kinetics. For those semiconductors more favourable catalytically to oxygen evolution, such as tungsten trioxide, an external bias voltage may be required to displace the Fermi level negative of the hydrogen evolution potential, a voltage which may be supplied by an auxiliary photovoltaic device [5].

Given the overpotentials involved, it is not surprising that alternative hole consumption processes are offered, the most significant of which is the anodic photocorrosion of the semiconductor itself. A typical example is a cadmium chalcogenide, which would otherwise offer appropriate valence and conduction band edge levels to sustain photoelectrolysis. For instance, cadmium sulfide with a badgap of 2.2 eV is not ideal but could still be relatively efficient (Figure 2). Instead in its photoanodic reaction cadmium ions dissolve in the electrolyte and the surface becomes sulphur coated:

$$CdS + 2h^+ \Longleftrightarrow Cd^{2+} + S \tag{3}$$

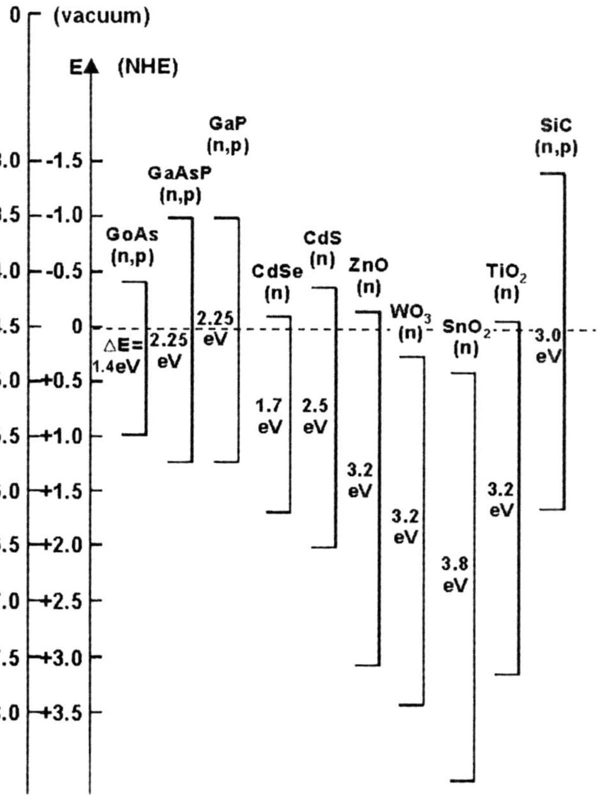

Figure 2 Bandgaps and band edge energies of compound semiconductors for photoelectrochemical applications. E_c^s is the conduction band edge at the semiconductor–electrolyte interface

The chemical stability of a compound against an oxidative photoprocess is therefore related to its bandgap, a reaction similar to Equation (3) being promoted for the narrower-bandgap materials. This is a process in competition even with the redox reaction in regenerative cells, so that while a redox mechanism may relatively stabilise a narrow-gap semiconductor, it is never totally inert against photocorrosion. Intensive research over the past three decades has led to an inescapable conclusion: stability is attained only at the cost of a wide band gap with insensitivity to the greater part of the solar spectrum and therefore low energy conversion efficiency. This recalcitrant problem has impeded the establishment of a credible electrochemical photovoltaic energy conversion system over this period; the resolution of this dilemma lies in the separation of light absorption from the charge separation function of the semiconductor, by sub-bandgap sensitisation to visible light of a wide bandgap stable material through visible light absorption in an electroactive dye.

3 Photoelectrochemistry, Photography and Sensitisation

The choice of a silver halide as the electrode material in Becquerel's experiments reflects the rapid growth in knowledge of photochemical effects at that time. Biot [6] was aware of the use of paper sensitised by silver chloride, being contemporary with the introduction of photography by Daguerre and Fox-Talbot. Although the art of formulating photographic emulsions only became a science a century later with the theoretical analysis of the process by Gurney and Mott in 1938 [7], there was constant empirical progress in extending the overall sensitivity of photographic emulsions, which had been particularly deficient for mid-spectral visible light and towards the red. The early 'orthochromic' materials had limited capability to register scenes observed in colour realistically on a grey scale due to the semiconductor nature of the silver halides with their band gaps between 2.7 and 3.2 eV and for which the photoresponse is negligible for wavelengths longer than 460 nm. It was noted, for example, that the origin of the gelatin used as the support medium for the alkali halide grains significantly modified the film spectral sensitivity. Only in the twentieth century was it demonstrated that an organosulphur compound present in calf skin gelatin was responsible [8], and which is now known to have its effect by inducing a superficial nanostructure of silver sulphide on each grain. This can now be recognised as the first sensitisation at a semiconductor heterojunction, silver sulphide to halide. Even more significant was the work of Vogel, professor of 'photochemistry, spectroscopy and photography' in Berlin. In 1873 [9] he established empirically that silver halide emulsions could be sensitised to red and even infrared light by suitably chosen dyes, thereby making possible the modern 'panchromatic' broad-spectrum black and white film, and more recently with the use of spectrally-selective dyes, colour photography.

The concept of dye enhancement was carried over already by 1887 from photography to the photoelectric effect by Moser [10] using the dye erythrosine, again on silver halide electrodes, and confirmed by Rigollot in 1893 [11]. (It is, incidentally, curious to note that Moser, writing before the formulation of the Einstein theory of the photoelectric effect, records his observations of dye enhancement in terms of potential (volts) rather than the more fundamental current (amps), despite referring in his title to a 'strengthened photoelectric *current* through optical sensitising'.) This parallel between sensitisation in photography and in photoelectrochemistry, both of which are charge transfer processes, still seems to come as a surprise to successive generations of chemists [12]. It continues to apply to such recent observations as two-electron sensitisation [13] where the photoexcited dye after transferring one electron to the halide, reacts in turn with a reducing agent in the emulsion producing a second charge-transfer step; such current-enhancement reactions have been investigated in photoelectrochemistry since 1984 [14].

That the same dyes were particularly effective for both processes was recognised among others by Namba and Hishiki [15] at the 1964 International Conference on Photosensitisation in Solids in Chicago, a seminal event in the history of dyes in the photosciences. It was also recognised there that the dye should be adsorbed on the semiconductor surface in a closely packed monolayer for maximum sensitisation efficiency [16]. On that occasion the theoretical understanding of the processes was clarified, since until then it was still disputed

Notiz über Verstärkung photoelektrischer Ströme durch optische Sensibilisirung.[1]

Von Dr. James Moser.

(Aus dem physikalisch-chemischen Laboratorium der Wiener Universität.)

(Vorgelegt in der Sitzung am 23. Juni 1887.)

Ich erlaube mir mitzutheilen, dass ich die von Herrn E. Becquerel entdeckten photoelektrischen Ströme erheblich dadurch verstärken konnte, dass ich die beiden chlorirten, jodirten oder bromirten Silberplatten in einer Farbstofflösung, z. B. Erythrosin, badete.

Beispielsweise war zwischen zwei chlorirten Silberplatten die elektromotorische Kraft im Sonnenlicht 0·02, zwischen zwei anderen in gleicher Weise behandelten, aber gebadeten Platten 0·04 Volt.

Bisher sind nur an jodirten Platten von Herrn Egoroff elektromotorische Kräfte beobachtet, und zwar bis $^1/_{15}$ Volt. Ich konnte bei jodirten und bromirten Platten durch Baden in Erythrosin $^1/_4$ Volt erreichen.

Ich halte es für meine Pflicht, schon an dieser Stelle Herrn Max Reiner, der mir bei diesen Versuchen assistirt, meinen verbindlichsten Dank auszusprechen.

[1] Akadem. Anzeiger Nr. XVI.

Figure 3 Report of the first dye-sensitised photoeffect observed on an illuminated semiconductor – Moser, Vienna, 1887.

whether the mechanism was a charge-transfer or an Auger-like energy-coupling process, as illustrated in Figure 4, taken from the proceedings of the Conference [17]. With the subsequent work of Gerischer and Tributsch [18, 19] on ZnO, there could be no further doubt about the mechanism, nor about the significance for photoelectrochemistry of the dye-sensitisation phenomenon, finally rendering compatible effective wideband visible spectral absorption with stability of a semiconductor substrate. It is now evident that the process involved the excitation of the dye from its charge-neutral ground state to an excited state by the absorption of the energy of a photon.

The chemical bond results from the interaction of the valence electrons associated with the atoms of a molecule. As in the semiconductor, the lowest-lying permitted energy levels are first filled, up to the highest occupied molecular level, the HOMO. Excitation, for example by the absorption of the energy of a photon, can promote an electron to the lowest unoccupied molecular level, the LUMO. Therefore as far as absorption of light is concerned, the HOMO–LUMO gap of a molecule is fully analogous to the band gap of a semiconductor. It defines the response to incident light and consequently the optical absorption spectrum. At the same time the absolute energy level of the LUMO can determine the energetics of the permitted relaxation processes of the excited molecule. When it lies above the conduction band edge of a semiconductor substrate, relaxation may take the form of emission of an electron from the dye into the semiconductor, leaving that molecule in a positively charged cationic state. In the photographic process the injected electron reduces a silver ion, and thereby establishes the latent image. In a photoelectrochemical cell, injected electrons provide a current in an external circuit, returning to the redox electrolyte through a cathode in contact with it. The uncharged ground state of the dye is then restored by electron transfer from the redox system, completing the circuit

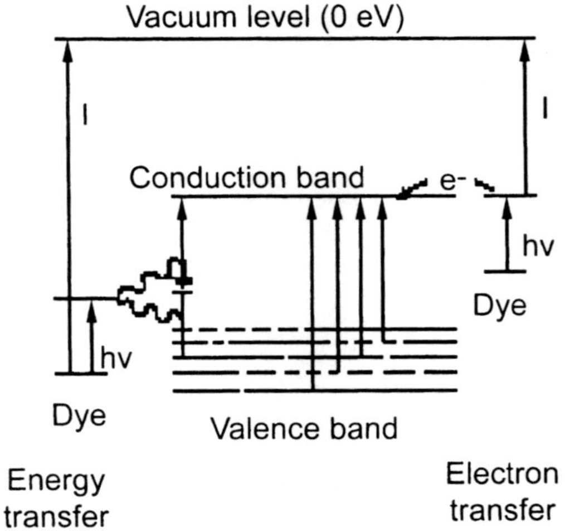

Figure 4 Disputed mechanism for sensitisation in photographic semiconductor colloid particles – here silver bromide – as presented by Bourdon in 1964 [17].

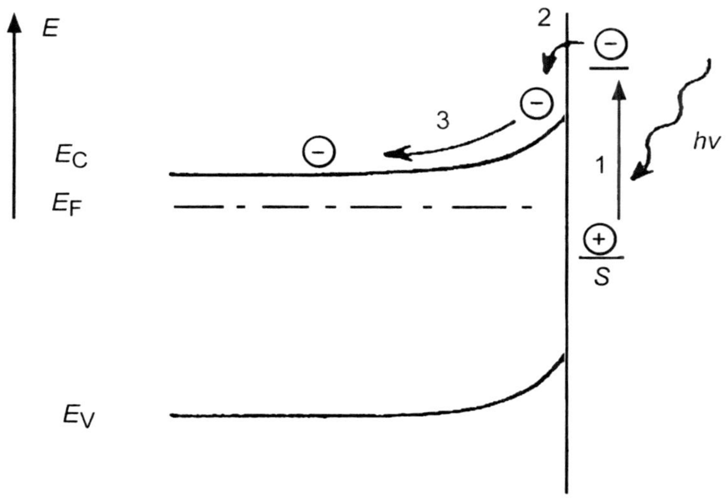

Figure 5 Definitive photosensitisation mechanism for a ZnO semiconductor as presented by Gerischer and Tributsch in 1968 [18].

and providing a regenerative cycle functionally comparable with other photovoltaic devices, as presented in Figure 6. It was also recognised that the major loss mechanism associated with semiconductors in photovoltaics, the recombination of the photoexcited charge carriers in the crystalline lattice, is strongly inhibited at a dye-sensitised photoelectrode [20]. Conventional photovoltaic junctions are essentially minority carrier devices, holes being generated in the n-type material, electrons in the p-type material, then transported to the interface. During this process carriers of each polarity can be lost due to recombination with the local majority carriers, the losses being minimised by attention to the bulk crystallinity and surface quality of the semiconductor. In the dye-sensitisation case, in contrast, no specific action is necessary, as the electron injected from the photoexcited dye finds itself within the semiconductor, already separated spatially, in a different phase, and by a potential barrier from the oxidised dye species and from the electrolyte (Figure 6). It is a majority carrier device, no holes being found in the n-type semiconductor valence band. Electron escape from the semiconductor and recapture by the dye cations or the redox electrolyte is therefore significantly suppressed, as will be seen later. This relieves the restrictions placed in the solid-state case by the requirement for low bulk and surface recombination velocities on the morphology, impurity content and crystallinity of the semiconductor. A rough and porous surface is therefore permissible, so that the dye is no longer a single monolayer on a plane area, with consequently weak optical absorption; on the rough surface with a greatly enhanced area, the optical density is such that the photoelectrode is opaque and incident light is efficiently absorbed [21].

The dye-sensitised photoelectrochemical cell integrates all these considerations, the molecular engineering of a suitable dye, its adsorption as a monomolecular

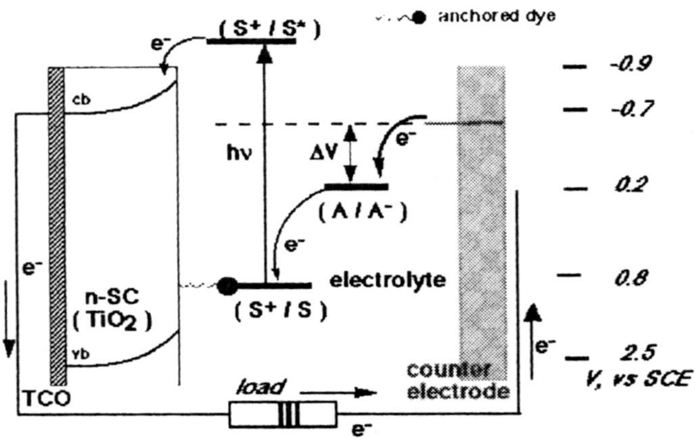

Figure 6 The regeneration cycle in a dye-sensitised solar cell. After relaxation of the photoexcited state of the dye by electron loss to the semiconductor substrate, the positive charge remaining is neutralised by the electrolyte, which in turn recovers the electron through the external circuit and the metallic cathode. A^- = redox anion, S = sensitising dye.

film on a rough wide bandgap stable semiconductor to obtain adequate optical absorption, an ohmic contact to deliver the resulting current to an external circuit, and finally establishment of a suitable regenerative system through a cathodic counterelectrode and redox electrolyte, in a single device [21]. The established semiconductor choice for this application is titanium dioxide, TiO_2, with a bandgap of 3.1 eV. It has many advantages for sensitised photochemistry and photoelectrochemistry, being a low cost, widely available, non-toxic and biocompatible material, and as such it is even used in health care products as well as in domestic applications such as paint pigmentation. Since light must enter the cell to photoexcite the dye- semiconductor composite, the ohmic contact is usually made to a transparent conducting oxide (TCO) such as an indium-tin oxide or zinc oxide supported on glass or a suitable polymer. Most research has used the iodine/iodide ($3I^- \Longleftrightarrow I_3^- + 2e^-$) redox system, with others, including transition metal complexes, now under development. Progress of the sensitised electrochemical photovoltaic device since 1991, with a conversion efficiency at that time of 7.1% under solar illumination, has been incremental, by optimising the synergy of structure, substrate roughness, dye photochemistry, counter-electrode kinetics and electrolyte redox chemistry. That evolution has continued progressively since then, with certified efficiency now over 10% (Figure 8).

4 Molecular Engineering of Electroactive Dyes

The constraints determining the dye selection have already been indicated, and the production of a suitable formulation is a demanding exercise in synthetic chemistry. Firstly and evidently there is the matter of an optical absorption

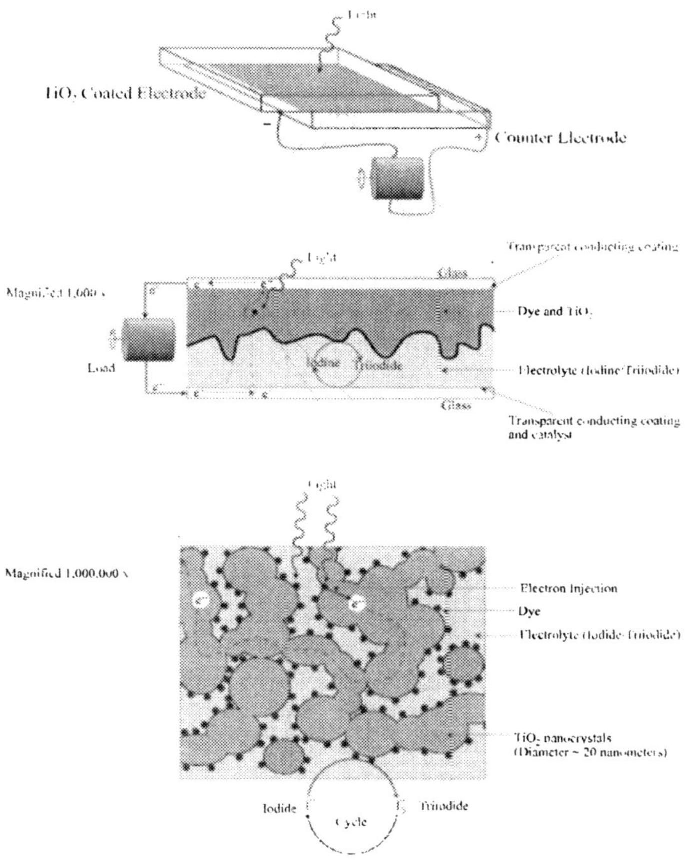

Figure 7 Structure of a dye-sensitised photoelectrochemical cell on different scales. Top: a complete cell (centimetric scale). Centre: electrodes and electrolyte, micron scale. Bottom: nanocrystalline semiconductor interconnected porous layer, with adsorbed monolayer of dye and entrained electrolyte (nanometric scale).

spectrum closely matched to the application of the photovoltaic device, whether it be to the solar spectrum for energy conversion or to artificial light sources for indoor use. With a high optical absorption coefficient across the visible spectrum, the excitation process should be rapid but subsequent relaxation slow, as already explained. For attachment to the semiconductor surface the molecule should adsorb strongly, by preference through a specific chemical bond, to the substrate, but avoid aggregation so that a monolayer coverage forms spontaneously on contact between the semiconductor and the dye in solution. The dye structures selected for intensive development arose originally from biomimetic considerations, given the role of photosynthesis in the natural world. The prototype energy-converting dye provided by nature is of course chlorophyll, a molecule consisting of a central magnesium atom surrounded by a nitrogen-containing porphyrin ring. Variations are due to minor modifications of certain side groups. Chlorophyll is in turn similar in structure to haemoglobin,

EPFL (Switzerland) nano-crystal dye sensitized cell

Sample: PL0710/2
Oct 30, 1998 10:41 AM
ASTM E 892-87 Global

Temperature = 25.0 °C
Area = 0.1863 cm ^{2}
Irradiance: 1000.0 Wm $^{-2}$

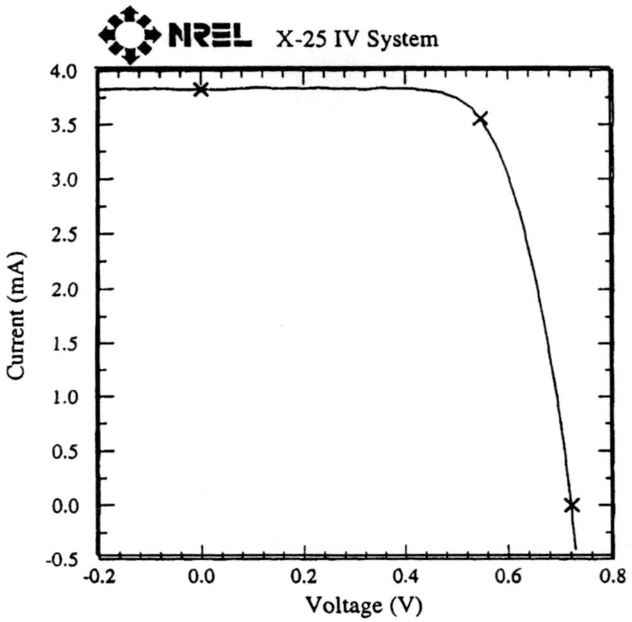

V_{oc} = 0.7210 V
I_{max} = 3.552 mA
J_{sc} = 20.53 mAcm $^{-2}$
Fill Factor = 70.41%

V_{max} = 0.5465 V
I_{sc} = 3.824 mA
P_{max} = 1.941 mW
Efficiency = 10.4 %

tape aperture, "black" dye

Figure 8 Certified I–V characteristic under simulated sunlight, of a dye-sensitised photoelectrochemical cell, showing a conversion efficiency > 10%.

the oxygen-carrying iron-based pigment found in blood. Given that the development of the dye-sensitised cell was associated with an interest in artificial photosynthesis, the adoption of porphyrin-like organometallic dyes as sensitisers was logical. However, whereas plant photosynthetic processes rely on chlorophyll the synthetic chemist can select from the whole range of complex-forming metals to design an appropriate metal–ligand charge transfer structure.

Typical is the use of ruthenium pyridyl complexes. Clark and Suttin had already used a tripyridyl ruthenium complex in 1977 to sensitise titanium dioxide to sub-bandgap illumination [22], but in solution only. Charge transfer could only occur only after diffusion of the photoexcited molecule to the semiconductor so the efficiency of the sensitisation was very low. By 1980 the idea of chemisorption of the dye, through an acid carboxylate group bonding to the metal oxide surface had already emerged [23] so that the sensitiser was immobilised and formed a monomolecular film on the semiconductor substrate, thereby facilitating charge transfer by electron injection. The carboxylated trisbipyridyl dye ('RuL_3') therefore became the prototype sensitiser for this type of electrochemical cell. The objective at that time was to photoelectrolyse water using sensitised electrodes. Although other compounds have since been assessed as sensitisers, such as zinc porphyrins and even Prussian blue analogues [24], the most suitable dyes today are still modifications of the ruthenium-based pyridyl complexes. The objective had also evolved to become a photovoltaic device, rather than photosynthesis.

The first priority for any PV cell is of course efficient conversion of incident radiation, requiring the spectral match of the optically absorbant component to the incident light, and in this regard the early dyes were inadequate. Development of dye molecules for extension of sensitivity, ultimately into the infrared, is dramatically illustrated in Figure 10. The progress in extension of the absorption spectrum of ruthenium complexes from the early tris-bipyridyl ('RuL_3') molecule, absorbent only in the blue region and therefore appearing red, to the present-day wide-spectrum material ('$RuL'(NCS)_3$'), panchromatic, absorbing right across the visible range and therefore presenting itself as a 'black' dye, has been a key step in cell development. The strategy has been to

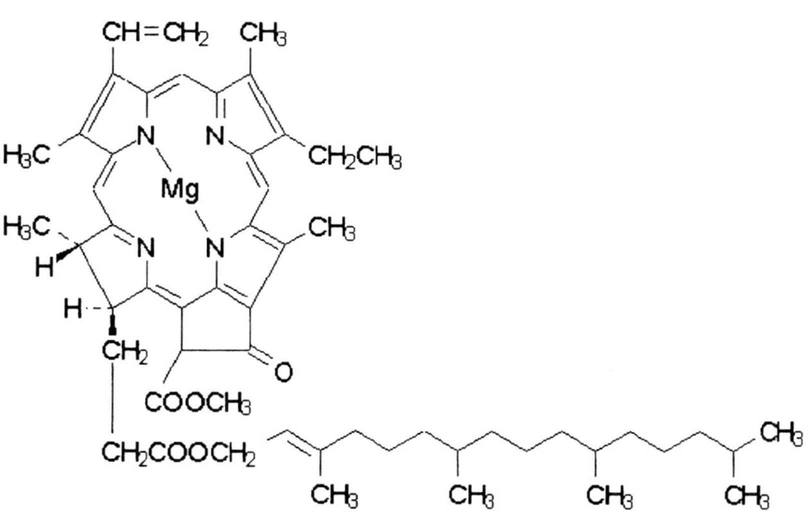

Figure 9 Structure of one of the principal variants of chlorophyll, the natural pigment converting solar energy in the photosynthetic process in plants.

retain ruthenium as the metallic component, with structural modification of the ligands to tune the energetics of the molecule. The LUMO level to which the photoexcited electron in promoted must remain above the conduction band edge of the semiconductor substrate if electron injection is to occur. Extension of the spectral sensitivity is then accomplished by raising the HOMO of the molecule, as in Figure 11. This molecular engineering process is of particular interest to physicists, as there is a close analogy to band gap modification in compound semiconductors.

A trimer structure, consisting of three ruthenium bis-biphridyl entities, cyanide bridged, and chemisorbed through carboxylate groups to the titania surface, gave the first significant extension of absorption spectrum, on the principle that an antenna structure with consequent charge transfer could be appropriate. However monomer structures are inherently simpler, and modification of pyridyl complexes by nucleophilic groups provided a more promising route. By thiocyanate substitution, this gave rise to the dye $RuL_2(NCS)_2$, (L = bipyridyl), for several years our standard chromophore and the basis for most of the system development work. Under the product name N3 it is now probably the best-understood photosensitisation molecule, particularly for its charge exchange kinetics [25]. In HOMO the metal ion is in oxidation state (II), but on excitation it loses a further electron to the organic ligand structure, a metal-to-ligand charge transfer. Injection of the electron through the bridging carboxylate to the semiconductor substrate occurs within picoseconds, faster than other relaxation processes and explaining the high quantum efficiency of the system sensitised with this dye.

Isomerism of the dye molecule can permit the identification of structures with variant optical properties. For example the standard N3 is a *cis*-isomer (both NCS thiocyanate groups on the same side of the molecule); the corresponding trans-isomer has an extended sensitivity into the infrared, but over time it tends to revert to the *cis*-structure. A *trans*-type configuration can be stabilised by replacing the L_2, bis-bipyridyl ligand system by a single tetrapyridyl coordination [26]. This example illustrates the sensitive dependence of the orbitals and their energetics on molecular structure. Addition of a third thiocyanate group, with the necessary reduction of pyridyl complexation from bis-bi- or tetra- to ter-pyridyl (L′) gives the structure of the state-of the art black dye, $RuL'(NCS)_3$, whose spectral sensitivity extends throughout the visible and into the infrared, approaching the ideal absorption edge position (1.4 eV) for optimal solar energy conversion. It can be expected that this type of molecular engineering will make available suitable dye formulations for a variety of applications, such as infra-red sensitive dyes with lower visible absorptivity, useful for 'transparent' cells, or dyes with selective absorption, the resulting colour presentation being determined for example by architectural options for building-integrated PV.

The interaction between the dye molecule and the adsorbant surface also influences the energetics of the dye and the kinetics of its chemical reactions. For example it is known that on adsorption of a carboxylated dye, a titania surface becomes positively charged, due to the deprotonation of the acid dye. The

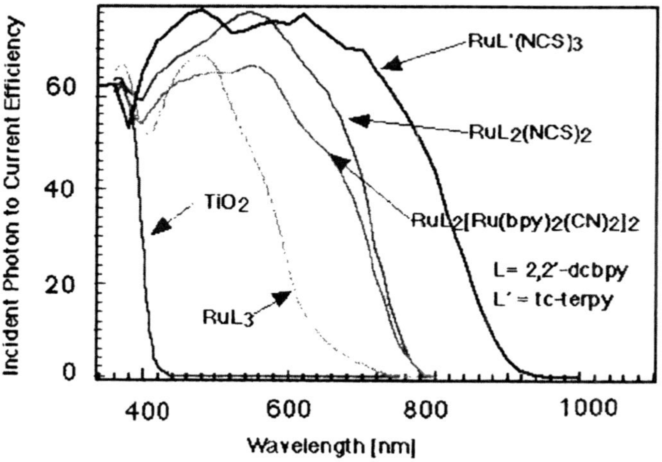

Figure 10 The photocurrent action spectrum of a cell containing various sensitisers, where the incident photon to current conversion efficiency is plotted as a function of wavelength.

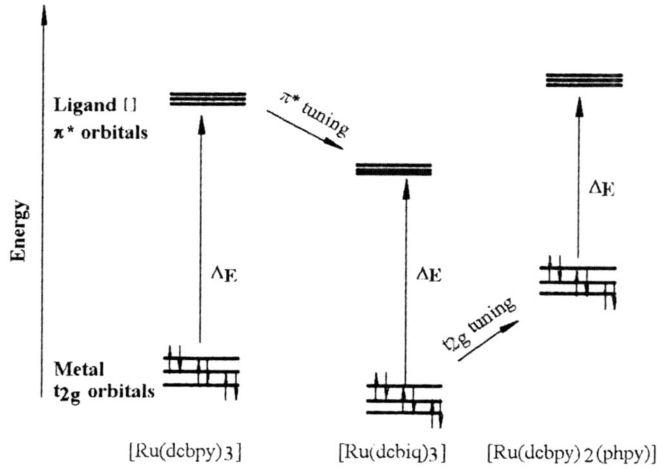

Figure 11 Tuning of HOMO (t_{2g}) and LUMO (π^*) orbital energy in various ruthenium complexes. (a) Ru(dcbpy)$_3$ = ruthenium tris(dicarboxyl bipyridyl) or RuL$_3$. (b) (dcbiq) = dicarboxybiquinoline. (c) Ru(dcbpy)$_2$(phpy) is a modified RuL$_2$ where phpy = phenylpyridine. In all cases the ruthenium atom is 6-fold coordinated with cyclic pyridyl-type ligands.

consequent interface potential gradient obviously enhances electron injection efficiency, leading to a higher photocurrent. However efficiency does not proportionately rise, since the positive shift of band edge is associated with a lower open-circuit voltage. Optimisation of the cell efficiency also requires a control of protonation of the dye. With this in mind several salts of N3 with organic cations such as tetrabutylammonium and imidazolium have been investigated [26]. These salts are also more soluble, facilitating purification by recrystallisation and thereafter adsorption to the semiconductor from the more concentrated solution. The thermal dehydration behaviour shows the association of only 1.5 molecules of water per Ru ion, for the optimum di-tetrabutylammonium salt, with a thermal stability thereafter to 190°C; at that point the organic cation tends to dissociate. When adsorbed on the semiconductor the dye stability is further enhanced, decarboxylation requiring over 300°C. This remarkable rise is attributed to the strength of the Ti ion to carboxylate bond between the semiconductor surface and the chemisorbed molecule. This confirmation of the thermal stability of the surface-bonded dye therefore opens the prospect of a very wide temperature window for processing the sensitised semiconductor during any manufacturing routine, simplifying bonding and sealing processes by permitting thermal treatment and curing of sealants even after the dye has been admitted to the cell.

Modification of the surface of the semiconductor to optimise adsorption and charge transfer behaviour of the dye is a further option. Several years ago, deposition of an outer intrinsic titania [27] or a solid solution layer [28] on each semiconductor grain was investigated in order to control the interface properties after dye adsorption. Evidently a nanostructure emulating the oxide or insulator layer in MOS or MIS solid state Schottky devices has a similar effect in defining the band edge characteristics in the photoelectrochemical case.

5 Semiconductor Layer Characteristics

If molecular design and engineering has underpinned the evolution of efficient stable sensitiser dyes, for the semiconductor substrate it is a matter of the materials science of nanoporous ceramic films. The nanoporous structure permits the specific surface concentration of the sensitising dye to be sufficiently high for total absorption of the incident light, necessary for efficient solar energy conversion, since the area of the monomolecular distribution of adsorbate is 2–3 orders of magnitude higher than the geometric area of the substrate. As already noted, this high roughness does not promote charge carrier loss by recombination, since the electron and the positive charge find themselves within picoseconds on opposite sides of the liquid-solid interface, much faster than any possible electron escape and redox or cation capture process. The original semiconductor structure used for early photosensitisation experiments was a fractal derived by hydrolysis of an organo-titanium compound. Later, suspensions of commercially available anatase titania powders were found to be equally effective. At present however hydrothermal techniques are employed for the synthesis of an optimised nanoparticulate anatase TiO_2 powder which is then used in suspension in a liquid medium [29].

A specific advantage of the hydrothermal technique is the ease of control of the particle size, and hence of the nanostructure and porosity of the resultant semiconductor substrate. The microstructure of the semiconductor is of course a compromise, to achieve an optimal optical absorption and photovoltaic performance. Nanosize grains give the greatest surface area, but pores must remain sufficiently large that the mobility of the charge carriers in the electrolyte, the redox ions, is not unduly inhibited. Also some degree of optical scattering by larger particles in the semiconductor film is also desirable, particularly for devices which function under indirect illumination such as vertical building facades. Processing parameters such as precursor chemistry, hydrothermal growth temperature for the titania powder, and sintering conditions are varied in the optimisation procedure. A flowsheet for hydrothermal processing is presented in Table 1 as an example of the required development work. Figure 12 shows the control of porosity of the final film which results from it, as determined by a nitrogen adsorption method. The procedure involves the hydrolysis of the titanium alkoxide precursor in an aqueous medium to produce a sol which is then subjected to the hydrothermal Ostwald ripening in an autoclave. The temperature of the hydrothermal treatment is decisive for the ultimate particle size. A standard sol, treated for 12 hours at 230°C has a mean particle diameter of the order of 10 nm. The colloidal suspension can be applied to the TCO support by one of several standard coating processes – tape casting, spraying or screen printing. A firing protocol to dry the film, then pyrolyse binders and organics ensures a coherence and low-resistance ohmic contact to the conducting underlayer. Figure 13 shows a micrograph of a semiconductor film suitable for sensitisation. Other oxide semiconductors which have been studied in the context of dye-sensitised photovoltaics include ZnO, SnO_2, Nb_2O_5, and $SrTiO_3$ [30–34].

For cell assembly, the anode as described is photosensitised by chemisorption of the dye from solution. The cathode or counter-electrode is usually a similar TCO supported on glass, treated to electrocatalyse the redox electrolyte

Table 1 Flow diagram for the preparation of TiO_2 colloids and mesoporous films (after Brooks et al. [29]).

Precursor Preparation	Modify Ti-isopropoxide with acetic acid
↓ Hydrolysis	Rapidly add precursor to water
↓ Peptization	Acidify with HNO_3, reflux
↓ Hydrothermal Growth	Autoclave: 12 hours at 230°C
↓ Concentrate Colloid	Rotovap: 45°C, 30mbar
↓ Solvent Exchange, Ethyl Cellulose Addition	Flocculation, centrifuging
↓ Homogenize Paste	3-roll mill, 15 minutes
↓ Screen Print Films	
↓ Dry and Anneal Films	Anneal: 450°C, 20 minutes

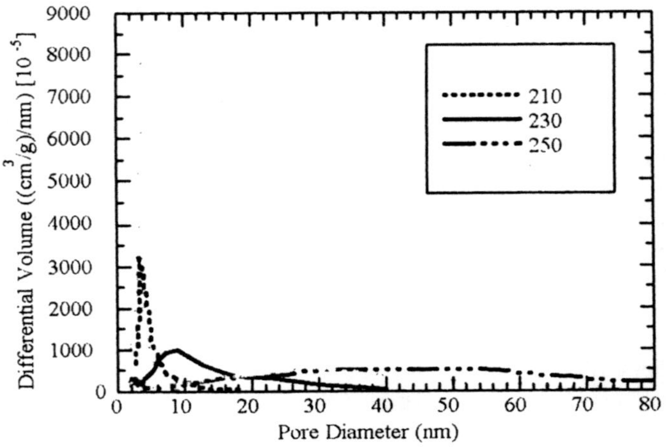

Figure 12 Control of semiconductor substrate porosity by temperature of hydrothermal processing.

regeneration, for example with highly dispersed platinum. The cell is completed with an edge seal to contain the electrolyte, usually a liquid, although gel or solid variants are also possible.

6 Dye Sensitisation in Heterojunctions

Given that a dye at the interface in a photoelectrochemical cell can sensitise the system to photons of lower energy that would not otherwise be absorbed by the semiconductor, the possibility presents itself that the effect is not restricted to the electrochemical device. Both conductors may in principle be in the solid state. Here we are not considering the special cases of gel or polymer electrolytes [35], because in these media charge transport is ionic, by the same mechanism as in a conventional liquid electrolyte. The key distinction is that a heterojunction is essentially a solid-state device, a contact between two chemically different semiconductors, organic or inorganic; they also differ in the conduction mechanism, for example with an n-type electron-conducting TiO_2 contacted by a p-type hole-conducting organic species. In an electrochemical device on the other hand the phase contacting the semiconductor substrate, as already stated, is an electrolyte with an ionic conduction mechanism. The difference is essentially in the nature of the mobile carrier, although the devices may be functionally the same. Since the sensitiser provides charge separation but not charge transport, and is distributed at an interface in the form of an immobilised molecular monolayer, it is evident that for charge transfer each dye molecule must bridge both conducting phases. This applies in the first instance to the wide bandgap porous semiconductor substrate into which the photoexcited chemisorbed molecules inject electrons. It is also evident that in the photoelectrochemical format of the sensitised cell the liquid electrolyte penetrates

Figure 13 SEM image of the surface of a mesoporous film prepared from the hydrothermal TiO_2 colloid.

into the porosity, thereby permitting the required intimate contact with the charged dye molecule necessary for charge neutralisation after the electron loss by exchange with the redox system in solution. It is not immediately evident that an interpenetrating network of two conducting inorganic solids can be established so effectively that an immobilised molecule at their interface can exchange charge carriers with both. This is a particular obstacle to the realisation of a sensitised nanostructured metal–semiconductor Schottky junction. However use of transparent organic semiconductors is a possibility, and initial results [36, 37] are promising. The materials science, fabrication procedures and photovoltaic performance of systems incorporating organic conductors are presented in a separate chapter. These heterojunction devices can now be expected to follow a similar learning curve to the electrochemical concept, which has significantly increased in efficiency since the initiation of its development [21], and as noted, the improvements in efficiency and stability are not the consequence of any radical breakthrough, but rather of gradual evolution.

7 Commercial Prospects

The status of the dye-sensitised device as the only verified alternative to solid-state junction devices has already been mentioned. However it must be recognised that the solid-state devices, particularly the silicon p-n junction cells, benefit from over 40 years of industrial and development experience, the technology transfer from the silicon-based electronics industry, and even the widespread availability of high quality silicon at low cost resulting from the expansion of that industry. The procedures for high-yield fabrication of silicon devices, both crystalline and amorphous, are well understood, with costing precisely established, based on decades of solid industrial experience. For the dye-sensitised cell, in contrast, fabrication procedures require development and costing is based on estimates of the requirements of chemical processes rather than those of the silicon metallurgy with its elevated temperatures and vacuum technology as required for conventional cells. This may in fact turn to the advantage of some variants of the dye-sensitised concept. Equally it is well known that the substitution of an established technology by an upcoming alternative requires that the new concept has definite advantages and no clear disadvantages. It is therefore noted with some satisfaction that several companies in Europe, Japan and Australia have taken up the challenge and are currently engaged under license in the venture of industrialisation and commercialisation of dye-sensitised PV cells. More, the existence of a credible challenger is a stimulus to the solid-state photovoltaic industry to improve its existing products and to remain open to new concepts.

8 Conclusions

The development of reproducible and stable photovoltaic devices adapted for manufacturing processes has proceeded in evolutionary steps, with each component optimised and verified for compatibility with system requirements. The dye-sensitised nanocrystalline solar cell, either the electrochemical device, or the closely related sensitised heterojunction, provides a scientifically verified photovoltaic option, challenging the conventional semiconductor junction solid-state cell. Time and the market will tell if it can compete successfully.

Acknowledgements

The team at EPFL greatly appreciates the vote of confidence represented by the licences taken up by industry for this type of solar cell. In the present work, the initiative and innovative spirit of the research team members, past and present, in the areas of dye synthesis, catalysis and electrochemistry, and semiconductor thin film fabrication are acknowledged.

References

[1] Becquerel, A.E., 1839. Memoire sur les effets électriques produits sous l'influence des rayons solaires. *C. R. Acad. Sci. Paris*, Vol. 9, pp. 561–567.

[2] Gerischer, H., 1969. Role of electrons and holes in surface reactions on semiconductors. *Surface Science*, Vol. 13, pp. 265–278.

[3] Fujishima, A. and Honda, K., 1972. Electrochemical photolysis of water on a semiconductor electrode. *Nature*, Vol. 238, pp. 37–38.

[4] Memming, R., 1988. Photoelectrochemical solar energy conversion. *Topics Current Chem.*, Vol. 143, pp. 79–112.

[5] Grätzel, M. and Augustynski, J., 2000. Tandem cell for water cleavage by visible light. *Int. Patent Appl.* PCT/EP00/06350.

[6] Biot, J.B., 1839. Memoires et communications. *C. R. Acad. Sci. Paris*, Vol. 9, pp. 169–173.

[7] Gurney, R.W. and Mott, N.F., 1938. Theory of photolysis of silver bromide and the photographic latent image. *Proc. Roy. Soc.*, Vol. A164, pp. 151–167.

[8] James, T.H., 1977. *The Theory of the Photographic Process*. 4th. Ed., Macmillan, New York, USA.

[9] West, W., 1974. First hundred years of spectral sensitisation, *Photogr. Sci. Eng.*, Vol. 18, pp. 35–48.

[10] Moser, J., 1887. Notiz über Verstärkung photoelektrischer Ströme durch optischer Sensibilierung, *Monatsh.Chem.*, Vol. 8, p. 373.

[11] Rigollot, H., 1893. Memoires et communications. *C. R. Acad. Sci. Paris*, Vol. 116, p. 873.

[12] Spitler, M.T., 1983. Dye photo-oxidation of semiconductor electrodes: a corollary to spectral sensitization in photography. *J. Chem. Ed.*, Vol. 60, pp. 330–332.

[13] Gould, I.R., Lenhard, J.R., Münter, A.A., Godelski, S.A. and Farid, S., 2000. Two-electron sensitisation: a new concept for silver halide photography. *J. Am. Chem. Soc.*, Vol. 122, pp. 11934–11943.

[14] Kato, T., Maekawa, E., Fujishima, A. and Honda, K., 1984. Current doubling effect generated by formate on the cadmium sulfide photoanode and characterisation of the photo-electrochemical cell. *Nippon Kagaku Kaishi*, pp. 233–238.

[15] Namba, S. and Hishiki, Y., 1965. Color sensitisation of zinc oxide with cyanine dyes. *J. Phys. Chem.*, Vol. 69, pp. 774–779.

[16] Nelson, R.C., 1965. Minority carrier trapping and dye sensitisation. *J. Phys. Chem.*, Vol. 69, pp. 714–718.

[17] Bourdon, J., 1965. Spectral sensitisation of chemical effects in solids, *J. Phys. Chem.*, Vol. 69, pp. 705–713.

[18] Gerischer, H. and Tributsch, H., 1968. Electrochemical studies on the spectral sensitisation of zinc oxide single crystals. *Ber. Bunsenges. Phys.Chem.*, Vol. 72, pp. 437–445.

[19] Tributsch, H., 1968. Ph.D. thesis, Techn. Hochschule München, Germany.

[20] McEvoy, A.J., 1988. Functional components for solar energy chemical photoconversion systems. In: Bloss, W.H. and Pfisterer, F., Eds., *Advances in Solar Energy*, Pergamon, Oxford, UK, pp. 2921–2926.

[21] O'Regan, B. and Grätzel, M., 1991. A low-cost high effiiciency solar cell based on dye-sensitised colloidal titanium dioxide films. *Nature*, Vol. 335, pp. 737–739.

[22] Clark, W.D.K. and Suttin, N., 1977. Spectral sensitisation of n-type titanium dioxide by polypyridine-ruthenium complexes. *J. Amer. Chem. Soc.*, Vol. 99, pp. 4676–4682.

[23] Dare-Edwards, M.P., Goodenough, J.B., Hamnett, A., Seddon, K.R. and Wright, R.D., 1980. Sensitisation of semiconductor electrodes with ruthenium-based dyes. *Faraday Discuss. Chem. Soc.*, Vol. 70, pp. 285–298.

[24] Vlachopoulos, N., Liska, P., McEvoy, A.J. and Grätzel, M., 1988. Efficient sensitisation of TiO_2 electrodes using transition metal charge transfer complexes. In: Bloss, W.H. and Pfisterer, P., Eds., *Advances in Solar Energy*, Pergamon, Oxford, UK, pp. 3003–3009.

[25] Haque, S.A., Tachibana, Y., Willis, R.L., Moser, J.E., Grätzel, M., Klug, D.R. and Durrant, J.A., 2000. Parameters influencing charge recombination kinetics in dye-sensitised nanocrystalline titanium dioxide films. *J. Phys. Chem. B*, Vol. 104, pp. 538–547.

[26] Nazeeruddin, M.K., Zakeeruddin, S.M., Humphrey-Baker, R., Jirouzek, M., Liska, P., Vlachopoulos, N., Shklover, V., Fischer, C.-H. and Grätzel, M., 1999. Acid–base equilibria of (2,2'-bipyridyl-4,4'-dicarboxylic acid)ruthenium(II) complexes and the effect of protonation on charge-transfer sensitisation of nanocrystalline titania. *Inorg. Chem.*, Vol. 38, pp. 6298–6305.

[27] Kay, A., O'Regan, B. and Grätzel, M., 1996. Method for the manufacture of a photo-electrochemical cell and a cell made by this method. US patent 5525440.

[28] O'Regan, B., Grätzel, M. and Nazeeruddin, M.K., 1994. Photovoltaic cells, US patent 5350644.

[29] Brooks, K.G., Burnside, S.D., Shklover, V., Comte, P., Arendse, F., McEvoy, A.J. and Grätzel, M., 1999. Mesoporous nanocrystalline thin films for electrochemical applications. *Proc. Am. Ceram. Soc.*, Indianapolis, USA, pp. 115–122.

[30] Bedja, I., Hotchandani, S., Kamat, P.V., 1994. Preparation and photoelectrochemical characterization of thin SnO_2 nanocrystalline semiconductor films and their sensitisation with bis(2,2'-bipyridine)(2,2'-bipyridine-4,4'-dicarboxylic acid)ruthenium(II) complex. *J. Phys. Chem.*, Vol. 98, pp. 4133–4140.

[31] Sayama, K., Sugihara, H. and Arakawa, H., 1998. Photoelectrochemical properties of a porous Nb_2O_5 electrode sensitised by a ruthenium dye. *Chem. Mater.* Vol. 10, pp. 3825–3832.

[32] O'Regan, B. and Schwartz, D.T., 1996. Efficient dye-sensitised charge separation in a wide-band-gap p-n heterojunction. *J. Appl. Phys.*, Vol. 80, pp. 4749–4754.

[33] Rensmo, H., Keis, K., Lindstrom, H., Sodergren, S., Solbrand, A., Hagfeldt, A., Lindquist, S.E., Wang, L.N. and Muhammed, M., 1997.

High light-to-energy conversion efficiencies for solar cells based on nanostructured ZnO electrodes. *J. Phys. Chem. B*, Vol. 101, pp. 2598–2601.
[34] Dabestani, R., Bard, A.J., Campion, A., Fox, M.A., Mallouk, T.E., Webber, S.E. and White, J.M., 1988. Sensitization of titanium dioxide and strontium titanate electrodes by ruthenium(II) tris(2,2'-bipyridine-4.4'-dicarboxylic acid) and zinc tetrakis(4-carboxyphenyl)porphorin: an evaluation of sensitisation efficiency for component photoelectrodes in a multipanel device. *J. Phys. Chem.*, Vol. 92, pp. 1872–1878.
[35] Mikoshiba, S., Sumino, H., Yonetsu, M. and Hayase, S., 2000. Highly efficient photo-electrochemical cell with novel polymer gel electrolytes. *Proc. 16th European Photovoltaic Solar Energy Conf.*, Glasgow, pp. 47–50.
[36] Bach, U., Lupo, D., Comte, P., Moser, J.-E., Weissörtel, F., Salbeck, J., Spreitzer, H. and Grätzel, M., 1998. Solid-state dye-sensitised mesoporous TiO_2 solar cells with high photon-to-electron conversion efficiencies. *Nature*, Vol. 395, pp. 583–585.
[37] Halls, J.J.M., Walsh, C.A., Greenham, N.C., Marseglia, E.A., Friend, R.H., Moratti, S.C. and Holmes, A.B., 1995. Efficient photodiodes from interpenetrating polymer networks. *Nature*, Vol. 376, pp. 498–500.

IIe-2

Organic and Plastic Solar Cells

Jenny Nelson, Centre for Electronic Materials and Devices,
Department of Physics, Imperial College London, UK

1 Introduction

The last three years have seen an unprecedented growth of interest in solar cells made from organic electronic materials. This is due partly to the rapid growth of the photovoltaic market, which has stimulated research into longer term, more innovative photovoltaic technologies, and partly to the development of organic electronic materials for display applications. Rapid progress in optoelectronic molecular materials has introduced a range of potential new photovoltaic materials, as well as an improved understanding of the capabilities of such materials and confidence in their application. For a review see [1].

Organic materials are attractive for photovoltaics primarily through the prospect of high throughput manufacture using processes such as reel-to-reel deposition. Additional attractive features are the possibilities for ultra thin, flexible devices which may be integrated into appliances or building materials, and tuning of colour through chemical structure. The field has made impressive progress since the late 1990s. At the time of writing, solar power conversion efficiencies of over 3% have been reported for four distinct classes of organic solar cell, a growing range of new photovoltaic materials have been studied and increasing numbers of research groups and companies have declared an interest in 'soft' solar cells [2–4].

Unlike other photovoltaic technologies featured in this handbook, organic or 'plastic' solar cells are far from commercialisation. It is not known which, if any, of several different materials systems and cell designs will prevail and there is no experience in module production. In this chapter, we will discuss the principles of organic electronic materials and the ways in which they differ from inorganic photovoltaic materials, the range of device concepts, the most efficient device designs and the main challenges in increasing efficiency. Possible production and processing technologies are discussed, but the practical details are largely speculative.

2 Organic Photovoltaic Materials

Organic electronic materials are conjugated solids where both optical absorption and charge transport are dominated by partly delocalised π and π^* orbitals. Candidates for photovoltaic applications include crystalline or polycrystalline films of 'small molecules' (conjugated molecules of molecular weight of a few 100), amorphous films of small molecules prepared by vacuum deposition or solution processing, films of conjugated polymers or oligomers processed from solution, and combinations of any of these either with other organic solids or with inorganic materials. A comprehensive discussion of the development of organic solids for photovoltaic applications is given by Halls [4].

Organic photovoltaic materials differ from inorganic semiconductors in the following important respects.

- Photogenerated excitations ('excitons') are strongly bound and do not spontaneously dissociate into charge pairs. (Dissociation requires an input of energy of ~100 meV compared to a few meV for a crystalline semiconductor.) This means that charge carrier separation generation does not necessarily result from the absorption of light.

- Charge transport proceeds by *hopping* between localised states, rather than transport within a band, which results in low mobilities.
- The spectral range of optical absorption is relatively narrow compared to the solar spectrum.
- Absorption coefficients are high ($\sim 10^5$ cm^{-1}) so that high optical densities can be achieved, at peak wavelength, with films less than 100 nm thick.
- Many organic materials are susceptible to degradation in the presence of oxygen or water.
- As one-dimensional semiconductors, their electronic and optical properties can be highly anisotropic. This is potentially useful for device design.

The first two features are due to the fact that the intermolecular van der Waals forces in organic solids are weak compared to bonds in inorganic crystals and much weaker than the intramolecular bonds. As a consequence all electronic states are localised on single molecules and do not form bands. Low mobility is aggravated by the high degree of disorder present in many organic solids. The optical excitations accessible to visible photons are usually π to π^* transitions. Most conjugated solids absorb in the blue or green; absorption in the red or infrared is harder to achieve. However, the absorption bandwidth depends on the degree of conjugation and wider spectral sensitivity can be achieved in highly conjugated dye molecules.

These properties impose some constraints on organic photovoltaic devices:

- A strong driving force should be present to break up the photogenerated excitons.
- Low charge carrier mobilities limit the useful thickness of devices.
- Limited light absorption across the solar spectrum limits the photocurrent.
- Very thin devices mean interference effects can be important.
- Photocurrent may be sensitive to temperature through hopping transport and thermal dissociation of excitons.

3 Principles of Operation and Device Concepts

3.1 Homojunctions

The simplest device structure is a layer of organic material sandwiched between two different conducting contacts, typically indium tin oxide (ITO) and a low work function metal such as Al, Ca or Mg (Figure 1). The difference in work function provides an electric field which drives separated charge carriers towards the respective contacts (in rough analogy to a p–i–n junction in

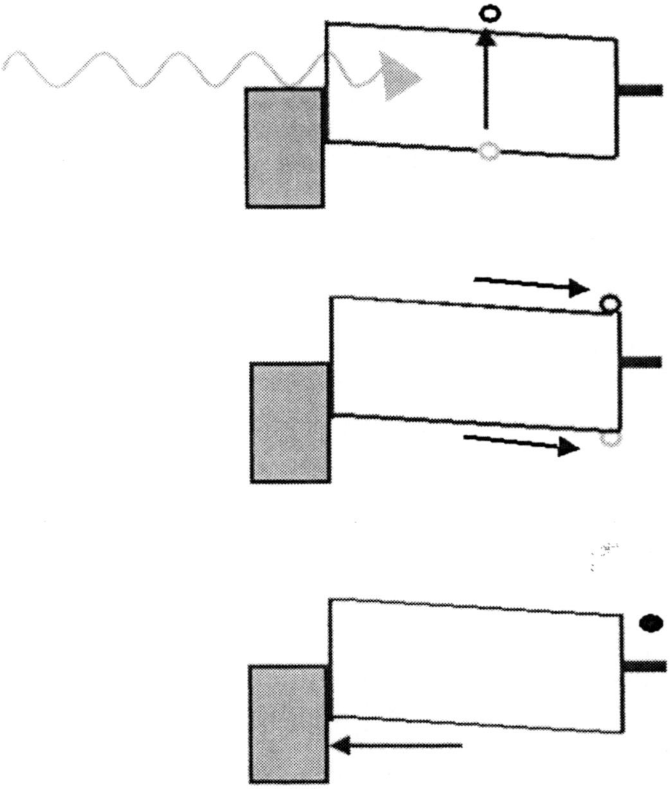

Figure 1 Schematic energy-band diagram of a simple device consisting of a single organic layer between two metal contacts. An electric field results from the difference in work functions of the contacts. Absorbed photons generate excitons which diffuse towards one or other contact where they may dissociate to yield charge pairs. Only the layer of organic material which lies within an exciton diffusion length of a contact can contribute to the photocurrent.

amorphous silicon). This electric field is seldom sufficient to break up the photogenerated exciton. Instead the exciton diffuses within the organic layer until it reaches a contact, where it may be broken up to supply separate charges, or recombine. Since exciton diffusion lengths are short, typically 1–10 nm, exciton diffusion limits charge carrier generation in such a device. Photocarrier generation is therefore a function not only of bulk optical absorption, but also of available mechanisms for exciton dissociation. Other loss factors are non-radiative recombination at the interfaces and non-geminate recombination at impurities or trapped charges.

Single layer solar cells of this type typically deliver quantum efficiencies (QE) of less than 1% and power conversion efficiencies of less than 0.1%. (QE is the ratio of electrons delivered to the external circuit per incident photon of a given wavelength, and is the figure of merit in organic photovoltaics. High QE is a necessary, though not sufficient, condition for high photovoltaic efficiency. In organic devices the value is still far from the values of 80–90% typical in inorganic solar cells.)

3.2 Heterojunctions

Most of the developments that have improved performance of organic photovoltaic devices are based on donor–acceptor heterojunctions. At the interface between two different materials, electrostatic forces result from the differences in electron affinity and ionisation potential. If both electron affinity and ionisation potential are greater in one material (the electron acceptor) than the other (the electron donor) then the interfacial electric field drives charge separation (Figure 2). These local electric fields are strong and may break up photogenerated excitons provided that the differences in potential energy are larger than the exciton binding energy. In a planar heterojunction, or 'bi-layer' device, the organic donor–acceptor interface separates excitons much more efficiently than the organic–metal interfaces in a single layer device and with very high purity materials, photovoltaic devices with high QE may be made.

3.3 Dispersed Heterojunctions

A revolutionary development in organic photovoltaics (and photodetectors) came in the mid 1990s with the introduction of a *dispersed* heterojunction, where an electron accepting and an electron donating material are blended

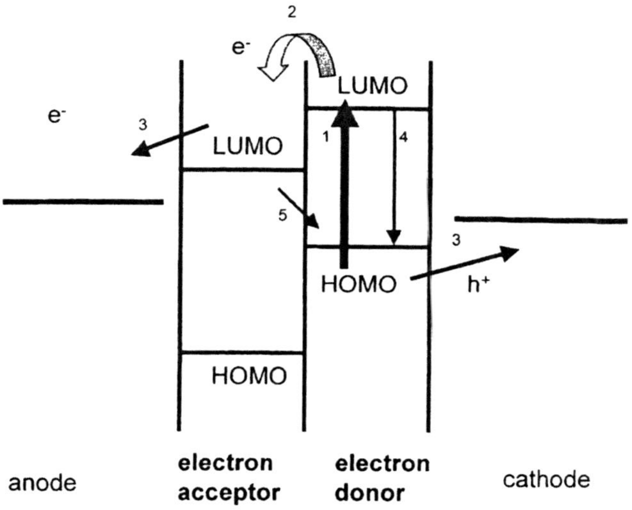

Figure 2 Schematic energy-band diagram of a donor-acceptor heterojunction. If both the excited state (LUMO) and ground state (HOMO) of the donor material lie at energies sufficiently higher than those of the acceptor material, then it is energetically favourable for an exciton reaching the interface to dissociate, leaving a positive polaron on the acceptor and a negative polaron on the donor. For efficient photocurrent generation, charge separation (2) should compete successfully with geminate recombination (4) after a photon absorption event (1), and transfer to contacts (3) should compete with interfacial recombination (5).

together. If the domain size in either material is similar to the exciton diffusion length, then wherever an exciton is photogenerated in that material, it is likely to diffuse to an interface and break up. If continuous paths exist in each material from the interface to the respective electrodes, then the separated charge carriers may travel to the contacts and deliver current to the external circuit (Figure 3(a)). This effect was reported independently by several groups [5–7] for a blend of two conjugated polymers. The blend improved QE to around 6–8% from less than 1% for either polymer alone. Around the same time, Yu and co-workers reported a QE of 29% for a blend of the hole transporter, poly-phenylene vinylene (PPV), with a derivative of C_{60} [8], where the C_{60} acts as the electron transporting component (Figure 3(b)).

This was followed by observations of enhanced QE in heterojunctions made from conjugated polymers with inorganic nanocrystals [9,10] and organic dye crystals [11]. The demonstration of improved QE with dispersed heterojunctions represents a departure from the device physics of conventional solar cells and has led to new device and materials designs. The principles of operation are shared by dye sensitised solar cells which are discussed in reference [12] and Chapter IIe-1.

4 Leading Device Designs

The early 2000s have seen developments in the synthesis of new photovoltaic materials; the combination of materials into new device architectures; studies of the effect of processing conditions and other factors on morphology and performance; and manipulation of materials at a molecular level, exploiting molecular self assembly and modification of interfaces. During this period power conversion efficiencies of over 3% have been reported in four distinct classes of device and QEs of over 20% achieved in several others. Several studies of issues relating to production, field performance and stability have been reported. Progress has been made with the understanding of photocurrent generation and charge separation, and the first approaches to developing a device physics for organic solar cells, have been made.

At time of writing, organic solar cell technology is young and the field is still wide open. Several very different device designs are competing as front runners and it is too early to guess which, if any, of these will emerge as a serious contender for commercialisation. Below we summarise the main features of the most promising options.

Power conversion efficiencies of over 3% have now been achieved in the three classes of device listed below and in Table 1. (High power conversion efficiencies have been reported [13,14] for a fourth class, a Schottky barrier device made from a doped molecular crystal in contact with an n-type a metal or n-type oxide layer, but these results are still unconfirmed.)

(1) Planar heterojunction devices made by vacuum deposition of thin films of small molecules have been studied by several groups, for application to LEDs as well as solar cells. An impressive result has been achieved at Princeton with a

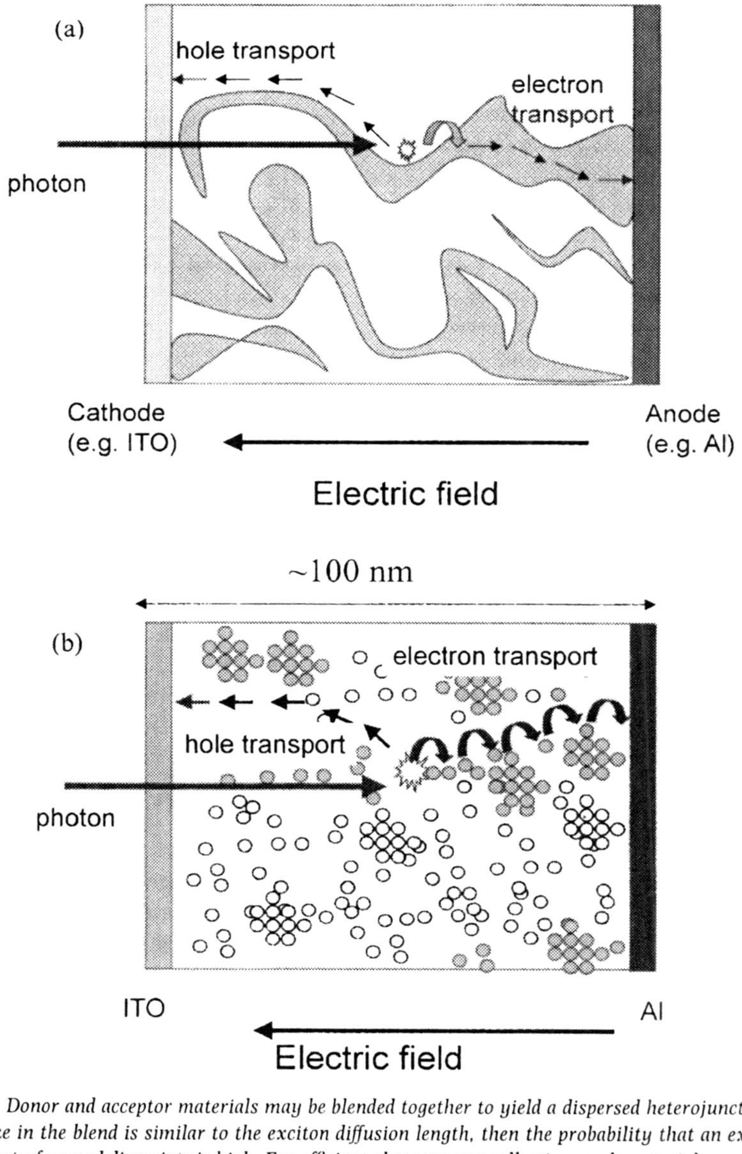

Figure 3 Donor and acceptor materials may be blended together to yield a dispersed heterojunction. If the domain size in the blend is similar to the exciton diffusion length, then the probability that an exciton will reach the interface and dissociate is high. For efficient photocurrent collection, each material must provide a continuous path for the transport of separated charge to the contacts. Isolated domains can trap charges, causing recombination. (a) A blend of two polymers. (b) A blend of one polymer with electron accepting nanoparticles or fullerenes. The concentration of nanoparticles should be sufficient to allow percolation.

four layer heterojunction, containing wide band gap hole transporting and electron transporting 'window' layers [15, 16]. These buffer layers function to block excitons from lossy metal contacts and to enhance optical field strength in photoactive layers via interference effects. Subsequent research on the same materials system has shown that greatly increased voltages can be achieved using tandem structures with tunnel junctions made from nanoparticulate silver. [17].

(2) Blends of polyphenylenevinylene derivatives and methanofullerenes are a well studied combination and are under intense development at Linz [3]. Photon absorption in the polymer is followed by electron transfer to the fullerene on a sub ps time scale. Current collection depends on charge percolation through the fullerene network and the polymer matrix and is therefore critically dependent upon the blend ratio and the degree of phase separation. The combination of poly(2-methoxy-5-(3',7'-dimethyloctyloxy)-1-4-phenylene vinylene), (MDMO-PPV) with 1-(3-methoxycarbonyl)-propyl-1-phenyl-(6,6)C_{61} (PCBM) appears to be a promising materials system. A breakthrough was achieved in 2001 by using chlorobenzene as a solvent in place of toluene, leading to QE of over 50% and power conversion efficiency of 2.5% [18]. The much improved performance is attributed to improved phase separation with chlorobenzene. Further development led to increased open circuit voltages by using LiF interfacial layers, leading to power conversion efficiencies exceeding 3% at 50°C [19].

(3) Solid-state dye-sensitised solar cells (DSSC) are the most promising amongst organic-inorganic composite devices to date. In the DSSC three active materials are used: an organic dye as light absorber, a nanocrystalline metal oxide film as electron transporter and liquid or organic hole transporting material (HTM) [12]. The original design used a redox active liquid electrolyte for hole transport, but a non volatile HTM is desirable for commercialisation. The ideal material should regenerate the photo-oxidised dye quickly and transport holes with high mobility. Candidates for solid state HTMs include doped arylamine based small molecules (OMeTAD) [20] and polythiophenes [21]. A power conversion efficiency of 2.5% using OMeTAD was reported by the EPFL group in 2001 [22], greatly improving on earlier studies of that material. The improvement is attributed to the suppression of electron–hole recombination at the metal oxide surface using adsorbed pyridine. The efficiency of this system was raised to over 3% by using silver ions to increase the density of adsorbed dye molecules on the TiO_2 surface, so increasing both J_{sc} and V_{oc} through increased photocurrent generation [23].

A number of other device types have achieved QE comparable with the devices in Table 1, though lower power conversion efficiencies. These include donor–acceptor polymer blend devices with well controlled morphology [24, 25], three layer donor–sensitiser–acceptor structures [26], inorganic–organic heterojunctions [27]; and liquid crystal–crystalline dye devices [28], all discussed below.

5 Challenges in Materials and Device Design

Table 1 shows that whilst organic solar cells produce quite respectable open-circuit voltages, the short circuit photocurrent and fill factor are much lower than those available from inorganic devices. The lower photocurrent is due to poorer light absorption as well as photocurrent generation and transport;

Table 1 The key performance characteristics for these cells as well as the peak QE. For comparison, the characteristics of the best amorphous and monocrystalline silicon solar cells are listed.

Material system	J_{sc} (ma cm^{-2})	V_{oc} (V)	Fill Factor	Efficiency (%)	Peak QE and wavelength (nm)	Reference
Cu phthalocyanine/C_{60} bi-layer cell	13[a]	0.53	0.52	3.6	18% at 620 nm; 35% at 400 nm	[16]
MDMO-PPV-PCBM	5.25	0.83	0.61	2.5; 3.3[b]	50% at 470 nm	[18], [19]
Dye-sensitised solar cell with OMeTAD	4.6	0.93	0.71	3.2	38% at 520 nm	[23]
Amorphous silicon	19.4	0.887	0.74	12.7	~90%	http://www.pv.unsw.edu.au/eff
Monocrystalline silicon	42.2	0.706	0.83	24.7	›90%	http://www.pv.unsw.edu.au/eff

[a] Estimated.
[b] Calculated, at 50°C.

the fill factor is due to poor transport and recombination. Most current research on organic photovoltaic materials and devices is therefore focussed on the following goals:

- Improving light harvesting.
- Improving photocurrent generation.
- Improving charge transport.
- Understanding device function and limits to performance.

Approaches to processing technology and stability are discussed in Sections 7 and 8.

5.1 Light Harvesting

A preferred strategy is to replace conducting polymers in devices with others which absorb further into the red. In the polymer-fullerene cell, lower optical gap replacements for the PPV derivatives, include polythiophene derivatives [29], polypyrrole/thiazadole copolymers [30] and thiophene/naphthene copolymers [31]. (Polythiophenes are of general interest as the hole transporting component in a blend, on account of both their red absorption and good hole mobilities when aligned.) Special red absorbing polymers may be developed for photovoltaic applications, including copolymers with mixed properties, such as a hole transporting backbone with red absorbing side groups.

Another approach is to replace the electron transporting polymer in a blend with conjugated crystalline dyes, such as anthracene or perylene, with wider absorption bands [32]. Extended red absorption can also be achieved using red absorbing *inorganic* semiconductor nanoparticles. Cadmium selenide nano-rods extend absorption to 700 nm [33]. Copper indium diselenide is particularly attractive on account of its very low band gap ($\sim$1 eV in the bulk) and is now being used in polymer-nanoparticle blends [34].

Dye sensitisation is a different strategy where a monolayer of a third material, usually an organic dye, is introduced between donor and acceptor to function as light absorber. Since light absorption and charge transport are carried out by different materials, the light absorber does not need to be a good bulk transporter of charge. Efforts to improve light harvesting in DSSCs include development of alternative dyes and combinations of dyes. A similar concept, of an all-organic donor-absorber-acceptor structure was proposed by Yoshino [6].

A different approach is to maximise the absorption of incident photons in the organic layers. From the perspective of light absorption, organic photovoltaic films are different from inorganic structures in the following important respects:

- Films are sufficiently thin that interference occurs within the layered device, allowing for the design of *light trapping* structures.
- Films are optically anisotropic.

Light trapping structures improve the capture of photons in the active layers through constructive interference. They allow thinner photovoltaic films to be used, and this improves the efficiency of charge transport and collection. The possibilities for light trapping for organic solar cells are discussed by Inganas [35]. Some of these have been investigated using embossed polymer layers for light trapping [36, 37] and exploiting interference effects inside cavities made from photoactive and optically transmitting organic layers [15, 38]. Embossed polymer light trapping structures are already used in thin film silicon solar cells.

Optical anisotropy is important for illumination at large angles, and is relevant to the use of plastic solar cells in diffuse light conditions. In the case of spin cast polymers, the conjugated backbone is likely to be lying in the plane of the film [39] with the result that light incident at large angles is less likely to be absorbed than in an isotropic material. For columnar structures such as discotic liquid crystals, anisotropy *enhances* absorption of light incident at wide angles.

Both effects need to be considered when designing organic photovoltaic devices. Calculation of optical absorption is non-trivial, and relies on good data for the optical functions of the films, usually provided by optical ellipsometry. Figure 4 shows how absorption of light in a thin organic film can be affected by interference.

5.2 Improving Charge Transport

Charge transport is limited by the low intrinsic mobilities of organic solids, and by the charge trapping effects of impurities and defects. In several recent studies, higher mobility polymers such as fluorene-triarylamine and thiophene copolymers have been used to replace MEH-PPV in blend devices [40, 41]. Materials with ordered phases offer high, though anisotropic, mobilities and in this respect liquid crystals [28, 42] and polymers with liquid crystal phases, such as polythiophenes [43], are interesting.

Because organic electron mobilities are generally very poor, an inorganic electron transporting component may be preferred. Thin film [27, 44, 45] and dispersed nanocrystalline [10] TiO_2 has been used in several approaches. TiO_2 is particularly interesting as a cheap and non-toxic material which is stable in oxygen, can be prepared as rigid nanostructured templates, can be used in light trapping structures, and benefits from a wide experience base due to dye sensitised solar cell research. Efficient charge transfer to TiO_2 from various polymers has been reported (MEH-PPV [10, 46], phenyl-amino-PPV [27], polythiophene [21]) and the sensitisation of other oxides such as SnO_2 with polymers has been demonstrated [47].

Elongated crystalline components are attractive as electron transporters if crystal size and orientation can be controlled. Blend devices using needle-like crystalline dyes [32, 48], CdSe nano-rods [33] and carbon nanotubes [49, 50] have been studied. Meanwhile, organic electron transporters with improved mobility and stability are being developed.

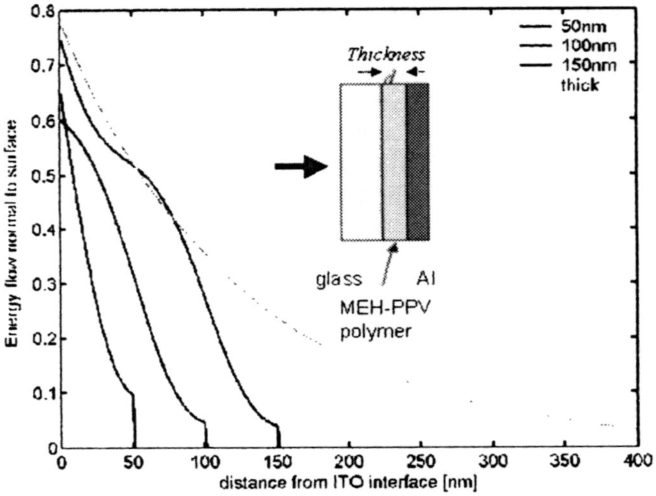

Figure 4 Interference effects influence light harvesting in layered structures. Shown is the fraction of 500 nm light absorbed in a layer of MEH-PPV of thickness d, *sandwiched between glass and aluminium contacts. Interference effects are clear in the shape of the attenuation curve for films of less than 200 nm thick. Notice that a film of 50nm is able to absorb 55% of the light, whilst if interference were neglected 200 nm of material would be needed to absorb the same fraction of the light.*

5.3 Control of Morphology

In a dispersed heterojunction device, both photocurrent generation and charge transport are functions of morphology. Photocurrent generation requires blending on the scale of the exciton diffusion length while transport requires continuous paths from interface to contacts. In polymer-nanocrystal or polymer-fullerene blends, the concentration of the particulate component should be sufficient for charge percolation [9, 51]. One attractive hypothetical configuration is a set of interdigitated electron and hole transporting channels, directed perpendicular to the contacts (this is one motivation for the study of rod-like nanocrystals). Another is a compositionally graded blend with an excess of donor type material on one side and acceptor type on the other. This concept has been demonstrated by Yoshino using vacuum deposited layers [6], and by Granstrom for a laminated assembly [24]. The concept was taken further by Takahashi and co-workers who reported QE of 49% for a three layer structure where electron and hole transporting layers are separated by a heterodimer light absorbing layer [26]. The absorbing layer is polarised upon light absorption to drive the charges towards the appropriate transport layers. This is essentially an all-organic version of the dye-sensitised solar cell. Some different morphologies are illustrated in Figure 5.

In practice many materials tend to segregate when blended, and much attention has focussed on ways of controlling the morphology of blends. Routes include:

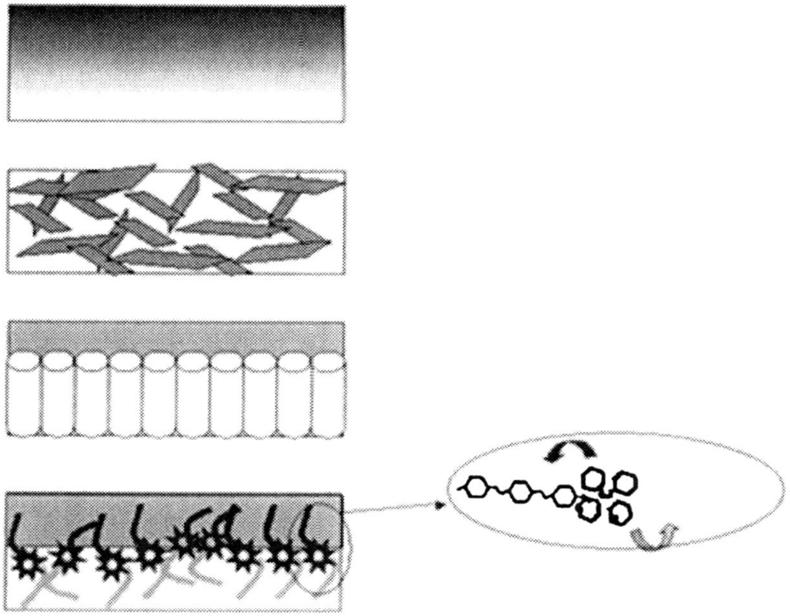

Figure 5 Some approaches to improved charge separation and collection in organic solar cells. In each figure the direction of photocurrent generation is from top to bottom and the electron accepting component is shaded. (a) A compositionally graded blend improves the collection of positive and negative charges near the respective contacts [6, 24]. (b) Elongated electron acceptors such as nano-rods or dye crystals can improve electron transport [32, 33], but for efficient collection they should be directed perpendicular to the contacts. (c) Self-organising discotic liquid crystals are one way of achieving preferential charge transport perpendicular to the contacts [28]. (d) A light absorbing dimer designed to drive positive and negative charges apart after photoexcitation can help to channel respective charges towards electron transporting and hole transporting layers [26].

- Control of blend morphology through processing conditions. Choice of solvent, atmosphere and substrate temperature strongly influence the morphology of polymer blends [40, 41]. Choice of solvent appears to influence segregation of fullerenes in MDMO-PPV [18].
- Self organisation. Self assembly by discotic liquid crystals [28], and by ionically or electrostatically interacting monolayers [52, 53] have been used to construct structured heterojunctions. Self-assembled monolayers can also be used to modify substrate surfaces to control the segregation of blend components [25].
- Synthesis of donor (D)–acceptor (A) copolymers (such as polymer with pendant fullerene groups [54]) and block copolymers. Positioning D and A groups on the same polymer backbone can ensure effective photoinduced D $\rightarrow$ A electron transfer under all conditions and avoids the problems of phase segregation. D–A copolymers may be designed to absorb longer wavelength photons than single polymers, so improving light harvesting, but charge extraction may be more difficult.

- Use of porous organic or inorganic films as templates (for example, [55]).
- Cosublimation of small molecules to form graded D–A heterostructures [56, 57].

Novel characterisation techniques have been developed to study blend morphology, including spatially resolved fluorescence [40] and confocal Raman spectroscopy [58] as well as AFM and SEM imaging techniques.

5.4 Understanding Function

Understanding of the device physics of organic solar cells is still at a primitive stage, compared to inorganic solar cells. As a basis for device physics, consider the steady-state continuity equation which describes the charge carrier population in any photonic device:

$$\frac{1}{q}\nabla.J_p + G - U = 0 \tag{1}$$

where p is the density of carriers of charge q, J_p the current density, G the volume generation rate and U the recombination rate. For crystalline inorganic p–n junction devices, G can be equated with the local photon absorption rate, and minority carrier recombination and current terms are linear in the carrier density. In organic devices, the description of all three terms is more complicated.

Charge carrier generation, G, can be related to exciton generation if the exciton lifetimes and diffusion lengths, and exciton dissociation efficiencies at interfaces, are known. In planar structures, charge carriers are effectively generated only at the interface (Figure 6) and photocurrent generation can be interpreted in terms of a filter effect [59]. This effect may cause higher photocurrents to result from thinner films, in contradiction to the usual case. In blends, photocurrent generation is a function of interface morphology. In the ideal blend, all photogenerated excitons are dissociated by nearby interfaces, and charge carrier generation is equivalent to light absorption. (Ultrafast transient spectroscopy has been used to demonstrate the fast (sub ps) rate and high yield of charge separation in well dispersed blends, such as polymer-fullerene structures [3].) In all cases interference should be accounted for in calculating the exciton generation profile.

Because organic materials are seldom heavily doped, charge recombination, U, is a bimolecular process. However the kinetics may be influenced by the trapping of charge at defects in the disordered material or competing recombination and transport pathways, and need to be measured for each new material system. Transient optical spectroscopy has proved useful for monitoring the rate of charge recombination in dye sensitised devices [60] and polymer-fullerene structures [61], as has electron spin resonance [62]. In the case of MDMO-PPV/PCBM structures under solar illumination, recombination appears

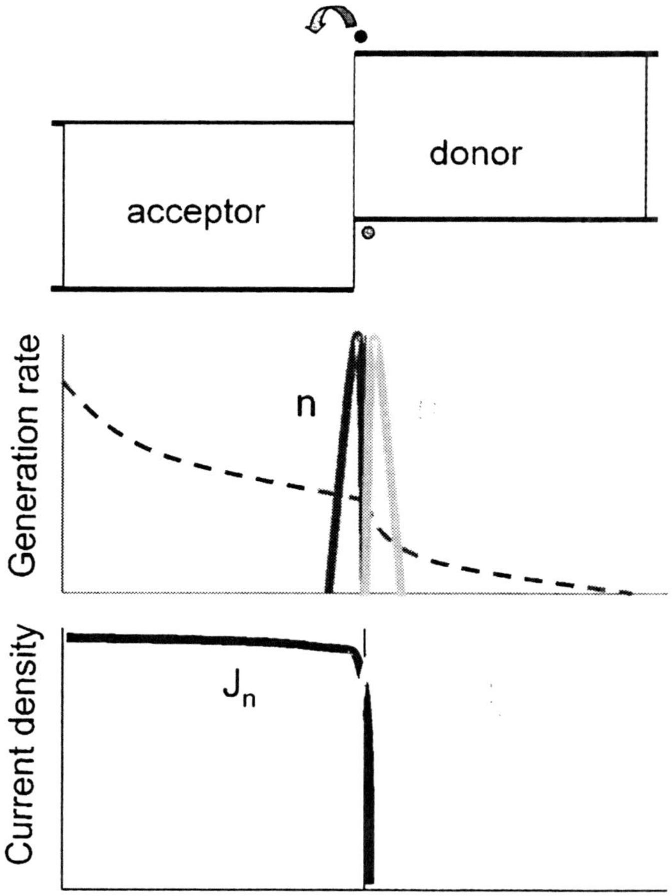

Figure 6 Schematic exciton generation, charge generation and current density profiles for bi-layer structure. Exciton generation follows the absorption of light in the two materials (dashed line). Charge carrier generation occurs only at the interface, leading to delta function profiles for negative charge generation in the acceptor material (black line) and positive charge generation in the donor material (grey line). These narrow charge generation region lead to constant, unipolar currents in the respective materials, and a bimolecular recombination zone located at the interface. The amplitude of the charge generation and therefore the current depends upon the exciton diffusion length and thickness of the front (acceptor) layer. For short exciton diffusion lengths, thinner layers result in higher photocurrents.

to be dominated by the release of positive polarons from trap states and U is a superlinear function of the polaron density.

The current term, J, depends upon the relative contributions of drift and diffusive currents and charge trapping, as well as upon the morphology of donor–acceptor heterostructures. Many organic conductors exhibit dispersive charge transport characteristics, so that both diffusive and drift currents are non-linear functions of carrier density. In blends, charge carrier mobilities are also dependent upon blend composition [63]. The relative importance of drift and diffusive currents at photovoltaic operating conditions is not well known, but recent work [64, 65] indicates that in planar heterojunctions the driving force due to charge carrier generation at the interface may dominate over the electrostatic driving force due to difference in work function of the contacts, and may dominate the open circuit voltage (Figure 6). This is in contrast with single layer devices where photovoltage is limited by the difference in work functions of the electrode materials, but in agreement with experimental studies of polymer-fullerene blends which show that the photovoltage depends upon the electron affinity of the acceptor rather than the work function of the cathode [66]. This indicates that photocurrent collection in donor-acceptor blends does not require a macroscopic electric field, a situation which is largely agreed to apply to dye sensitised solar cells [67]. Charge trapping further influences current through its influence on the electric field distribution.

On the whole, the device physics of organic solar cells is complicated by uncertainties in material parameters, the effects of heterojunctions and interfaces on materials, and the effects of light, bias and ageing. Further understanding and the development of appropriate models will require further fundamental experimental studies on model systems.

6 Organic Photovoltaic Modules

Plastic solar cells are first likely to be commercialised for low-power, consumer applications, which currently take up about 10% of the market. In this they will compete (most directly) with amorphous silicon (a-Si), and to a lesser extent with emerging thin film technologies. Plastic solar cells may offer advantages over a-Si, in lower production cost, in ease of deposition onto flexible substrates or as conformal coatings on to building materials, or in colour, and these may open up new areas in the consumer photovoltaics market.

Examples of applications might be power supplies for displays and monitors, control electronics, mobile communications, battery chargers, thermoelectric coolers, various consumer products, and photovoltaic windows and other architectural materials. The modules could be integrated into appliances or building materials. In such applications the module would be designed for the specific power output required and would consist of a small number of cells in series, depending on the operating voltage required. Note that since organic solar cells deliver relatively high voltages, any given operating voltage may be achieved with fewer cells than for inorganic materials. Whether cells need to be connected in parallel depends upon the current output required. The high series resistance of ITO coated flexible substrates (10–100 Ω cm^2) place a limit on the useful cell area before unacceptable power loss to series resistance. With a short circuit current density of 10 mA cm^{-2} at one sun (a reasonable goal for organic photovoltaics), this limit might be a few tens of cm^2. However, initial applications for plastic solar cells are likely to be designed for low light levels, such as indoors. The lower illumination levels (indoor illumination is typically less than one tenth of AM1.5 intensity) lead to lower current densities and lower series resistance penalties, and may avoid the need for parallel connections for current ratings of less than, say, 100 mA.

Modules would be produced directly during cell production (discussed below) whether a batch or a continuous process. Cell interconnects can be integrated into the modules by etching the conductive substrate, scribing the photoactive layers and pattering the top contacts during production, much as is done for inorganic thin films. After metallisation the cell material must be encapsulated in a barrier coating to protect against moisture and oxygen, and in an outer coating appropriate to the intended environment. Outer plastic encapsulants may be used as passive optical concentrators to enhance the amount of light absorbed in the thin active layers the cell. In the case of building integrated modules such as windows or roofing tiles, the deposition of contacts, photoactive layers and cell interconnections could be integrated into the production process for that component. Again, cell sizing and connections will be a function of the expected current output and power requirements.

Although there is as yet no commercial production of organic solar cells, prototype plastic solar modules have been produced at the University of Linz, consisting of several cells in series, with a cell area of a few tens of cm^2. Studies of the effects of increasing cell area [68] show that polymer-fullerene cells can be scaled up to $150 cm^2$ without large losses in performance.

Production of standard 12 or 24 V dc modules for power generation using plastic photovoltaic materials is still remote. The low efficiency of organic solar cells leads to larger area requirements for power generation, and higher relative costs for the balance of system components. The lower lifetime expectations also increase the cost per kWh. For these reasons organic solar cells are not foreseen to be competitive with inorganic materials for high power applications at least until the obstacles to efficiency and lifetime are overcome.

7 Production Issues

The main driver for organic solar cell development has been the prospect of fabricating large area devices at low cost using production techniques developed for conventional plastics. However, there is as yet no experience in large scale organic cell production, and the following discussion is largely speculative. We discuss below established organic device production technologies which could be applied to photovoltaics, several other approaches which have been explored only in the laboratory, and hypothetical production schemes which are so far untested.

7.1 General Issues

Issues for organic solar cell production are:

- *Flexible substrates.* Conductivity, transmissivity and impermeability to oxygen and water are required. The most widely used material is ITO coated poly(ethylene terephthalate) (PET) which is already used as a substrate for flexible amorphous silicon solar cells. ITO is limited by a transmissivity of around 80–90% in the visible and a relatively high resistivity (typically 10–50 Ω cm^2) which leads to series resistance limitations, as mentioned above. Film forming properties and electrical contact are improved by coating the ITO with a layer of the doped conducting polymer PEDOT:PSS, though this introduces an additional processing stage. New organic transparent conducting coatings are being studied, but do not yet compete with ITO.

- *Ease of deposition.* In principle, deposition of films from solution is easier and cheaper than that from the gas phase, therefore solution processible materials are preferred. 'One pot, one shot' approaches to deposition of donor-acceptor blend films, where both components are deposited from a single solution, are attractive and enable much easier processing than the sequential deposition of separate layers. In such approaches the use of self assembling materials such as discotic liquid crystals [28] or solvent controlled blend segregation (e.g. [25]) can help control the resulting blend morphology.
- *Chemical stability.* Many conjugated organic solids are unstable in the presence of oxygen and light, and may also be sensitive to water. Devices for research are preferentially fabricated in an inert and dry atmosphere, but maintaining such conditions for large-scale production is costly. Costs could be reduced by finding materials which are sufficiently stable so that devices can be prepared in ambient conditions before encapsulation.
- *Encapsulation.* Coatings are needed which are sufficiently impermeable to moisture and oxygen. This is a problem with organic substrates such as PET which are intrinsically permeable. Recent progress in development of barrier coatings for flexible OLEDs, usually using multiple deposition of organic and inorganic layers to achieve the required combination of flexibility and impermeability [69], has enabled plastic OLED lifetimes approaching 4000 hours [70]. Such coatings may be used for plastic photovoltaic devices.
- *Processing temperature.* High temperature processing stages are costly and damaging to organic films. This may be an issue for solid state dye sensitised solar cells, where electrical contact must be achieved within the metal oxide film without the usual high temperature sintering stage.

The production technology depends upon the materials system used. For dry deposition of photoactive layers, a batch process similar to that used for organic light emitting diodes (OLEDs) could be used. For deposition from solution, batch or continuous processes are in principle possible, although there is as yet no industrial experience in this.

7.2 Production Processes for Molecular Films

Batch or semi-batch processes are already routinely employed for production of OLEDs. The organic materials used are layers of 'small molecules' which can be evaporated on to the conducting glass substrate, followed by top metal electrodes. The devices are encapsulated in a stainless steel can using an epoxy resin before separation into individual devices. In this process the most expensive stage is the vacuum evaporation, since many expensive vacuum components are required in order to maintain high vacuum ($\sim 10^{-6}$ torr). Batches of area $\sim 370 \times 470$ mm^2 are processed this way. Solar cells or modules could be produced by the same techniques. The process for solar cells should be much simpler since (a) the device unit area (pixel) is much larger and the fine interconnections needed to address OLEDs are not needed, and (b) the number of molecular layers needed is lower (for a colour OLED, more than three electro-active materials are needed as well as contact layers). Nevertheless the process would still be expensive for solar cell production, on account of the high capital and maintenance costs of the vacuum chamber.

A typical batch process for deposition on glass is summarised below in Table 2. The procedures on plastic substrates would be different, and include additional encapsulation stages.

An alternative to vacuum deposition for small molecules is organic vapour-phase deposition (OVPD). Molecules are carried in a flowing gas at high temperature (~ 300°C) and condense on to a cool substrate. Compared to vacuum deposition, the technique does not require high vacuum nor high maintenance costs, it can enable large area coating and it may be compatible with continuous processing. Films of a few tens to hundreds of nm can be produced and surface roughness can be minimised by controlling the flow rate. The applicability of the technique to multilayer OLED fabrication has been demonstrated [71].

7.3 Batch Production Processes for Polymer Films

When the active materials are available in solution, new deposition techniques are available which may greatly reduce costs. Deposition from solution is more viable for solar cells than for LEDs because of the larger device unit area and facile interconnections.

Table 2 Steps in a typical batch process for organic LED deposition on glass.

Pre process	– Clean substrate
	– Etching of bottom contact patterns
	– Building of passivation layer and cathode separator by photolithography
Post process	– Evaporation of molecular layers
	– Evaporation of top contacts
Encapsulation process	– Attachment of desiccant film
	– Alignment of stainless steel casing.
	– Bonding of can to substrate.
Module process	– Scribing into individual devices
	– Connection to flexible printed circuit board
	– Reliability test

For research devices, spin coating is commonly used, but this is uneconomical for large areas because of the material waste. Some of the following production techniques have been considered:

- Doctor blading or 'wire' blading. A rod is drawn over a line of the polymer solution, to spread it out into a thin sheet. This process works well for films of 1 μm thickness, but for thicknesses of a few 100 nm, uniform thicknesses are harder to achieve. Thinner films may be achieved using a rod loosely wrapped with wire of a given thickness. The wire supports the rod and determines the thickness of the resulting polymer film. These approaches are being investigated at Linz.
- Screen printing. A fine mesh screen loaded with polymer solution is brought into temporary contact with the substrate and polymer solution flows across. This technique has been applied to (small area) organic solar cells made from polymer-fullerene blends, producing thin and uniform films (active layer thickness of 40 nm with a roughness of about 2 nm) [72]. The screen printing technique is readily applied to large areas and is widely used in conventional photovoltaic cell production for contact printing. It is also being studied for depositing the metal oxide paste in dye-sensitised solar cells. However, for large area thin films, the mesh is likely to introduce thickness variations.
- Inkjet printing. A fine jet of polymer solution is sprayed on to the substrate in a required pattern. The technique has huge potential for organic electronic devices where fine circuit structures are required, and has been used successfully for organic thin-film transistors [73]. It offers precise contact definition and an easy way to deposit multiple layers. The technique could be applied to organic photovoltaic devices, although processes to achieve uniform film thickness over large areas will need to be developed (the polymer solution needs to wet the substrate surface rather than bead up). A simpler approach is printing polymer and contact materials using an embossed stamp [74]. This may be compatible with reel-to-reel processing.

Some of these techniques are illustrated in Figure 7.

7.4 Continuous Production Processes for Polymer Films

Since the device structure and contact designs for organic solar cells and modules are relatively simple, the modules could be produced from solution by a continuous reel-to-reel process similar to the processes currently used to produce plastic sheeting. Such a process is particularly appropriate for solar cell applications (compared to LEDs) because large areas are required yet the contact patterns and device structures are simple. In principle, well understood production technologies can be adapted and existing plant may be redeployed to the new purpose as a way of reducing production costs. However, reel-to-reel processing has not yet been attempted for any application of organic electronic materials and experience is extremely limited.

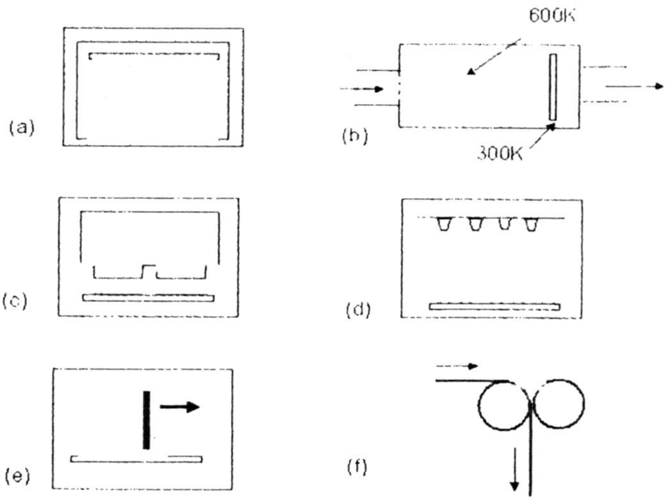

Figure 7 Several possible techniques for deposition of organic photovoltaic films. (a) Molecular deposition in high vacuum. (b) Organic vapour phase deposition. Molecules carried in a heated gas are condensed on to cooled substrates. (c) Printing. Patterned stamps can be used to print polymer films of defined area. (d) Inkjet printing. This allows very fine contact definition. (e) Doctor blading. (f) Compression between rollers.

The stages involved could be as follows (see also Figure 8):

- patterned conducting coating on flexible substrate,
- scribe the conducting coating,
- deposit active layers of 100–200 nm thickness,
- scribe the active layers,
- deposit metal contacts by in-line sputtering at low pressure,
- low-temperature annealing,
- add the encapsulating layer by reel-to reel contact,
- dry and cut.

The key stage in such a process is the deposition of the active layers. The material could be deposited as a blend from a single dispenser, or from a second roller. Dispersed donor acceptor blends can be deposited from a single solution, or laminated donor–acceptor 'bi-layer' structures could be deposited by contact between rollers. As yet these processes are untried. Achieving uniform film thicknesses to within 10 nm over areas of hundreds of cm^2 represents a major challenge.

An important consideration is the sensitivity of the materials to air and moisture whilst in solution. For sensitive materials the entire process could be carried out in a dry and inert atmosphere, but this adds to the cost of the plant.

A number of other industrial techniques are relevant to the development and production of organic photovoltaics. An example is the use of combinatorial techniques to screen different materials combinations and select the best ones [44].

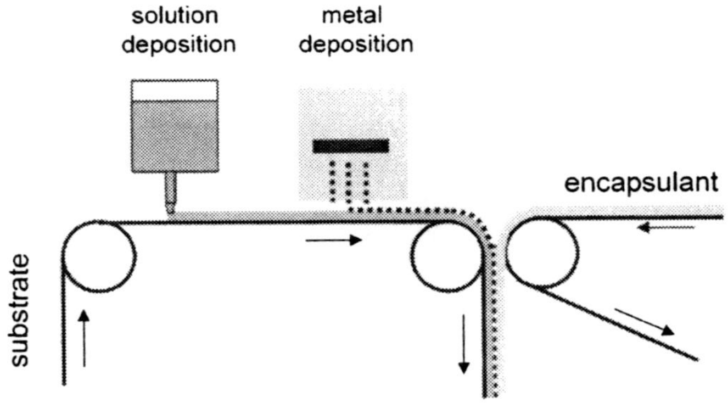

Figure 8 A hypothetical reel-to-reel process. A thin layer of polymer solution is deposited on a conducting substrate, followed by deposition of top metal contacts by a process such as in-line sputtering. Barrier coatings can be applied by a second roller. Scribing of layers to interconnect cells into modules can be integrated into the process (not shown).

8 Stability and Performance

Organic solar cells suffer from instability in two main respects. First, many conjugated materials are unstable in the presence of oxygen and light. The photogenerated exciton is capable of reducing oxygen, generating reactive products (such as superoxide) which react with the molecular material and chemically degrade it. The second is the mechanical instability of donor–acceptor blends. On casting the components are frozen into a non-equilibrium configuration. Over time, the components may segregate, reducing the degree of blending and the effectiveness of charge separation. The latter is a particular problem with polymer-fullerene blends, where the fullerenes tend to cluster.

As yet there is no general solution to the problem of the photostability of conjugated solids. One option is to develop or to use more stable components. Some conjugated dye molecules, liquid crystals and inorganic nanocrystals offer better stability. Until more air stable materials can be developed, devices need to be encapsulated using barrier coatings, as discussed above. An alternative approach is to incorporate sacrificial oxygen scavenging components into the device, such as TiO_2 or other metal oxides. The TiO_2 in solid-state dye-sensitised cells may help to protect the organic components from degradation. Another lesson from dye sensitised solar cell research is that the photoactive components in a solar cell do not have to be extremely stable in isolation. The ruthenium based dye which is used in DSSCs is more stable in the cell configuration than in isolation [75]. This can be attributed to the very short lifetimes of the reactive photogenerated excited state and cation state when in the cell configuration. A similar effect is observed in polymer/fullerene blends, where the photodegradation of the polymer is slowed down by blending with fuilevenes, and again can be attributed to the removal of the reactive exciton by rapid charge separation [76].

The most detailed stability and performance studies of organic solar cells have been carried out on polymer/fullerene systems. These indicate that under normal operation unprotected cells degrade after a few hours, encapsulated cells maintain over 50% of their initial performance for several months [76], degradation is accelerated by high temperatures both in the laboratory [77] and in the field in hot ambient conditions [78]. However, these devices exhibit improved performance with increasing temperature, an advantage over inorganic devices which is attributed to temperature dependent mobility, [79] and shared by polymer blend devices [39] and DSSCs.

Structural instability due to phase segregation in donor-acceptor blends is a function of materials, temperature and time. To some extent segregation may be minimised by choice of solvent and control of interfacial forces (as discussed in Section 5). The problem can be avoided by the use of rigidly connected porous templates such as inorganic semiconductor or metal oxide nanostructures.

A common problem with new device designs of degradation of QE with increasing light intensity, so that cells perform well only under low illumination. Such behaviour is a signature of recombination mechanisms which may be due to impurities, and must be eliminated for the cells to be useful in solar conditions.

Finally it is worth commenting on the measurement and characterisation of organic solar cells. On account of their poor spectral range and poor performance under high light intensities, organic photovoltaic device performance is often reported at a single wavelength and low light intensity. For new materials combinations, the figure of merit is often the QE at peak wavelength rather than the power conversion efficiency. Since wavelength and light intensity vary between reported measurements, as well as ambient conditions (vacuum or air), the comparison of device performance is difficult.

Since organic solar cells are likely to be used in quite different contexts to conventional photovoltaics, for instance, under low or diffuse light levels, or in windows where partial transparency is required, the standard test condition may not be the most relevant. There is a case for establishing new measurement standards more relevant for organic solar cell measurement and calibration.

9 Conclusions

Progress with organic photovoltaic materials and devices in recent years has been impressive. Power conversion efficiencies over 3% have been achieved in at least three different device structures, varying from high quality, vacuum deposited multilayer molecular films to dispersed heterojunctions in spin cast soluble polymers. All are based on the concept of a donor–acceptor system where photogenerated excitons are split by forces at the donor–acceptor interface. Higher efficiency requires improvements in absorption of red light, in charge transport and in material stability. Recent research focuses on the synthesis and testing of new photovoltaic materials in established device structures, and the development of new structures where morphology is controlled through self-assembly and processing conditions. Experience with commercial scale devices is still limited, as is the theoretical understanding of device function. Based on current trends, efficiencies of 5–10% appear to be within reach, although stability remains an obstacle.

Acknowledgements

The author is grateful to Youngkyoo Kim, Stephen Forrest, Ralph Gottschalg and James Durrant for helpful discussions, and to Felix Braun for help with Figure 4. The support of the Engineering and Physical Sciences Research Council and the Greenpeace Environmental Trust is acknowledged.

References

[1] Heeger, A.J., 2001. Semiconducting and metallic polymers: The fourth generation of polymeric materials. *J. Physical Chemistry B*, Vol. 105(36), pp. 8475–8491.

[2] Wallace, G.G. et al., 2000. Conjugated polymers: New materials for photovoltaics. *Chemical Innovation*, Vol. 30 (4), pp. 15–22.

[3] Brabec, C.J., Sariciftci, N.S. and Hummelen, J.C., 2001. Plastic solar cells. *Advanced Functional Materials*, Vol. 11(1), pp. 15–26.

[4] Halls, J.J.M. and Friend, R.H., 2001. Organic Photovoltaic Devices. In: Archer, M.D. and Hill, R.D., Eds., *Clean Electricity from Photovoltaics*, Imperial College Press, London, pp. 377–445.

[5] Yu, G. and Heeger, A.J., 1995. Charge Separation and Photovoltaic Conversion in Polymer Composites with Internal Donor–Acceptor Heterojunctions. *J. Appl. Phys.* Vol. 78(7), pp. 4510–4515.

[6] Yoshino, K. et al., 1997. Novel photovoltaic devices based on donor–acceptor molecular and conducting polymer systems. *IEEE Transactions on Electron Devices*, Vol. ED-44(8), pp. 1315–1324.

[7] Halls, J.J.M. et al., 1995. Efficient Photodiodes from Interpenetrating Polymer Networks. *Nature*, Vol. 376(6540), pp. 498–500.

[8] Yu, G. et al., 1995. Polymer Photovoltaic Cells – Enhanced Efficiencies Via a Network of Internal Donor–Acceptor Heterojunctions. *Science*, Vol. 270(5243), pp. 1789–1791.

[9] Greenham, N.C., Peng, X.G. and Alivisatos, A.P., 1996. Charge separation and transport in conjugated polymer/semiconductor-nanocrystal composites studied by photoluminescence quenching and photoconductivity. *Physical Review B*, Vol. 54(24), pp. 17628–17637.

[10] Salafsky, J.S., 1999. Exciton dissociation, charge transport, and recombination in ultrathin, conjugated polymer-TiO2 nanocrystal intermixed composites. *Physical Review B*, Vol. 59(16), pp. 10885–10894.

[11] Petritsch, K. et al., 2000. Dye-based donor/acceptor solar cells. *Solar Energy Materials and Solar Cells*, Vol. 61(1), pp. 63–72.

[12] Hagfeldt, A. and Gratzel, M., 2000. Molecular photovoltaics. *Accounts of Chemical Research*, Vol. 33(5), pp. 269–277.

[13] Schon, J.H. et al., 2000. Efficient organic photovoltaic diodes based on doped pentacene. *Nature*, Vol. 403(6768), pp. 408–410.

[14] Schon, J.H., Kloc, C. and Batlogg, B., 2001. Pentacene based photovoltaic devices. *Synthetic Metals*, Vol. 124(1), pp. 95–97.

[15] Peumans, P., Bulovic, V. and Forrest, S.R., 2000. Efficient photon harvesting at high optical intensities in ultrathin organic double-heterostructure photovoltaic diodes. *Applied Physics Letters*, Vol. 76(19), pp. 2650–2652.

[16] Peumans, P. and Forrest, S.R., 2001. Very-high-efficiency double-heterostructure copper phthalocyanine/C-60 photovoltaic cells. *Applied Physics Letters*, Vol. 79(1), pp. 126–128.

[17] Yakimov, A. and Forrest, S.R., 2002. High photovoltage multiple-heterojunction organic solar cells incorporating interfacial metallic nanoclusters. *Applied Physics Letters*, Vol. 80(9), pp. 1667–1669.

[18] Shaheen, S.E. et al., 2001. 2.5% efficient organic plastic solar cells. *Applied Physics Letters*, Vol. 78(6), pp. 841–843.

[19] Brabec, C.J. et al., 2002. Effect of LiF/metal electrodes on the performance of plastic solar cells. *Applied Physics Letters*, Vol. 80(7), pp. 1288–1290.

[20] Bach, U. et al., 1998. Solid-state dye-sensitized mesoporous TiO2 solar cells with high photon-to-electron conversion efficiencies. *Nature*, Vol. 395(6702), pp. 583–585.

[21] Spiekermann, S. et al., 2001. Poly(4-undecyl-2,2'-bithiophene) as a hole conductor in solid state dye sensitized titanium dioxide solar cells. *Synthetic Metals*, Vol. 121(1–3), pp. 1603–1604.

[22] Kruger, J. et al., 2001. High efficiency solid-state photovoltaic device due to inhibition of interface charge recombination. *Applied Physics Letters*, Vol. 79(13), pp. 2085–2087.

[23] Kruger, J. et al., 2002. Improvement of the photovoltaic performance of solid-state dye- sensitized device by silver complexation of the sensitizer cis-bis(4,4'-dicarboxy-2,2'-bipyridine)-bis(isothiocyanato) ruthenium(II). *Applied Physics Letters*, Vol. 81(2), pp. 367–369.

[24] Granstrom, M. et al., 1998. Laminated fabrication of polymeric photovoltaic diodes. *Nature*, Vol. 395(6699), pp. 257–260.

[25] Arias, A.C. et al., 2002. Vertically segregated polymer blend photovoltaic thin film structures through surface mediated solution processing. *Applied Physics Letters*, in press.

[26] Takahashi, K. et al., 2000. Three-layer organic solar cell with high-power conversion efficiency of 3.5%. *Sol. Energy Mater. Sol. Cells*, Vol. 61(4), pp. 403–416.

[27] Arango, A.C. et al., 2000. Efficient titanium oxide/conjugated polymer photovoltaics for solar energy conversion. *Advanced Materials*, Vol. 12(22), p. 1689.

[28] Schmidt-Mende, L. et al., 2001. Self-organized discotic liquid crystals for high-efficiency organic photovoltaics. *Science*, Vol. 293(5532), pp. 1119–1122.

[29] Brabec, C.J. et al., 2001. Influence of disorder on the photoinduced excitations in phenyl substituted polythiophenes. *Journal of Chemical Physics*, Vol. 115(15), pp. 7235–7244.

[30] Dhanabalan, A. et al., 2001. Synthesis and characterization of a low bandgap conjugated polymer for bulk heterojunction photovoltaic cells. *Advanced Functional Materials*, Vol. 11(4), pp. 255–262.

[31] Shaheen, S.E. et al., 2001. Low band-gap polymeric photovoltaic devices. *Synthetic Metals*, Vol. 121(1–3), pp. 1583–1584.

[32] Dittmer, J.J., Marseglia, E.A. and Friend, R.H., 2000. Electron trapping in dye/polymer blend photovoltaic cells. *Advanced Materials*, Vol. 12(17), 2000, p. 1270.

[33] Huynh, W.U., Dittmer, J.J. and Alivisatos, A.P., 2002. Hybrid nanorod-polymer solar cells. *Science*, Vol. 295(5564), pp. 2425–2427.

[34] Arici, E. et al., 2001. CIS Plastic Solar Cells. *Proc. 17th European Photovoltaic Solar Energy Conf.*

[35] Inganas, O. et al., 2001. Recent progress in thin film organic photodiodes. *Synthetic Metals*, Vol. 121(1–3), pp. 1525–1528.

[36] Niggemann, M. et al., 2001. Trapping Light in Organic Plastic Solar Cells with integrated Diffraction Gratings. *Proc. 17th European Photovoltaic Solar Energy Conf.*

[37] Roman, L.S. et al., 2000. Trapping light in polymer photodiodes with soft embossed gratings. *Advanced Materials*, Vol. 12(3), p. 189.

[38] Rostalski, J. and Meissner, D., 2000. Photocurrent spectroscopy for the investigation of charge carrier generation and transport mechanisms in organic p/n junction solar cells. *Sol. Energy Mater. Sol. Cells*, Vol. 63(1), pp. 37–47.

[39] Ramsdale, C.M. and Greenham, N.C., 2002. Ellipsometric determination of anisotropic optical constants in electroluminescent conjugated polymers. *Advanced Materials*, Vol. 14(3), p. 212.

[40] Halls, J.J.M. et al., 2000. Photodiodes based on polyfluorene composites: Influence of morphology. *Advanced Materials*, Vol. 12(7), p. 498.

[41] Arias, A.C. et al., 2001. Photovoltaic performance and morphology of polyfluorene blends: A combined microscopic and photovoltaic investigation. *Macromolecules*, Vol. 34(17), pp. 6005–6013.

[42] Struijk, C.W. et al., 2000. Liquid crystalline perylene diimides: Architecture and charge carrier mobilities. *Journal of the American Chemical Society*, Vol. 122(45), pp. 11057–11066.

[43] Videlot, C., El Kassmi, A. and Fichou, D., 2000. Photovoltaic properties of octithiophene-based Schottky and p/n junction cells: Influence of molecular orientation. *Sol. Energy Mater. Sol. Cells*, Vol. 63(1), pp. 69–82.

[44] Thelakkat, M., Schmitz, C. and Schmidt, H.W., 2002. Fully vapor-deposited thin-layer titanium dioxide solar cells. *Advanced Materials*, Vol. 14(8), pp. 577–581.

[45] Fan, Q. et al., 2001. A solid state solar cell using sol-gel processed material and a polymer. *Chemical Physics Letters*, Vol. 347(4–6), pp. 325–330.

[46] Savenije, T.J., Warman, J.M. and Goossens, A., 1998. Visible light sensitisation of titanium dioxide using a phenylene vinylene polymer. *Chemical Physics Letters*, Vol. 287(1–2), pp. 148–153.

[47] Anderson, N.A. et al., 2002. Subpicosecond photoinduced electron transfer from a conjugated polymer to SnO_2 semiconductor nanocrystals. *Physica E*, Vol. 14, pp. 215–218.

[48] Cabanillas-Gonzales, J. et al., 2003. Effect of aggregation on photocurrent generation in polyfluorene doped with violanthrone. *Synthetic Metals*, Vol. 137, pp. 1471–1472.

[49] Ago, H. et al., 1999. Composites of carbon nanotubes and conjugated polymers for photovoltaic devices. *Advanced Materials*, Vol. 11(5), pp. 1281.

[50] Lee, S.B. et al., 2001. Electrical and optical properties of conducting polymer-C-60 carbon nanotube system. *Synthetic Metals*, Vol. 12(1–3), pp. 1591–1592.

[51] Brabec, C.J. et al., 1999. Photovoltaic properties of conjugated polymer/methanofullerene composites embedded in a polystyrene matrix. *J. Appl. Phys.*, Vol. 85(9), pp. 6866–6872.

[52] Baur, J.W. et al., 2001. Photovoltaic interface modification via electrostatic self- assembly. *Synthetic Metals*, Vol. 121(1–3), pp. 1547–1548.

[53] Schroeder, R. et al., 2001. Control of excited state dynamics in ionically self-assembled monolayers of conjugated molecules. *Synthetic Metals*, Vol. 121(3), pp. 1521–1524.

[54] Ramos, A.M. et al., 2001. Photoinduced electron transfer and photovoltaic devices of a conjugated polymer with pendant fullerenes. *Journal of the American Chemical Society*, Vol. 123(27), pp. 6714–6715.

[55] de Boer, B. et al., 2001. Synthesis and self-organization of PPV-based block copolymers for photonic applications. *Synthetic Metals*, Vol. 121(1–3), pp. 1541–1542.

[56] Murgia, M. et al., 2001. Intedigitated p–n junction: a route to improve the efficiency in organic photovoltaic cells. *Synthetic Metals*, Vol. 121(1–3), pp. 1533–1534.

[57] Pfeiffer, M., et al., 2000. Controlled p-doping of pigment layers by cosublimation: Basic mechanisms and implications for their use in organic photovoltaic cells. *Sol. Energy Mater. Sol. Cells*, Vol. 63(1), pp. 83–99.

[58] Stevenson, R., et al., 2001. Raman microscopy determination of phase composition in polyfluorene composites. *Applied Physics Letters*, Vol. 79(14), pp. 2178–2180.

[59] Harrison, M.G., Gruner, J. and Spencer, G.C.W., 1997. Analysis of the photocurrent action spectra of MEH-PPV polymer photodiodes.*Physical Review B*, Vol. 55(12), pp. 7831–7849.

[60] Nogueira, A.F. et al., 2001. Electron transfer dynamics in dye sensitized nanocrystalline solar cells using a polymer electrolyte. *Journal of Physical Chemistry B*, Vol. 105(31), pp. 7517–7524.

[61] Montanari, I. et al., 2002. Transient optical studies of charge recombination dynamics in a polymer/fullerene composite at room temperature. *Applied Physics Letters*, Vol. 81(16), pp. 3001–3003.

[62] Dyakonov, V. et al., 2001. Spectroscopy on polymer-fullerene composites and photovoltaic cells. *Synthetic Metals*, Vol. 121(1–3), pp. 1529–1532.

[63] Pacios, R. et al., 2003. Efficient polyfluorene based solar cells. *Synthetic Metals*, Vol. 137, pp. 1469–1470.

[64] Ramsdale, C.M. et al., 2002. The origin of the open-circuit voltage in polyfluorene-based photovoltaic devices. *Journal of Applied Physics*, Vol. 92(8), pp. 4266–4270.

[65] Gregg, B.A. and Hanna, M.C., 2003. Comparing organic to inorganic photovoltaic cells: Theory, experiment and simulation. *Journal of Applied Physics*, Vol. 93, pp. 3605–3614.

[66] Brabec, C.J. et al., 2001. Origin of the open circuit voltage of plastic solar cells. *Advanced Functional Materials*, Vol. 11(5), pp. 374–380.

[67] Cahen, D. et al., 2000. Nature of photovoltaic action in dye-sensitized solar cells. *Journal of Physical Chemistry B*, Vol. 104(9), pp. 2053–2059.

[68] Padinger, F. et al., 2000. Fabrication of large area photovoltaic devices containing various blends of polymer and fullerene derivatives by using the doctor blade technique. *Opto-Electronics Review*, Vol. 8(4), pp. 280–283.

[69] Burrows, P.E. et al., 2001. Ultra barrier flexible substrates for flat panel displays. *Displays*, Vol. 22(2), pp. 65–69.

[70] Weaver, M.S. et al., 2002. Organic light-emitting devices with extended operating lifetimes on plastic substrates. *Applied Physics Letters*, Vol. 81(16), pp. 2929–2931.

[71] Shtein, M. et al., 2001. Material transport regimes and mechanisms for growth of molecular organic thin films using low-pressure organic vapor phase deposition. *J. Appl. Phys.*, Vol. 89(2), pp. 1470–1476.

[72] Shaheen, S.E. et al., 2001. Fabrication of bulk heterojunction plastic solar cells by screen printing. *Applied Physics Letters*, Vol. 79(18), pp. 2996–2998.

[73] Sirringhaus, H. et al., 2000. High-resolution inkjet printing of all-polymer transistor circuits. *Science*, Vol. 290, pp. 2123–2126.

[74] Kim, C., Shtein, M. and Forrest, S.R., 2002. Nanolithography based on patterned metal transfer and its application to organic electronic devices. *Applied Physics Letters*, Vol. 80(21), pp. 4051–4053.

[75] Nazeeruddin, M.K. et al., 1993. Conversion of Light to Electricity by Cis-X2bis(2,2'-Bipyridyl-4,4'-Dicarboxylate) Ruthenium(II) Charge-Transfer Sensitizers ($X = Cl^-$, Br^-, I^-, Cn^-, and Scn^-) on Nanocrystalline TiO_2 Electrodes. *Journal of the American Chemical Society*, Vol. 115(14), pp. 6382–6390.

[76] Neugebauer, H. et al., 2000. Stability and photodegradation mechanisms of conjugated polymer/fullerene plastic solar cells. *Sol. Energy Mater. Sol. Cells*, Vol. 61(1), pp. 35–42.
[77] Padinger, F. et al., 2001. Degradation of bulk heterojunction solar cells operated in an inert gas atmosphere: a systematic study. *Synthetic Metals*, Vol. 121(1–3), pp. 1605–1606.
[78] Tuladhar, S.M. et al., 2001. Out-door photovoltaic characterization of plastic solar cells. Presented at the *International Workshop on Nanostructures in Photovoltaics*, Dresden, unpublished.
[79] Katz, E.A. et al., 2001. Temperature dependence for the photovoltaic device parameters of polymer-fullerene solar cells under operating conditions. *J. Appl. Phys.*, Vol. 90(10), pp. 5343–5350.

IIf

Testing, Monitoring and Calibration

IIf-1

Standards, Calibration and Testing of PV Modules and Solar Cells

Carl R. Osterwald, National Renewable Energy Laboratory, Golden, Colorado, USA

1 PV Performance Measurements

1.1 Introduction

When referring to the performance of a PV cell or module, the most important parameter is, of course, the maximum power point P_{max}, which is usually determined by varying the forward bias voltage across the device under test while illuminated. The light I–V curve is then recorded by measuring the current and voltage. Unfortunately, determination of P_{max} is complicated because it is a function of the total and spectral irradiance incident upon the device, the spatial and temporal uniformity of the irradiance, and the temperature of the device. It can even be a function of the voltage sweep rate and direction, as well as the voltage biasing history. Another important parameter, the power conversion efficiency η, defined as the power out divided by the power in, is proportional to P_{max}, and also to radiometric (the total incident irradiance, G) and physical (the device area, A) quantities: $\eta = P_{max}/AG$. Accurate determination of PV performance requires knowledge of the potential measurement problems and how these problems are influenced by the specific device to be tested. This section covers common PV measurement techniques and show how potential problems and sources of error are minimised.

Because the first solar cell applications were for satellite power systems, it was important for designers to know how much power could be expected from an individual solar cell in Earth orbit (i.e., when illuminated by the extraterrestrial solar irradiance). This could not be determined exactly for two reasons: (1) the precise nature of the extraterrestrial irradiance could only be estimated, and (2) sunlight at the Earth's surface is filtered by the atmosphere. Therefore, space solar cell performance could not be measured in a laboratory. By the early 1970s, a similar situation existed for terrestrial applications, but for a different reason – the total and spectral irradiance vary continuously due to the effects of the Earth's atmosphere. Thus, in both cases, it was essentially impossible to independently verify the efficiency measurements made by any laboratory [1].

1.2 Radiometry

The problems with measurement comparisons led to the concept of standard reporting conditions (SRC, also referred to as standard test conditions, STC), which consist of the device temperature, the total irradiance, and the spectral irradiance under which PV performance measurements are made or corrected to [1, 2]. Using SRC allows performance comparisons to be made within the error uncertainty limits of the measurements involved [2]. Table 1 lists the SRCs for space and terrestrial applications, and Figure1 is a plot of the extraterrestrial and global spectral irradiances, which are the result of measurements made by solar resource satellites, high-altitude balloons, and aircraft, combined with ground measurements and atmospheric models [4–11].

Table 1 Standard Reporting Conditions (SRC)

Application	Reference spectral irradiance	Total irradiance (W/m^2)	Temperature (°C)
Low-Earth orbit	ASTM E 490 [1]	1366.1	28
Terrestrial global (non-concentrator)	IEC 60904-3 [3] or ASTM G 159 [2]	1000	25

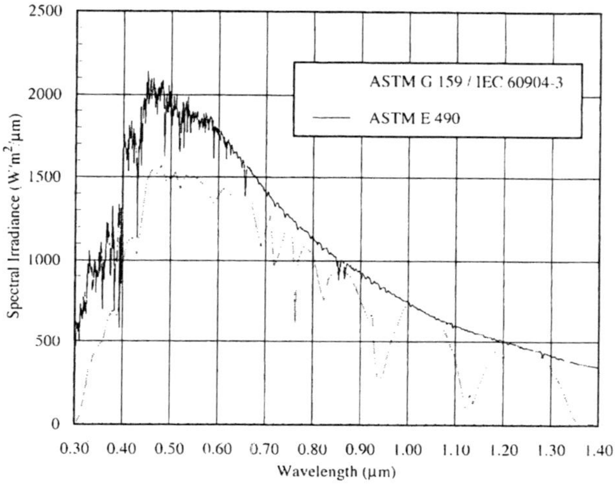

Figure 1 Reference spectral irradiances for Air Mass 0 (AM0) and Air Mass 1.5 hemispherical (global) [1–3].

For space applications, it is easy to identify a reference spectral irradiance, which is the extraterrestrial solar irradiance at a distance of one astronomical unit from the sun. Identification of a reference spectral irradiance for terrestrial applications is a much more difficult subject because of the effects of the atmosphere on sunlight, which are a function of the path length through the atmosphere [7]. A way of describing an atmospheric path is called the relative optical path length, which is commonly simplified to air mass (AM). To a first approximation, the air mass is the secant of the solar zenith angle. Thus, AM1 indicates the sun is directly overhead, and AM1.41 is a 45° zenith angle. Because air mass is actually a relative path length referenced to sea level, it can be less than one as the altitude increases above sea level. The extraterrestrial irradiance is commonly called air mass zero (AM0) because at the top of the atmosphere, the path length is zero.

Because solar cells convert light to electricity, radiometry is a very important facet of PV metrology. Radiometric measurements have the potential to introduce large errors in any given PV performance measurement because radiometric instrumentation and detectors can have total errors of up to 5% even with careful calibration [12, 13]. Other errors can be introduced through means as subtle as misunderstanding the objective of a particular measurement or its potential limitations. Broadband radiometers, such as pyranometers and pyrhelimeters, can be difficult to use in PV measurements because of the large spectral response differences. The calibration and use of radiometric instruments such as pyrheliometers, spectroradiometers, and pyranometers is a subject that is beyond the scope of this chapter, but these instruments are important for PV measurements. A number of standards for radiometric instrumentation are available [14–19].

It should be emphasised that the standardised spectral irradiances in Figure 1 cannot be reproduced exactly in the laboratory. In addition, although light sources such as xenon arc lamps can approximate these curves (i.e. solar simulators), a method of setting the total irradiance is needed. These limitations have been overcome by the so-called reference cell method. Using this method, an unknown device is tested in a solar simulator for which the total irradiance has been set with a calibrated reference cell that has the same or similar spectral response as the test device. The output level of the simulator is adjusted until the short-circuit current, I_{sc}, of the reference cell is equal to its calibration. With the total irradiance established, the reference cell is replaced with the device to be tested and the performance measurement can then be made [20–22].

When considering the reference cell method, another problem becomes immediately obvious: How can it be used to test an unknown cell if a matched reference cell is not available? This problem was solved by calculating the error in a measurement of I_{sc} caused by spectral response differences between the reference and unknown devices, and by spectral irradiance differences of the solar simulator from the desired standard spectrum [23–25]; this error is now called the spectral mismatch error, M. It is important to note that once M is known, it can be used to correct PV current measurements for spectral error by dividing by M. The PV performance standards mentioned above all rely on the reference cell method with spectral mismatch corrections [20–22].

The measurement problem is then reduced to obtaining calibrated reference cells. Historically, reference cells calibrated in sunlight have been called primary cells, and cells calibrated in solar simulators are called secondary cells. Historically, for space applications, primary reference cell calibrations have relied on I_{sc} measurements under spectral conditions as close to AM0 as possible with locations such as high-altitude balloons and aircraft, and manned spacecraft [26–28]. Primary terrestrial calibrations require stable, clear sky conditions with total irradiances measured with an absolute cavity radiometer [29, 30]. Other primary terrestrial calibration methods are used by national laboratories worldwide [31]. Reference [32] is a standard for secondary reference cell calibrations, and referenec [33] describes calibrations and use of reference modules.

1.3 Instrumentation and Solar Simulation

Figure 2 is an electrical block diagram that illustrates how PV current-voltage measurements are made. A four-wire (or Kelvin) connection to the device under test allows the voltage across the device to be measured by avoiding voltage

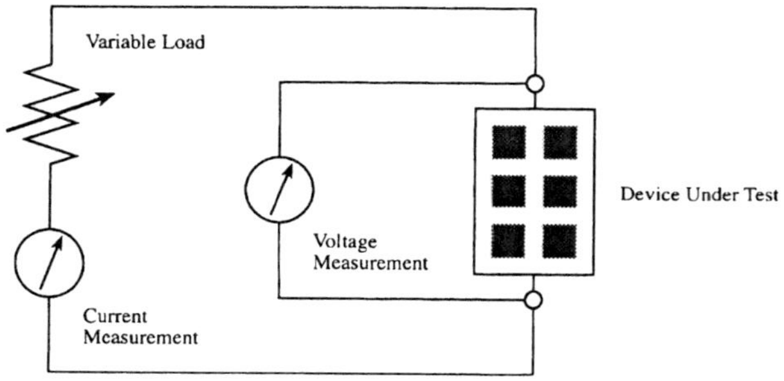

Figure 2 Block diagram of the four-wire Kelvin technique for I–V curve measurements of PV devices.

drops along the wiring in the current measurement loop. The device under test is illuminated, the load is varied, and the operating point of the device under test changes, which allows the current and voltage points to be captured along the I–V curve [3]. Typically, the entire measurement process is computer controlled.

There are two factors that greatly influence the design choices of an I–V measurement system: the type of the devices to be tested and the illumination source. For example, it is much easier to design a variable load if all test devices are similar in size and output, such as testing modules at the end of a production line. On the other hand, if a measurement system will be required to handle a wide variety of module sizes and outputs, or individual solar cells, the instrumentation must have the necessary voltage and current ranges. The need to test multijunction devices can greatly complicate the problems that must be solved in a measurement system design.

Any light source used (i.e. a solar simulator or natural sunlight) will have spatial non-uniformities, temporal instabilities, drift of its spectral irradiance, and an illumination time [34]. Outdoor measurements normally have uniform illumination in the test plane (less than 1% spatial variation) and stable irradiance for time periods up to several minutes. A Xe long-arc flash solar simulator can also achieve good spatial uniformities, but these lamps can have measurement times in the 1–20 ms range. Other flash simulators use a continuously pulsing source and measure one point on the I–V curve at each pulse. Continuous Xe simulators are commercially available for a wide range of sample sizes, from 50 mm square up to as much as 2 m square. Solar simulators with longer measurement times typically have poorer spatial uniformity, from several percent up to as much as 10 or 20%. Two standards have been developed that quantify the performance of solar simulators and can be used as aids in selection and use [35, 36].

Using these constraints, the equipment needed for I–V measurements in Figure 2 can be considered [3]. Voltage measurements are easily obtained with commercial voltmeters or high-speed analogue-to-digital (A/D) conversion cards. These same instruments can be connected across the sense leads of a four-wire resistor intended for current measurements. In high-current situations, such as source strings in an array, magnetic current probes are very convenient. The variable load can be as simple as a variable resistor, but such loads cannot be used to trace the I–V curve outside of the power quadrant. Computer programmable DC power supplies, on the other hand, can be obtained in nearly any voltage and current range needed, and bipolar supplies can operate a PV device in reverse bias or beyond open-circuit voltage (V_{oc}).

In higher power applications, such as larger modules or whole arrays, a popular load for I–V measurements is a capacitor. The capacitor is initially uncharged and connected in series with the PV device to be tested, and any current produced by the device is stored in the capacitor. When the device is illuminated, the voltage across the device is close to zero. As the capacitor charges, the capacitor voltage and the device voltage both increase, thereby sweeping through the I–V curve. The charging stops when the current goes to zero at V_{oc}. Although it cannot operate the device outside of the power quadrant, this technique has the advantage of not dissipating large amounts of power, and it can sweep at high rates. Commercial capacitive sweep systems are available that are portable and can be used to obtain I–V curves of arrays or array segments up to about 100 kW in size.

1.4 Temperature

The I–V curve of a PV device under illumination is a strong function of temperature, which must be accounted for in performance measurements [37]. Typically, I_{sc} has the smallest temperature dependence, caused by the semiconductor bandgap shifting to longer wavelengths with higher temperatures. V_{oc} and P_{max}, on the other hand, degrade rapidly with increasing temperature [38, 39]. These strong dependencies are the reason a fixed temperature is used for SRC.

Temperature measurements of PV devices can be difficult. In general, measured temperatures should be those of the actual semiconductor junctions, but usually the only temperature that can be measured is at the rear surface of a cell. For modules, it not possible to contact individual cells and one is forced to apply temperature sensors to the rear surfaces. This leads to an error because the rear surface will normally be cooler than the cells laminated inside. Reference [40] states that at 1000 W/m^2, this error is an average of 2.5°C for typical crystalline Si modules. It is possible to correct I–V measurements to a reference temperature if temperature coefficients are known [41]. Another problem with modules is that the temperature can vary from cell to cell, requiring some sort of average if a single value is needed. The same problem is exacerbated in an array, where larger variations can be expected [42]. Temperature coefficients can also be difficult to apply correctly because they can be functions of irradiance and other factors [43].

1.5 Multijunction Devices

A multijunction PV device consists of several individual semiconductor junctions stacked together (also called subcells) and connected in series to obtain higher performance. With two subcells, a multijunction solar cell is commonly referred to as a tandem cell. In such devices, however, both the I_{sc} and the fill factor (FF) are functions of the incident spectral irradiance, greatly complicating the determination of device performance at SRC [1]. Because all subcells are in series, the same current must flow through each, and the output current is usually limited by the junction generating the smallest current (called the limiting subcell). A special case occurs when all subcells generate the same amount of current, which is termed current matched. The reason the FF can be a function of the spectral irradiance is that the current generated by a limiting subcell can cause a different subcell to operate in reverse bias and result in a stepped I–V curve in forward bias [1, 44–46].

The only way to avoid measuring the wrong I_{sc} and FF is to ensure the currents generated in each subcell under the test light source are equal to those that would be generated under the desired reference spectral irradiance. In general, this can only be accomplished if the spectral response for each subcell is known and if the spectral irradiance of the test light source can be adjusted [1, 44, 45]. Such a simulator suitable for standard multijunction performance is called a spectrally adjustable solar simulator. There are a number of ways an adjustable simulator can be realised, including multiple independently adjustable light sources and selective filtering [44]. A number of procedures can be used to perform spectral matching in adjustable simulators. One procedure is an extension of the reference cell method with spectral mismatch corrections. An iterative process is used in which the spectral mismatch M for each subcell in the test is obtained, following a measurement of the simulator spectral irradiance. Based on these results, the simulator spectral irradiance is adjusted and the process is repeated. When all values of M are within an acceptable tolerance of one (usually 2–3%), the process is stopped and the I–V measurements are made [44, 45, 47]. This procedure will be standardised in the near future [48]. Another procedure relies on a multiple source simulator in which each source can be adjusted independently without changing its spectral distribution. A system of linear equations is then used to obtain the current outputs of one or more reference cells that indicate when the simulator is correctly adjusted [49].

It should be noted that the spectral adjustment of the simulator does not reproduce the reference spectral irradiance; instead, it sets the simulator spectral irradiance so that the test device is operating as if it were illuminated by the reference spectrum.

1.6 Other Performance Ratings

Although performance at SRC is an important parameter that allows measurement comparisons between different laboratories and different PV devices, it is not a realistic indication of the performance that can be expected outdoors, where devices see a range of irradiance conditions and almost never operate at 25°C when the irradiance is greater than 900 W/m^2. Another performance rating method is called Performance Test Conditions (PTC), the conditions for which are listed in Table 2 [50, 51]. This method is usually used to produce a rating for a system over a period of time (typically one month) when

Table 2 Performance Test Conditions (PTC)

Application	Irradiance (W/m^2)	Ambient temperature (°C)	Wind speed (m/s)
Flat-plate fixed tilt	1000 (global)	20	1
Concentrators	850 (direct normal)	20	1

conditions are close to the rating conditions. Data filtering and a regression fit to a linear equation are then used to obtain the performance at PTC. These ratings can be used to monitor the performance of a system over time, and it can be used for DC as well as AC ratings if the system is grid-connected. Another system rating uses translations to SRC based on a number of parameters measured outdoors on individual modules [52]. This method employs an empirical air mass versus irradiance correction factor that was called a measure of 'the systematic influence of solar spectrum' [53], but it should be noted that this factor is not a spectral correction to the reference spectral irradiance and is site specific [54].

There has also been interest in rating modules based on energy production, rather than output power at a single fixed condition [55]. A proposed energy rating establishes five reference days of hourly weather and irradiance data: hot-sunny, cold-sunny, hot-cloudy, cold-cloudy, and nice. Hourly module power outputs for each of the five days are obtained and then integrated to obtain the energy output for each condition [56]. Module energy ratings have been obtained from laboratory measurements as well as outdoor performance data [57–60].

1.7 Potential Problems and Measurement Uncertainty

Measurement uncertainty analysis is a formal process of identifying and quantifying possible errors, and combining the results to obtain an estimate of the total uncertainty of a measurement [61]. An inherent part of this process is understanding the potential problems that can affect the results of any given PV performance measurement. Although an exhaustive discussion of such problems cannot be included here, there are a number of common problems and pitfalls that can be identified. Many of these result from the instrumentation and apparatus used, the characteristics of the device to be tested, or both.

The voltage bias rate and direction can have profound effects on measured I–V data [62–64]. Test samples can exhibit hysteresis when swept toward forward bias and toward reverse bias, resulting in two different values of FF. Note that many I–V measurement systems are unable to sweep in both directions; especially capacitive loads, which can completely hide hysteresis problems. Thin-film devices such as $CuInSe_2$ and CdTe have mid-bandgap states that make them especially vulnerable to errors caused by fast sweep rates and prior light and voltage bias conditions [62]. Electronic loads used with flash simulators can greatly improve or degrade measurement results [65]. Mundane subjects such as area measurements, temperature measurements, and contacts and wiring cannot be ignored [66]. Amorphous Si modules exhibit a dependence of the FF on the incident spectral irradiance, which can affect the results of outdoor performance measurements [67].

A number of uncertainty analyses of PV measurements have been published for general I–V measurements [68, 69], spectral corrections [70], and reference cell calibrations [29, 71]. Reference [70] concluded that the magnitude of uncertainty in spectral corrections is directly proportional to the size of the spectral mismatch factor.

2 Diagnostic Measurements

A number of diagnostic measurements are widely used in photovoltaic research and development. Two of these date to the earliest days of PV devices: dark I–V and spectral response (also called quantum efficiency). Dark I–V shows how a device operates as a p–n junction and can be used to obtain series resistance, shunt resistance, and diode quality factor [38, 39]. Although complicated by multiple individual cells connected in series, dark I–V can also be useful for module diagnostics [72]. Many of the same parameters can be derived from light I–V measurements as well [73, 74], and with careful design the same measurement system illustrated in Figure 1 can be used for both light and dark I–V measurements. Shunt resistance of individual cells is an important factor for the performance of monolithic thin-film modules such as a-Si, and reference [75] describes a technique for measuring individual shunt resistances in modules.

Spectral response is a fundamental property of solar cells, and it can provide information about optical losses such as reflection, and give insights into carrier recombination losses [39]. As previously stated, it is an important parameter for performance measurements because it is used as the basis for spectral mismatch corrections [24, 25]. Because of this importance, two standards for spectral response measurements have been developed [76, 77]. A wide variety of measurement schemes have been used for spectral response measurements [78–81]. Reference [82] discusses several sources of error that are commonly encountered. Note that special light- and voltage-biasing techniques are required to measure the spectral response of series-connected multijunction devices [47, 48, 83].

A very useful diagnostic for cells and modules is known as Laser (or Light) Beam Induced Current (LBIC), which produces a map of a device's response by rastering a laser spot across the front surface and measuring the resultant current [84, 85]. These maps can easily identify locations of reduced output, which can greatly simplify diagnostic investigations, such as looking for cracks in polycrystalline Si cells. LBIC can be used to measure photocurrents and shunt resistances of individual cells in crystalline Si modules [84]. When used with monolithic thin-film modules, LBIC will show regions of reduced photocurrent that might be caused by deposition non-uniformities.

Another technique that has become increasingly popular in recent years is thermal mapping using commercial infrared imaging cameras [86]. These cameras easily measure temperature variations across the surfaces of both modules and arrays, and they are especially useful for locating hot spots, which can be caused by a number of conditions. A localised shunt path within a crystalline Si cell can produce small 20°C hot spots when a module is operated in reverse bias. In PV arrays, it is not uncommon for individual cells or a single module to be forced into reverse bias (thereby absorbing rather than generating power) if its output current has been reduced for some reason and the array design does not have adequate bypass diodes. These conditions will result in hot spots, 20–40°C higher than the surrounding cells or modules, which are easily detected in an infrared image.

Ultrasonic imaging is a non-destructive test that can identify voids and delamination in encapsulants. It is useful for studying the condition of solder joints in crystalline Si modules [87].

3 Commercial Equipment

There are a number of companies worldwide that market PV instrumentation, solar simulators, and complete PV measurement systems. Products are available for testing everything from small individual cells to large modules and small-sized arrays, for both space and terrestrial applications. These companies include:

- Berger Lichttechnik, Baierbrunn, Germany
- Beval S.A. (Pasan), Valangin, Switzerland
- Daystar, Inc., Las Cruces, NM, USA
- EKO Instruments Trading Co., Ltd, Tokyo, Japan
- Energy Equipment Testing Service Ltd, Cardiff, UK
- The Eppley Laboratory, Inc., Newport, RI, USA
- h.a.l.m elektronik GmbH, Frankfurt/Main, Germany
- Kipp & Zonen B.V., Delft, The Netherlands
- K.H. Steuernagel Lichttechnik, Morfelden-Walldorf, Germany
- LI-COR Environmental Research and Analysis, Lincoln, NE, USA
- NewSun, Nepean, ON, Canada
- NPC Inc., Tokyo, Japan
- Optosolar GmbH, Merdingen, Germany
- Photo Emission Tech., Inc., Newbury Park, CA, USA
- PV Measurements, Boulder, CO, USA
- Shanghai Jiaoda GoFly Green Energy Co., Ltd, Shanghai, China
- Spectrolab, Inc., Sylmar, CA, USA
- Spire Corp., Bedford, MA, USA
- Telecom-STV Co., Ltd, Moscow, Russia
- Thermo Oriel, Stratford, CT, USA
- Thermosenorik GmbH, Erlangen, Germany
- Vortek Industries, Ltd, Vancouver, BC, Canada
- Wacom Electric Co., Ltd, Tokyo, Japan

4 Module Reliability and Qualification Testing

4.1 Purpose and History

The 'holy grail' of module reliability that many people ask for is a single test, which if passed, indicates that a certain module design will last *x* number of years in use (typically 20 or 30 years). No such test exists, nor can it be developed [88]. The reason for this is that every possible failure mechanism has to be known and quantified. This condition is impossible to meet because some failures may not show themselves for many years and because manufacturers continually introduce new module designs and revise old designs.

Instead of a search for a single test, module reliability testing aims to identify unknown failure mechanisms and determine whether modules are susceptible to known failure mechanisms. Accelerated testing is an important facet of reliability testing, but accelerated tests need to be performed in parallel with real-time tests to show that a certain failure is not caused by the acceleration factor and will never appear in actual use. Qualification tests are accelerated tests, usually of short duration, that are known to produce known failure mechanisms, such as delamination. Thus, passing a qualification test is not a guarantee of a certain lifetime, although extending the duration of a qualification can provide added confidence that a module design is robust and durable [89].

The first PV module qualification tests were developed by the Jet Propulsion Laboratory (JPL) as part of the Low-Cost Solar Array program funded by the US Department of Energy [90–93]. Elements of the Block V qualification sequence included:

- Temperature cycling
- Humidity-freeze cycling
- Cyclic pressure loading
- Ice ball impact
- Electrical isolation (hi-pot)
- Hot-spot endurance
- Twisted-mounting surface test

Following the qualification sequences, test modules were compared with baseline electrical performance tests and visual inspection results to determine whether the design passed or failed. These tests have served as the starting point for all the qualification sequences that have been developed since.

The next development in module qualification was the adoption of the European CEC 502 sequence [94], which was significantly different from Block V and added several new tests:

- UV irradiation
- High-temperature storage

- High-temperature and high-humidity storage
- Mechanical loading

Another significant difference was that CEC 502 lacked humidity-freeze cycling. At the same time, Underwriters Laboratories (UL) developed the UL 1703 safety standard, which has become a requirement for all modules in the USA [95]. It incorporated the humidity-freeze, thermal cycling, and hi-pot tests from Block V, plus a large number of other safety-related tests. Note that as a safety standard, UL 1703 does not require a module to retain its performance at a certain level; rather, it simply must not become hazardous as a result of the test sequences.

With the development of commercially available a-Si modules, an investigation into the applicability of Block V to these devices resulted in publication of the so-called Interim Qualification Tests (IQT) [96]. The IQT were similar to the JPL tests but added two tests from UL 1703: surface cut susceptibility and ground continuity. A wet insulation resistance test was included as a check for electrochemical corrosion susceptibility.

Again building on prior work, Technical Committee 82 (TC-82) of the International Electrotechnical Commission (IEC) produced an international standard for qualification of crystalline Si modules, IEC 1215 (the designation was later changed to 61215), that included elements of all the prior tests [97]. It followed the sequences of JPL's Block V but added many of the new elements in CEC 502. Probably the most significant addition was the 1000-h damp heat test that replaced the high-temperature and high-humidity storage tests in CEC 502. Before IEC 61215 became a standard, a nearly identical draft version was adopted in Europe as CEC 503 [98]. Since its adoption, IEC 61215 has been a major influence on the reputation of excellent reliability that crystalline Si modules currently have. It has proven to be an invaluable tool for discovering poor module designs before they are sold on the open market [99].

In the USA, another significant development was that of IEEE 1262 [100], which was motivated by the lack of a US qualification standard. In 1989, IEC 61215 was only a draft standard that was several years away from adoption. Also, the IEC document was not suitable for a-Si modules because the UV and outdoor exposures resulted in the initial light-induced degradation inherent in these devices, which makes determination of performance losses due to the qualification sequences very difficult. Therefore, the goal was a document that was applicable to both technologies, which was accomplished with the addition of short thermal annealing steps to remove the light-induced degradation. IEEE 1262 includes a wet hi-pot test that is nearly impossible to pass if a module has insulation holes or flaws. Otherwise, IEEE 1262 is very similar to IEC 61215. TC-82 solved the problem of a-Si degradation by issuing a separate standard specifically for a-Si modules, IEC 61646 [101]. Rather than thermal annealing, however, IEC 61646 uses light soaking to condition the test modules prior to the qualification testing. In practice, light soaking has been shown to be lengthy and expensive for testing, so it is likely that light soaking will be replaced with thermal annealing in a future version.

4.2 Module Qualification Tests

TC-82 has been working on a revision of IEC 61215 for module qualification that is due to proceed to the Committee Draft for Voting (CDV) stage at the time of this writing [102]. This is an important document for the PV industry, so a brief description of the main test sequences is included below.

The new draft includes a wet insulation test, adopted from IEEE 1262, that all test modules must pass prior to and following the test sequences. A module is immersed in a surfactant solution and the insulation resistance between the solution and the shorted module leads is measured at 500 V. This test reveals insulation flaws that could be safety hazards in use, and also exposes possible paths of moisture intrusion that might result in performance degradation.

4.2.1 Thermal Cycling Sequence

One thermal cycle consists of a −40°C freeze for a minimum of 10 min, followed by an excursion to +85°C, again with a 10 min hold. The rate of change of module temperature must not exceed 100°C/h, and a complete cycle should not last longer than 6 h. This sequence is repeated until 200 cycles are completed.

According to reference [89], thermal cycling 'is intended to accelerate thermal differential-expansion stress effects so that design weaknesses associated with encapsulant system, cells, interconnects, and bonding materials can be detected as a direct result of the test'. An important change from the previous version of IEC 61215 is the added requirement that a forward-bias current equal to the output current at SRC P_{max} be passed through the test modules when the temperature is above 25°C. Current bias simulates the stresses that solder joints experience in actual use, and the biasing has been shown to reveal poor soldering [103].

4.2.2 Damp Heat Sequence

This sequence begins with the standard damp heat, which is 1000 h of exposure to 85% relative humidity (RH) at 85°C. Damp heat is followed by the 25 mm ice ball impact and 2400 Pa mechanical loading tests. Operating conditions as severe as 85% RH and 85°C never occur in actual use, but damp heat stresses the encapsulation and can result in delamination. Note that because these extremes, it is conceivable that damp heat could produce failures that would not be seen in the field.

4.2.3 UV Exposure, Thermal Cycling, and Humidity-Freeze Sequence

Prior to any stress tests, a short UV exposure of 60 MJ/m^2 (15 kWh/m^2) preconditions the modules. The UV exposure is followed by 50 thermal cycles and then 10 humidity-freeze cycles. Module designs that have tapes inside the encapsulation will typically delaminate during the subsequent thermal or humidity-freeze cycles. This sequence concludes with a robustness of termination test.

4.3 Reliability Testing

As noted previously, qualification testing uses accelerated tests, but it should not be considered reliability testing. However, this does not mean the specific tests that are part of a standard qualification sequence cannot be used for reliability testing. An example is extending the duration of the standard thermal cycle test by increasing the total number of cycles, or until failure of the module. Another is combining a qualification test with an added stress, such as damp heat with high voltage bias [103]. If possible, accelerated tests should be combined with real time tests. Reference [104] is a standard for solar weathering that uses techniques adapted from the materials industry. This standard uses total UV exposure doses to quantify both real-time and accelerated tests.

4.4 Module Certification and Commercial Services

Product certification is a formal process involving accredited testing laboratories and certification agencies that issue licenses to manufacturers indicating their products have been tested and are in conformance [105]. Certification is well known in other industries, especially for product safety, but it is a fairly recent development for photovoltaics. PV module safety certifications have been available for almost 20 years, but efforts have been made within the past 10 years to initiate certifications based on qualification testing. An example of this is the Global Approval Program for Photovoltaics (PV GAP) [106].

There are a number of independent PV laboratories that offer testing services, either as part of a formal certification program or upon request. These include:

- ASU Photovoltaic Testing Laboratory, Mesa, AZ, USA
- European Commission Joint Research Centre, Environment Institute, Renewable Energies Unit, Ispra, Italy
- TÜV Rheinland, Berlin/Brandenburg, Germany
- Underwriters Laboratories, Inc., Northbrook, IL, USA
- VDE Testing and Certification Institute, Offenbach, Germany

5 Module Degradation Case Study

To conclude this chapter, a diagnosis of a degraded polycrystalline Si module is presented as an example that uses many of the techniques outlined here. A small 20 W module was subjected to real-time outdoor weathering with a resistive

Table 3 Polycrystalline Si module I–V parameters

	V_{oc} (V)	I_{sc} (A)	FF	P_{max} (W)	R_s (Ω)
Initial	21.1	1.18	0.733	18.2	0.60
Exposed	20.8	1.15	0.687	16.4	0.43

load, and light and dark I–V measurements were made prior to exposure and at several intervals over a period of four years. At the end of this period, P_{max} under STC as measured with a pulsed Xe simulator had declined by 9.8%, as listed in Table 3. Most of this loss has occurred in the FF, which was down by 6.3%. Series resistance (R_s) values determined from dark I–V measurements did not show large increases, so a gradual degradation of the solder joints or the cell metallisation was not responsible. Examination of the light I–V curves, Figure 3, shows a 'stair step', which in a series-connected PV device typically indicates a mismatch in current output between one or more individual cells.

An attempt to locate such a problem was made with an IR camera by placing the module in forward bias (1 A and 22 V) and allowing the cells to heat for several minutes. The resulting IR image, Figure 4, was made of the module rear

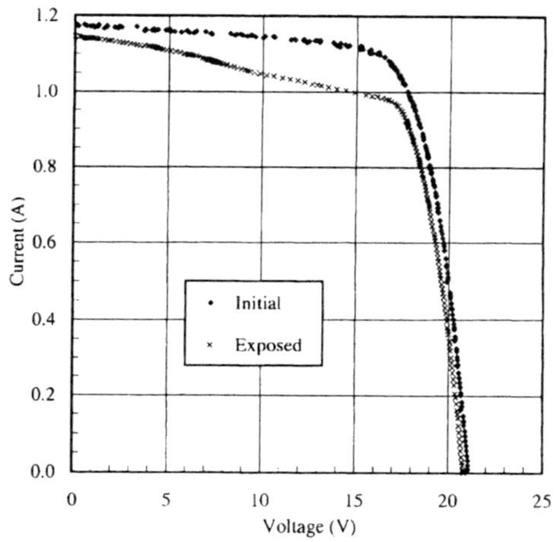

Figure 3 Light I–V curves at SRC of a polycrystalline Si module weathered outdoors for four years.

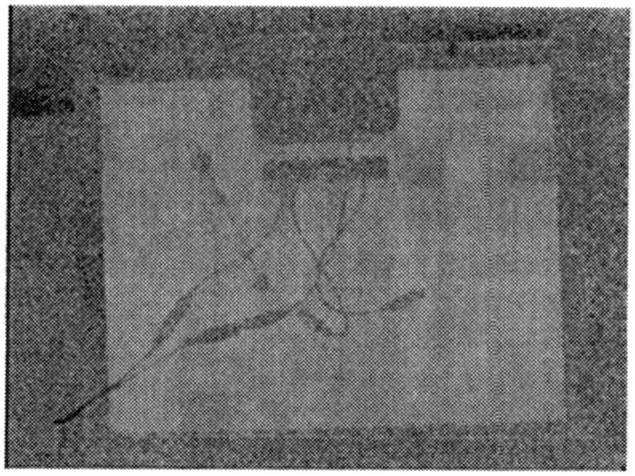

Figure 4 IR thermal image of the rear surface of the module from Figures 3 and 4. The module was placed at 1 A, 22 V in forward bias and allowed to heat for several minutes.

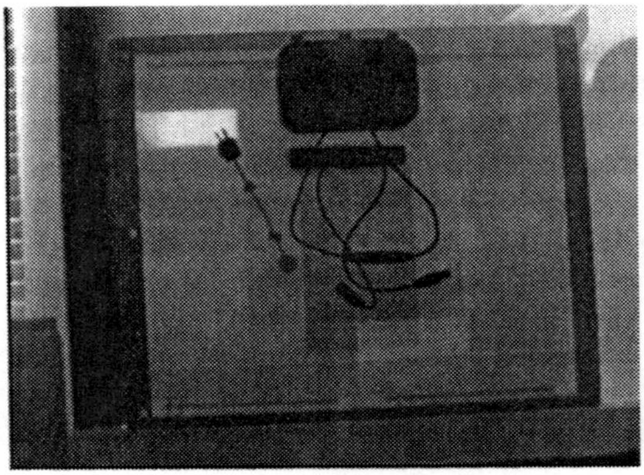

Figure 5 Another IR thermal image of the same module from Figure 3. The module was shorted for several minutes in sunlight before the image was captured.

surface through the polymeric backsheet (shadows from the junction box and the wiring are visible) and shows temperature rises of only about 1°C above the ambient. One or two cells appear to have slight temperature variations from the rest of the module, so this test was inconclusive at best. As it turns out, forward bias is not a very stressful condition for crystalline Si, and it is much easier to see hot spots when a module is shorted while illuminated in sunlight. After just a few minutes with the module in this condition, one cell developed a hot spot greater than 6°C, which was easily visible with an IR camera (Figure 5). This cell is most likely being forced into reverse bias. Looking at the hot cell in Figure 5, it appears one corner is not heating at all, and a visual examination of this cell revealed a crack in this location. Therefore, this cell has an output current about 10% lower than all the others, which caused the loss of fill factor.

Acknowledgements

Preparation of this chapter was supported by the US Department of Energy under contract No. DE-AC36-99-G010337. Steve Rummel, Allan Anderberg, Larry Ottoson, and Tom McMahon performed and assisted with the measurements used in the module degradation case study.

Permissions

This submitted manuscript has been offered by Midwest Research Institute (MRI) employees, a contractor of the US Government Contract No. DE-AC36-99-GO10337. Accordingly, the US Government and MRI retain non-exclusive, royalty-free license to publish or reproduce the published form of this contribution, or allow others to do so, for US Government purposes.

References

[1] Emery, K.A. and Osterwald, C.R. 1988. Efficiency measurements and other performance-rating methods. In: Coutts, T.J. and Meakin, J.D., Eds., *Current Topics in Photovoltaics*, Vol. 3, Academic Press, London.

[2] Emery, K. 1999. The rating of photovoltaic performance. *IEEE Transactions on Electron Devices*, Vol. ED-46, pp. 1928–1931.

[3] Emery, K.A. and Osterwald, C.R. 1986. Solar cell efficiency measurements. *Solar Cells*, Vol. 17, pp. 253–274.

[4] ASTM standard E 490: Standard Solar Constant and Zero Air Mass Solar Spectral Irradiance Tables. In: *ASTM Annual Book of Standards*, Vol. 12.02, ASTM International, West Conshohocken, PA, USA, 2002.

[5] ASTM standard G 159: Standard tables for references solar spectral irradiance at air mass 1.5: direct normal and hemispherical for a 37° tilted surface. In: *ASTM Annual Book of Standards*, Vol. 14.04, ASTM International, West Conshohocken, PA, 2002.

[6] IEC standard 60904-3: Measurement principles for terrestrial photovoltaic (PV) solar devices with reference spectral irradiance data. Photovoltaic Devices, International Electrotechnical Commission, Geneva, Switzerland, 1998.

[7] Iqbal, M. 1985. *An Introduction to Solar Radiation*, Academic Press Canada, Ontario, Canada.

[8] Riordan, C., Cannon, T., Myers, D. and Bird, R. 1985. Solar irradiance models, data, and instrumentation for PV device performance analyses. *Proc. 18th IEEE Photovoltaic Specialists Conf.*, Las Vegas, pp. 957–962.

[9] Osterwald, C.R. and Emery, K.A. 2000. Spectroradiometric sun photometry. *Journal of Atmospheric and Oceanic Technology*, September, pp. 1171–1188.

[10] Myers, D., Emery, K. and Gueymard, C. 2002. Proposed reference spectral irradiance standards to improve photovoltaic concentrating system design and performance evaluation. *Proc. 29th IEEE Photovoltaic Specialist Conf.*, New Orleans.

[11] Myers, D., Emery, K. and Gueymard, C. 2002. Terrestrial solar spectral modeling tools and applications for photovoltaic devices. *Proc. 29th IEEE Photovoltaic Specialist Conf.*, New Orleans.

[12] Myers, D.R., Emery, K.A. and Myers, D.R. 1989. Uncertainty estimates for global solar irradiance measurements used to evaluate PV device performance. *Solar Cells*, Vol. 27, pp. 455–464.

[13] Myers, D.R., Myers, D.R. and Reda, I. 2002. Recent progress in reducing the uncertainty in and improving pyranometer calibrations. *Journal of Solar Energy Engineering*, Vol. 124, pp. 44–50.

[14] ASTM standard G 130: Standard test method for calibration of narrow- and broad-band ultraviolet radiometers using a spectroradiometer. In: *ASTM Annual Book of Standards*, Vol. 14.04, ASTM International, West Conshohocken, PA, USA, 2002.

[15] ASTM standard G 138: Standard test method for calibration of a spectroradiometer using a standard source of irradiance. In: *ASTM Annual Book of Standards*, Vol. 14.04, ASTM International, West Conshohocken, PA, USA, 2002.

[16] ASTM standard E 816: Standard test method for calibration of secondary reference pyrheliometers and pyrheliometers for field use. In: *ASTM Annual Book of Standards*, Vol. 14.04, ASTM International, West Conshohocken, PA, USA, 2002.

[17] ASTM standard E 824: Standard test method for transfer of calibration from reference to field radiometers. In: *ASTM Annual Book of Standards*, Vol. 14.04, ASTM International, West Conshohocken, PA, USA, 2002.

[18] ASTM standard E 913: Standard method for calibration of reference pyranometers with axis vertical by the shading method. In: *ASTM Annual Book of Standards*, Vol. 14.04, ASTM International, West Conshohocken, PA, USA, 2002.

[19] ASTM standard E 941: Standard test method for calibration of reference pyranometers with axis tilted by the shading method. In: *ASTM Annual Book of Standards*, Vol. 14.04, ASTM International, West Conshohocken, PA, USA, 2002.

[20] ASTM standard E 948: Test method for electrical performance of photovoltaic cells using reference cells under simulated sunlight. In: *ASTM Annual Book of Standards*, Vol. 12.02, ASTM International, West Conshohocken, PA, USA, 2002.

[21] ASTM standard E 1036: Test methods for electrical performance of nonconcentrator terrestrial photovoltaic modules and arrays using reference cells. In: *ASTM Annual Book of Standards*, Vol. 12.02, ASTM International, West Conshohocken, PA, USA, 2002.

[22] IEC standard 60904-1: Measurement of photovoltaic current-voltage characteristics. *Photovoltaic Devices.* International Electrotechnical Commission, Geneva, Switzerland, 1987.

[23] Seaman, C.H. 1982. Calibration of solar cells by the reference cell method – the spectral mismatch problem. *Solar Energy*, Vol. 29, pp. 291–298.

[24] ASTM standard E 973: Test method for determination of the spectral mismatch parameter between a photovoltaic device and a photovoltaic reference cell. In: *ASTM Annual Book of Standards*, Vol. 12.02, ASTM International, West Conshohocken, PA, USA, 2002.

[25] IEC standard 60904-7: Photovoltaic devices. Part 7: computation of spectral mismatch error introduced in the testing of a photovoltaic device. International Electrotechnical Commission, Geneva, 1987.

[26] Anspaugh, B. 1987. A verified technique for calibrating space solar cells. *Proc. 19th IEEE Photovoltaic Specialist Conf.*, New Orleans, pp. 542–547.

[27] Brandhorst, H.W. 1971. Calibration of solar cells using high-altitude aircraft. In: *Solar Cells*, Gordon and Breach, London, UK.

[28] Bücher, K. 1997. Calibration of solar cells for space applications. *Progress in Photovoltaics: Research and Applications*, Vol. 5, pp. 91–107.

[29] Osterwald, C.R., Emery, K.A., Myers, D.R. and Hart, R.E. 1990. Primary reference cell calibrations at SERI: history and methods. *Proc. 21st IEEE Photovoltaic Specialists Conf.*, Kissimmee, pp. 1062–1067.

[30] ASTM standard E 1125: Test method for calibration of primary non-concentrator terrestrial photovoltaic reference cells using a tabular spectrum. In: *ASTM Annual Book of Standards*, Vol. 12.02, ASTM International, West Conshohocken, PA, USA, 2002.

[31] Osterwald, C.R., Anevsky, S., Bücher, K., Barua, A.K., Chaudhuri, P., Dubard, J., Emery, K., Hansen, B., King, D., Metzdorf, J., Nagamine, F., Shimokawa, R., Wang, Y.X., Wittchen, T., Zaaiman, W., Zastrow, A. and Zhang, J. 1999. The world photovoltaic scale: an international reference cell calibration program. *Progress in Photovoltaics: Research and Applications*, Vol. 7, pp. 287–297.

[32] ASTM standard E 1362: Test method for calibration of non-concentrator photovoltaic secondary reference cells. In: *ASTM Annual Book of Standards*, Vol. 12.02, ASTM International, West Conshohocken, PA, USA, 2002.

[33] IEC standard 60904-6: Requirements for reference solar modules. Photovoltaic Devices, International Electrotechnical Commission, Geneva, Switzerland 1994.

[34] Emery, K., Myers, D. and Rummel, S. 1988. Solar simulation – problems and solutions. *Proc. 20th IEEE Photovoltaic Specialists Conf.*, Las Vegas, pp. 1087–1091.

[35] ASTM standard E 927: Specification for solar simulation for terrestrial photovoltaic testing. In: *ASTM Annual Book of Standards*, Vol. 12.02, ASTM International, West Conshohocken, PA, USA, 2002.

[36] IEC standard 60904-9: Solar simulator performance requirements. *Photovoltaic Devices*, International Electrotechnical Commission, Geneva, Switzerland, 1995.

[37] Osterwald, C.R., Glatfelter, T. and Burdick, J. 1987. Comparison of the temperature coefficients of the basic I–V parameters for various types of solar cells. *Proc. 19th IEEE Photovoltaic Specialists Conf.*, New Orleans, pp. 188–193.

[38] Sze, S.M. 1981. *Physics of Semiconductor Devices*. Wiley, New York, USA.

[39] Fahrenbruch, A.L. and Bube, R.H. 1983. *Fundamentals of Solar Cells*. Academic Press, New York, USA.

[40] Whitfield, K. and Osterwald, C.R., 2001. Procedure for determining the uncertainty of photovoltaic module outdoor electrical performance. *Progress in Photovoltaics: Research and Applications*, Vol. 9, pp. 87–102.

[41] Osterwald, C.R. 1986. Translation of device performance measurements to reference conditions. *Solar Cells*, Vol. 18, pp. 269–279.

[42] King, D.L., Kratochvil, J.A. and Boyson, W.E. 1997. Temperature coefficients for PV modules and arrays: measurement methods, difficulties, and results. *Proc. 26th IEEE Photovoltaic Specialists Conf.*, Anaheim, pp. 1183–1186.

[43] Whitaker, C.M., Townsend, T.U., Wenger, H.J., Iliceto, A., Chimento, G. and Paletta, F. 1991. Effects of irradiance and other factors on PV temperature coefficients. *Proc. 22nd IEEE Photovoltaic Specialists Conf.*, Las Vegas, pp. 608–613.

[44] Glatfelter, T. and Burdick, J. 1987. A method for determining the conversion efficiency of multiple-cell photovoltaic devices. *Proc. 19th IEEE Photovoltaic Specialists Conf.*, New Orleans, pp. 1187–1193.

[45] Emery, K.A., Osterwald, C.R., Glatfelter, T., Burdick, J. and Virshup, G. 1988. A comparison of the errors in determining the conversion efficiency of multijunction solar cells by various methods. *Solar Cells*, Vol. 24, pp. 371–380.

[46] King, D.L., Hansen, B.R., Moore, J.M. and Aiken, D.J. 2000. New methods for measuring performance of monolithic multi-junction solar cells. *Proc. 28th IEEE Photovoltaic Specialists Conf.*, Anchorage, pp. 1197–1201.

[47] Emery, K., Meusel, M., Beckert, R., Dimroth, F., Bett, A. and Warta, W. 2000. Procedures for evaluating multijunction concentrators. *Proc. 28th IEEE Photovoltaic Specialists Conf.*, Anchorage, pp. 1126–1130.

[48] ASTM draft standard: Standard test method for measurement of electrical performance and spectral response of nonconcentrator multijunction photovoltaic cells and modules. *ASTM Annual Book of Standards*, Vol. 12.02, ASTM International, West Conshohocken, PA, USA, 2002, in press.

[49] Meusel, M., Adelhelm, R., Dimroth, F., Bett, A.W. and Warta, W. 2002. Spectral mismatch correction and spectrometric characterization of monolithic III–V multi-junction solar cells. *Progress in Photovoltaics: Research and Applications*, Vol. 10, pp. 243–255.

[50] Smith, S., Townsend, T., Whitaker, C. and Hester, S. 1989. Photovoltaics for utility-scale applications: project overview and data analysis. *Solar Cells*, Vol. 27, pp. 259–266.

[51] Whitaker, C.M., Townsend, T.U., Newmiller, J.D., King, D.L., Boyson, W.E., Kratochvil, J.A., Collier and Osborn, D.E. 1997. Application and validation of a new PV performance characterization method. *Proc. 26th IEEE Photovoltaic Specialists Conf.*, Anaheim, pp. 1253–1256.

[52] King, D.L., Kratochvil, J.A., Boyson, W.E. and Bower, W. 1998. Field experience with a new performance characterization procedure for photovoltaic arrays. *Proc. 2nd World Conf. on Photovoltaic Solar Energy Conversion*, pp. 1947–1952.

[53] Myers, K., del Cueto, J.A. and Zaaiman, W. 2002. Spectral corrections based on optical air mass. *Proc. 29th IEEE Photovoltaic Specialists Conf.*, New Orleans, in press.

[54] King, D.L., Kratochvil, J.A. and Boyson, W.E. 1997. Measuring solar spectral and angle-of-incidence effects on photovoltaic modules and solar irradiance sensors. *Proc. 26th IEEE Photovoltaic Specialists Conf.*, Anaheim, pp. 1113–1116.

[55] Kroposki, B, Emery, K., Myers, D. and Mrig, L. 1994. A comparison of photovoltaic module performance evaluation methodologies for energy

ratings. *Proc. First World Conf. on Photovoltaic Energy Conversion*, Hawaii, pp. 858–862.

[56] Marion, B., Kroposki, B., Emery, K., del Cueto, J., Myers, D. and Osterwald, C. 1999. Validation of a photovoltaic module energy ratings procedure at NREL. National Renewable Energy Laboratory Technical Report NREL/TP-520-26909, available from the Nationa Technical Information Service, Springfield, VA, USA.

[57] Chianese, D., Rezzonico, S., Cereghetti, N. and Realini, A. 2001. Energy rating of PV modules. *Proc. 17th European Photovoltaic Solar Energy Conf.*, Munich, pp. 706–709.

[58] Cereghetti, N., Realini, A., Chianese, D. and Rezzonico, S. 2001. Power and energy production of PV modules. *Proc. 17th European Photovoltaic Solar Energy Conf.*, Munich, pp. 710–713.

[59] Anderson, D., Sample, T. and Dunlop, E. 2001. Obtaining module energy rating from standard laboratory measurements. *Proc. 17th European Photovoltaic Solar Energy Conf.*, Munich, pp. 832–835.

[60] del Cueto, J.A. 2002. Comparison of energy production and performance from flat-plate photovoltaic modules deployed at fixed tilt. *Proc. 29th IEEE Photovoltaic Specialists Conf.*, New Orleans, in press.

[61] *Guide to the Expression of Uncertainty in Measurement*. International Electrotechnical Commission, Geneva, Switzerland 1995.

[62] Emery, K.A. and Osterwald, C.R. 1990. PV performance measurement algorithms, procedures, and equipment. *Proc. 21st IEEE Photovoltaic Specialists Conf.*, Kissimmee, pp. 1068–1073.

[63] Ossenbrink, H.A., Zaaiman, W. and Bishpop, J. 1993. Do multi-flash solar simulators measure the wrong fill factor? *Proc. 23rd IEEE Photovoltaic Specialists Conf.*, Louisville, pp. 1194–1196.

[64] Emery, K.A. and Field, H. 1994. Artificial enhancements and reductions in the PV efficiency. *Proc. First World Conf. on Photovoltaic Energy Conversion*, Hawaii, pp. 1833–1838.

[65] Mantingh, E.G., Zaaiman, W. and Ossenbrink, H.A. 1994. Ultimate transistor electronic load for electrical performance measurement of photovoltaic devices using pulsed solar simulators. *Proc. First World Conf. on Photovoltaic Energy Conversion*, Hawaii, pp. 871–873.

[66] Osterwald, C.R., Anevsky, S., Barua, A.K., Dubard, J., Emery, K., King, D., Metzdorf, J., Nagamine, F., Shimokawa, R., Udayakumar, N., Wang, Y.X., Wittchen, T., Zaaiman, W., Zastrow, A. and Zhang, J. 1996. Results of the PEP'93 intercomparison of reference cell calibrations and newer technology performance measurements. *Proc. 25th IEEE Photovoltaic Specialists Conf.*, Washington DC, pp. 1263–1266.

[67] Rüther, R., Kleiss, G. and Reiche, K. 2002. Spectral effects on amorphous silicon solar module fill factors. *Sol. Energy Mater. Sol. Cells*, Vol. 71, pp. 375–385.

[68] Emery, K.A., Osterwald, C.R. and Wells, C.V. 1987. Uncertainty analysis of photovoltaic efficiency measurements. *Proc. 19th IEEE Photovoltaic Specialists Conf.*, New Orleans, pp. 153–159.

[69] Heidler, K. and Beier, J. 1988. Uncertainty analysis of PV efficiency measurements with a solar simulator: spectral mismatch, non-uniformity, and other sources of error. *Proc. 8th European Photovoltaic Solar Energy Conf.*, Florence, pp. 554–559.

[70] Field, H. and Emery, K.A. 1993. An uncertainty analysis of the spectral correction factor. *Proc. 23rd IEEE Photovoltaic Specialists Conf.*, Louisville, pp. 1180–1187.

[71] King, D.L., Hansen, B.R. and Jackson, J.K. 1993. Sandia/NIST reference cell calibration procedure. *Proc. 23rd IEEE Photovoltaic Specialists Conf.*, Louisville, pp. 1095–1101.

[72] King, D.L., Hansen, B.R., Kratochvil, J.A. and Quintana, M.A. 1997. Dark current-voltage measurements on photovoltaic modules as a diagnostic or manufacturing tool. *Proc. 26th IEEE Photovoltaic Specialists Conf.*, Anaheim, pp. 1125–1128.

[73] Chegaar, M., Ouennoughi, Z. and Hoffman, A. 2001. A new method for evaluating illuminated solar cell parameters. *Solid-State Electronics*, Vol. 45, pp. 293–296.

[74] del Cueto, J.A. 1998. Method for analyzing series resistance and diode quality factors from field data of photovoltaic modules. *Sol. Energy Mater. Sol. Cells*, Vol. 55, pp. 291–297.

[75] McMahon, T.J., Basso, T.S. and Rummel, S.R. 1996. Cell shunt resistance and photovoltaic module performance. *Proc. 25th IEEE Photovoltaic Specialists Conf.*, Washington DC, pp. 1291–1294.

[76] IEC standard 60904-8: Measurement of spectral response of a photovoltaic (PV) device. *Photovoltaic Devices*, International Electrotechnical Commission, Geneva, Switzerland, 1998.

[77] ASTM standard E 1021: Test method for measuring spectral response of photovoltaic cells. In: *ASTM Annual Book of Standards*, Vol. 12.02, ASTM International, West Conshohocken, PA, USA 2002.

[78] Hartman, J.S. and Lind, M.A. 1983. Spectral response measurements for solar cells. *Solar Cells*, Vol. 7, pp. 147–157.

[79] van Steenwinkel, R. 1987. Measurements of spectral responsivities of cells and modules. *Proc. 7th European Photovoltaic Solar Energy Conf.*, p. 325.

[80] Budde, R., Zaaiman, W. and Ossenbrink, H.A. 1994. Spectral response calibration facility for photovoltaic cells. *Proc. First World Conf. on Photovoltaic Energy Conversion*, Hawaii, pp. 874–876.

[81] Osterwald, C.R., Anevsky, S., Barua, A.K., Bücher, K., Chauduri, P., Dubard, J., Emery, K., King, D., Hansen, B., Metzdorf, J., Nagamine, F., Shimokawa, R., Wang, Y.X., Wittchen, T., Zaaiman, W., Zastrow, A. and Zhang, J. 1998. The results of the PEP'93 intercomparison of reference cell calibrations and newer technology performance measurements: final report. National Renewable Energy Laboratory Technical Report NREL/TP-520-23477, available from the National Technical Information Service, Springfield, VA, USA.

[82] Field, H. 1997. Solar cell spectral response measurement errors related to spectral band width and chopped light waveform. *Proc. 26th IEEE Photovoltaic Specialists Conf.*, Anaheim, pp. 471–474.

[83] Burdick, J. and Glatfelter, T. 1986. Spectral response and I–V measurements of tandem amorphous-silicon alloy photovoltaic devices. *Solar Cells*, Vol. 18, pp. 301–314.

[84] Eisgruber, I.L. and Sites, J.R. 1996. Extraction of individual-cell photocurrents and shunt resistances in encapsulated modules using large-scale laser scanning. *Progress in Photovoltaics: Research and Applications*, Vol. 4, pp. 63–75.

[85] Agostinelli, G., Friesen, G., Merli, F., Dunlop, E.D., Acciarri, M., Racz, A., Hylton, J., Einhaus, R. and Lauinger, T. 2001. Large area fast LBIC as a tool for inline PV module and string characterization. *Proc. 17th European Photovoltaic Solar Energy Conf.*, Munich, pp. 410–413.

[86] King, D.L., Kratochvil, J.A., Quintana, M.A. and McMahon, T.J. 2000. Applications for infrared imaging equipment in photovoltaic cell, module, and system testing. *Proc. 28th IEEE Photovoltaic Specialists Conf.*, Anchorage, pp. 1487–1490.

[87] King, D.L., Quintana, M.A., Kratochvil, J.A., Ellibee, D.E. and Hansen, B.R. 2000. Photovoltaic module performance and durability following long-term field exposure. *Progress in Photovoltaics: Research and Applications*, Vol. 8, pp. 241–256.

[88] McMahon, T.J., Jorgensen, G.J., Hulstrom, R.L., King, D.L. and Quintana, M.A. 2000. Module 30 year life: what does it mean and is it predictable/achievable? *Proceedings of the NCPV Program Review Meeting*, available from the National Technical Information Service, Springfield, VA, USA.

[89] Wohlgemuth, J.H. 1994. Reliability testing of PV modules. *Proc. First World Conf. on Photovoltaic Energy Conversion*, Hawaii, pp. 889–892.

[90] Hoffman, A.R. and Ross, R.G. 1979. Environmental qualification testing of terrestrial solar cell modules. *Proc. 13th IEEE Photovoltaic Specialists Conf.*, Washington DC, pp. 835–842.

[91] Hoffman, A.R., Griffith, J.S. and Ross, R.G. 1982. Qualification testing of flat-plate photovoltaic modules. *IEEE Transactions of Reliability*, Vol. R-31, pp. 252–257.

[92] Block V solar cell module design and test specification for intermediate load applications – 1981. Jet Propulsion Laboratory report 5101-161 1981, Pasadena, CA, USA.

[93] Smokler, M.I., Otth, D.H. and Ross, R.G. 1985. The block program approach to photovoltaic module development. *Proc. 18th IEEE Photovoltaic Specialists Conf.*, pp. 1150–1158.

[94] Qualification test procedures for photovoltaic modules. Specification No. 502, Issue 1, Commission of the European Communities, Joint Research Center, Ispra Establishment, 1984.

[95] Standard for flat-plate photovoltaic modules and panels. ANSI/UL 1703-1987, American National Standards Institute, New York, USA, 1987.

[96] DeBlasio, R., Mrig, L. and Waddington, D. 1990. Interim qualification tests and procedures for terrestrial photovoltaic thin-film flat-plate modules. *Solar Energy Research Institute Technical Report* SERI/TR-213-3624, available from the National Technical Information Service, Springfield, VA, USA.
[97] IEC standard 61215: Crystalline silicon terrestrial photovoltaic (PV) modules – design qualification and type approval. International Electrotechnical Commission, Geneva, Switzerland, 1993.
[98] Ossenbrink, H., Rossi, E. and Bishop, J. 1991. Specification 503 – implementation of PV module qualification tests at ESTI. *Proc. 10th European Photovoltaic Solar Energy Conf.*, Lisbon, pp. 1219–1221.
[99] Bishop, J. and Ossenbrink, H. 1996. Results of five years of module qualification testing to CEC specification 503. *Proc. 25th IEEE Photovoltaic Specialists Conf.*, Washington DC, pp. 1191–1196.
[100] IEEE standard 1262: IEEE recommended practice for qualification of photovoltaic (PV) modules. Institute of Electrical and Electronic Engineers, New York, USA, 1995.
[101] IEC standard 61646: Thin-film terrestrial photovoltaic (PV) modules – design qualification and type approval. International Electrotechnical Commission, Geneva, Switzerland, 1996.
[102] Wohlgemuth, J.H., BP Solar, private communication.
[103] Wohlgemuth, J.H., Conway, M. and Meakin, D.H. 2000. Reliability and performance testing of photovoltaic modules. *Proc. 28th IEEE Photovoltaic Specialists Conf.*, Anchorage, pp. 1483–1486.
[104] ASTM standard E 1596: Test methods for solar radiation weathering of photovoltaic modules. *ASTM Annual Book of Standards*, Vol. 12.02, ASTM International, West Conshohocken, PA, USA, 2002.
[105] Osterwald, C.R., Hammond, R., Zerlaut, G. and D'Aiello, R. 1994. Photovoltaic module certification and laboratory accreditation criteria development. *Proc. First World Conf. on Photovoltaic Energy Conversion*, Hawaii, pp. 885–888.
[106] Osterwald, C.R., Varadi, P.F., Chalmers, S. and Fitzgerald, M. 2001. Product certification for PV modules, BOS components, and systems. *Proc. 17th European Photovoltaic Solar Energy Conf.*, Munich, pp. 379–384.

IIf-2

Calibration, Testing and Monitoring of Space Solar Cells

Emilio Fernandez Lisbona, ESA-Estec, Noordwijk, The Netherlands

1 Introduction

Solar energy is the main power source technology for most spacecraft since the 1960s. A total failure of the solar array (SA) performance will lead to complete mission loss. SA behaviour in the space environment has to be predicted in order to assure endurance during mission life.

The SA electrical performance is a basic parameter that needs to be predicted for mission life, tested on ground and monitored continuously in space. Electrical performance (EP) at beginning of life conditions is measured on ground to check power output prediction, based on performance measurements of single solar cells and before their integration on the SA. These measurements are performed with solar simulators, having adjusted their light intensity to standard AM0 illumination conditions with suitable reference solar cells. Reference cells are space calibrated using different methods that will be described in Section 2.

Endurance of the SA to the space environment has to be simulated by ground environmental testing. Different mechanical and environmental tests, together with electrical tests for degradation assessment, are performed at the different steps of development, manufacture and integration of SA components and intermediate assemblies. An overview of these tests is given in Section, mainly focused on tests at solar cell levels.

Monitoring the performance of the SA in orbit is essential to assess the predicted behaviour during the mission and this provides valuable data for verification of ground testing and further SA design improvements. Section 4 deals with the monitoring of spacecraft SA in orbit. Flight experiments are conducted to assess the performance and behaviour in space of novel solar cell or SA integration technologies. Two other important sources of data are unpredicted anomalies in orbit and investigations carried out on returned from space SAs.

2 Calibration of Solar Cells

Standard solar cells are used to set the intensity of solar simulators to standard illumination conditions, in order to electrically characterise solar cells with similar spectral response. Space calibration methods of solar cells can be extraterrestrial when performed outside the atmosphere or synthetic if they are carried out on the ground, using natural sunlight or simulated indoor illumination [1]. To prevent continuous handling operations of the expensive extraterrestrial/synthetic cells, so-called secondary working standard solar cells are calibrated for routine electrical performance testing in industry and testing laboratories.

2.1 Extraterrestrial Methods

Two calibration methods are the main suppliers of extraterrestrial standards: the high altitude balloon and the high altitude aircraft. Both methods require minimum data correction due to the small residual air mass at the altitude where the calibration is performed.

2.1.1 High Altitude Balloon

Calibrations are performed on board stratospheric balloons flying at altitudes of around 36 km, where the illumination sun conditions are very close to AM0. Cells to be calibrated are directly exposed to the sun, mounted on supports with sun trackers. Currently, two institutions, JPL-NASA in the USA [2] and CNES in France [3, 4], are conducting, on a yearly basis, these calibration campaigns. The main differences between the two calibration institutes are the position of the cells, which in the case of JPL-NASA is mounted on the balloon apex and in the case of CNES, is a gondola hanging from the balloon. Both institutes correct calibrated data taking into account the effect of temperature and the variation of illumination due to the Earth–Sun distance variation over the year. CNES also corrects its calibrated data, taking into account the effect of the residual atmosphere.

2.1.2 High Altitude Aircraft

Calibrations are performed on board of an aircraft capable of flying at altitudes of 15–16 km. Cells are mounted at the end cap of a collimating tube on a temperature controlled plate. NASA Glenn Research Centre is currently conducting more than 25 flights per calibration campaign using a Gates Learjet 25 equipped even with a spectroradiometer to measure the solar spectrum at that altitude. Data are corrected for the ozone absorption, the geocentric distance and extrapolated to the air mass value of zero [5].

2.1.3 Space Methods

The most realistic environment on which calibration of solar cells can be performed is indeed outside the atmosphere. The first constraint of these methods is their relatively high cost compared with the other two extraterrestrial methods and their lower level of maturity.

- Space shuttle: On board the space shuttle, the Solar Cell Calibration Experiment (SCCE) was conducted in two flights in 1983/84, where solar cells from different agencies, institutions and space solar cell industries around the world were calibrated and returned back to Earth [6].
- Photovoltaic Engineering Testbed: This is a NASA-proposed facility to be flown in the International Space Station, where after exposure and calibration of cells in the space environment, they are returned back to Earth for laboratory use [7].
- Lost Twin: This is an ESA-proposed method, based on the flight of several solar cells on a non-recoverable spacecraft. Cells nearly identical to the flight ones are kept on Earth. The orbiting cells are calibrated and these calibrated values are given to their respective twin cells.

2.2 Synthetic Methods

There are two methods carried out under natural sunlight conditions.

2.2.1 Global Sunlight

The cells to be calibrated and a pyranometer are placed on a horizontal surface, where simultaneous readings of spectral irradiance over the sensitivity range of the pyranometer and short circuit current of the cells are recorded in global sunlight. The calibration site environmental conditions need to fulfil several requirements relating to global and diffuse irradiance levels, solar elevation, unobstructed view over a full hemisphere, etc. The calibrated short circuit current of the cell is calculated by means of the following formula:

$$I_{sc} = I_{sg} \frac{\int (k_2 E_{g\lambda}) d\lambda}{E_{glob}} \frac{\int (k_1 s_\lambda) E_{s\lambda} d\lambda}{\int (k_1 s_\lambda)(k_2 E_{g\lambda}) d\lambda}$$

where $k_1 s_\lambda$ is the absolute spectral response of the cell, $k_2 E_{g\lambda}$ the absolute spectral irradiance of the sun at the calibration site, $E_{s\lambda}$ the AM0 spectral irradiance, E_{glob} the pyranometer irradiance reading and I_{sg} the measured short circuit current of the cell.

The final calibration value is the average of three calibrations of three different days. The former RAE (UK) performed for several years global sunlight calibrations at Cyprus [8] and presently INTA-SPASOLAB (Spain) is performed on a yearly basis and at Tenerife [9, 10].

2.2.2 Direct Sunlight

The cells to be calibrated are placed on the bottom plate of a collimation tube, a normal incidence pyrheliometer and a spectroradiometer are kept pointing to direct sunlight while measurements of short-circuit current, total irradiance and spectral irradiance are recorded. Several conditions need to be fulfilled by the calibration site and its environment, i.e. certain irradiance level, stable cell short-circuit readings, ratio of diffuse to direct irradiance, etc. The calibrated short circuit current of the cell is calculated by means of the following formula:

$$I_{sc} = \frac{I_{sd} \int E_{d\lambda} d\lambda \int E_{s\lambda} s_\lambda d\lambda}{E_{dir} \int E_{d\lambda} s_\lambda d\lambda}$$

where I_{sd} is the measured short circuit current, E_{dir} is the total solar irradiance, $E_{d\lambda}$ is the spectral solar irradiance, $E_{s\lambda}$ is the AM0 spectral irradiance and s_λ is the relative spectral response of the cell to be calibrated.

The calibrated short circuit current value is the average of three calibrations performed in three different days. CAST (China) presently performs calibrations following this method [11].

The following two methods are carried out under simulated sunlight.

2.2.3 Solar Simulator

The cell to be calibrated is illuminated by means of a steady-state solar simulator adjusted to 1 AM0 solar constant with a previously calibrated cell or a suitable detector. The spectral irradiance of the solar simulator is measured with a spectroradiometer and the relative spectral response of the cell is measured separately. The calibrated short circuit current of the cell is calculated as follows:

$$I_{sc} = I_{sm} \frac{\int Es\lambda s_{\lambda} d\lambda}{\int Em\lambda s\lambda d\lambda}$$

where I_{sm} the short circuit current and $E_{m\lambda}$ the spectral irradiance, both measured under the solar simulator. NASDA (Japan) regularly performs calibrations following this method [12].

2.2.4 Differential Spectral Response

The calibrated short circuit current of the cell is calculated with its absolute spectral response together with the reference AM0 solar spectral irradiance. The absolute spectral response is obtained as follows: first, the relative spectral response of the cell to be calibrated and then for certain wavelengths the absolute differential spectral response, is determined by the ratio of the cell short circuit current to irradiance measured by a standard detector. This method was developed and is frequently presently used by PTB (Germany) for solar cell calibration [13].

2.3 Secondary Working Standards

Secondary working standard (SWS) solar cells are used to set intensity of solar simulators to standard conditions for routine measurements of identical (same spectral response) solar cells during acceptance or qualification testing. For the EP characterisation of SA, panels or coupons, SWSs are preferred for reference. SWSs are calibrated using standards obtained by the methods defined above and a continuous or pulsed light source. The measured data are corrected by means of the spectral response of both cells and the spectral irradiance of the light source and the standard AM0 spectrum, following the spectral mismatch correction method [14]. This secondary calibration method also gives relations between calibrated solar cells by different methods [15].

3 Testing of Space Solar Cells and Arrays

In order to assess the behaviour of solar cells and solar arrays for a specific space mission or environment, several tests need to be conducted at different hardware levels and phases of a project.

- Solar cells:
  - Development: To know their performance, their endurance to the space environment and therefore decide on the most appropriate solar cell candidate for specific application.

- Design: Measured solar cell data at different environmental conditions is necessary for an accurate power prediction during the mission and therefore a suitable sizing of the solar array.
- Qualification: Verify that the solar cells manufactured in the production line meet a set of requirements defined by the specific space mission [17].
- Acceptance: To provide cell performance and physical data; essential for their further integration in the solar array electrical network.

- Higher levels of solar array components integration: The so-called photovoltaic assemblies (test specimens with all the components existing and integrated as in the solar array electrical network) are also tested in the development and qualification phases.
- Solar array level: Tests are performed in development phases and in the qualification phase of the flight hardware. These tests are required to see whether or not the solar array is integrated with the spacecraft body.

Tests on solar cells and solar arrays can be split in three types: Electrical, Environmental and Mechanical/Physical characteristics. The following paragraphs deal with these types of tests, focusing chiefly for their application to solar cells assemblies (SCAs), however, when relevant, their application to higher levels of solar array integration or other solar array components is described.

3.1 Electrical Tests

3.1.1 Electrical Performance

The objective of this test is to assess the corresponding electrical parameters of the solar cells and to provide data for solar generator design. The electrical current of solar cells under 1 Solar Constant AM0 equivalent illumination shall be measured and recorded at a certain voltage. A solar cell test set up consists basically of a continuous or pulsed light source, a load connected across the cell's terminals and electrical current and voltage measurement equipment. During the measurement, the temperature of the cell junction is kept at a constant temperature (25°C) and a four-point probe measurement of the cell is used in order to minimise the effects of lead and contact resistances [16].

Solar simulators need to meet certain requirements on their light beam spectrum, uniformity and stability for optimum EP measurements of photovoltaic devices [1, 17]:

- Spectrum: Maximum allowable deviations of spectral energy in certain wavelength regions of the standard AM0 spectrum define the solar simulator spectral quality classification. The spectral irradiance is measured with spectro-radiometers [18] or special filtered solar cells [19].
- Uniformity: Uniformity of the irradiance on the test area is a critical parameter for accurate measuring of panels or SA.

- Stability: The light beam stability has to be maintained under certain values, especially when no simultaneous correction is done when measuring the EP.

Continuous or pulsed light sources are used to simulate solar illumination in laboratories or test facilities:

- Continuous solar simulators are mostly based on xenon short arc lamps where the beam is filtered and collimated to achieve the above-mentioned requirements. They are mainly used for the electrical characterisation of solar cells and small coupons. Large area continuous solar simulators based on Argon discharge lamps are used to electrically characterised solar cells or panels [20]. Multi-source solar simulators are required for measuring multijunction (Mj) solar cells, in order to set equivalent AM0 illumination conditions on each sub-cell [21].
- Pulsed solar simulators are based on xenon large arc lamps where the beam usually is not filtered to meet the above requirements on the test plane. Either solar cells or large panels can be electrically characterised, being not heated during the test, but special techniques are needed for measuring slow response cells [22, 23]. When measuring Mj solar cells, a better matching of the AM0 is needed, precisely in the near infrared spectral range, where xenon large arc lamps have less radiant energy [24].

Reference cells, either primary or secondary standards, are used to set the intensity of solar simulators to standard illumination conditions. For Mj solar cells, either so-called component cells (Mj cell structures with only one active junction) [25] or methods based on mismatch factor are followed to set standard illumination conditions on each cell junction [26].

Under standard illumination conditions and constant temperature the current voltage curve of the photovoltaic device is traced by polarising at different voltages. The shape and magnitude of the I–V curve depends on the junction characteristics, shunt and series resistance, and on total radiant energy converted, regardless of wavelength composition [27]. However, for Mj solar cells, wavelength composition of the radiant energy affects the shape of the I–V curve [28].

Temperature coefficients of solar cell electrical parameters can be calculated from experimental data, by measuring the device EP at different temperatures [29, 30] (Table 1).

The solar cell EP behaviour under different angles of incidence is of most importance for SA designs with curved substrates and operation of planar SA at high tilt levels. The potential angle of incidence-dependent effects are the cosine function, Fresnel reflectivity, coverglass coatings and filters, solar cell multi-layer anti-reflecting coating, extreme angle effects and end-of-life (EOL) behaviour. Assessments of these effects for each SCA component combinations are needed for SA performance prediction [31].

Table 1 Typical EP parameters and temperature coefficients of some space solar cells. Abbreviations: Sj = single junction; Dj = double junction; Tj = triple junction

Solar cell technology	I_{sc} (mA/cm^2)	V_{oc} (mV)	P_{max} (mW/cm^2)	η (%)	$dIsc/dT$ (mA/cm^2/°C)	$dVoc/dT$ (mV/°C)	dP_{max}/dT (mW/cm^2/°C)
Si BSR	37.0	595	17.5	13.0	0.02	−2.20	−0.080
Si BSFR	39.0	610	19.0	14.0	0.03	−2.00	−0.075
Sj GaAs/Ge	32.0	1030	26.5	19.5	0.02	−1.85	−0.050
Dj GaInP/GaAs/Ge	16.3	2350	31.5	23.0	0.01	−5.50	−0.065
Tj GaInP/GaAs/Ge	16.5	2560	41.5	26.0	0.01	−6.50	−0.085

3.1.2 Relative Spectral Response

Relative Spectral Response is the short-circuit current density generated by unit of irradiance at a particular wavelength as a function of wavelength. Relative spectral response provides valuable data for improving solar cells under development, for the calculation of performance measurement errors and for solar simulator verifications. It is measured by illuminating with a narrow bandwidth (monochromator or narrow band filters) light source (pulsed or continuous) the solar cell [32], at different wavelengths in its sensitivity range, while measuring the cell short circuit current and the irradiance with a sensor. A cell with known spectral response can be used as reference, replacing the irradiance sensor [1].

To measure spectral response of Mj solar cells, each junction needs to be characterised separately by light biasing (filtered light or variable intensity lasers) of the non-measured junctions and by voltage biasing, to measure in short circuit conditions the sub-cell junction under test [33] (Figure 1).

3.1.3 Reverse Characterisation

The reverse voltage behaviour of solar cells is needed for the prediction of shadowing and hot-spot phenomena on solar cell strings. Reverse-biased cells may experience excessive heating, permanent loss of minor power output, or permanent short-circuit failure [34]. Generally, single and Mj gallium arsenide solar cells are more sensitive to reverse bias than silicon cells [35] as seen in Figure 2, driving to implement by-pass diodes on each cell for effective protection. Testing apparatus and procedures are similar to the EP ones, but current and power limitations are needed to avoid cell breakdown.

3.1.4 Capacitance Characterisation

The dynamic behaviour of solar cells may introduce specific requirements on the subsequent solar array regulator. Therefore, the capacitance of solar cells needs to be characterised following two different methods:

- Small signal or frequency domain method: This is the measured high frequency impedance around a certain bias point. Tests are performed with voltage biasing and in darkness [36].

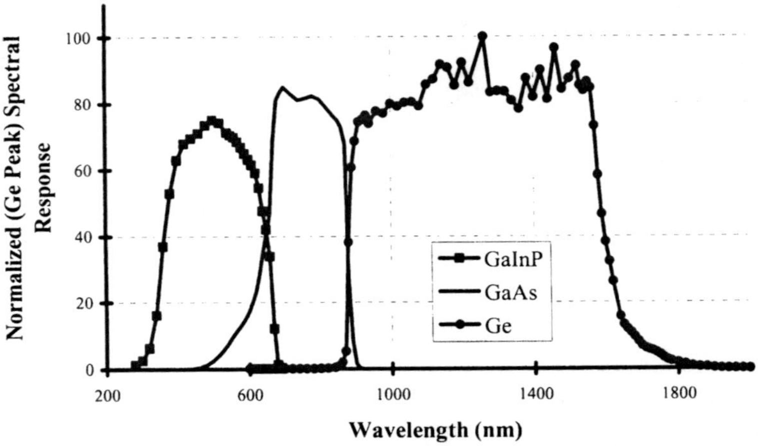

Figure 1 Spectral response of a proton-irradiated Tj solar cell.

- Large signal or time domain method: The rise of solar cell voltage between two operational points gives the solar cell capacitance by applying the formula $C = I_{sc}(t_2 - t_1)/(V_2 - V_1)$ [37], where t_2, t_1 and V_2, V_1 are the time and voltages associated with these operational points.

3.2 Environmental Tests

Environmental tests are performed to check solar array endurance to the different surroundings to which it is exposed during its complete lifetime. The most damaging environments are depicted here:

- Ground operations: Solar arrays are exposed to possible physical damage during manufacturing, integration, handling and transportation activities. During long storage periods, solar array components maybe corroded by humidity. Tests are performed at component, solar cell and SCA levels.
- Launch: Vibration, shocks, acceleration and acoustic fields affect the solar array in this phase, producing high mechanical stress levels that could produce physical damage either just after testing or in orbit. Vibration, shock and acoustic tests are performed usually at higher levels of solar array integration; panel, wing and spacecraft level.
- Space: Particles, temperature, vacuum and micrometeoroids are the main factors degrading solar arrays in space. Each factor affects different solar array components and interfaces. Tests are mainly conducted at solar cell, SCA level and coupon level.

EP and visual inspection tests are performed before and after exposure of photovoltaic devices to any environmental tests. The main environments affecting solar array performance are described below in more detail.

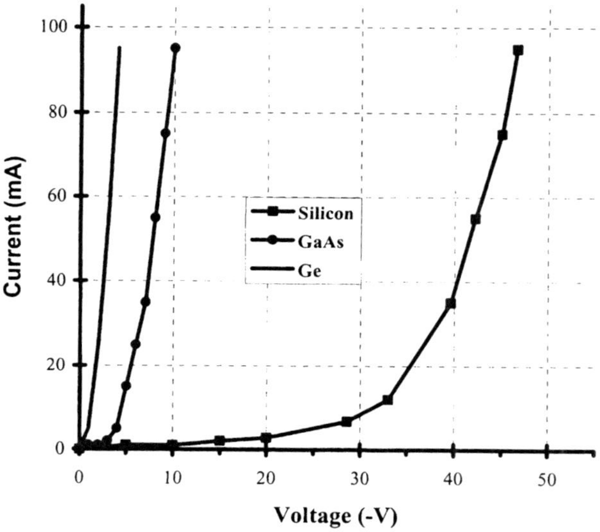

Figure 2 Reverse characteristics of some space solar cells.

3.2.1 Radiation Testing

The radiation environment in space basically comprises electrons and protons of different spectral energies. Solar cells are permanently damaged by these particles; displacement damage is produced in the cells' crystalline structure, reducing the minority carrier diffusion length and lifetimes in the cells' base region, driving a degradation of the cells' electrical parameters. For medium/high radiation environment missions, solar cell particle degradation is the key parameter for solar array sizing. Coverglasses and adhesives can be darkened by radiation reducing the array performance, by transmission losses and operational temperature increases.

Two methods are followed to predict the performance of solar cells under the space radiation environment: JPL method based on reducing all proton/electron energies from a certain space environment to an equivalent normal incidence and mono-energetic irradiation, usually 1MeV electrons [38–40] and NRL model based in the displacement damage dose methodology [41] (see Figure 3).

Solar cell radiation testing is performed on solar cell or SCAs at electron and proton irradiation facilities:

- Electrons are produced by Van der Graaff generators. Typical electron energies range from 0.6 up to 2.5 MeV and flux between 10^9 up to 1.5×10^{12} $e^-/cm^2/s$. Cells are irradiated under vacuum or inert gas conditions.
- Low energy protons (< 2 MeV) are produced by hydrogen ionising chambers and mass separators. Tandem Van der Graaff generators produced protons with energy from 2 to 10 MeV and cyclotrons and synchro-cyclotrons from 10 to 50 MeV and 50 to 155 MeV respectively. Cells are always irradiated under vacuum conditions.

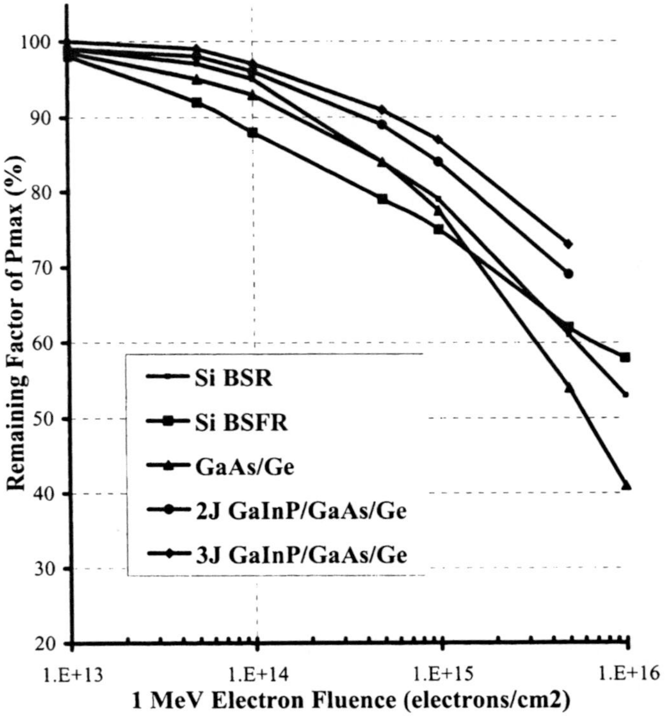

Figure 3 Power degradation of space solar cells under 1 MeV electron particles.

In general, the crystalline damage and performance degradation of irradiated solar cells is not stable for certain types. Recovery or further degradation phenomena are observed after annealing at temperatures higher than 20°C and exposure to sunlight [42, 43], suggesting performance of photon irradiation and annealing testing after particle irradiation.

3.2.2 Ultraviolet Radiation

Ultraviolet radiation can darken certain types of coated solar cell coverglass and adhesives, reducing the sunlight transmission to the solar cells, increasing solar array temperature and therefore lowering its EP [44]. Cracks on solar cell coverglasses may increase cell current degradation by 2% more for EOL [45]. Tests are conducted in vacuum chambers on coverglasses, SCAs or solar array coupons, at solar cell operational temperature in orbit, using UV light sources based on xenon arc, high-pressure mercury or halogen arc discharge lamps [46].

3.2.3 Atomic Oxygen (ATOX)

For low earth orbits (between 180 and 650 km), the presence of ATOX is a main cause of erosion of silver solar cell interconnectors [47, 48] and the kapton foil glued to the support structure outer layer [49]. ATOX durability testing on components or solar array coupons is performed on plasma asher chambers, being air raw material that becomes a plasma of atomic oxygen and other particles [50]. ATOX dose is determined using uncoated kapton samples whose erosion is known from flying data.

3.2.4 Thermal Cycling

The temperature cycling experienced by solar arrays in orbit (eclipses) is the cause with time of fatigue cracking of harnesses, bus-bars, interconnector material and interconnector solder/weld joints [51], and also the cause of increased series resistance in the solar cell and interconnector interface. Temperature cycling tests are performed following two methods on either solar array coupons or SCAs:

- Vacuum: So-called *thermal vacuum* or *vacuum thermal cycling* provides a good simulation of the space environment, but not of the temperature rate decay in orbit [52].
- Ambient pressure: Fast temperature change rates are achieved with circulating inert gas chambers. Cost and test duration are considerably reduced compared to vacuum chambers, but failures tend to happen earlier than with vacuum chambers [53].

3.2.5 Vacuum

A space vacuum might vaporise metals (Mg, Cd and Zn) and also volatile materials like adhesives. Thermal vacuum is a standard test performed at component and up to solar array level, for endurance testing of components and interfaces. Chambers as described in [52] are commonly available in the space photovoltaic industry and test houses. Failures coming from wrong manufacturing process or contamination of materials are also quickly revealed with these tests [54].

3.1.6 Micrometeoroids

More frequent impacts from micrometeoroids and space debris (between 10^{-6} and 10^{-3} g) mainly erode coverglass and solar array exposed coatings, with small solar array performance degradation due to optical losses. Predictions are in agreement with in-orbit degradation [55] and permanent loss of solar array sections by impacts on harnesses, though these are rare [56]. Hypervelocity impacts of particles are simulated with plasma drag accelerators [57, 58] and light gas guns [59].

3.2.7 Electrostatic Discharge

Dielectric solar array surfaces, mainly solar cell coverglass and kapton layers, are subject to electrostatic charging due to geomagnetic substorm activity or by the spacecraft surrounding plasma. Subsequent sudden electrostatic discharge (ESD) effects may permanently damage solar array components [60]. Coverglasses are coated with conductive coatings (i.e. ITO) and grounded [61] to lighten charging and to give an equi-potential surface for scientific field measurements. Tests are conducted at coupon level to check adequacy of components and interfaces [62] and at component level for the survival of the conductive coating to the mission environment [17].

3.2.8 Humidity

Accelerated humidity/temperature testing of solar cells is conducted to check the stability of solar cell contacts and anti-reflection coatings for long storage periods [63]. GaAs solar cells with AlGaAs window layers are submitted to this test in order to assure the effective protection of the anti-reflection coating to the corrosion of this window layer [64, 65].

3.3 Physical Characteristics and Mechanical Tests

Several tests are presented in this section, not only tests to check mechanical characteristics as adhesion of coatings, contacts or interconnectors, but also measurements of some physical characteristics needed for solar array sizing or essential inputs to other solar array analysis (Mass Budget, Thermal Analysis, etc). A summary is depicted in Table 2.

4 Monitoring of Space Solar Cells and Arrays

4.1 Flight Experiments

Several flight experiments have been conducted with solar cells/coupons in order to verify their endurance to the space environment and ground radiation testing assessment. Most of the flight experiments measure main electrical parameters of the cells and coupons (I_{sc}, V_{oc}, and power at certain voltage, full I–V curve), sun aspect angle and the operational temperature. In Table 3 some of the most relevant recent flight experiments are listed together with the publication reference, dates of data acquisition, orbit (apogee/perigee), cell/coupon types and main conclusions achieved.

Some flight experiments are, at the time of writing, in preparation:

- Mars array technology experiment (MATE): Several solar cell technologies shall be sent to Mars surface for checking their performance and endurance, together with instrumentation for the Mars surface (sun spectrum, dust, temperature, etc.) characterisation [79].
- Concentrator solar cell array technology flight experiment: Assessment of the performance of reflective concentrators with Mj solar cells [80].

4.2 Monitoring of Solar Array Performance in Space

Monitoring of solar arrays in space is mainly needed to verify that their performance meets the spacecraft power requirements for planned operations

Table 2 Mechanical/physical characterisation tests on space solar cells

Test name or phys./charact.	Purpose	Test method	Requirements
Visual inspection	Find solar cell or component obvious defects	Unaided eye or low magnification (5×–10×)	Several defects are not allowed at component level, relaxation criteria for higher levels of integration exist.
Interconnector adherence	Interconnector weld and cell contact adhesion	Pull test	Maximum pull force value and breakage mode.
Coating adherence	Contact and coatings adhesion	Tape peel test	Percentage of delaminated area below certain value.
Solar cell dimensions	External dimensions and contacts disposition	Microscope	Maximum dimensions provided by solar array electrical network or cell manufacturers.
Solar cell weight	Data for solar array weight budget	Balance	Maximum weight provided by solar array or cell manufacturers.
Solar cell flatness	Solar cell flatness	Profile microscope	Maximum bow provided by solar array electrical network manufacturer.
Contact thickness	Data for interconnector integration	X-ray spectroscopy	Maximum thickness provided by solar array electrical network or cell manufacturers.
Contact surface roughness	Data for interconnector integration	Roughness tester	Maximum thickness provided by solar array electrical network or cell manufacturers.
Hemispherical emittance	Emitted energy by the cell	Infrared spectrophotometer	Maximum emittance provided by solar array or cell manufacturers.
Solar absorptance	Absorbed/incident energy to the cell	Solar spectrometer	Maximum absorptance provided by solar array or cell manufacturers.

and that the design performance predictions for the complete mission are met. Reliable pre-flight data based on ground performance measurements, solar cell qualification tests and power budget calculations, based on qualification tests, is needed initially for an accurate performance evaluation in orbit. For flight data acquisition, temperature sensors, operational and short circuit sensors, and operational and open circuit voltage sensors are required, together with precise attitude and orbit data. Their quantity and precision drives the flight data quality [81]. Flight data is converted to standard conditions (1 Solar Constant and 25°C) for comparison with predicted data. In-orbit performance of some recent spacecraft is shown in Table 4, which includes relevant literature references, dates of evaluated data, orbit, SA type (array layout, power and solar cell type) and main conclusions achieved.

Table 3 Summary of recent flight experiments

Experiment	Ref.	Dates	Orbit	Cell/coupon types	Main conclusions for each cell/coupon type
Equator-S	[66], [67]	Dec. 97, May 98	Equatorial 500 km/ 67,000 km High radiation	1. GaAs/Ge 2. MBE Dj $GaInP_2$/GaAs/GaAs 3. UT[a] GaAs 4. Si NRS[b]/BSF 5. CIGS 6. CIS	1. Degradation according to modelling. 2. High radiation tolerance. 3. High radiation tolerance-thick cover. 4. Confirms ground radiation-testing data. 5. Improvement performance by light soaking effect in orbit. Degradation according to modelling. 6. Low energy protons heavily damaged uncovered cells. Degradation according to modelling.
PASP-Plus	[68–70]	Aug. 94, Aug. 95	70° Elliptic 362 km/ 2552 km High radiation	1. Si Planar & ISS[c] 2. GaAs 3. Dj AlGaAs/GaAs 4. Dj GaAs/CIS 5. InP 6. a-Si	1. Less degradation than predicted with modelling (*). 2. Degradation according to modelling (*). 3. GaAs degradation according to modelling. 4. GaAs degradation less than predicted with modelling. 5. Less degradation than predicted with modelling. 6. Positive P_{max} temp/coeff. with increasing temp.
ETS-VI (SCM)	[71]	Nov. 94, Jun. 96	Elliptic 8550 km/ 38,700 km High radiation	1. Sj GaAs/Si 2. GaAs/GaAs 3. Si BSFR 4. Si BSR	1. and 2. GaAs/Si cells more resistant to radiation damage than GaAs/GaAs.
ETS-V (SCM)	[72]	Sep. 87, Sep. 97	GEO 150°E	1. GaAs LPE 2. GaAs MOCVD 3. Si BSFR 4. Si BSR 5. Si NRS/BSFR	3. and 4. BSR are more resistant to radiation than BSFR, as in ground tests.

Table 3 (*continued*)

Experiment	Ref.	Dates	Orbit	Cell/coupon types	Main conclusions for each cell/coupon type
EURECA (ASGA)	[73]	Aug. 92, Jun. 93	LEO 510 km Circular	1. GaAs/Ge MOCVD 2. GaAs LPE 3. GaAs MOCVD	1. GaAs/Ge cells showed higher operation temperature than GaAs/GaAs. 2. No degradation of solar cells during the flight confirmed with post-flight ground measurements. 3. Post-flight analysis studied the effects on coupon components (interconnectors) of LEO environment (Large number of thermal cycles and ATOX).
LIPS-III	[74–77]	May 87, Aug. 93	LEO 1100 km 60° circular	1. Si BSR and BSFR 2. a-Si:H 3. $CuInSe_2$ 4. GaAs MOCVD and LPE 5. Sj AlGaAs/GaAs	1. BSR more resistant than BSFR. Degradation according to modelling. 2. Photodegradation main cell degradation mechanism (40% in power). 3. Extremely high radiation resistant. 4. and 5. Degradation according to modelling.
UoSAT-5	[78]	Jul. 91, Jul. 95	770 km polar Sun-sync. Low radiation	1. InP 2. ITO/InP 3. GaAs/Ge 4. Si High Eta	1. Small degradation according to modelling. 2. Anomalous degradation in voltage. 3. Small degradation according to modelling. 4. Higher degradation than 1. and 3. but fits models.

(*) Parasitic current collected by these coupons correlates with ground testing and prediction models.
[a] UT = ultra thin.
[b] NRS = non-reflective.
[c] Silicon cells of the International Space Station Array
For other abbreviations and standard solar cell terms see text or Chapter IIa-1.

Table 4 In-orbit performance of recent spacecraft.

Spacecraft (design life)	Ref.	Dates	Orbit	SA type	Main conclusions
SOHO (2.5 years)	[82]	Dec. 95, Dec. 01	LG1[a] 1.5×10^6 km from Earth	2 wings × 2 rigid panels EOL[b]power 1.4 kW Si 2 Ω cm BSR	1. Solar array design and good margin between working and P_{max} point allowed mission extension. 2. Less SA radiation degradation than predicted. 3. SA recovery after sun flares degradation.
SPOT 1 (3 years)	[83]	Feb. 86, Feb. 98	LEO[c] Sun-synchronous	2 wings × 1 flexible panel EOL[b] power 1 kW Si 1 Ω cm	1. Several loss factors have been over evaluated on the design, allowing a longer SA life. 2. SPOT 1 database shall improve EOL[b] performance predictions of coming LEO[c] spacecraft.
HS 601 HP C1 (15 years)	[84]	Aug. 97, Mar. 98	GEO[d]	2 wings × 3 rigid panels BOL[e] power 9.5 kW Dj $GaInP_2$/GaAs/Ge	1. SA power in orbit is 1.1% less than predicted. 2. Systematic errors may be the source of this discrepancy: Calibration of flight balloon standard and the calibration of ground performance testing.
INTELSAT-V (7 years)	[85]	Dec. 81, Dec. 94	GEO[d]	2 wings × 3 rigid panels EOL[b] power 1.5 kW Si BSR	1. SA power is 8–10% higher than predicted (solar flares) and 4–6% higher than predicted (no solar flares) for 13 spacecraft.
CS-3A (7 years)	[86]	Feb. 88, Nov. 88	GEO[d]	2 body mounted panels BOL[e] power 0.85 kW Sj GaAs	1. SA power is 1–5% higher than predicted.
HIPPARCOS (3 years)	[87]	Feb. 90, Jun. 93	GTO[f]	3 deployed panels EOL[b] power 325 W Si BSR 10 Ω cm	1. SA power is according to radiation degradation modelling.
IRS-1A (3 years)	[88]	Mar. 88, Mar. 95	LEO[c]	6 deployed panels EOL[b] power 0.7 kW Si BSR 10 Ω cm	1. The silver mesh interconnector survived more than 35000cy. 2. Effects of ATOX negligible on interconnectors. 3. Power degradation due to radiation matches with modelling.

Table 4 (*continued*)

Spacecraft (design life)	Ref.	Dates	Orbit	SA type	Main conclusions
Space Telescope SA (5 years)	[89]	Apr. 90, Dec. 93	LEO[c] 600 km	2 flexible wings × double roll-out 2 year life power 4.4 kW Si BSFR 10 Ω cm	1. Degradation performance is in agreement with most design loss factors. 2. Radiation fluence below initial prediction. 3. Random failures main degradation SA mechanism.
JCSAT (10 years)	[90]	Mar. 89, May 91	GEO[d]	2 telescopic cylindrical EOL[b] power 1.7 kW K7 and K3 Si	1. 1–2% less BOL[e] in-orbit performance than predicted probably due to reference standards for performance ground testing. 2. Increase performance over time probably due to radiation model more severe than in orbit. 3. Some solar flares did not produce any damage on the SA. No explanation.

[a] La Grangian Point 1.
[b] End of life.
[c] Low Earth orbit.
[d] Geosynchronous orbit.
[e] Beginning of life.
[f] Geosynchronous transfer orbit.
For other abbreviations and standard solar cell terms see text or Chapter IIa-1.

4.3 Spacecraft Solar Array Anomalies in Orbit

Another source of data for improving solar array design comes regrettably from anomalies faced by spacecraft SA in orbit. Investigations of the failure mechanism in-orbit are much more complicated due to the small quantities of data often available. However, some anomalies in orbit could be acceptably explained; a few of them are depicted here below:

4.3.1 European Communication Satellite (ECS) and Maritime European Communication Satellite (MARECS)

After 1.5 years in GEO both SA (virtually identical, two wings of three rigid panels each with silicon solar cells) started to suffer partial loss of power [60]. The failures seemed to be short-circuits between the cell network and panel structure. These failures continued intermittently until the end both missions, however, for ECS the power losses were recovered. Several potential failure modes were identified: imperfections of the Kapton insulation layer or embedded particles in between layers, insulation breakdown by electrostatic discharge, thermal cycling, corona effects, micrometeoroids or a combination of all of them. None of the potential failure modes could be identified as being responsible for the ECS and MARECS anomalies, however, several weak points in the SA design were identified, investigations continued in the direction of the most probable failure mode (ESD) [91] and some improvements were proposed aiming to lower the risk of these failures: designs should be adapted to incorporate sufficient margins in areas where uncertainties exist, parallel cell strings sections instead of single string sections and more stringent tests in manufacturing and acceptance for early failure detection.

4.3.2 X-ray Timing Explorer (XTE)

The XTE spacecraft was launched in December 1995. SA is composed of two wings of three rigid panels each, with silicon solar cells. Shortly after launch, the array showed discontinuous current drops, consistent with the loss of a part of a cell, when coming out from eclipse. The failure mechanism seems to be cell cracks not detected in ground inspections that became open in orbit due to the temperature gradients. These cracks were probably produced during the extensive tap tests, performed to detect SCA to substrate delaminations. During testing on the ground, following the same activities as for the flight SA, the qualification panel showed these effects, giving high confidence to this theory [92].

4.3.3 GPS Navstars 1–6

Six GPS Navstars satellites were placed in 20,000 km circular orbits from 1980. Mission lifetime for each spacecraft was five years and silicon solar cells K4 or K6 were in the SAs. After two years in orbit all spacecraft suffered an unexpected additional degradation of 2.5%. Investigations carried out in optical reflectors surfaces of one of the spacecraft revealed traces of contamination covering all spacecraft external surfaces. These contaminants mainly come from the outgassing of materials from the spacecraft, leading to reflectivity degradation of the coverglasses [93].

4.3.4 Pioneer Venus Orbiter SA

Pioneer Venus orbiter was a spin-stabilised (5 rpm) cylindrical spacecraft that operated in a high eccentric near polar orbit around Venus for more than eight years. After two years orbiting, power drops correlated with string losses were observed depending on the vehicle rotating angle. This suggested failures on strings due to reverse bias of cells (no shunt diodes protected the strings) produced by cyclic shadows made by the magnetometer boom cast, not predicted and unavoidable for the mission success. Ground tests were not conclusive that the cyclic reverse bias operation ended in cell breakdowns. Therefore, other interactions, as the ATOX environment in the Venus upper atmosphere, could favour the SA degradation [94].

4.4 Post-flight Investigations on Returned Solar Arrays

Returned SAs from space are valuable opportunities to assess their predicted behaviour in the space environment. Few SAs have been returned to Earth and a brief summary of their investigation programmes and the major conclusions are outlined in subsequent paragraphs.

4.4.1 Hubble Space Telescope Solar Array 1

One wing of the Hubble Space Telescope SA was retrieved from space in December 1993, after more than 3.5 years operating in a low earth orbit. The SA of the Hubble Space Telescope consisted of two wings of a double roll-out concept using two flexible solar cell blankets on each wing. The 48760 Silicon BSFR solar cells should provide the required 4.4 kW after two years in operation.

The post-flight investigation programme carried out between 1994 and 1995 [95] had the following main objectives:

- Assess the effect of different LEO interaction and environments as: thermal fatigue, ATOX, meteoroid and space debris damage, contaminations, UV, etc.
- Explain the anomalies experienced in orbit.

During the investigation programme the SA was submitted to several tests like detailed visual inspections, EP and health checks, wipe testing etc. The SA mechanisms were also mechanically tested to study their deploy/retract performance and finally the SA was totally disassembled for detailed investigation of all its components.

The main conclusions of the post-flight investigation programme related to the SA blankets are the following:

- SA performance: 5% more power than predicted after 3.6 years in orbit, despite several anomalies (string shorts) that reduced the power by 6.7%.

Random failures are the main contributors to SA degradation. SA overall degradation excluding failures was less than predicted, mainly as radiation model used was pessimistic.

- Solar cell interconnectors: No fatigue effects on interconnection loops were detected as expected from pre-flight qualification data.
- Harness: Fatigue effects were evident on flexible data harnesses, but no full detachments were found.
- Adhesives for ATOX protection: Darkening due to UV could increase SA operational temperature.
- Micrometeoroids: More than 4000 impacts were detected on the SA, but none of them produced permanent short circuits. The loss factor applied in the design is in full agreement with the results of the observed degradation (1.8%).

4.4.2 EURECA

The European Retrievable Carrier (EURECA) was launched in July 1992 (500 km orbit) and completely retrieved in July 1993 by the Space Shuttle. The SA consists of two interchangeable wings of five rigid panels ($\sim$100 m^2) each providing initially 5 kW. Silicon BSFR 10 Ω cm solar cells of two sizes were used to manufacture the charge and load array networks. The solar array post flight investigation programme had the objectives of studying LEO environment effects and mainly the anomalies faced during the mission [96]. Main conclusions are depicted here:

- Failures by fatigue (inadequate bend radii in the stress relief loop) in the Wiring Collecting Panels (WCPs) were responsible for open circuits on solar cell strings. WPCs were never tested in a flight representative configuration, as it was not possible to detect in advance the weakness of this design.
- A short circuit on the load array produced current from the battery during eclipse to the solar array (no blocking diodes were placed between the SA and battery circuits). A large burn mark was found at the suspected location of the short circuit after retrieval.
- Kapton FEP (Fluorinated Ethylene-Propylene) coatings of the cable insulations were completely eroded in X-ray/UV direction.
- The exposed side of MoAg interconnectors was oxidised and eroded by ATOX.
- Adhesives for ATOX protection were also darkened (top surfaces converted to SiO_2) and all surfaces investigated showed contamination of carbon or silicone.

4.4.3 MIR Solar Array

In January 1998, a segment of the MIR solar array was retrieved by the space shuttle. The segment, composed of eight panels, spent 10.5 years in a 380 km orbit. The panel design is exclusive; a laminated sandwich of coverglass, glass cloth, silicon solar cells (11% efficiency), glass cloth and optical solar reflectors

(OSRs) [97]. Two post-flight investigation programmes have been conducted in the USA and Russia. The main conclusions are the following:

- Hot spots are the main reason for the 50% power degradation of the solar array. By-pass diodes were not installed on the panel, relying especially on solar cell screening for handling full reverse currents. High temperatures during the hot spots destroyed separate commutation bundles in the circuits of serial connected solar cells [98, 99].
- Large SiO_x contaminations were found on all exposed surfaces of the panel, due to outgassing of silicone adhesives, resulting in a total power loss of only 0.72%.
- The meteoroid and space debris impact produced less than 1% power loss.
- The temperature increase over life was 7°C, due to an increase of the emittance and decrease of the absorptance.
- Solar cells not influenced by the hot spots had only 10–15% power degradation.

Acknowledgements

To my wife, Maria Jesus, for her constant support and patience. My colleagues C. Signorini and R. Crabb (ESA-Estec) and T.J. Gomez (Spasolab) for their good advice and helpful comments about the contents of this chapter. All my colleagues at Estec, especially of the solar generator section, Spasolab and the space photovoltaic community for fruitful discussions about these subjects. I am grateful to ESA for its support and permission to publish this work.

References

[1] ISO/DIS 15387: Space Systems – Single-Junction Space Solar Cells – Measurement and Calibration Procedures.

[2] Anspaugh, B.E. et al. 2002. Results of the 2001 JPL balloon flight solar cell calibration program. *JPL Publication* 02-004.

[3] Pichetto, V. et al. 2002. Casolba calibration of solar cells using balloon flight. *Proc. 29th IEEE Photovoltaic Specialists Conf.*, New Orleans.

[4] Roussel, M. et al. 1984. Calibration de cellules solaires hors atmosphère. *Proc. 4th European Space Power Conf.*, ESA SP-210, pp. 257–264.

[5] Jenkins, P. et al. 1997. Uncertainty analysis of high altitude aircraft air mass zero solar cell calibration. *Proc. 26th IEEE Photovoltaic Specialists Conf.*, Anaheim, pp. 857–860.

[6] Suppa, E.G. 1984. Space Calibration of Solar Cells. Results of 2 shuttle flight missions. *Proc. 17th IEEE Photovoltaic Specialists Conf.*, Orlando, pp. 301–305.

[7] Landis, G.A. et al. 2001. Calibration and measurement of solar cells on the international space station: A new test facility. *Proc. 36th Intersociety Energy Conversion Conf.*, pp. 229–231.

[8] Davies, M.A.H. and Goodbody, C. 1991. The calibration of solar cells in terrestrial sunlight. *Proc. 2nd European Space Power Conf.*, ESA SP-320, pp. 583–587.

[9] Garcia-Cervantes, L. et al. 1998. Ground level sunlight calibration of space solar cells. *Proc. 5th European Space Power Conf.*, ESA SP-416, pp. 615–620.

[10] Garcia, L. et al. 2001. Uncertainty analysis for ground level sunlight calibration of space solar cells at Tenerife. *Proc. 17th European Photovoltaics Solar Energy Conf.*, Munich, pp. 2259–2262.

[11] Yiqiang, Y. et al. 1996. Calibration of AM0 reference solar cells using direct normal terrestrial sunlight. *Proc. 9th Asia/Pacific Photovoltaic Science and Engineering Conf.*

[12] Kawasaki, O. et al. 1994. Study of solar simulator method and round robin calibration plan of primary standard solar cell for space use. *Proc. 1st World Conf. on Photovoltaic Energy Conversion*, Hawaii.

[13] Metzdorf, J. et al. 1986. Absolute indoor calibration of large area solar cells. *Proc. 5th European Symp. on Photovoltaic Generators in Space*, ESA SP-267, pp. 397–402.

[14] ASTM E973M-96: Test method for determination of the spectral mismatch parameter between a Photovoltaic device and a Photovoltaic reference cell.

[15] Gras, A. et al. 2000. Terrestrial secondary calibration analysis. *Proc. 16th European Space Power Conf.*, pp. 1011–1014.

[16] Gras, A. et al. 1993. Generic test procedure for solar cell testing. *Proc. 3rd European Space Power Conf.*, ESA WPP-054, pp. 743–748.

[17] ESA PSS-01-604 Generic specification for silicon solar cells, 1988.

[18] Seaman, C.H. et al. 1980. The spectral irradiance of some solar simulators and its effect on cell measurements, *Proc. 14th IEEE Photovoltaic Specialists Conf.*, San Diego, pp. 494–499.

[19] Goodelle, G.S. et al. 1981. Simulator spectral characterization using balloon calibrated solar cells with narrow band pass filters. *Proc. 15th IEEE Photovoltaic Specialists Conf.*, Orlando, pp. 211–217.

[20] Thrum, T. et al. 2000. Characterizing state of the art solar panels – A new approach for large area testing. *Proc. 28th IEEE Photovoltaic Specialists Conf.*, Anchorage, pp. 1320–1323.

[21] Kilmer, L.C. 1995. A more accurate, higher fidelity dual source AM0 solar simulator design. *Proc. 4th European Space Power Conf.*, ESA SP-369, pp. 671–675.

[22] Sturcbecher, J.J. et al. 1994. The mini-flasher: a solar array test system. *Solar Energy Materials and Solar Cells*, Vol. 36, pp. 91–98.

[23] Lukschal, W. et al. 1989. A pulsed solar simulator for electrical performance tests of space solar cells/arrays. *Proc. 1st European Space Power Conf.*, ESA SP-294, pp. 689–693.

[24] Granata, J.E. et al. 2000. Triple-junction $GaInP_2$/GaAs/Ge solar cells, production status, qualification results and operational benefits. *Proc. 28th IEEE Photovoltaics Specialists Conf.*, Anchorage, pp. 1181–1184.
[25] Gras, A. et al. 2002. Analysis for multi-junction solar cell measurements at Spasolab. *Proc. 6th European Space Power Conf.*, ESA SP-502, pp. 577–580.
[26] Emery, K. et al. 2000. Procedures for evaluating multi-junction concentrators. *Proc. 28th IEEE Photovoltaic Specialists Conf.*, Anchorage, pp. 1126–1130.
[27] Rauschenbach, H.S. 1980. *Solar Cell Array Design Handbook*, Litton Educational Publishing.
[28] Adelhelm, R. et al. 2000. Matching of multi-junction solar cells for solar array production. *Proc. 28th IEEE Photovoltaic Specialists Conf.*, Anchorage, pp. 1336–1339.
[29] King, D.L. et al. 1997. Temperature coefficients for PV modules and arrays: Measurement methods, difficulties and results. *Proc. 26th IEEE Photovoltaic Specialists Conf.*, Anaheim, pp. 1183–1186.
[30] Adelhelm, R. et al. 1997. Temperature coefficients of tandem solar cells under appropriate spectra. *Proc. 14th European Photovoltaics Solar Energy Conf.*, Barcelona.
[31] Burger, D.R. et al. 1996. Angle of incidence corrections for GaAs/Ge solar cells with low absorptance coverglass. *Proc. 25th IEEE Photovoltaic Specialists Conf.*, Washington DC, pp. 243–246.
[32] Larue, J.C. 1979. Pulsed measurement of solar cell spectral response. *Proc. 2nd European Photovoltaic Solar Energy Conf.*, West Berlin, pp. 477–486.
[33] King, D.L. et al. 2000. New methods for measuring performance of monolithic Mj solar cells. *Proc. 28th IEEE Photovoltaic Specialists Conf.*, Anchorage, pp. 1197–1201.
[34] Rauschenbach, H.S. et al. 1972. Breakdown phenomena on reverse biased silicon solar cells. *Proc. 9th IEEE Photovoltaic Specialists Conf.*, Silver Springs, pp. 217–225.
[35] Baron, W.R. et al. 1987. GaAs solar cell reverse characteristics. *Proc. 19th IEEE Photovoltaic Specialists Conf.*, New Orleans, pp. 457–462.
[36] Schwander, D. 2002. Dynamic solar cell measurement techniques: New small signal measurement techniques. *Proc. 6th European Space Power Conf.*, ESA SP-502, pp. 603–608.
[37] Rueda, P. et al. 2002. Mj GaAs solar cell capacitance and its impact upon solar array regulators. *Proc. 6th European Space Power Conf.*, ESA SP-502, pp. 29–34.
[38] Tada, H.Y. et al. 1982. *The Solar Cell Radiation Handbook*. JPL publication 82-69.
[39] Anspaugh, B.E. 1996. *GaAs Solar Cell Radiation Handbook*. JPL publication 96-9.

[40] Marvin, D.C. 2000. Assessment of Mj solar cell performance in radiation environments. *Aerospace Report TOR-2000 (1210)-1*. The Aerospace Corporation.

[41] Walters, R.J. et al. 2000. Analysis and modelling of the radiation response of Mj space solar cells. *Proc. 28th IEEE Photovoltaic Specialists Conf.*, Anchorage, pp. 1092–1097.

[42] Crabb, R. 1973. Photon induced degradation of electron and proton irradiated silicon solar cells. *Proc. 10th IEEE Photovoltaic Specialists Conf.*, Palo Alto, pp. 396–403.

[43] Fischer, H. et al. 1973. Investigation of photon and thermal induced changes in silicon solar cells. *Proc. 10th IEEE Photovoltaic Specialists Conf.*, Palo Alto, pp. 404–411.

[44] Goodelle, G.S. et al. 1975. High vacuum UV test of improved efficiency solar cells. *Proc. 11th IEEE Photovoltaic Specialists Conf.*, Scottsdale, pp. 184–189.

[45] Meulenberg, A. et al. 1997. Evidence for enhanced UV degradation to cracked coverslides. *XV Space Photovoltaic Research and Technology*, pp. 213–218.

[46] Matcham, J. et al. 1998. Effects of simulated solar-UV radiation on solar cell efficiency and transparent cell components. *Proc. 5th European Space Power Conf.*, ESA SP-416, pp. 643–650.

[47] Gerlach, L. et al. 1985. Advanced solar generator technology for the Eureca low earth orbit. *Proc. 18th IEEE Photovoltaic Specialists Conf.*, Las Vegas, pp. 78–83.

[48] Dunnet, A. et al. 1991. Assessment of ATOX erosion of silver interconnects on Intelsat VI, F3. *Proc. 2nd European Space Power Conf.*, ESA SP-320, pp. 701–706.

[49] Banks, B.A. et al. 1985. Protection of solar array blankets from attack by low earth orbital atomic oxygen. *Proc. 18th IEEE Photovoltaic Specialists Conf.*, Las Vegas, pp. 381–386.

[50] Ruthledge, S.K. et al. 1991. Atomic oxygen effects on SiO_x coated kapton for photovoltaic arrays in low earth orbit. *Proc. 22nd IEEE Photovoltaic Specialists Conf.*, Las Vegas, pp. 1544–1547.

[51] Richard, D. 1981. A rational approach to design and test a space photovoltaic generator. *Proc. 15th IEEE Photovoltaic Specialists Conf.*, Orlando, pp. 554–559.

[52] Ley, W. 1976. DFVLR facility for thermal cycling tests on solar cells panel samples under vacuum conditions. *Proc. 12th IEEE Photovoltaic Specialists Conf.*, Baton Rouge, pp. 406–412.

[53] Larue, J.C. et al. 1978. Accelerated thermal cycling of solar array samples. *Proc. 1st European Symposium on Photovoltaic Generators in Space*, ESA SP-140, pp. 57–64.

[54] Norris Blake III, L. 1996. Lessons learned about fabrication of space solar arrays from thermal cycle failures. *Proc. 25th IEEE Photovoltaic Specialists Conf.*, Washington DC, pp. 329–332.

[55] Gerlach, L. et al. 1995. HST-SA1: Electrical performance evaluation. *Hubble Space Telescope Solar Array Workshop*, ESA WPP-77, pp. 257–264.

[56] Murray, J.F. et al. 1991. Space environment effects on a rigid panel solar array. *Proc. 22nd IEEE Photovoltaic Specialists Conf.*, Las Vegas, pp. 1540–1543.

[57] Paul, K.G. et al. 1995. Post-Flight particle impacts on HST solar cells. *Hubble Space Telescope Solar Array Workshop*, ESA WPP-77, pp. 493–500.

[58] Brandhorst, Jr., H.W. et al. 2002. Hypervelocity impact testing of stretched lens array modules. *Proc. 6th European Space Power Conf.*, ESA SP-502, pp. 585–590.

[59] Schneider E. 1986. Micrometeorite impact on solar panels. *Proc. 5th European Symposium on Photovoltaic Generators in Space*, ESA SP-267, pp. 171–174.

[60] Bogus, K. et al. 1985. Investigations and conclusions on the ECS Solar Array in orbit power anomalies. *Proc. 18th IEEE Photovoltaic Specialists Conf.*, Las Vegas, pp. 368–375.

[61] Stern, T.G. et al. 2000. Development of an electrostatically clean solar array panel. *Proc. 28th IEEE Photovoltaic Specialists Conf.*, Anchorage, pp. 1348–1351.

[62] Bogorad, A. et al. 1991. Electrostatic discharge induced degradation of solar arrays. *Proc. 22nd IEEE Photovoltaic Specialists Conf.*, Las Vegas, pp. 1531–1534.

[63] Bishop, C.J. 1970. The fundamental mechanism of humidity degradation in silver-titanium contacts. *Proc. 8th IEEE Photovoltaic Specialists Conf.*, Seattle, pp. 51–61.

[64] Iles, P.A. et al. 1985. The role of the AlGaAs window layer in GaAs heteroface solar cells. *Proc. 18th IEEE Photovoltaic Specialists Conf.*, Las Vegas, pp. 304–309.

[65] Mitsui, K. et al. 1984. A high quality AR coating for AlGaAs/GaAs solar cells. *Proc. 17th IEEE Photovoltaic Specialists Conf.*, Orlando, pp. 106–110.

[66] La Roche, G. et al. 2000. Evaluation of the flight data of the Equator-S mini-modules. *Proc. 16th European Photovoltaic Solar Energy Conf.*, Glasgow, pp. 945–950.

[67] Messenger, S.R. et al. 2000. A displacement damage dose analysis of the Comets and Equator-S space solar cell experiments. *Proc. 16th European Photovoltaic Solar Energy Conf.*, Glasgow, pp. 974–977.

[68] Curtis, H. et al. 1996. Final results from the PASP-Plus flight experiment. *Proc. 25th IEEE Photovoltaic Specialists Conf.*, Washington DC, pp. 195–198.

[69] Davis, V.A. et al. 1995. Parasitic current collection by PASP PLUS solar arrays. *XIV Space Photovoltaic Research and Technology*, NASA CP-10180, pp. 274–285.

[70] Guidice, D.A. 1995. High voltage space-plasma interactions measured on the PASP Plus test arrays. *XIV Space Photovoltaic Research and Technology*, NASA CP-10180, pp. 286–295.

[71] Imaizumi, M. et al. 2000. Flight degradation data of GaAs-on-Si solar cells mounted on highly irradiated ETS-VI. *Proc. 28th IEEE Photovoltaic Specialists Conf.*, Anchorage, pp. 1075–1078.
[72] Aburaya, T. et al. 2001. Analysis of 10 years' flight data of solar cell monitor on ETS-V. *Solar Energy Materials and Solar Cells*, Vol. 68, pp. 15–22.
[73] Flores, C. et al. 1994. Post-flight investigation of the ASGA solar cell experiment on Eureca. *Proc. 1st World Conf. on Photovoltaic Energy Conversion*, Hawaii, pp. 2076–2081.
[74] Burgess, R.M. et al. 1993. Performance analysis of $CuInSe_2$ and GaAs solar cells aboard the LIPS-III flight Boeing lightweight panel. *Proc. 23rd IEEE Photovoltaic Specialists Conf.*, Louisville, pp. 1465–1468.
[75] Woodyard, J.R. et al. 1996. Analysis of LIPS-III satellite a-Si:H alloy solar cell data. *Proc. 25th IEEE Photovoltaic Specialists Conf.*, Washington DC, pp. 263–266.
[76] Severns, J.G. et al. 1988. LIPS-III. A solar cell test bed in space. *Proc. 20th IEEE Photovoltaic Specialists Conf.*, Las Vegas, pp. 801–807.
[77] Kulms, H. et al. 1990. Results of the MBB LIPS-III experiment. *Proc. 21st IEEE Photovoltaic Specialists Conf.*, Orlando, pp. 1159–1163.
[78] Goodbody, C. et al. 1996. The UoSAT-5 solar cell experiment – Over 4 years in orbit. *Proc. 25th IEEE Photovoltaic Specialists Conf.*, Washington DC, pp. 235–238.
[79] Scheiman, D.A. et al. 2000. Mars array technology experiment (MATE). *Proc. 28th IEEE Photovoltaic Specialists Conf.*, Anchorage, pp. 1362–1365.
[80] Jain, J.K. et al. 2002. Concentrator solar array technology flight experiment. *Proc. 29th IEEE Photovoltaic Specialists Conf.*, New Orleans, pp. 1362–1365.
[81] Bogus, K. et al. 1993. Comparative evaluation of the in-orbit performance of ESA's satellite solar generators. *Proc. 3rd European Space Power Conf.*, ESA WPP-054, pp. 529–535.
[82] Rumler, P. et al. 2002. SOHO power system performance during 6 years in orbit. *Proc. 6th European Space Power Conf.*, ESA SP-502, pp. 141–146.
[83] Jalinat, A. et al. 1998. In orbit behaviour of SPOT 1,2 and 3 solar arrays. *Proc. 5th European Space Power Conf.*, ESA SP-416, pp. 627–631.
[84] Fodor, J.S. et al. 1998. In-orbit performance of Hughes HS 601 solar arrays. *Proc. 2nd World Conf. on Photovoltaic Energy Conversion*, Vienna, pp. 3530–3533.
[85] Ozkul, A. et al. 1994. In-orbit performance characteristics of Intelsat-V solar arrays. *Proc. 1st World Conf. on Photovoltaic Energy Conversion*, Hawaii, pp.1994–1997.
[86] Takata, N. et al. 1989. In-orbit performance of CS-3A spacecraft GaAs solar array. *Proc. 1st European Space Power Conf.*, ESA SP-294, pp. 823–828.

[87] Crabb, R.L. and Robben, A.P. 1993. In-flight Hipparcos solar array performance degradation after three and a half years in GTO. *Proc. 3rd European Space Power Conf.*, ESA WPP-054, pp. 541–549.

[88] Puthanveettil, S.E. et al. 1995. IRS-1A Solar array – In-orbit performance. *Proc. 4th European Space Power Conf.*, ESA SP-369, pp. 583–585.

[89] Gerlach, L. et al. 1995. Hubble Space Telescope and EURECA solar generators a summary of the post flight investigations. *Proc. 4th European Space Power Conf.*, ESA SP-369, pp. 5–20.

[90] Gelb, S.W. et al. 1991. In-orbit performance of Hughes HS 393 solar arrays. *Proc. 22nd* IEEE *Photovoltaic Specialists Conf.*, Las Vegas, pp. 1429–1433.

[91] Levy, L. et al. 1986. MARECS & ECS anomalies: Attempt for insulation defect production in kapton. *Proc. 5th European Symposium on Photovoltaic Generators in Space*, ESA SP-267, pp. 161–169.

[92] Gaddy, E.M. et al. 1997. The Rossi X-Ray timing explorer XTE solar array anomaly. *XV Space Photovoltaic Research and Technology*, pp.144–153.

[93] Marvin, D.C. et al. 1988. Anomalous solar array performance on GPS. *Proc. 20th IEEE Photovoltaic Specialists Conf.*, pp. 913–917.

[94] Goldhammer, L.J. et al. 1987. Flight performance of the Pioneer Venus orbiter solar array. *Proc. 19th IEEE Photovoltaic Specialists Conf.*, New Orleans, pp. 494–499.

[95] *Proceedings of the Hubble Space Telescope SA workshop*. ESA WPP-77, 1995.

[96] EURECA The European retrievable carrier. *Technical Report ESA* WPP-069, 1994.

[97] Pinkerton, R.J. 2001. MIR returned solar array. *Proc. 36th Intersociety Energy Conversion Conf.*, pp. 217–222.

[98] Letin, V.A. 2002. Optical, radiation and thermal cycling losses of power solar array returned from orbital station MIR after 10.5 years of operation. *Proc. 6th European Space Power Conf.*, ESA SP-502, pp. 713–718.

[99] Grabov, A.B. et al. 2002. A terrestrial investigation of material's degradation mechanisms in silicon solar cells, which returned from MIR space station after ten years exploitation. *Proc. 6th European Space Power Conf.*, ESA SP-502, pp. 733–740.

Appendices

APPENDIX A

Constants, physical quantities and conversion factors

Name	Symbol	Value
Astronomical unit (mean distance between the Sun and the Earth)	R_{SE}	$1.496\ 10^{11}$ m
Avogadro's number	N_{Av}	$6.023\ 10^{23}$ molecules/mol
Boltzmann constant	k_B	$1.381\ 10^{-23}$ J/K
Electron charge	q	$1.602\ 10^{-19}$ C
Electronvolt	eV	$1.602\ 10^{-19}$ J
Energy of 1 μm photon		1.240 eV
Free electron mass	m_O	$9.109\ 10^{-31}$ kg
Permitivity of free space	$\varepsilon_O = 10^7/4\pi c^2$	$8.854\ 10^{-12}$ F/m
Permeability of free space	$\mu_O = 4\pi\ 10^{-7}$	$1.257\ 10^{-6}$ H/m
Plank's constant	h	$6.625\ 10^{-34}$ J.s
	$\hbar = h/2\pi$	$1.055\ 10^{-34}$ J.s
Radius of the Sun	R_S	$6.96\ 10^{8}$ m
Thermal voltage at 300 K	$V_T = k_B T/q$	25.9 mV
Solid angle subtended by the Sun	ω_S	$6.85\ 10^{-5}$ sterad
	$f_\omega = \omega_S/\pi$	$2.18\ 10^{-5}$
Solar constant (mean irradiance outside Earth's atmosphere)	I_o	1367 W/m^2
Speed of light in vacuum	c	$2.998\ 10^{10}$ m/s
Stefan-Boltzmann constant	σ	$5.670\ 10^{-8}$ W/m^2 K^4
Wavelength of 1 eV photon		1.240 μm

APPENDIX B

List of principal symbols

	Quantity	Subscripted quantity		Usual units
		Symbol	Name	
A	Area			m^2, cm^2
B	Beam irradiance (without subscript)			Wm^{-2}
B	Beam irradiation[a]	B_h	Hourly beam irradiation	MJm^{-2}, Whm^{-2} [b]
		B_d	Daily beam irradiation	
		B_m	Monthly mean beam irradiation	
B	Radiative recombination constant			cm^3sec^{-1}
C	Concentration ratio			–
C	Auger recombination constant	C_{no}	... for electrons	cm^6sec^{-1}
		C_{no}	... for holes	
C_A	Normalised array size			–
C_n	Battery capacity (in energy units)			Wh
C_S	Number of days of autonomy			days
D	Diffusion constant	D_n	Electron ...	cm^2sec^{-1}
		D_p	Hole ...	
		D_a	Ambipolar ... $\{=2D_n D_p/(D_n+D_p)\}$	
D	Density of localised states (in amorphous semiconductor)	D_{Ct}	... in conduction band	cm^{-3}
		D_{Vt}	... in valence band	
		D_{DB}	... of dangling bonds	

(Appendix B continued on next page)

Appendix B (*continued*)

	Quantity	Subscripted quantity: Symbol	Subscripted quantity: Name	Usual units
D	Diffuse irradiance (without subscript)			Wm^{-2}
D	Diffuse irradiation			$MJ\,m^{-2}$, $Wh\,m^{-2}$ [b]
		D_h	Hourly diffuse irradiation	
		D_d	Daily diffuse irradiation	
		D_m	Monthly mean diffuse irradiation	
d	Thickness (of antireflection coating)			μm
E	Energy			eV
		E_c	... of the edge of the conduction band	
		E_v	... of the edge of the valence band	
		E_g	Energy gap, band gap ($= E_c - E_v$)	
		E_F	Fermi energy	
		E_{Fn}	Electron quasi-Fermi level	
		E_{Fp}	Hole quasi-Fermi level	
		E_a	Activation energy	
		E_A	Energy generated by PV array	Wh or J
		E_{AC}	AC energy	Wh or J
EQE	External quantum efficiency			–
EOT	Equation of time			h or min
$\mathcal{E}$	Electric field			$V\,m^{-1}$
FF	Fill factor	FF_O	Fill factor of ideal solar cell characteristic	–
$F_{1/2}$	$\frac{2}{\sqrt{\pi}}\int_0^\infty \frac{x^{1/2}}{1+\exp(x-z)}dx$			
G	Global irradiance (without subscript)			Wm^{-2}
G	Global irradiation [a]			MJm^{-2}, Whm^{-2} [b]
		G_h	Hourly global irradiation	
		G_d	Daily global irradiation	
		G_m	Monthly mean global irradiation	
		G_o	Extraterrestrial irradiation (with further suffix indicating time interval)	
G	Carrier generation rate per unit volume			$cm^{-3}sec^{-1}$
g	Carrier generation function (generation rate per unit distance)			$cm^{-1}sec^{-1}$
H	Daily irradiation in the plane of the array [a]			$kWh\,m^{-2}$ [b]
		H_I	Daily global irradiation in the plane of the array	
		H_d	Daily direct irradiation in the plane of the array	

Appendix B *(continued)*

	Quantity	Subscripted quantity		Usual units
		Symbol	Name	
I	Current	I_{sc}	Short-circuit current	A
		I_{ph}	Photogenerated current	
		I_o	Diode dark saturation current	
		I_m	Current at the maximum power point	
		I_{DC}	Nominal DC current (in PV system)	
IQE	Internal quantum efficiency			–
J	Current density	J_{sc}	Short-circuit current density	Am^{-2}
		J_{ph}	Photogenerated current density	
		J_o	Diode dark saturation current density	
K	Damage constant	K_L	Diffusion length damage constant	–
		K_τ	Lifetime damage constant	cm^2sec^{-1}
k	Wave vector			cm^{-1}
KT	Clearness index	KT_h	Hourly clearness index $(=G_h/G_{oh})$	–
		KT_d	Daily clearness index $(=G_d/G_{od})$	
		KT_m	Monthly mean clearness index $(=(G_d)_m/(G_{od})_m)$	
L	Diffusion length	L_n	Electron diffusion length	μm
		L_p	Hole diffusion length	
L	Load (daily) energy consumption			Wh
L	Losses (system)	L_c	Array capture losses	kWh/(kWp*day)
		L_s	System losses	
L_D	Debye length			μm
ℓ	Drift (collection) length	ℓ_n	Electron drift length	μm
		ℓ_p	Hole drift length	
m	Electron mass	m_o	Free-electron mass	kg
		m_c	Density-of-states effective mass at the bottom of conduction band	
		m_v	Density-of-states effective mass at the top of valence band	

(*Appendix B continued on next page*)

Appendix B (*continued*)

	Quantity	Subscripted quantity		Usual units
		Symbol	Name	
N	Dopant or defect concentration	N_A	Acceptor concentration	cm^{-3}
		N_D	Donor concentration	
		N_B	$= N_A N_D/(N_A+N_D)$	
		N_{eff}	Effective dopant concentration	
		N_t	Concentration of carrier traps or recombination centres	
		N_{dop}	Dopant concentration	
N	Number of modules in a PV array	N_s	... connected in series	–
		N_p	... connected in parallel	
$NOCT$	Nominal operating cell temperature			K
$\mathcal{N}$	Density of states	$\mathcal{N}_C$	... in the conduction band	cm^{-3}
		$\mathcal{N}_V$	... in the valence band	
n	Electron concentration in semiconductor	n_O	....at equilibrium	cm^{-3}
		n_i	Intrinsic carrier concentration	
n_{id}	Diode ideality factor			–
$\mathbf{n}$	Refractive index	$\mathbf{n}_O$	.... of material surrounding solar cell	–
		$\mathbf{n}_{ar}$	.... of antireflection coating	
		$\mathbf{n}_{sc}$	... of a semiconductor	
P	Power	P_{max}	Power at the maximum power point	W
		P_A	Actual power produced by array	
		P_{eff}	Effective power rating of array (in stand-alone systems)	
		P_o	Nominal power of array	
		P_{AC}	AC power output	
		P_I	Nominal AC power of inverter	
PR	Performance ratio $(=Y_f/Y_r)$			–
PSH	Peak Solar Hours			h
p	Hole concentration in semiconductor	p_O	.... at equilibrium	cm^{-3}
QE	Quantum efficiency			–

Appendix B *(continued)*

	Quantity	Subscripted quantity		Usual units
		Symbol	Name	
R	Resistance	R_s	Series resistance	Ω
		R_p	Parallel (shunt) resistance	
$\mathcal{R}$	Reflection coefficient			–
r	Fresnel reflection coefficient			–
r	Normalised resistance	r_s	$= R_s I_{sc}/V_{oc}$	–
		r_p	$= R_p I_{sc}/V_{oc}$	
S	Surface recombination velocity			$cm\ sec^{-1}$
S	Sunshine duration	S_m	Monthly mean ...	h
SR	Spectral response			$A\ W^{-1}$
T	Temperature	T_s	Black body temperature of the Sun	K
		T_c	Cell temperature	
		T_a	Ambient temperature	
		T_n	$= T/300$	
$\mathcal{T}$	Transmission coefficient			–
t	Time			sec, hours
U	Recombination rate per unit volume	U_{rad}	Radiative ...	$cm^{-3}sec^{-1}$
		U_{Auger}	Auger ...	
		U_{SHR}	SHR ... (at defect)	
V	Voltage	V_{oc}	Open circuit voltage	V
		V_m	Voltage at the maximum power point	
		V_{bi}	Built-in voltage of a p-n junction	
		V_{bat}	Battery voltage	
		V_{DC}	Nominal voltage of PV system	
v	Velocity	v_{sat}	Saturation velocity	$cm\ sec^{-1}$
		v_{th}	Thermal velocity	
v_{oc}	$= qV_{oc}/k_BT$ or $qV_{oc}/n_{id}k_BT$			–

(*Appendix B continued on next page*)

Appendix B (*continued*)

	Quantity	Subscripted quantity: Symbol	Subscripted quantity: Name	Usual units
W	Thickness of a region in solar cell	W_j	Junction width	μm
		W_e	Emitter width	
		W_b	Base width	
		W_i	Width of intrinsic region	
x	Space coordinate			m
x_g	$= E_g/k_B T_s$			–
Y	Yield	Y_A	Array yield	kWh/(kWp*day)
		Y_f	Final yield	
		Y_r	Reference yield	
α	Absorption coefficient			cm^{-1}
α_s	Solar azimuth			degrees or radians
γ_s	Solar altitude angle			degrees or radians
χ	Electron affinity			eV
δ	solar declination			degrees or radians
ε	Static dielectric constant			–
ε	Irradiance correction to mean solar distance			–
Φ	Photon flux			$cm^{-2}sec^{-1}$
ϕ	Quasi-Fermi level potential	ϕ_n	Electron ...	V
		ϕ_p	Hole	
ϕ	Potential barrier (of heterojunction)			eV
ϕ	Particle fluence			$cm^{-2}sec^{-1}$
ϕ	Latitude angle			degrees
η	Efficiency	η_C	Carnot efficiency	–
		η_{CA}	Curzon - Ahlborn efficiency	
		η_L	Landsberg efficiency	

Appendix B (*continued*)

	Quantity	Subscripted quantity: Symbol	Subscripted quantity: Name	Usual units
		η_{PT}	Photothermal efficiency	
		η_{Amean}	Mean array efficiency	
		η_I	Inverter efficiency	
		η_{tot}	Overall PV plant efficiency	
ϑ	Collection efficiency			–
κ	Extinction coefficient			–
λ	Wavelength			μm, nm
μ	Carrier drift mobility			cm^2 V sec
		μ_n	Electron mobility	
		μ_p	Hole mobility	
ν	Frequency			sec^{-1}
ρ	Charge density			cm^{-3}
ρ_g	Ground albedo			
σ	Carrier capture cross section			cm^2
σ	Photoconductance			Siemens
τ	Minority carrier lifetime			sec
		τ_n	Electron lifetime	
		τ_p	Hole lifetime	
		τ_{rad}	Radiative lifetime	
		τ_{Auger}	Auger lifetime	
		τ_{SHR}	SHR lifetime	
		τ_{eff}	Effective	
ψ	electrostatic potential			V
ω	hour angle	ω_S	Sunset or sunrise ...	h
ω	solid angle	ω_S	... subtended by the Sun.	sterad

[a] *B* and *G* are the conventional symbols for the direct (beam) and global irradiation, as recommended by the International Solar Energy Society. IEC standard 61724:1998 recommends the symbols H_I and H_d for the daily global and direct irradiation in the plane of the array.

[b] Units $Whm^{-2}h^{-1}$ and $Whm^{-2}day^{-1}$ (or $MJm^{-2}h^{-1}$ and $MJm^{-2}day^{-1}$) are also used for the hourly and daily irradiation

APPENDIX C

Abbreviations and acronyms

AC	Alternating current
A/D	Analogue to digital
AFM	Atomic force microscopy
AGM	Absorbent glass mat
AR	Antireflection
ARC	Antireflection coating
a-Si	(Hydrogenated) amorphous silicon
AM	Air mass
AOI	Angle of incidence
APCVD	Atmospheric pressure chemical vapour deposition
ASTM	American Society for Testing and Materials
ATOX	Atomic oxygen
BCSC	Buried contact solar cell
BIPV	Building-integrated photovoltaics
BMBF	German Ministry for Research, Technology and Education
BOL	Beginning of life
BOS	Balance of system
BR	Bragg reflector
BSF	Back-surface field
BSFR	Back surface field and reflector
BSR	Back surface reflector
BST	British standard time
CAST	Chinese Academy of Space Technology
CBD	Chemical bath deposition
CDV	Committee draft for voting
CEI	Comitato Elettrotecnico Italiano (Italy)
CGA	Compressed Gas Association (USA)
CHP	Combined heat and power, co-generation
CIBSE	Chartered Institution of Building Services Engineers (UK)
CIS	Copper indium di-selenide
CIGS	Copper indium gallium di-selenide
CNES	Centre National d'Etudes Spatiales (France)

CVD	Chemical vapour deposition
CZ	Czochralski
DC	Direct current
DEV	Deutsches Einheitsverfahren (Germany)
DIN	Deutsches Institut für Normung (Germany)
Dj	Double junction
DOD	Depth of discharge
DOE	Department of Energy (USA)
DOS	Density of states
DR	Directional solidification
DTA	Differential thermal analysis
DTI	Department of Trade and Industry (UK)
DSSC	Dye sensitised solar cell
EB	Electron beam
EBIC	Electron-beam induced current
EFG	Edge-defined film fed growth
EHL	Environmental Health Laboratories
EHS	Environmental, health and safety
EMC	Electromagnetic continuous casting
ENS	Einrichtung zur Netzüberwachung mit zugerordnetem allpoligem Schaltorgan
EOL	End of life
EOT	Equation of time
EP	Electrical performance
EPA	Environmental Protection Agency (USA)
EPBT	Energy pay-back time
EPIA	European Photovoltaic Industry Association
EQE	External quantum efficiency
ESA	European Space Agency
ESD	Electrostatic discharge
ESRA	European Solar Radiation Atlas
ESTI	Eidgenossische Starkstrominspektorat (Switzerland)
ETSU	Energy Technology Support Unit for the DTI (UK)
EURECA	European Retrievable Carrier
EVA	Ethylene vinyl acetate
FF	Fill factor
FSF	Front surface field
FZ	Floating zone
GAP	Global Approval Programme (see also PV GAP)
GEO	Geostationary orbit
GDP	Gross domestic product
GMT	Greenwich mean time
HIT	Heterojunction with intrinsic thin layer (cell)
HTM	Hole transporting material
HOMO	Highest occupied molecular orbital
HV	High voltage
IBC	Interdigitated back contact
IEA	International Energy Agency
IEC	International Electrotechnical Commission
IEE	Institution of Electrical Engineers (UK)
IEEE	Institute of Electrical and Electronics Engineers (USA)
INTA	Instituto Nacional de Técnica Aeroespacial (Spain)
IPP	Independent power producers (Portugal)
IQE	Internal quantum efficiency
ISO	International Standards Organisation

ISS	International Space Station
ITO	Indium-tin oxide
JIS	Japan Industry Standard
JPL	Jet Propulsion Laboratory (USA)
JQA	Japan Quality Assurance Agency
LASS	Low angle silicon sheet
LAT	Local apparent time
LBIC	Light-beam induced current
LBSF	Local back-surface field
LCA	Life cycle assessment
LCR	Load coverage rate
LED	Light emitting diode
LEO	Low earth orbit
LLP	Loss-of-load probability
LMT	Local mean time
LOLP	Loss-of-power probability
LPCVD	Low pressure chemical vapour deposition
LPE	Liquid phase epitaxy
LPSC	Loss of power supply probability
LUMO	Lowest unoccupied molecular orbital
LV	Low voltage
MATE	Mars array technology experiment
MBE	Molecular beam epitaxy
mc	Multicrystalline
MDMO-PPV	Poly(2-methoxy-5-(3′,7′-dimethyloctyloxy) 1,4-phenylene vinylene
MEH-PPV	Poly(2-methoxy-5-2′-ethyl-hexyloxy)1,4-phenylene vinylene)
MEO	Mid-earth orbit
MG	Metallurgical grade
MINP	Metal-insulator np junction (cell)
MIS	Metal-oxide-semiconductor
MITI	Ministry of Trade and Industry (Japan)
Mj	Multijunction
MOCVD	Metalorganic chemical vapour deposition
MOS	Metal oxide semiconductor
MOVPE	Metalorganic vapour phase epitaxy
MPP	Maximum power point
MPPT	Maximum power point tracker
MSD	**M**ains monitoring units with allocated all-pole **s**witching **d**evices connected in series (also known as ENS)
MV	Medium voltage
NASA	National Aeronautics and Space Administration (USA)
NASDA	National Space Development Agency (Japan)
NEDO	New Energy and Industrial Technology Development Organisation (Japan)
NFPA	National Fire Protection Association (USA)
NIR	Near infrared
NREL	National Renewable Energy Laboratory (USA)
NRL	Naval Research Laboratory
NRS	Nonreflective
NSRDB	National Solar Radiation Data Base (USA)
O&M	Operation and maintenance
ODC	Ordered defect compound
OECD	Organisation for Economic Cooperation and Development
OECO	Obliquely evaporated contact
OLED	Organic light-emitting diode
OCVD	Open-circuit voltage decay

OMeTAD	2,2′,7,7′-tetrakis(N,N-di-p-methoxyphenylamine)-9,9′-spirobifluorene
ÖVE	Östereichische Verband für Elektrotechnik (Austria)
OVPD	Organic vapour phase deposition
PC	Point contact
PCD	Photoconductive decay
PCBM	1-(3-methoxycarbonyl)-propyl-1-phenyl-(6,6)C_{61}
PECVD	Plasma-enhanced chemical vapour deposition
PEDOT:PSS	Poly(3,4-ethylenedioxythiopene):poly(styrene sulfonate)
PERL	Passivated emitter, rear locally diffused (cell)
PESC	Passivated emitter solar cell
PET	Poly(ethylene terephalate)
PPV	Poly-phenylene vinylene
PR	Performance ratio
PRT	Platinum resistance thermometer
PSH	Peak solar hours
PTC	Performance test conditions
PV	Photovoltaic, photovoltaics
PVD	Physical vapour deposition
PVGAP	PV Global Approval Programme
PVPS	Photovoltaic Power Systems (Programme)
PVUSA	Photovoltaics for Utility Scale Applications
QE	Quantum efficiency
QSSPC	Quasi-steady-state photoconductance (method)
QSSVoc	Quasi-steady state open circuit voltage (method)
R&D	Research and development
RES	Renewable energy sources
RF	Radio frequency
RGS	Ribbon growth on substrate
RH	Relative humidity
ROCOF	Rate of change of frequency
ROW	Rest of the world
RTCVD	Rapid thermal chemical vapour deposition
RTD	Research and technology development
RTP	Rapid thermal process
SA	Solar array
SCA	Solar cell assembly
SCCE	Solar cell calibration experiment
SDL	Surface defect layer
SELF	Solar Electric Light Fund
SEM	Scanning electron microscopy
SG	Specific gravity
SRH	Shockley-Read-Hall
SHS	Solar home system(s)
SIPOS	Semi-insulating polysilicon
Sj	Single junction
SLI	Starting, lighting and ignition
SMUD	Sacramento Municipal Utility District
SOC	State of charge
SR	String ribbon
SR	Spectral response
SRC	Standard reporting conditions
SSP	Space solar power
STC	Standard test conditions
STAR	Surface texture and enhanced absorption with back reflector
SRV	Surface recombination velocity

SWE	Staebler-Wronski effect
SWS	Secondary working standard
S-Web	Supporting web
TJ	Tandem junction
Tj	Triple junction
TC	Technical committee
TCO	Transparent conducting oxide
TEC	Thermal expansion coefficient
TEM	Transmission electron microscopy
TPV	Thermophotovoltaics
TTV	Total thickness variation
UCPTE	Union for the Coordination of Production and Transmission of Electricity
UL	Underwriters Laboratory (USA)
UT	Ultrathin
UV	Ultraviolet
VDE	Verband der Elektrotechnik, Elektronik und Informationstechnik (Germany)
VdEW	Verband der Elektrizitätswirtschaft (Germany)
VHF	Very high frequency
VSE	Verband Schweizerische Elektrizitätswerke (Switzerland)
WCP	Wiring collecting panels
WET	West european time
WMO	World Meteorological Organisation
XRD	X ray diffraction
X-Si	Crystalline silicon
XTE	X-ray timing explorer
YAG	Yttrium aluminium garnet
ZMR	Zone-melt recrystalization

APPENDIX D

Bibliography

Books about solar cells, photovoltaic systems and applications

Aberle, A.G. *Crystalline Silicon Solar Cells-Advanced Surface Passivation & Analysis*, University of New South Wales, Sydney.

Andreev, V.M., Grilikhes, V.A.and Rumyantsev, V.D. *Photovoltaic Conversion of Concentrated Sunlight*, John Wiley & Sons, Chichester, 1997.

Archer, M.D. and Hill, R. *Clean Electricity from Photovoltaics*, Imperial College Press, London, 2001.

Brabec, C.J., Dyakonov, V., Parisi, J. and Sariciftci, N.S. Eds. *Organic Photovoltaics: Concepts and Realization*, Springer, Heidelberg, 2003.

Brendel, R. *Thin-Film Crystalline Silicon Solar Cells*, Wiley, Weinheim, 2003.

Bube R.H. *Photovoltaic Materials*, Imperial College Press, London, 1998

Bubenzer, A and Luther, J. *Photovoltaics Guidebook for Decision Makers*, Springer, Heidelberg, 2003.

Castañer, L. and Silvestre, S. *Modelling Photovoltaic Systems Using Pspice*, John Wiley & Sons, Chichester, 2002

Coutts, T.J. and Meakin, J.D. Eds. *Current Topics in Photovoltaics*, Academic Press, London, Vols. 1 – 4, 1985 – 1990.

Davidson, J. *The New Solar Electric Home, The Photovoltaics How-To Handbook*, Aatec Publishers, Michigan, 1987.

Derrick, A., Barlow, R.W., McNelis, B. and Gregory, J.A. *Photovoltaics: A Market Overview*, James & James Science Publishers, 1993.

Derrick, A., Francis, C. and Bokalders, V. *Solar Photovoltaic Products : A Guide For Development Workers*, Intermediate Technology Publications, London, 1991.

Fahrenbruch, A.L. and Bube, R.H. *Fundamentals of Solar Cells. Photovoltaic Solar Energy Conversion*, Academic Press, New York, 1983.

Fonash, S.J. *Solar Cell Device Physics*, Academic Press, New York, 1981.

Goetzberger, A., Knobloch J. and Voss, B. *Crystalline Silicon Solar Cells*, John Wiley & Sons, Chichester, 1998.

Green, M.A. *Solar Cells: Operating Principles, Technology and Practice*. Prentice Hall, New York, 1982.

Green, M.A. *High Efficiency Silicon Solar Cells*, Trans Tech Publications, 1987.

Green, M.A. *Silicon Solar Cells: Advanced Principles and Practice*. Centre for Photovoltaic Devices and Systems, University of New South Wales, 1995.

Green, M.A. *Third Generation Photovoltaics: Advanced Solar Energy Conversion*, Springer, Heidelberg, 2003.

Haberlin, H. *Photovoltaik*, Aarau, Switzerland, 1991 (in German).

Hill, R. Ed. *Applications of Photovoltaics*, Adam Hilger, Bristol, 1988.

Hille, G., Roth, W. and Schmidt, H. *Photovoltaic Systems (seminar coursebook)*, Frauenhofer Institute for Solar Energy Systems, Freiburg, 1995.

Hovel, H.J. Semiconductor solar cells, in: *Semiconductors and Semimetals*, R.K. Willardson and A.C. Beer, editors), Vol. 11, Academic Press, New York, 1975.

Humm, O. and Togweiler, P. *Photovoltaics in Architecture : The Integration of Photovoltaic Cells in Building Envelopes*, Birkhäuser, Basel, 1993

Iamura, M.S. , Helm, P. and Palz, W. Eds. *Photovoltaic System Technology: A European Handbook*, H.S. Stephens & Associates, Bedford, 1992.

Various chapters in Johansson, T.B., Kelly, H., Reddy, A.K.N. and Williams, R.H. Eds. *Renewable Energy: Sources for Fuel and Electricity*, Earthscan, London, 1993.

Johnston, W.D. *Solar Voltaic Cells*, Marcel Dekker Inc., New York, 1980

Komp, R.J. *Practical Photovoltaics: Electricity from Solar Cells*, Aatec Publishers, Michigan, 2002.

Lasnier, F. and Ang, T.G. *Photovoltaic Engineering Handbook*, Adam Hilger, Bristol, 1990.

Lorenzo, E. *Solar Electricity: Engineering of Photovoltaic Systems*, Progensa, Seville, 1994 (translation from the Spanish)

Luque, A. *Solar Cells and Optics for Photovoltaic Concentration*, Adam Hilger, Bristol, 1989.

Luque, A. and Araùjo, G.L. Eds. *Physical Limitations to Photovoltaic Energy Conversion*, Adam Hilger, Bristol, 1990.

Luque, A. and Hegedus, S. *Handbook of Photovoltaic Science and Engineering*, John Wiley & Sons, Chichester, 2003.

Markvart, T. Ed. *Solar Electricity (2nd edition)*, John Wiley & Sons, Chichester, 2001.

Maycock, P. and Stirewalt, E. *A Guide to the Photovoltaic Revolution. Sunlight to Electricity in One Step*, Rodale Press, Emmaus, 1985.

Mazer, J.A. *Solar Cells: An Introduction to Crystalline Photovoltaic Technology*, Kluwer Academic Publishers, Dordrecht, 1997

Messenger, R.A., and Ventre, J.G. *Photovoltaic Systems Engineering*, CRC Press, Boca Raton, USA, 1999.

Möller, H.J. *Semiconductors for Solar Cells*, Artech House, Boston, 1993.

Nelson, J. *The Physics of Solar Cells: Photons In, Electrons Out*, Imperial College Press, London, 2003

Neville, R.C. *Solar Energy Conversion (2nd edition)*, Elsevier, 1995

Novem – The Netherland Agency of Energy and the Environment, *Building With Photovoltaics*, James & James Science Publishers, London, 1997.

Overstraeten, R. van and Mertens, R.P. *Physics, Technology and Use of Photovoltaics*, Hilger, Bristol, 1986.

Partain, L.D. Ed. *Solar Cells and their Applications*, John Wiley & Sons, New York, 1995.

Perlin, J. *From Space to Earth : The Story of Solar Electricity*, Harvard University Press, Cambridge Mass., 2002.

Pulfrey, D.L. *Photovoltaic Power Generation*, Van Nostrand Reinhold Co., New York, 1978.

Randall, T. Ed. *Photovoltaics and Architecture*, Routledge – Spon Press, London, 2001.

Rauschenbach, H.S. *Solar Cell Array Design Handbook : The Principles and Technology of Photovoltaic Energy Conversion*, Van Nostrand Reinhold, New York, 1980.

Roberts, S. *Solar Electricity: A Practical Guide to Designing and Installing Small Photovoltaic Systems*, Prentice Hall, New Jersey, 1991.

Ross, M. and Royer, J. *Photovoltaics in Cold Climates*, James & James Publishers, London, 1998.

Sandia National Laboratories, *Stand-Alone Photovoltaic Systems. A Handbook of Recommended Design Practices*, US Department of Energy, 1990. Updated 1995.

Scheer, H. *A Solar Manifesto*, James & James Publishers, London, 2001.

Schropp, R.E.I and Zeman, M. *Amorphous and Microcystalline Silicon Solar Cells: Modelling, Materials and Device Technology*, Kluwer, Boston, 1998.

Smestad, G.P. *Optoelectronics of Solar Cells*, Society of Photo-Optical Instrumentation Engineers, Bellingham, 2002.

Takahashi, K. and Konagai, M. *Amorphous Silicon Solar Cells*, John Wiley & Sons, 1986.

Tao, G. *Optical Modeling and Characterization of Hydrogenated Amorphous Silicon Solar Cells*, Coronet Books, Philadelphia, 1994.

Treble, F.D. Ed. *Generating Electricity from the Sun*, Pergamon Press, Oxford, 1991

Turner, R.P. *Solar Cells and Photocells*, Howard W. Sams & Co, Indianapolis, 1975

Wenham, S.R., Green, M.A. and Watt, M.E. *Applied Photovoltaics*, Centre for Photovoltaic Devices and Systems, University of New South Wales, 1994.

Würfel, P. *Physik der Solarzellen (2nd edition)*, Spektrum, Heidelberg – Berlin, 2000 (in German; an English translation is in preparation)

Zweibel, K. and Hersch, P. *Basic Photovoltaic Principles and Methods*, Van Nostrand Reinhold Company, New York, 1984.

Books of interest which focus on topics related to photovoltaics

Born, M. and Wolf, E. *Principles of Optics (7th ed)*, Cambridge University Press, Cambridge, 1999

Brozel, M.R. and Stillman, G.E. Eds. *Properties of Gallium Arsenide (3rd edition)*, IEE/INSPEC, The Institution of Electrical Engineers, London, 1996.

Bube, R.T. *Photoconductivity of Solids*, Wiley, New York – London, 1960

Duffie, J.A. and Beckman, W.A. *Solar Engineering of Thermal Processes (2nd edition)*, John Wiley & Sons, New York, USA, 1991.

Fritzsche, H. Ed. *Amorphous Silicon and Related Materials*, University of Chicago, Chicago, 1989.

Keiser, B.E. *Principles of Electromagnetic Compatibility (5th edition)*, Artech House Inc, Ma, USA, 1985.

Lander, C.W. 1993. *Power Electronics (3rd Edition)*, McGraw-Hill, London, 1993

Landsberg, P.T. *Recombination in Semiconductors*, Cambridge University Press, Cambridge, 1991.

Levinstein, M., Rumyantsev, S. and Shur, M. Eds. *Handbook Series on Semiconductor Parameters*, Vols. 1 and 2, World Scientific, London, 1996, 1999.

Mohan, N., Undeland, T.M. and Robbins, W.P. *Power Electronics, Converters, Applications and Design (2nd Edition)*, John Wiley & Sons, New York, 1995

Pankove, J.I. *Optical Processes in Semiconductors*, Dover, New York, 1975

Palik, E.D. Ed. *Handbook of Optical Constants of Solids*, Academic Press Handbook Series, Orlando, 1985.

Palik, E.D. Ed. *Handbook of Optical Constants of Solids II*, Academic Press, San Diego, 1991.

Palz, W. Ed. *European Solar Radiation Atlas (2nd edition)*, Vols. 1 and 2, Verlag TÜV Rheinland, Köln, 1984.

Pearsall, T.P. Ed. *Properties, Processing and Applications of Indium Phosphide*, IEE/INSPEC, The Institution of Electrical Engineers, London, 2000.

Rashid, M.H. *Power Electronics, Circuits, Devices and Applications (2nd Edition)*, Prentice Hall, 1993.

Scharmer, K. and Grief, J. Co-ordinators. *European Solar Radiation Atlas (4th Edition)*, Les Presses de l' École des Mines de Paris, Paris, 2000.

Searle, T. Ed. *Properties of Amorphous Silicon and its Alloys* (EMIS Data Reviews No. 19), IEE/INSPEC, The Institution of Electrical Engineers, London, 1998.

Selberherr, S. *Analysis and Simulation of Semiconductor Devices*, Springer, Vienna – New York, 1984

Shur, M. *Physics of Semiconductor Devices*, Prentice Hall, Englewood Cliffs, 1990.

Smith, R.A. *Semiconductors (2nd edition)*, Cambridge University Press, Cambridge, 1978.

Street, R.A. *Hydrogenated amorphous silicon*, Cambridge University Press, Cambridge, 1991

Sze, S.M. *Physics of Semiconductor Devices (2nd edition)*, John Wiley & Sons, New York,, 1981.

deVos, A. *Endoreversible Thermodynamics of Solar Energy Conversion*, Oxford University Press, 1992.

Welford, W.T. and Winston, R. *The Physics of Non-imaging Concentrators*, Academic Press, New York, 1978.

APPENDIX E

International and US Standards with Relevance to Photovoltaics

Standards published by the International Electrotechnical Commission. *IEC Central Office, 3, rue de Varembé, P.O. Box 131, CH - 1211 GENEVA 20, Switzerland. http://www.iec.ch/*

IEC 60364-7-712:2002. Electrical installations of buildings. Part 7-712: Requirements for special installations or locations – Solar photovoltaic (PV) power supply systems

IEC 60891:1987. Procedures for temperature and irradiance corrections to measured I-V characteristics of crystalline silicon photovoltaic devices.

IEC 60904-1:1987. Photovoltaic devices. Part 1: Measurement of photovoltaic current-voltage characteristics.

IEC 60904-2: 1989. Photovoltaic devices. Part 2: Requirements for reference solar cells.

IEC 60904-3: 1998. Photovoltaic devices. Part 3: Measurement principles for terrestrial photovoltaic (PV) solar devices with reference spectral irradiance data.

IEC 60904-5: 1996. Photovoltaic devices. Part 5: Determination of the equivalent cell temperature (ECT) of photovoltaic (PV) devices by the open-circuit voltage method.

IEC 60904-6: 1994. Photovoltaic devices. Part 6: Requirements for reference solar modules.

IEC 60904-7: 1987. Photovoltaic devices. Part 7: Computation of spectral mismatch error introduced in the testing of a photovoltaic device.

IEC 60904-8: 1998. Photovoltaic devices. Part 8: Measurement of spectral response of a photovoltaic (PV) device.

IEC 60904-9: 1995. Photovoltaic devices. Part 9: Solar simulator performance requirements.

IEC 60904-10: 1998. Photovoltaic devices. Part 10: Methods of linearity measurement.

IEC 61173:1992. Overvoltage protection for photovoltaic (PV) power generating systems. Guide.

IEC 61194:1992. Characteristic parameters of stand-alone photovoltaic (PV) systems

IEC 61215: 1993. Crystalline silicon terrestrial photovoltaic (PV) modules – design qualification and type approval.

IEC 61277:1995. Terrestrial photovoltaic (PV) power generating systems. General and guide.

IEC 61345:1998. UV test for photovoltaic (PV) modules

IEC 61427:1999. Secondary cells and batteries for solar photovoltaic energy systems. General requirements and methods of test.

IEC 61646: 1996. Thin-film terrestrial photovoltaic (PV) modules – design qualification and type approval.

IEC 61683:1999. Photovoltaic systems. Power conditioners. Procedure for measuring efficiency

IEC 61701:1995. Salt mist corrosion testing of photovoltaic (PV) modules

IEC 61702:1995. Rating of direct coupled photovoltaic (PV) pumping systems

IEC 61721:1995. Susceptibility of a photovoltaic (PV) module to accidental impact damage (resistance to impact test)

IEC 61724:1998. Photovoltaic system performance monitoring. Guidelines for measurement, data exchange and analysis.

IEC 61725:1997. Analytical expression for daily solar profiles

IEC 61727:1995. Photovoltaic (PV) systems. Characteristics of the utility interface.

IEC 61829:1995. Crystalline silicon photovoltaic (PV) array. On-site measurement of I-V characteristics

IEC/TR2 61836:1997. Solar photovoltaic energy systems – Terms and symbols

IEC/PAS 62111:1999. Specifications for the use of renewable energies in rural decentralised electrification

Draft IEC Standards

IEC 61215 Ed. 2.0 Crystalline silicon terrestrial photovoltaic (PV) modules – Design qualification and type approval

IEC 62116. Testing Procedure of Islanding Prevention Measures for Grid Connected Photovoltaic Power Generating Systems

IEC 61727 Ed. 2.0 Characteristics of the utility interface for photovoltaic (pv) systems

IEC 61730-1 Ed. 1.0 Photovoltaic module safety qualification – Part 1: Requirements for construction

IEC 61730-2 Ed. 1.0 Photovoltaic module safety qualification – Part 2: Requirements for testing

IEC 61836 TR Ed. 2.0 Solar photovoltaic energy systems – Terms and symbols

IEC 61853 Ed. 1.0 Performance testing and energy rating of terrestrial photovoltaic (PV) modules

IEC 62093 Ed. 1.0 Balance-of-system components for photovoltaic systems – Design qualification natural environments

IEC 62108 Ed. 1.0 Concentrator photovoltaic (PV) receivers and modules – Design qualification and type approval

IEC 62109 Ed. 1.0 Electrical safety of static inverters and charge controllers for use in photovoltaic (PV) power systems

IEC 62116 Ed. 1.0 Testing procedure – Islanding prevention measures for power conditioners used in grid connected photovoltaic (PV) power generation systems

IEC 62124 Ed. 1.0 Photovoltaic (PV) stand alone systems – Design verification

IEC 62145 Ed. 1.0 Crystalline silicon PV modules – Blank detail specification

IEC 62234 Ed. 1.0 Safety guidelines for grid connected photovoltaic (PV) systems mounted on buildings

IEC 62253 Ed. 1.0 Direct coupled photovoltaic pumping systems – Design qualification and type approval

IEC 62257 Ed. 1.0 Recommendations for the use of renewable energies in rural decentralised electrification

IEC 62257-1 TS Ed. 1.0 Recommendations for small renewable energy and hybrid systems for rural electrification – Part 1: General introduction to rural electrification

IEC 62257-2 TS Ed. 1.0 Recommendations for small renewable energy and hybrid systems for rural electrification – Part 2: From requirements to a range of electrification systems

PNW 82-304 Ed. 1.0 Photovoltaic module safety qualification – Part 1: Requirements for construction

PNW 82-306 Ed. 1.0 Photovoltaic module safety qualification – Part 2: Requirements for testing

PNW 82-314 Ed. 1.0 Procedures for establishing the traceability of the calibration of photovoltaic reference devices

PWI 82-1 Ed. 1.0 Photovoltaic electricity storage systems

Standards published by ASTM International (formerly American Society for Testing and Materials). *100 Barr Harbor Drive, PO Box C700, West Conshohocken, Pennsylvania, USA 19428-2959. http://www.astm.org*

E 490-00a Standard Solar Constant and Zero Air Mass Solar Spectral Irradiance Tables

E 816-95 Standard test method for calibration of secondary reference pyrheliometers and pyrheliometers for field use.

E 824-94 Standard test method for transfer of calibration from reference to field radiometers.

E 913-82(1999) Standard method for calibration of reference pyranometers with axis vertical by the shading method.

E 941-83(1999) Standard test method for calibration of reference pyranometers with axis tilted by the shading method.

E1362-99 Standard Test Method for Calibration of Non-Concentrator Photovoltaic Secondary Reference Cells

E1125-99 Standard Test Method for Calibration of Primary Non-Concentrator Terrestrial Photovoltaic Reference Cells Using a Tabular Spectrum

E948-95(2001) Standard Test Method for Electrical Performance of Photovoltaic Cells Using Reference Cells Under Simulated Sunlight

E973-02 Standard Test Method for Determination of the Spectral Mismatch Parameter Between a Photovoltaic Device and a Photovoltaic Reference Cell

E973M-96 Standard Test Method for Determination of the Spectral Mismatch Parameter Between a Photovoltaic Device and a Photovoltaic Reference Cell [Metric]

E1040-98 Standard Specification for Physical Characteristics of Nonconcentrator Terrestrial Photovoltaic Reference Cells

E1036M-96e2 Standard Test Methods for Electrical Performance of Nonconcentrator Terrestrial Photovoltaic Modules and Arrays Using Reference Cells

E1039-99 Standard Test Method for Calibration of Silicon Non-Concentrator Photovoltaic Primary Reference Cells Under Global Irradiation

E1830-01 Standard Test Methods for Determining Mechanical Integrity of Photovoltaic Modules

E1799-02 Standard Practice for Visual Inspections of Photovoltaic Modules

E1596-99 Standard Test Methods for Solar Radiation Weathering of Photovoltaic Modules

E1524-98 Standard Test Method for Saltwater Immersion and Corrosion Testing of Photovoltaic Modules for Marine Environments

E1171-01 Standard Test Method for Photovoltaic Modules in Cyclic Temperature and Humidity Environments

E1597-99 Standard Test Method for Saltwater Pressure Immersion and Temperature Testing of Photovoltaic Modules for Marine Environments

E1143-99 Standard Test Method for Determining the Linearity of a Photovoltaic Device Parameter with Respect To a Test Parameter

E1021-95(2001) Standard Test Methods for Measuring Spectral Response of Photovoltaic Cells

E1802-01 Standard Test Methods for Wet Insulation Integrity Testing of Photovoltaic Modules

E1462-00 Standard Test Methods for Insulation Integrity and Ground Path Continuity of Photovoltaic Modules

E1038-98 Standard Test Method for Determining Resistance of Photovoltaic Modules to Hail by Impact with Propelled Ice Balls

E2047-99 Standard Test Method for Wet Insulation Integrity Testing of Photovoltaic Arrays

E1328-99 Standard Terminology Relating to Photovoltaic Solar Energy Conversion

E927-91(1997) Standard Specification for Solar Simulation for Terrestrial Photovoltaic Testing

E2236-02 Standard Test Methods for Measurement of Electrical Performance and Spectral Response of Nonconcentrator Multijunction Photovoltaic Cells and Modules

E1036-02 Standard Test Methods for Electrical Performance of Nonconcentrator Terrestrial Photovoltaic Modules and Arrays Using Reference Cells

E782-95(2001) Standard Practice for Exposure of Cover Materials for Solar Collectors to Natural Weathering Under Conditions Simulating Operational Mode

E881-92(1996) Standard Practice for Exposure of Solar Collector Cover Materials to Natural Weathering Under Conditions Simulating Stagnation Mode

E822-92(1996) Standard Practice for Determining Resistance of Solar Collector Covers to Hail by Impact With Propelled Ice Balls

G113-01 Standard Terminology Relating to Natural and Artificial Weathering Tests of Nonmetallic Materials

G 130-95 Standard test method for calibration of narrow- and broad-band ultraviolet radiometers using a spectroradiometer.

G 138-96 Standard test method for calibration of a spectroradiometer using a standard source of irradiance.

G 159-98 Standard tables for references solar spectral irradiance at air mass 1.5: direct normal and hemispherical for a 37° tilted surface.

Work item WK558 Reference Solar Spectral Irradiances: Direct Normal and Hemispherical on 37 Tilted Surface

Standards published by the Institution of Electrical and Electronic Engineers. *IEEE Standards Association, PO Box 1331, 445 Hoes Lane, Piscataway NJ 08855-1331, USA. http://standards.ieee.org/*

IEEE Std 1145-1990 IEEE recommended practice for installation and maintenance of nickel-cadmium batteries for photovoltaic (PV) systems

IEEE Std 1013-1990 IEEE recommended practice for sizing lead-acid batteries for photovoltaic (PV) systems

IEEE Std 1262-1995 IEEE recommended practice for qualification of photovoltaic (PV) modules

IEEE Std 1144-1996 IEEE recommended practice for sizing nickel-cadmium batteries for photovoltaic (PV) systems

IEEE Std 1374-1998 IEEE guide for terrestrial photovoltaic power system safety

IEEE Std 1145-1999 IEEE recommended practice for installation and maintenance of nickel-cadmium batteries for Photovoltaic (PV) systems

IEEE Std 929-2000 IEEE Recommended Practice for Utility Interface of Photovoltaic (PV) Systems

IEEE Std 937-2000 IEEE Recommended Practice for Installation and Maintenance of Lead-Acid Batteries for Photovoltaic (PV) Systems

IEEE Std 1013-2000 IEEE recommended practice for sizing lead-acid batteries for photovoltaic (PV) systems

IEEE Std 1513-2001 IEEE recommended practice for qualification of concentrator photovoltaic (PV) receiver sections and modules

Draft European Standards

prEN 50312-1:1999 Photovoltaic systems. Solar home systems. Part 1. Safety. Test requirements and procedures

prEN 50312-2:1999 Photovoltaic systems. Solar home systems. Part 2. Performance. Test requirements and procedures

prEN 50313-1:1999 Photovoltaic systems. Solar modules. Part 1. Safety. Test requirements and procedures

prEN 50313-2:1999 Photovoltaic systems. Solar modules. Part 2. Performance. Test requirements and procedures

prEN 50314-1:1999 Photovoltaic systems. Charge regulators. Part 1. Safety. Test requirements and procedures

prEN 50314-2:1999 Photovoltaic systems. Charge regulators. Part 2. EMC. Test requirements and procedures

prEN 50314-3:1999 Photovoltaic systems. Charge regulators. Part 3. Performance. Test requirements and procedures

prEN 50315-1:1999 Accumulators for use in photovoltaic systems. Part 1. Safety. Test requirements and procedures

prEN 50315-2:1999 Accumulators for use in photovoltaic systems. Part 1. Performance. Test requirements and procedures

prEN 50316-1:1999 Photovoltaic lighting systems. Part 1. Safety. Test requirements and procedures

prEN 50316-2:1999 Photovoltaic lighting systems. Part 2. EMC. Test requirements and procedures

prEN 50316-3:1999 Photovoltaic lighting systems. Part 3. Performance. Test requirements and procedures

prEN 50322-1:1999 Photovoltaic systems. Part 1. Electromagnetic compatibility (EMC). Requirements for photovoltaic pumping systems

prEN 50330-1:1999 Photovoltaic semiconductor convertors. Utility interactive fail safe protective interface for PV-line commutated converters. Design qualification and type approval

prEN 50331-1:1999 Photovoltaic systems in buildings. Part 1. Safety requirements

prEN 50380:2001 Datasheet and nameplate information for photovoltaic modules

Other standards and guidelines

Qualification test procedures for photovoltaic modules. Specification No. 502, Issue 1, Commision of the European Communities, Joint Research Centre, Ispra, Italy, 1984. http://www.jrc.org/

ANSI/UL 1703-1987 Standard for flat-plate photovoltaic modules and panels. American National Standards Institute, New York, USA, 1987. http://www.ansi.org/

UL 1741-1999 Inverters, Converters, and Controllers for Use in Independent Power Systems. Underwriters Laboratories, Inc. http://www.ul.com/

NFPA 70. National Electrical Code (NEC). 1999. National Fire Protection Association, Quincy, MA 02269, USA. http://www.nfpa.org/

ISO/DIS 15387:2002 Space systems. Space solar cells. Requirements, measurements and calibration procedures. International Organization for Standardization, Geneva, Switzerland, 2002. http://www.iso.ch/

APPENDIX F

Useful web sites, journals and newsheets

There is a large and growing number of web sites, journals and newsletters with renewable energy content. This Appendix contains a cross section of those which contain useful informaion relating to photovoltaics and solar cells.

Web sites

http://www.pvpower.com/
Contains a wealth of information including PV glossary, bibliography, system design software, PV standards, units and conversion factors, and environmental safety and health information

http://www.solarbuzz.com
PV industry information with a comprehensive list of manufacturers of solar cells and PV system components.

http://www.solaraccess.com/
A renewable energy website; includes information about companies, products, jobs, education events and services.

http://www.solarenergy.org/
Web site of Solar Energy International – includes information about workshops and projects.

http://www.iea-pvps.org/
Web site of the Photovoltaic Power Systems Programme of the International Energy Agency. A wealth of information and IEA reports; many can be downloaded from the site. Newsletter of the IEA PVPS programme can be found at http://www.oja-services.nl/iea-pvps/pvpower/home.htm

http://www.pvresources.com
A (personal ?) website with much useful information

http://www.censolar.es
Web site of the Spanish Solar Energy Training Centre (in Spanish)

http://www.ises.org/ises.nsf
The web site of the International Solar Energy Society

http://www.nrel.gov/
Web site of the U.S. National Renewable Energy Laboratory

http://www.seia.org/
Web site of Solar Energy Industries Association (SEIA)

http://www.pv.unsw.edu.au/
Web site of Key Centre for Photovoltaic Engineering, UNSW. Information about its activities and courses. AM1.5 spectrum can be downloaded from http://www.pv.unsw.edu.au/am1.5.html

http://www.eurosolar.org
Web site of Eurosolar; a European renewable energy agency.

http://www.soda-is.com
Web site of the SoDa Project. Solar radiation data for a range of applications including PV, energy in buildings, vegetation, oceanography and health.

http://www.satel-light.com
Solar radiation data for Europe with a processing service for a wide range of applications. Includes an extensive glossary of terms.

http://rredc.nrel.gov/solar/
A wealth of information with solar radiation data, spectra, standards, computer programs and glossary

http://www.pv.bnl.gov
Web site of the Photovoltaic Environmental, Health and Safety Assistance Center, Brookhaven National Laboratory.

http://www.jxj.com/yearbook/index.html
Database of renewable energy suppliers and installers

http://www.ioffe.rssi.ru/SVA/NSM/Semicond/
A web sites with semiconductor data. See also M. Levinstein et al, Handbook Series on Semiconductor Parameters, in Bibliography.

http://www.semiconductors.co.uk/
A web site with general information and data concerning semiconductors

http://www.solarweb.net
A Spanish renewable energy web site. A useful summary of the regulations for grid connection in Spain can be found at http://www.solarweb.net/fotovoltaica.php

Journals

Progress in Photovoltaics published by Wiley. http://www3.interscience.wiley.com/

Solar Energy Materials and Solar Cells published by Elsevier. http://www.elsevier.nl/locate/solmat

Photon International.. Solar Verlag GmbH, Wilhelmstrasse 34, D-52070 Aachen, Germany. http://www.photon-magazine.com/

Solar Energy published by Pergamon. Official Journal of the International Solar Energy Society. http://www.elsevier.com/locate/solener

Newsletters

PV News: editor: Paul Maycock; PV Energy Systems, 4539 Old Auburn Road, Warrenton, VA 20187 USA; Phone/Fax: 540-349-4497; e-mail pvenergy@crosslink.net; http://www.pvenergy.com.

Photovoltaic Insider's Report: editor: Richard Curry; 1011 W. Colorado Blvd., Dallas, TX 75208 USA; phone and fax 214-942-5248; e-mail rcurry@pvinsider.com; http://www.pvinsider.com.

PV Network News; editor: Paul Wilkins; 2303 Cedros Circle #l; Santa Fe NM 87505 USA; Phone: 505-473-1067; e-mail: pvpaulset@aol.com

The Solar Letter; editor: Allan L. Frank; ALFA Publishing, 9124 Bradford Rd., Silver Spring, MD 20901 USA; phone 301-565-2532; fax 301-565-3298.

Editorial Index

D

E

Q

R

Printed in the United States
124432LV00001B/70-72/A

9 781856 174572